METHODS IN MOLECULAR BIOLOGY™

Series Editor
John M. Walker
School of Life Sciences
University of Hertfordshire
Hatfield, Hertfordshire, AL10 9AB, UK

For other titles published in this series, go to
www.springer.com/series/7651

METHODS IN MOLECULAR BIOLOGY™

Gene Knockout Protocols

Second Edition

Edited by

Ralf Kühn

Institute for Developmental Genetics, Helmholtz Center Munich - German Research Center for Environmental Health, Munich, Germany; Technical University Munich, Munich, Germany

Wolfgang Wurst

Institute for Developmental Genetics, Helmholtz Center Munich - German Research Center for Environmental Health, Munich, Germany; Max-Planck-Institute of Psychiatry, Munich, Germany; Technical University Munich, Munich, Germany

Humana Press

Editors
Ralf Kühn
Institute for Developmental Genetics
Helmholtz Center Munich –
German Research Center
for Environmental Health
Technical University, Munich
Ingolstädter Landstr. 1
85764 Neuherberg, Munich, Germany
ralf.kuehn@helmholtz-muenchen.de

Wolfgang Wurst
Institute for Developmental Genetics
Helmholtz Center Munich –
German Research Center
for Environmental Health
Max-Planck-Institute of Psychiatry
Technical University, Munich
Ingolstädter Landstr. 1
85764 Neuherberg, Munich, Germany
wurst@helmholtz-muenchen.de

ISSN 1064-3745
ISSN 1940-6029 (eBook)
ISBN 978-1-62703-834-8
ISBN 978-1-59745-471-1 (eBook)
DOI 10.1007/978-1-59745-471-1

Softcover re-print of the Hardcover 2nd edition 2009

Printed on acid-free paper

springer.com

Preface

Following the completion of the mouse and human genome sequences, a major challenge is the functional characterization of every mammalian gene and the deciphering of their molecular interaction network. The mouse offers many advantages for the use of genetics to study human biology and disease, unmatched among other mammals. Its development, body plan, physiology, behavior, and diseases have much in common, based on the fact that 99% of the human genes have a mouse ortholog. The investigation of gene function using mouse models is based on many years of technological development. In the two decades since gene targeting in murine embryonic stem (ES) cells was first described by Mario Capecchi and colleagues, more than 3000 predesigned mouse mutants have been developed. To date, a variety of mouse mutagenesis techniques, either gene- or phenotype-driven, are used as systematic approaches. The availability of the genome sequence supports gene-driven approaches such as gene-trap and targeted mutagenesis in ES cells, allowing efficient and precise gene disruption. In combination with the use of site-specific DNA recombinases, in particular the Cre/loxP system, gene disruption can be directed to specific cell types in conditional mouse mutants. Furthermore, chemical and transposon mutagenesis of the mouse genome enables us to perform phenotype-driven screens for the unbiased identification of phenotype–genotype correlations involved in models of human disease. Over the next several years, the mouse genome will be systematically altered, and the techniques for achieving predesigned manipulations will be constantly developed further and improved.

The second edition of *Gene Knockout Protocols* brings together distinguished contributors with extensive experience in the gene targeting and mouse genetics fields. In line with the successful format of *Methods in Molecular Biology*, the volume provides a comprehensive collection of step-by-step protocols of use not only for the beginner in the field but also for experienced scientists. The new edition particularly emphasizes the range of new mutagenesis techniques developed over the last seven years, but also covers the basic methods relevant to researchers performing classical gene targeting experiments. The 25 chapters of this volume are organized into four sections on gene modification in ES cells, stem cell manipulation, the generation of genetically engineered mice, and mutant phenotype analysis. The contents reflect the diversification of mutagenesis approaches that now include, besides classical gene targeting, gene modification by oligonucleotides, gene trap mutagenesis, RNAi-mediated knockdown, transposon, and ENU mutagenesis. Conditional gene inactivation through Cre/loxP recombination is covered by chapters on the construction of conditional vectors for gene targeting, gene trap, gene knockdown, and chromosome engineering, complemented by chapters on the generation of constitutive and inducible Cre transgenic mice and the Cre mouse strain database. While most of the chapters describe methods to generate new mutants or transgenic mice the content is completed by techniques relevant for the preservation and phenotyping of mutants. These include sperm freezing, ES cell line establishment, ES cell in vitro differentiation, mouse pathology, mutant phenotyping, and the influence of genetic background on phenotypes.

We hope that this new edition of *Gene Knockout Protocols* that provides a unique collection of bench protocols written by experts will be a valuable resource for all scientists in the field and will further stimulate research on mouse genetics.

Ralf Kühn
Wolfgang Wurst

Contents

Contributors

MARIEKE AARTS • *The Netherlands Cancer Institute, Division of Molecular Biology, Amsterdam, The Netherlands*
JERZEY ADAMSKI • *Institute of Experimental Genetics, Helmholtz Zentrum München, Neuherberg, Germany; Lehrstuhl für Experimentelle Genetik, Technische Universität München, Munich, Germany*
THURE ADLER • *Institute for Medical Microbiology, Immunology and Hygiene, Technische Universität München, Munich, Germany*
ANTONIO AGUILAR PIMENTEL • *Clinical Research Division of Molecular and Clinical Allergotoxicology, Technische Universität München, Munich, Germany*
BERND AIGNER • *Institute of Molecular Animal Breeding and Biotechnology, Gene Center, Ludwig-Maximilians-Universität München, Munich, Germany*
KEIKO AKAGI • *Mouse Cancer Genetics Program, NCI-Frederick, Frederick, MD, USA*
JOACHIM ALTSCHMIED • *Department of Molecular Hematology, University of Frankfurt, Frankfurt am Main, Germany*
WOJTEK AUERBACH • *Regeneron Pharmaceuticals, Inc., Tarrytown, NY, USA*
LORE BECKER • *Friedrich-Baur-Institut, Department of Neurology, Medical School, Ludwig-Maximilians-Universität München, Munich, Germany*
HEIDRUN BEHRENDT • *Division of Environmental Dermatology and Allergy TUM/HMGUGSF, ZAUM-Center for Allergy and Environment, Technische Universität München, Munich, Germany*
AUKE BOERSMA • *Institute of Experimental Genetics, Helmholtz Zentrum München, German Research Center for Environmental Health, Munich, Germany*
INES BOLLE • *Institute for Inhalation Biology, Helmholtz Zentrum München, Neuherberg, Germany*
ALLAN BRADLEY • *Wellcome Trust Sanger Institute, Wellcome Trust Genome Campus, Hinxton, Cambridgeshire, United Kingdom*
RODERICK T. BRONSON • *Department of Biomedical Sciences, Pathology, Tufts University, Massachusetts, MA, USA*
DIRK H. BUSCH • *Institute for Medical Microbiology, Immunology and Hygiene, Technische Universität München, Munich, Germany*
JULIA CALZADA-WACK • *Institute of Pathology, Helmholtz Zentrum München, Neuherberg, Germany*
VINCENZO COPPOLA • *Mouse Cancer Genetics Program, NCI-Frederick, Frederick, MD, USA*
CLAUDIA DALKE • *Institute for Developmental Genetics, Helmholtz Zentrum München – German Research Center for Environmental Health, Munich, Germany*
THOMAS M. DECHIARA • *Regeneron Pharmaceuticals, Inc., Tarrytown, NY, USA*
MARLEEN DEKKER • *The Netherlands Cancer Institute, Division of Molecular Biology, Amsterdam, The Netherlands*

ROB DEKKER • *The Netherlands Cancer Institute, Division of Molecular Biology, Amsterdam, The Netherlands*
SABIT DELIC • *Institute for Developmental Genetics, Helmholtz Zentrum München – German Research Center for Environmental Health, Munich, Germany*
THOMAS DOETSCHMAN • *BIO5 Institute, University of Arizona, Tucson, AZ, USA*
NICOLE EHRHARDT • *Biology Faculty, Department of Animal Physiology, Philipps-Universität, Marburg, Germany*
DAVID ENGBLOM • *Molecular Biology of the Cell I, German Cancer Research Center, Heidelberg, Germany*
JACK FAVOR • *Institute of Human Genetics, Helmholtz Zentrum München, Neuherberg, Germany*
ROBERT FEIL • *Interfakultäres Institut für Biochemie, Universität Tübingen, Tübingen, Germany*
SUSANNE FEIL • *Interfakultäres Institut für Biochemie, Universität Tübingen, Tübingen, Germany*
BARBARA FERWAGNER • *Institute of Experimental Genetics, Helmholtz Zentrum München, Neuherberg, Germany*
DAVID FRENDEWEY • *Regeneron Pharmaceuticals, Inc., Tarrytown, NY, USA*
HELMUT FUCHS • *Institute of Experimental Genetics, Helmholtz Zentrum München, Neuherberg, Germany*
VALÉRIE GAILUS-DURNER • *Institute of Experimental Genetics, Helmholtz Zentrum München, Neuherberg, Germany*
JOACHIM GRAW • *Institute for Developmental Genetics, Helmholtz Zentrum München – German Research Center for Environmental Health, Munich, Germany; Lehrstuhl für Entwicklungsgenetik, Technische Universität München, Munich*
ANNA-KATERINA HADJANTONAKIS • *Mount Sinai Hospital, Samuel Lunenfeld Research Institute, Toronto, Ontario, Canada*
WOLFGANG HANS • *Institute of Experimental Genetics, Helmholtz Zentrum München, Neuherberg, Germany*
GERHARD HELDMAIER • *Biology Faculty, Department of Animal Physiology, Philipps-Universität Marburg, Germany*
CHRISTIANE HITZ • *Institute for Developmental Genetics, Helmholtz Zentrum München - German Research Center for Environmental Health, Munich, Germany*
ANDREAS HOFMANN • *Institut für Pharmakologie und Toxikologie, Universität Bonn, Bonn, Germany*
SABINE M. HÖLTER • *Institute for Developmental Genetics, Helmholtz Zentrum München – German Research Center for Environmental Health, Munich, Germany; Lehrstuhl für Entwicklungsgenetik, Technische Universität München, Munich, Germany*
GABRIELE HÖLZLWIMMER • *Institute of Pathology, Helmholtz Zentrum München, Neuherberg, Germany*
MARION HORSCH • *Institute of Experimental Genetics, Helmholtz Zentrum München, Neuherberg, Germany*
MARTIN HRABÉ DE ANGELIS • *Institute of Experimental Genetics, Helmholtz Zentrum München, Neuherberg, Germany; Lehrstuhl für Experimentelle Genetik, Technische Universität München, Munich, Germany*

BORIS IVANDIC • *Department of Medicine III, Div. of Cardiology, University of Heidelberg, Germany*
ANAHITA JAVAHERI • *Clinical Research Division of Molecular and Clinical Allergotoxicology, Technische Universität München, Munich, Germany*
MAGDALENA KALLNIK • *Institute for Developmental Genetics, Helmholtz Zentrum München – German Research Center for Environmental Health, Munich, Germany*
HUGO KATUS • *Department of Medicine III, Div. of Cardiology, University of Heidelberg, Germany*
HEIDRUN KERN • *Department of Applied Genetics, TaconicArtemis GmbH, Cologne, Germany*
SATOSHI KISHIGAMI • *Center for Developmental Biology RIKEN Kobe, Kobe, Japan*
EVA KLING • *Friedrich-Baur-Institut, Department of Neurology, Medical School, Ludwig-Maximilians-Universität München, Munich, Germany*
MARTIN KLINGENSPOR • *Zentralinstitut für Ernährungs- und Lebensmittelforschung, Technische Universität München, Munich, Germany*
THOMAS KLOPSTOCK • *Friedrich-Baur-Institut, Department of Neurology, Medical School, Ludwig-Maximilians-Universität München, Munich, Germany*
RALF KÜHN • *Institute for Developmental Genetics, Helmholtz Zentrum München – German Research Center for Environmental Health, Munich, Germany; Lehrstuhl für Entwicklungsgenetik, Technische Universität München, Munich, Germany*
DAVID A. LARGAESPADA • *University of Minnesota, Department of Genetics, Cell Biology and Development, Minneapolis, MN, USA*
SONG-CHOON LEE • *Wellcome Trust Sanger Institute, Hinxton, Cambridge, UK*
CHRISTOPH LENGGER • *Institute of Experimental Genetics, Helmholtz Zentrum München, Neuherberg, Germany*
PENTAO LIU • *Wellcome Trust Sanger Institute, Hinxton, Cambridge, UK*
LYNN MAR • *Mount Sinai Hospital, Samuel Lunenfeld Research Institute, Toronto, Ontario Canada*
SUSAN MARSCHALL • *Institute of Experimental Genetics, Helmholtz Zentrum München – German Research Center for Environmental Health, Munich, Germany*
HARALD VON MELCHNER • *Department of Molecular Hematology, University of Frankfurt, Frankfurt am Main, Germany*
ILONA MOSSBRUGGER • *Institute of Pathology, Helmholtz Zentrum München, Neuherberg, Germany*
CORINNA MÖRTH • *Institute of Molecular Animal Breeding and Biotechnology, Gene Center, Ludwig-Maximilians-Universität München, Munich, Germany*
ROBERT MUNROE • *College of Veterinary Medicine, Cornell University, Ithaca, NY, USA*
ANDRAS NAGY • *Mount Sinai Hospital, Samuel Lunenfeld Research Institute, Toronto, Ontario, Canada*
BEATRIX NATON • *Institute of Experimental Genetics, Helmholtz Zentrum München, Neuherberg, Germany*
MARKUS OLLERT • *Clinical Research Division of Molecular and Clinical Allergotoxicology, Technische Universität München, Munich, Germany*
MARY ELLEN PALKO • *Mouse Cancer Genetics Program, NCI-Frederick, Frederick, MD, USA*

ALEXANDER PFEIFER • *Institut für Pharmakologie und Toxikologie, Universität Bonn, Bonn, Germany*
WILLIAM T. POUEYMIROU • *Regeneron Pharmaceuticals, Inc., Tarrytown, NY, USA*
CORNELIA PREHN • *Institute of Experimental Genetics, Helmholtz Zentrum München, Neuherberg, Germany*
OLIVER PUK • *Institute for Developmental Genetics, Helmholtz Zentrum München – German Research Center for Environmental Health, Munich, Germany*
LETICIA QUINTANILLA-MARTINEZ • *Institute of Pathology, Helmholtz Zentrum München, Neuherberg, Germany*
JOHN SCHIMENTI • *College of Veterinary Medicine, Cornell University, Ithaca, NY, USA*
BIRGIT RATHKOLB • *Institute of Molecular Animal Breeding and Biotechnology, Gene Center, Ludwig-Maximilians-Universität München, Munich, Germany*
SUSAN W. REID • *Mouse Cancer Genetics Program, NCI-Frederick, Frederick, MD, USA*
HEIN TE RIELE • *The Netherlands Cancer Institute, Division of Molecular Biology, Amsterdam, The Netherlands*
JAN RODRIGUEZ PARKITNA • *Molecular Biology of the Cell I, German Cancer Research Center, Heidelberg, Germany*
ALEXANDRA ROLLETSCHE • *In Vitro Differentiation Group, Leibniz Institute of Plant Genetics and Crop Plant Research, Gatersleben, Germany*
JAN ROZMAN • *Biology Faculty, Department of Animal Physiology, Philipps-Universität Marburg, Germany*
PATRICIA RUIZ • *Center for Cardiovascular Research, Charité – Universitätsmedizin Berlin, Berlin, Germany*
FRANK SCHNÜTGEN • *Department of Molecular Hematology, University of Frankfurt, Frankfurt am Main, Germany*
ANJA SCHREWE • *Institute of Experimental Genetics, Helmholtz Zentrum München, Neuherberg, Germany; Department of Medicine III, Division of Cardiology, University of Heidelberg, Germany*
INSA S. SCHROEDER • *In Vitro Differentiation Group, Leibniz Institute of Plant Genetics and Crop Plant Research, Gatersleben, Germany*
HOLGER SCHULZ • *Institute for Inhalation Biology, Helmholtz Zentrum München, Neuherberg, Germany*
GÜNTHER SCHÜTZ • *Molecular Biology of the Cell I, German Cancer Research Center, Heidelberg, Germany*
CHARLES SHAW-SMITH • *Wellcome Trust Sanger Institute, Wellcome Trust Genome Campus, Hinxton, Cambridgeshire, UK*
EILEEN SOUTHON • *Mouse Cancer Genetics Program, NCI-Frederick, Frederick, MD, USA*
PATRICIA STEUBER-BUCHBERGER • *Institute for Developmental Genetics, Helmholtz Zentrum München – German Research Center for Environmental Health, Munich, Germany*
MIKA TANAKA • *Mount Sinai Hospital, Samuel Lunenfeld Research Institute, Toronto, Ontario, Canada*

LINO TESSAROLLO • *Mouse Cancer Genetics Program, NCI-Frederick, Frederick, MD, USA*
FRANK THIELE • *Institute of Experimental Genetics, Helmholtz Zentrum München, Neuherberg, Germany*
THUY T. TRUONG • *In Vitro Differentiation Group, Leibniz Institute of Plant Genetics and Crop Plant Research, Gatersleben, Germany*
DAVID M. VALENZUELA • *Regeneron Pharmaceuticals, Inc., Tarrytown, NY, USA*
NADEJDA VALTCHEVA • *Interfakultäres Institut für Biochemie, Universität Tübingen, Tübingen, Germany*
KRISTINA VINTERSTEN • *Mount Sinai Hospital, Samuel Lunenfeld Research Institute, Toronto, Ontario, Canada*
SANDRA DE VRIES • *The Netherlands Cancer Institute, Division of Molecular Biology, Amsterdam, The Netherlands*
SAYAKA WAKAYAMA • *Center for Developmental Biology RIKEN Kobe, Kobe, Japan*
TERUHIKO WAKAYAMA • *Center for Developmental Biology RIKEN Kobe, Kobe, Japan*
ANJA VAN DER WAL • *The Netherlands Cancer Institute, Division of Molecular Biology, Amsterdam, The Netherlands*
WEI WANG • *Wellcome Trust Sanger Institute, Hinxton, Cambridge, UK*
GRAHAM WATTS • *Mount Sinai Hospital, Samuel Lunenfeld Research Institute, Toronto, Ontario, Canada*
LOUISE VAN DER WEYDEN • *Wellcome Trust Sanger Institute, Wellcome Trust Genome Campus, Hinxton, Cambridgeshire, UK*
EVA WIELDERS • *The Netherlands Cancer Institute, Division of Molecular Biology, Amsterdam, The Netherlands*
CORNELIA WIESE • *In Vitro Differentiation Group, Leibniz Institute of Plant Genetics and Crop Plant Research, Gatersleben, Germany*
ANNA M. WOBUS • *In Vitro Differentiation Group, Leibniz Institute of Plant Genetics and Crop Plant Research, Gatersleben, Germany*
ECKHARD WOLF • *Institute of Molecular Animal Breeding and Biotechnology, Gene Center, Ludwig-Maximilians-Universität München, Munich, Germany*
WOLFGANG WURST • *Institute for Developmental Genetics, Helmholtz Zentrum München – German Research Center for Environmental Health & Max-Plank-Institute for Psychiatry, Munich, Germany; Lehrstuhl für Entwicklungsgenetik, Technische Universität München, Munich, Germany*
GEORGE D. YANCOPOULOS • *Regeneron Pharmaceuticals, Inc., Tarrytown, NY, USA*
BRANKO ZEVNIK • *Department of Applied Genetics, TaconicArtemis GmbH, Cologne, Germany*
SILKE DE-ZOLT • *Department of Molecular Hematology, University of Frankfurt, Frankfurt am Main, Germany*

Color Plates

Color Plate 1 — Protocol for mES cell-derived cardiac differentiation. Five-day EBs were plated onto gelatin- or laminin-coated plates and cultured in IMDM+20%FCS supplemented with L-glutamine, NEAA, and MTG for up to 24 days. Multilineage progenitors at the intermediate stage 2 co-express nestin and desmin, while terminally differentiated cardiac clusters (stage 3) show well-organized sarcomeric staining of Z-disk epitopes of titin. Beating frequency measured from a beating cluster (phase contrast) by the LUCIA HEART imaging system is shown at the *right*, bar = 50 μm (*see* discussion on p. 229)

Color Plate 2 — Protocol for mES cell-derived neuronal differentiation. ES cells were cultured as EBs for 4 days. After plating onto gelatin, cells were cultured in B1 supplements and FCS-containing medium for 24 h (*). After medium change (at day 4+1), EB outgrowths were cultured until day 4+8 without FCS to select for neural progenitors. At day 4+8, EBs were dissociated and replated onto poly-L-ornithine/laminin until day 4+14, when differentiation of mature neurons was induced by "Neurobasal" medium, B27 supplement, and SPFs ("survival promoting factors"). The table shows the media, additives, and substrates used with this protocol. Differentiation led to nestin-positive neural progenitors (stage 2) followed by β-III-tubulin-expressing neuronal cells at stage 3 (4+14 d) and dopaminergic neurons expressing tyrosine hydroxylase at stage 4. A phase contrast picture shows the morphology of the ES cell-derived neurons at stage 4 (*right*) (*see* discussion on p. 230)

Color Plate 3 — Protocol for mES cell-derived pancreatic differentiation. Scheme displays media, additives, and substrates used during the differentiation process. Five-day EBs were plated onto gelatin for spontaneous differentiation in IMDM containing 20% FCS, L-Glut, NEAA, and MTG. At day 5+9, EBs were dissociated and replated onto poly-L-ornithine/laminin and subjected to differentiation by adding the differentiation factors niacinamide (NA), laminin, insulin, sodium selenite, transferrin, progesterone, and putrescine (and FCS for 24 h after plating). After medium change (at day 5+10), differentiation was continued (without FCS) until day 5+28. During spontaneous differentiation, nestin/CK19 co-expressing multilineage progenitors were formed (stage 2). Directed differentiation resulted in C-peptide/nestin-positive committed progenitors (stage 3) and insulin/C-peptide co-expressing islet-like clusters (stage 4; images from (68)). Morphology by phase contrast is shown (*right*) from (77). Cell nuclei were visualized by Hoechst 33342 (*blue*). Bars = 20 μm (*see* discussion on p. 231)

Color Plate 4 Protocol for mES cell-derived hepatic differentiation. Scheme displays media, additives, and substrates used during the differentiation process. Five-day EBs were plated onto gelatin for spontaneous differentiation in IMDM containing 20% FCS, L-Glut, NEAA, and MTG. At day 5+9, differentiation into the hepatic lineage was induced by dissociation of the EBs and replating onto collagen I. Cells were cultured in differentiation medium (HCM) containing 10% FCS until day 5+9+30. Spontaneous differentiation led to nestin/AFP-positive multilineage progenitors (stage 2). Differentiation resulted in albumin/AFP co-expressing committed progenitors at stage 3, and albumin- and AAT-positive, partially binucleated hepatocyte-like cells (stage 4, images from (53)) with cuboidal morphology (phase contrast, *right*, from (77)) at day 5+9+30. Cell nuclei were visualized by Hoechst 33342 (*blue*). Bars = 20 μm (*see* discussion on p. 232)

Chapter 1

Overview on Mouse Mutagenesis

Ralf Kühn and Wolfgang Wurst

Abstract

In this chapter we give an overview of mutagenesis methods in the mouse as they evolved over the last two decades, an outlook of ongoing and future developments and advice for choosing a mutagenesis strategy. Where appropriate, reference is given to relevant chapters of this book, key original articles and links of web-based resources for mouse mutagenesis.

Key words: Knockout mice, conditional mutagenesis, Cre/loxP, RNAi knockdown, ENU mutagenesis, chimaeric mice.

1. Mutagenesis Strategies

Gene Knockout Protocols is considered an information resource for beginners in the field of mouse mutagenesis. This chapter gives a short review of mutagenesis methods as they evolved over time and an outlook of the ongoing developments to provide an integrated view of the specialized protocols in this volume.

Strategies for mutagenesis can be classified into the reverse or forward genetics approach. Forward genetics is a phenotype-driven approach whereby large numbers of mutations are induced at random and new mutants are identified through specific phenotype screens (**Fig. 1.1**). In mice, experimentally induced forward genetics methods include irradiation, chemical mutagenesis with ethylnitrosourea (ENU) and transposon-based mutagenesis. Depending on the nature of the mutagen mainly chromosomal aberrations, point mutations or insertions are induced. Since no prior assumption is made about the underlying genes the forward genetics approach represents an unbiased way for the identification

Ralf Kühn, Wolfgang Wurst (eds.), *Gene Knockout Protocols: Second Edition, vol. 530*

DOI 10.1007/978-1-59745-471-1_1 Springerprotocols.com

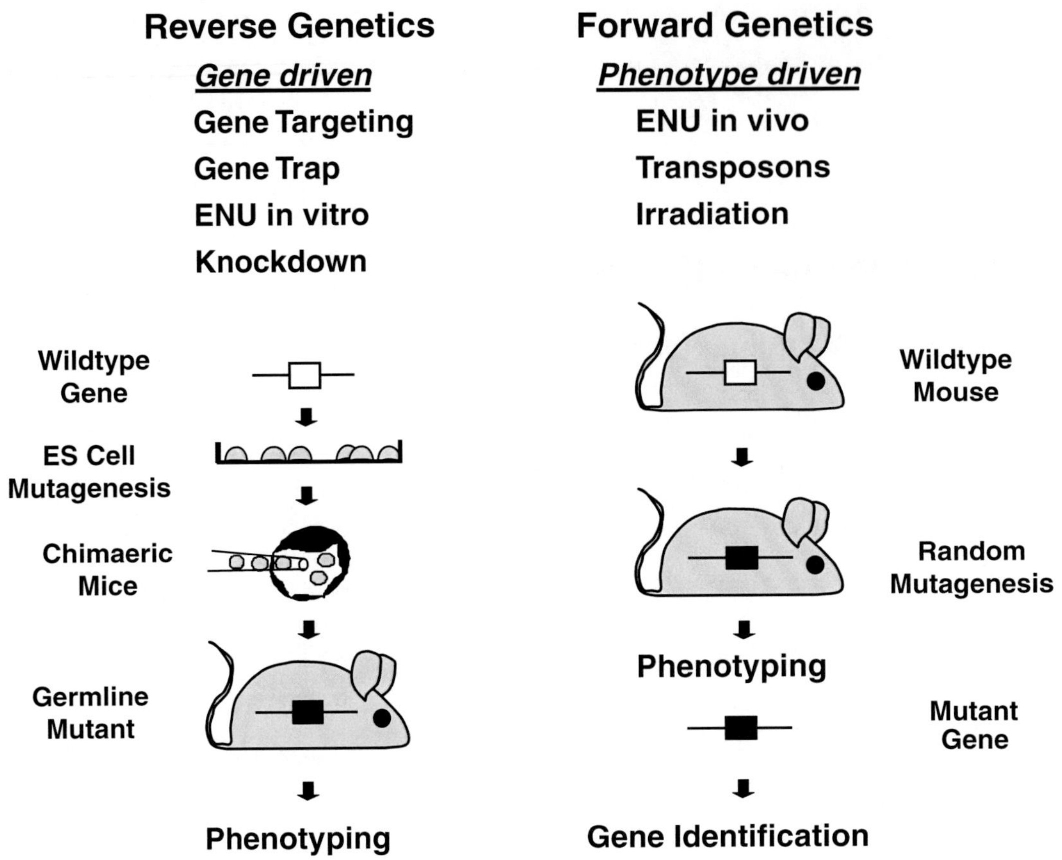

Fig. 1.1. **Comparison of reverse and forward mutagenesis strategies.** For the reverse genetics approach a preselected gene is manipulated in vitro in embryonic stem (ES) cells that are used to generate a mutant mouse strain. The phenotypic characterization of the mutant should reveal the essential, nonredundant function of the targeted gene. By contrast, forward genetics randomly mutagenizes a large number of genes in vivo. Upon breeding of the founder generation relevant mutant offspring are identified by phenotypic screening. The altered gene must be identified by genetic mapping and sequence analysis.

of genes and genetic pathways involved in biological processes. However, since large numbers of mice must be raised and handled over years such screens are usually performed only by large research centres. Besides chapters on transposon mutagenesis (**Chapter 20**) and chemical mutagenesis in embryonic stem (ES) cells (**Chapter 7**), this book also focuses on methods of reverse mouse genetics that are more suited for individual researchers and smaller research units.

The reverse genetics approach requires knowledge about the sequence and structure of a target gene and aims to characterize its in vivo function by the generation of a mutant mouse strain, the phenotype of which is compared to wild-type controls. This strategy, which includes the production of knockout mice by gene targeting, gene-trap mutagenesis, chemical ES cell mutagenesis and RNA interference (RNAi)-mediated knockdown (**Fig. 1.1**), relies on the use of murine ES cell lines. ES cell lines exhibit unique properties such as the ability, once established from the inner cell mass of a mouse blastocyst, to renew indefinitively in cell culture while retaining their early pluripotent differentiation state. This property enables to grow ES cells in large numbers and to select, since most mutagenesis methods are inefficient, rare genetic variants that are expanded into a pure stem cell clone that harbours a specific genetic alteration in the target gene (**Fig. 1.2**). Upon introduction of ES cells into mouse blastocysts and embryo transfer these cells contribute to all cell types of the developing chimaeric embryo, including the germline. By mating of germline chimaeras to normal mice the engineered genetic modification is inherited to their offspring and thereby transferred into the mouse germline (**Fig. 1.2**).

The technical basis for reverse mouse genetics was initially established in the decade of 1980–1990 in three steps and the basic scheme is followed since then, essentially unchanged. The

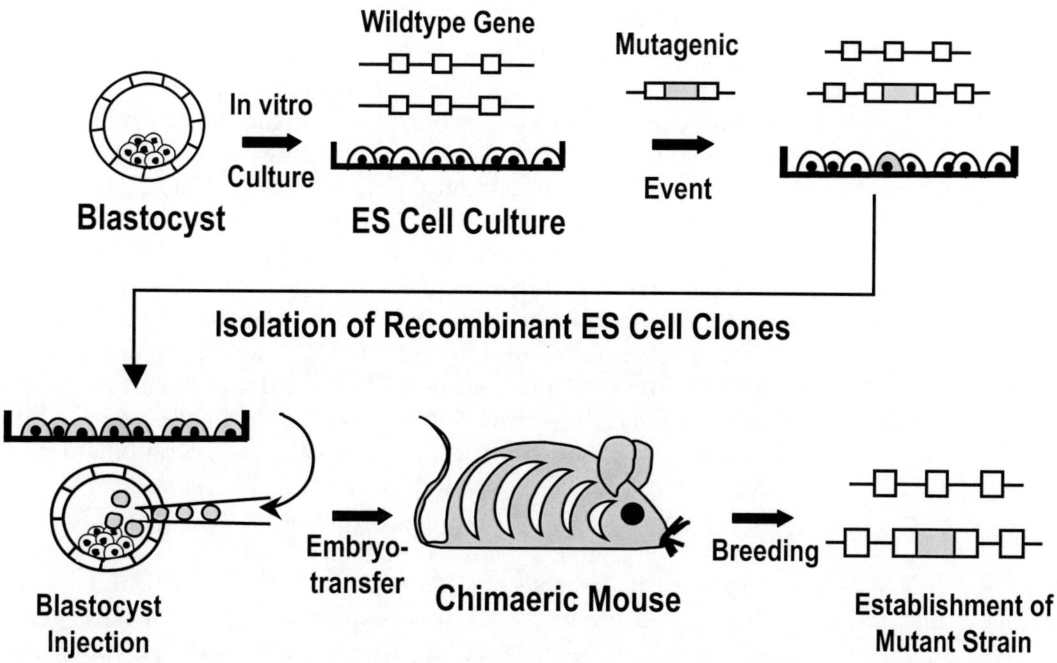

Fig. 1.2. **Generation of knockout mouse mutants by gene targeting in embryonic stem (ES) cells.** ES cells, initially isolated from mouse blastocysts, can be expanded in vitro to large numbers and used to induce rare genetic variants. Gene-targeting vectors are for homologous recombination with the wild-type gene are used as mutagenic event. A mutant, recombinant ES cell clone must be isolated to generate germline chimaeric mice through microinjection into blastocysts. Further breeding of the chimaeras allows to transfer the mutant allele to the next generation and the establishment of a new mutant strain.

first of these steps was the establishment of ES cell lines from cultured murine blastocysts and of culture conditions that maintain their pluripotent differentiation state in vitro *(1, 2)*. A few years later it was first reported that ES cells, upon microinjection into blastocysts, are able to colonize the germline in chimaeric mice *(3, 4)*. The third step concerns the technology to introduce preplanned, inactivating mutations into target genes in ES cells by homologous recombination between a gene-targeting vector and endogenous loci. Initially demonstrated for the directly selectable HPRT gene *(5)*, gene targeting was soon adopted for many other genes *(6)*, and the establishment of the first knockout mouse strain was reported in 1989 *(7)*. The reverse genetics approach proved very successful and led to the generation of more than 3000 knockout mouse lines that provided a wealth of information on in vivo gene functions. The basic technology for the production of knockout mice is covered in this book by six chapters on gene targeting in ES cells, ES cell manipulation, ES cell line establishment and the production of chimaeras through blastocyst injection (**Chapters 8–10 and 14–16).**

However, even the impressive number of 3000 mutant strains that have been generated within the last two decades represents only a small fraction of all genes contained in the mouse genome. One reason for this situation was the incomplete knowledge of the mouse genome sequence before 2002. Another reason for slow progress lies in the efforts of 1–2 years of benchwork required to establish a new knockout strain following the classical gene-targeting protocols. Therefore, a high demand exists to develop and implement new and more efficient procedures for vector construction, ES cell mutagenesis and chimaera production, many of which are included in this volume. Since just the construction of a gene-targeting vector for homologous recombination by PCR and standard cloning methods often requires up to 6 months, streamlined procedures have been developed that allow to assemble such vectors within a matter of days to weeks by homologous recombination in *Escherichia coli*. Two related protocols, ET cloning *(8)* and recombineering *(9)* (**Chapter 2**) that start with genomic BAC clones, were first described in 1999 and 2001 and further developed in later years *(10, 11)*. Since the whole mouse genome is available in the form of sequenced BAC clones all genes are readily accessible to these methods. Besides the more efficient construction of gene-targeting vectors, several new strategies for mutagenesis in ES cells were added to the basic scheme in later years that do not rely on homologous recombination and do not require the construction of individual gene-targeting vectors.

Among the four technologies of gene trapping, chemical mutagenesis, oligonucleotide targeting, and RNAi-based knockdown, the first two, gene-trap and chemical mutagenesis, can be performed as a systematic, large-scale process. As described in **Chapter 3** gene-trap mutagenesis is based on the random integration of an insertion mutagenesis vector across the genome of ES cells and the disruption of trapped genes through vector-specific elements. Gene-trap vectors simultaneously mutate a gene at the site of insertion, provide a sequence tag for the identification of the disrupted gene, and indicate the expression of the tagged gene by a reporter gene. Since a single DNA or retroviral vector can be used to hit a large number of genes, gene trapping is a high-throughput insertional mutagenesis approach that enables to establish libraries of mutant ES cell clones rapidly and at low costs *(12)*. The resulting databases of mutant genes provide the basis for the establishment of mutant mouse strains through germline chimaeras raised from selected ES cell clones. Based on several national gene-trap projects the International Gene Trap Consortium combines the local resources and provides access to a large number of mutant ES cell clones (**Table 1.1**).

A method that does not require the use of vectors for gene mutagenesis in ES cells relies on the use of chemical mutagens like EMS or ENU *(13)*. As described in **Chapter 7**, a library of mutagenized ES cells is produced in vitro and serves as the basis for the production of germline chimaeras and mutagenized offspring that are screened for phenotypic alterations. Alternatively, the library of mutagenized ES cells can be screened by RT-PCR and sequencing to identify clones that harbour mutations in a specific gene *(14)*. A collection of mutant clones allows to establish allelic series of mutants and to map functionally important residues of a protein at high resolution. The introduction of a preselected point mutation into a gene, for example, to mimic a human disease allele, has traditionally been a labour-intensive task that requires the construction of a specific gene-targeting vector. However, it has recently been found that synthetic oligonucleotides can act in ES cells as targeting vector surrogate if DNA repair mechanisms are transiently suppressed *(15)*. This method of oligonucleotide targeting (**Chapter 5**) simplifies the introduction of point mutations into the genome and greatly facilitates the production of specific disease models. Another recently emerging mouse mutagenesis method is RNAi-mediated gene silencing. RNAi has developed into a routine method to knock down genes in cultured cell lines but has also been found useful for silencing gene expression in embryos and adult mice *(16)*. In this book a method for RNAi mediated gene silencing through shRNA vector transgenesis is described in Chapter 6.

Table 1.1
Web-based resources for mouse mutagenesis

Mouse Genome Databases	
Ensembl Mouse Genome Server	http://www.ensembl.org/Mus_musculus/index.html
NCBI	http://www.ncbi.nlm.nih.gov/genome/guide/mouse/
Perlegen	http://mouse.perlegen.com/mouse/index.html
Jackson Lab Mouse Genome Informatics	http://www.informatics.jax.org/
Large-Scale Projects	
EUCOMM	http://www.eucomm.org/
KOMP	http://www.knockoutmouse.org/
NORCOMM	http://norcomm.phenogenomics.ca/index.htm
Gene-Trap Libraries	
International Gene Trap Consortium	http://www.genetrap.org/
German Gene Trap Consortium	http://www.genetrap.de/
Sanger Institute Gene Trap Resource	http://www.sanger.ac.uk/PostGenomics/genetrap/
Bay Genomics	http://baygenomics.ucsf.edu/
Mutant/Transgenics Databases	
Jackson Lab Mouse Genome Informatics	http://www.informatics.jax.org/
Cre/Floxed Mouse Database	http://www.mshri.on.ca/nagy/
Tetmouse Base	http://www.zmg.uni-mainz.de/tetmouse/index.htm
ENU Mutagenesis	
Harwell Mutagenesis Programme	http://www.mgu.har.mrc.ac.uk/
Munich ENU Project	http://www.gsf.de/ieg/groups/genome/enu.html

(continued)

Table 1.1 (continued)

Transposon Mutagenesis	
Mouse Transposon Insertion Database	http://mouse.ccgb.umn.edu/transposon/
Mutant Repositories and Archives	
Mouse Mutant Regional Resource Centre	http://www.mmrrc.org/
Canadian Mouse Mutant Repository	http://www.fimre.org
European Mouse Mutant Archive	http://www.emmanet.org
Jackson Induced Mutant Resource	http://www.jax.org/imr/notes.html
Phenotyping Centres/ Projects	
Mouse Clinical Institute	www-mci.u-strasbg.fr
German Mouse Clinic	http://www.gsf.de/ieg/gmc
Interphenome	http://www.interphenome.org/
Eumodic	http://www.eumodic.org/

2. Conditional Mutagenesis

Through "classical" gene targeting, germline knockout mice are obtained that harbour the mutation in all cells throughout development. This strategy identifies the first essential function of a gene during ontogeny. If the gene product fulfils an important role in development its inactivation can lead to embryonic lethality precluding further analysis in adult mice. To avoid embryonic lethality and to study gene function only in specific cell types, Gu et al. *(17, 18)* introduced a modified, conditional gene-targeting scheme that allows to restrict gene inactivation to specific cell types or developmental stages (**Fig. 1.3**). In a conditional mutant gene inactivation is achieved by the insertion of two 34-bp recognition (loxP) sites of the site-specific DNA recombinase Cre into introns of the target gene. Cre-mediated recombination between two loxP sites results in the deletion of loxP-flanked exons and gene inactivation. Conditional mutants initially require the generation of two mouse strains: one strain harbouring a loxP-flanked gene segment obtained by gene targeting in ES cells and a second,

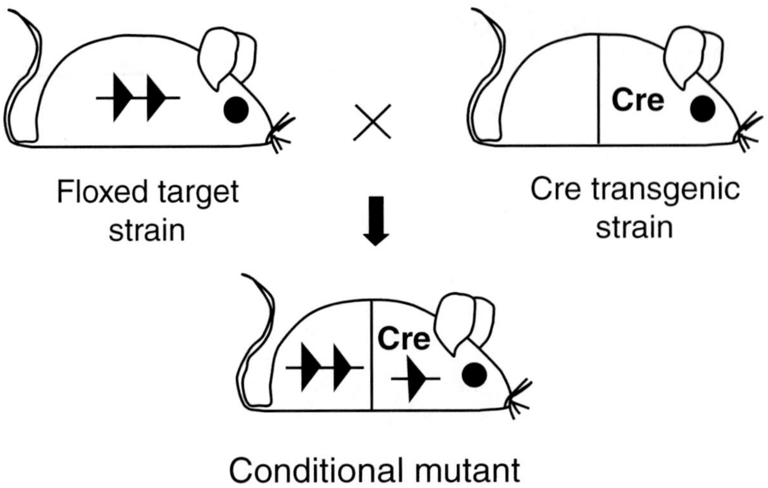

Fig. 1.3. **Principle of conditional gene targeting using the Cre/loxP recombination system.** In the floxed mouse strain, one or more exons of the target gene have been flanked with a pair of 34-bp Cre recombinase (loxP) recognition sites (*filled triangles*). Upon breeding of floxed mice with a transgenic strain that expresses Cre under control of a cell type-specific promoter, the loxP-flanked gene segment is excised from the floxed allele. The loss of an essential gene segment leads to the production of an unfunctional protein in conditional mutant mice.

transgenic strain expressing Cre recombinase in a cell type-specific manner. The conditional mutant is generated by crossing these two strains such that target gene inactivation occurs in a spatially and temporally restricted manner, according to the pattern of recombinase expression of the Cre transgenic strain (**Fig. 1.3**). Methods for conditional mutagenesis are covered by several chapters in this book, describing the construction of conditional gene-targeting vectors (**Chapter 2**), the production of Cre transgenic mice (**Chapter 17**), the use of inducible Cre mice (**Chapter 18**), and a database of Cre transgenic mouse lines (**Chapter 19**). In addition, other mutagenesis methods have adopted the Cre/loxP recombination system to generate conditional mutations, as described in **Chapter 3** for conditional gene-trap vectors and in **Chapter 6** for conditional gene knockdown. Conditional alleles have been generated for more than 100 genes that lead to embryonic lethality as germline knockouts *(19)*. Besides the avoidance of embryonic lethality a conditional mutant can reveal information about the function of a widely expressed gene in different tissues by combination with various Cre lines. More than 150 tissue-specific Cre transgenic strains have been reported and allow to select for tissue-specific gene inactivation out of a number of cell types (**Table 1.1**, **Chapter 19**). In addition to the use of Cre/loxP for gene inactivation site-specific recombination has been used to achieve other types of genome manipulation in ES cells or mice. These include chromosomal engineering, gene replacement, recombinase-mediated cassette exchange, and the inversion of gene segments *(20, 21)*. The engineering of chromosomes in ES cells by Cre/loxP recombination is covered in **Chapter 4** of this volume.

3. Choosing a Mutagenesis Strategy

Before starting a mouse mutant project it is certainly helpful to first define the priorities of the work in relation to the specifics of the target gene and then to select the technique that is most appropriate for mutant generation. In the following we give some thoughts that may help to make a good decision. First, there is no definitive rule to decide whether for a particular experiment conventional or conditional gene targeting is more appropriate since this depends on the specific biological question and the peculiarities of the gene studied. In fact, the answer to this question may not be predictable in many cases but will require the performance of the actual experiment. Therefore, we recommend by default to generate a conditional (loxP-flanked) allele of the target gene since it can be easily converted into a germline knockout by mating of the conditional mutant with a Cre "deleter" strain that expresses recombinase in the germline. This allows one to study first the knockout mutants, but in case of embryonic lethality or specific questions on adult gene function, the additional option for conditional mutagenesis can be easily realized by mating with a tissue-specific Cre line. However, as compared to the direct production of a germline knockout the construction of conditional gene-targeting vectors requires additional work, and matings to Cre lines require additional time. Thus, if time is of high priority and the work focused on embryonic development the direct production of germline knockout mutants may be the better choice. The same applies to genes that are of interest in the adult and exhibit a very restricted expression pattern making it unlikely to lead to embryonic lethality. Another time-saving option for the body-wide suppression of gene function is the production of RNAi-based knockdown mice. It must be taken into account, however, that gene knockdown rarely reaches 100% silencing and a risk of obtaining the phenotype only partly remains.

Regarding the practical aspects of mutant generation it should be first checked whether the planned mutant or a similar one already exists as an established strain or mutant ES cell clone. The most comprehensive database for published mutant strains is the Mouse Genome Informatics (MGI) resource of the Jackson Laboratory (**Table 1.1**). If the mutant of interest is not found at MGI one should further search through the databases of gene-trap libraries and the large-scale projects – Knockout Mouse Project (KOMP), European Conditional Mouse Mutagenesis Programme (EUCOMM) and North American Conditional Mouse Mutagenesis Project (NorCOMM) (**Table 1.1**) – to check whether mutant ES cell clones for the target gene are already available. If such clones are not found or a strategy is followed that is not covered by these projects (e.g. specific point mutations) there is no other way than generating the envisioned mutant yourself by construction of a gene-targeting vector, isolation of mutant ES cells, and production of germline chimaeras. At the beginning one of the mouse genome databases (**Table 1.1**) should be consulted to analyze the genomic structure of the target gene and to design an appropriate targeting vector. For the further steps this book offers numerous chapters on the practical aspects of vector construction and mutant production.

Before the desired mutant is actually obtained a plan for phenotype analysis should be made and fixed. There are no definitive rules on the extent of phenotyping since this depends on the particular question and the specifics of the gene studied. Many laboratories have a few favourite assays that are most relevant to their working area established in-house but are not experienced in most other fields. Phenotype analysis of mutants is an extensive field on its own that splits up into so many specialized assays that no single academic group could take care of them *(22)*. To provide a central service for the analysis of mutants several phenotyping centres have been built up that work on a collaborative basis (**Table 1.1**). Within this book, which focuses on mutagenesis protocols, we can provide only an overview on phenotyping represented by two chapters: **Chapter 24** covers the pathologic analysis of mutants while **Chapter 25** gives a broad overview on first-line phenotyping assays in the mouse.

4. Large-Scale Mutagenesis Projects

Gene targeting in its first two decades has largely progressed in a one-by-one manner by the contributions of a large number of laboratories. Since the generation of a single mutant requires several years of work for vector construction, ES cell culture, mouse breeding and analysis, gene targeting has been traditionally a low-throughput technology; in contrast, gene trapping generates large numbers of insertion mutations with less efforts because a generic vector mutagenizes any gene expressed in ES cells. Thus efforts have begun to build a functional map of the mammalian genome. Despite these efforts, the absolute number of ~4000 mouse mutants to date represents only a fraction of the genes in the mouse genome. To improve this situation several initiatives towards a genome-wide project for the systematic mutational and functional analysis of all mouse genes have recently emerged *(23)*. The European Conditional Mouse Mutagenesis Programme (EUCOMM) puts priority on the development of conditional gene trap and conditional gene-targeting strategies. The US-based Knockout Mouse Project (KOMP) also proposes to saturate the genome by a combination of gene trapping and gene targeting, with less emphasis on conditional mutagenesis. The North American Conditional Mouse Mutagenesis Project (NORCOMM) plans for genome-wide mutagenesis in ES cells using versatile gene-trap vectors that allow post-insertional modifications through Cre or transposon-mediated recombination. Altogether, the international mouse gene knockout programmes plan for the production of ES cell collections that include 22,000 unique gene-trap vector insertions and 18,500 mutant clones generated by gene targeting until 2010 *(23)*. In addition, more than 900 mutant mouse strains will be generated from ES cell clones derived from this resource and analyzed in phenotyping centres. The vectors, ES cell clones and mouse strains will be distributed to the research community through repository centres. When these resources become available we anticipate that the field of mouse mutagenesis shifts its focus from the production of new mutants more towards the analysis of the mutants' phenotypes. It can therefore be expected that these initiatives will lay the ground to build a more complete functional map of the mammalian genome over the next decade and promise a bright future for mouse genetics.

References

1. Evans MJ, Kaufman MH. Establishment in culture of pluripotential cells from mouse embryos. Nature 1981; 292:154–6.
2. Martin GR. Isolation of a pluripotent cell line from early mouse embryos cultured in medium conditioned by teratocarcinoma stem cells. Proc. Natl. Acad. Sci. USA 1981; 78:7634–8.
3. Bradley A, Evans M, Kaufman MH, Robertson E. Formation of germ-line chimaeras from embryo-derived teratocarcinoma cell lines. Nature 1984; 309:255–6.
4. Gossler A, Doetschman T, Korn R, Serfling E, Kemler R. Transgenesis by means of blastocyst-derived embryonic stem cell lines. Proc. Natl. Acad. Sci. USA 1986; 83:9065–9.
5. Thomas KR, Capecchi MR. Site-directed mutagenesis by gene targeting in mouse embryo-derived stem cells. Cell 1987; 51:503–12.
6. Capecchi MR. The new mouse genetics: altering the genome by gene targeting. Trends Genet 1989; 5:70–6.
7. Schwartzberg PL, Goff SP, Robertson EJ. Germ-line transmission of a c-abl mutation produced by targeted gene disruption in ES cells. Science 1989; 246:799–803.

8. Angrand PO, Daigle N, van der Hoeven F, Scholer HR, Stewart AF. Simplified generation of targeting constructs using ET recombination. Nucleic Acids Res. 1999; 27:e16.
9. Lee EC, Yu D, Martinez de Velasco J, et al. A highly efficient *Escherichia coli*-based chromosome engineering system adapted for recombinogenic targeting and subcloning of BAC DNA. Genomics 2001; 73:56–65.
10. Testa G, Vintersten K, Zhang Y, Benes V, Muyrers JP, Stewart AF. BAC engineering for the generation of ES cell-targeting constructs and mouse transgenes. Methods Mol. Biol. 2004; 256:123–39.
11. Liu P, Jenkins NA, Copeland NG. A highly efficient recombineering-based method for generating conditional knockout mutations. Genome Res. 2003; 13:476–84.
12. Stanford WL, Cohn JB, Cordes SP. Gene-trap mutagenesis: past, present and beyond. Nat. Rev. Genet. 2001; 2:756–68.
13. Munroe RJ, Bergstrom RA, Zheng QY, et al. Mouse mutants from chemically mutagenized embryonic stem cells. Nat. Genet. 2000; 24:318–21.
14. Vivian JL, Chen Y, Yee D, Schneider E, Magnuson T. An allelic series of mutations in Smad2 and Smad4 identified in a genotype-based screen of *N*-ethyl-*N*-nitrosourea-mutagenized mouse embryonic stem cells. Proc. Natl. Acad. Sci. USA 2002; 99:15542–7.
15. Aarts M, Dekker M, de Vries S, van der Wal A, te Riele H. Generation of a mouse mutant by oligonucleotide-mediated gene modification in ES cells. Nucleic Acids Res. 2006; 34:e147.
16. Kühn R, Streif S, Wurst W. RNA interference in mice. Handb. Exp. Pharmacol. 2007:149–76.
17. Gu H, Marth JD, Orban PC, Mossmann H, Rajewsky K. Deletion of a DNA polymerase beta gene segment in T cells using cell type-specific gene targeting. Science 1994; 265:103–6.
18. Rajewsky K, Gu H, Kuhn R, et al. Conditional gene targeting. J. Clin. Invest. 1996; 98:600–3.
19. Kwan KM. Conditional alleles in mice: practical considerations for tissue-specific knockouts. Genesis 2002; 32:49–62.
20. Branda CS, Dymecki SM. Talking about a revolution: the impact of site-specific recombinases on genetic analyses in mice. Dev. Cell 2004; 6:7–28.
21. Torres RM, Kühn R. Laboratory Protocols for Conditional Gene Targeting. Oxford: Oxford University Press, 1997:167p.
22. Hrabé de Angelis M, Chambon P, Brown S. Standards of Mouse Model Phenotyping. Weinheim: Wiley-VCH, 2006:332p.
23. Collins FS, Rossant J, Wurst W. A mouse for all reasons. Cell 2007; 128:9–13.

Part I

Gene Modification in ES Cells

Chapter 2

Construction of Gene-Targeting Vectors by Recombineering

Song-Choon Lee, Wei Wang, and Pentao Liu

Abstract

Recombineering is a technology that utilizes the efficient homologous recombination functions encoded by λ phage to manipulate DNA in *Escherichia coli*. Construction of knockout vectors has been greatly facilitated by recombineering as it allows one to choose any genomic region to manipulate. We describe here an efficient recombineering-based protocol for making mouse conditional knockout targeting vectors.

Key words: Conditional knockout, recombineering, vector, gene targeting, mouse, *E. coli*.

1. Introduction

The development of mouse knockout (gene targeting) technology is based on two scientific discoveries in the 1980s. First, mouse embryonic stem (ES) cells derived from wild-type mouse embryos were found to be able to propagate in Petri dishes *(1, 2)*, and were subsequently demonstrated to contribute to the germline in chimaera mice produced using ES cells *(3)*. Second, homologous recombination in mammalian cells enabled precise manipulation of a locus whereby a selection marker flanked by two homology arms could insert into a pre-determined locus *(4–6)*. Today, with the improved protocols for establishing and culturing germline-robust mouse ES cells and the advances in DNA manipulation, one can generate essentially all types of genetic mutations in ES cells and in the mouse *(7)*. To decode gene functions genome-wide, large mutagenesis programmes in the mouse have recently taken off aiming to mutate all coding genes in the mouse genome primarily through gene targeting

Ralf Kühn, Wolfgang Wurst (eds.), *Gene Knockout Protocols: Second Edition, vol. 530*

DOI 10.1007/978-1-59745-471-1_2 Springerprotocols.com

(8). The success of these large projects will undoubtedly generate enormous genetic resources for the research communities for decades to come.

The first generation of mouse knockout (KO) alleles were either deletion of an exon or simply insertion of a selection marker into a locus to disrupt the expression of the gene *(9)*. For many genes, owing to their essential roles in embryonic development, the knockout mutants die in uteri, thereby precluding studying their functions in late development or in specific tissues of adult mice. In some extreme cases, even loss of one copy of a gene is not tolerated in development (VEGF) *(10)*. The embryonic lethality problem is circumvented by using conditional knockout (cko) approaches that two *lox*P sites flank the critical exon(s) of a gene so that upon spatial and temporal Cre recombinase expression, this critical exon is deleted *(11)*. Operationally, construction of cko targeting vectors is laborious since one has to find perfect restriction digestion–ligation strategies to put together usually six to seven pieces of DNA fragments. Indeed, it had been a bottleneck for producing cko mice until recombineering technology became available.

Recombineering is a technology developed in the late 1990s that utilizes the efficient homologous recombination functions encoded by λ phage or cryptic phage Rac to perform DNA manipulation in *Escherichia coli (12–14)*. Recombining linear double-stranded (ds) DNA requires three λ phage proteins, Gam, Beta and Exo, which are collectively called λ Red proteins. The λ Gam protein inhibits RecBCD and SbcCD exonuclease activities of *E. coli*, thus preserving linear dsDNA and allowing it to be used as a substrate for recombination *(15)*. Beta protein is a single-stranded DNA (ssDNA)-binding protein that promotes annealing of complementary DNA strands *(14, 16)*. Beta can bind stably to ssDNA longer than 35 nucleotides *(16)*, and can protect ssDNA overhangs from single-strand nuclease degradation. This property of Beta protein makes it possible to use only about 50 nucleotides of homology for efficient recombination. The short homology can be conveniently supplied by incorporating the 50-nt genomic DNA sequence into PCR primer oligonucleotides *(12, 17)*. Exo is a dsDNA-dependent 5′-3′ exonuclease that processes linear dsDNA and generates a 3′ ssDNA overhang at each end, the substrate that Beta protein binds *(18)*.

Several efficient recombineering systems have been developed that utilize the λ Red proteins *(12, 17, 19–21)*. The system developed in the laboratories of Don Court and Neal Copeland appears to be commonly used for making conditional targeting vector *(22, 23)*. In this system, initially, recombineering was performed in *E. coli* strains that harbour a defective λ prophage *(17)*, which retains *pL* operon, where the Red genes are located, and the critical features of transcriptional control of this operon. Transcription of *pL* is under control of CI857 repressor which shuts down *pL* transcription at 32°C. At 42°C, CI857 repressor becomes inactive, so Red genes are transcribed and Red proteins are accumulated. Owing to the strong *pL* promoter, growth of *E. coli* at 42°C for 10–15 min is enough for efficient and clean homologous recombination in these cells. The Court/Copeland recombineering system is very efficient possibly due to the fact that the three Red genes are expressed from their natural operon with the strong but tightly regulated *pL* promoter, and, therefore, the three proteins are in appropriate molar ratio as they form a complex in vivo. The original recombineering strains were further modified to express Cre (EL350) or Flpe recombinases (EL250) upon arabinose induction. These new features make the Court/Copeland strains popular for making complicated DNA constructs that have *lox*P or *FRT* sites *(22, 23)*.

Using recombineering, we and other colleagues have previously described methods for constructing targeting vectors *(19, 23–27)*. However, due to some unknown reasons, some BAC DNAs are difficult to transform into EL350 or EL250 recombineering-competent strains, or have rearrangements after transformation. We have recently described new reagents that deliver recombineering functions directly into the cells hosting BACs or PACs *(27)*.

In this chapter, we describe a protocol for making conditional knockout targeting vectors using pSim18 plasmid, which carries the three Red genes under the control of *pL* promoter that is in turn regulated by the temperature-sensitive CI857 repressor *(27)*. By simple plasmid transformation, heat-inducible recombineering functions are delivered to BACs. The conditional targeting vectors constructed in this protocol have a *lacZ* reporter incorporated into the targeted allele.

2. Materials

2.1. Mobile Recombineering Reagents

1. pSim18 has the pSC101 replication origin which is low copy and temperature sensitive. This plasmid contains the three Red genes, *exo*, *bet* and *gam*, together with *CI857* repressor and *pL* promoter. Additionally, pSim18 plasmid has the coding sequence of a Hygromycin-resistant cassette inserted between *CI857* and *pL* promoter *(27)*. Consequently, transformants of pSim18 are selected with Hygromycin (*see* **Note 1**).

2. The retrieval vector, PL611, is derived from pBR322, which can host a relatively large piece of mammalian genomic DNA. PL613 is the *lacZ* reporter cassette plasmid. *I-SceI-Bsd-I-CeuI* and *loxP-F3-Neo-F3* are the selection cassettes for introducing *lox*P sites into the BACs. *Neo* is *PGK-EM7-Neo* so it is selectable in both *E. coli* and in ES cells. Plasmid *Cm-MC1TK* has a Chloramphenicol resistance cassette (*Cm*) and the negative selection marker in ES cells, *MC1TK*. *Cm-MC1TK* cassette is flanked by two 600-bp homology regions to PL611; therefore, *Cm-MC1TK* is added to the targeting vector backbone by recombineering. Details of these plasmids can be found in a recent publication *(27)*.

2.2. Primers for PCR Amplification of Selection Cassettes and the Retrieval VectorBackbone

Design rules for the long oligos can be found in a recent publication *(27)* (*see* **Note 2**). The underlined sequences are mouse genomic DNA used as homology in recombineering.

1. Primers for amplifying the *Bsd* cassette:

   For-5′-*TCTAGCCTCACATAGGGGAGAAAGTGTATTT CTCAGTTATACTTTAAGCCCTGGCATTTTTTTAAAG TGTCTGGGACATTCTAGGGATAACAGGGTAATG*;

   Rev-5′-*TAACCAACAGTTTACCAGCCAACTGCAACA TTTAAGGATGTAGAAGAGACAATGGCCTAGGGAC AAGGATGAGTCTAGCTTCGCTACCTTAGGACCGTTA*.

2. Primers for amplifying the *Neo* cassette:

   For-5′-*AAAGTTTTATTTTATTTTATTTTTAAATGGTTA TCAAATTGAATGTGAAATGTGCAAAGGCCCTGGAA TGTGATGAAATATGTAAAACGACGGCCAGTGA*;

   Rev-5′-*TCTACTTCTTTTGGCGCCAGAATTTCATTAA ATGCATCATTTTAAACAAGTATTGTCACAAGATG AACTTCTTGCTAATGAGGAAACAGCTATGACCATG*.

3. Primers for amplifying retrieval plasmid backbone:
   For-5′-*GAGACTTGGTTCAAGAAACAAATATGTGTC CCTTTTGTTGTTGTGCTAAATTGGGAGTGAGGTTT AAAAAAAAAATCAGATACGACTCACTATAGGGAG*;

   Rev-5′-*TAATGCCTTTTATCCAAAGCCAGGAGACTT TTATCTTTTTAAGCATCGGCAAAGTAAGGTGTTTG GCTCTTACTTTTATTTTAGTGAGGGTTAATTATCG*.

2.3. PCR Amplification

1. Plasmids PL611 (retrieval vector), *I-SceI-Bsd-I-CeuI* cassette and *loxP-F3-Neo-F3* cassette are digested with *Eco*RI/*Bam*HI, and the digestion products, 3.1 kb (*PL611*), 0.6 kb (*Bsd*) and 2.1 kb (*Neo*) are purified from the gel using Qiagen gel extraction kit (*see* **Note 3**).

2. A quantity of 1.0 ng of each of the purified DNA fragments is used as PCR templates. PCR amplification is carried out using Extensor Hi-Fidelity PCR Master Mix 2 (2X, ABgene). A volume of 25 μL of the Master Mix was added to 1 μL of the template (1.0 ng), 2 μL of each primer (10 μM) and 20 μL of PCR-grade water.
3. PCR was performed using PTC-225 PCR machine (Peltier Thermal Cycler) with the following settings: 94°C for 4 min, followed by 35 cycles of 94°C for 30 s, 60°C for 30 s, 68°C for 1 min (*Bsd* cassette) or 2–3 min (*Neo* and retrieval cassette). This is then followed by 68°C for 5 min.
4. After PCR reactions, 1.0 μL of *Dpn*I and 0.5 μL of exonuclease I (New England Biolabs) are added per 50 μL PCR products and incubated at 37°C for 1 h followed by heat inactivation at 80°C for 20 min. The PCR products are then purified using Qiagen mini-preparation columns and eluted in 50 μL of PCR-grade water (*see* **Note 4**).

2.4. Antibiotics and LB

1. Antibiotics are used at the following concentrations: Ampicillin (Amp), 50 μg/mL; Chloramphenicol (Cm), 12.5 μg/mL; Kanamycin (Kan), 20 μg/mL; and Hygromycin (Hygro), 75 μg/mL.
2. Fast-Media and agar with Blasticidin (Bsd) (fas-bl-s), Puromycin (Puro) (fas-pr-s) are purchased from InvivoGen. Prepare agar and TB as per instructions on packets, taking care not to overheat the mixture. Cool media rapidly in ice slurry for a few minutes after heating in a microwave. Rapid cooling of the heated mixture reduces degradation of the antibiotics. Puro selection in *E. coli* cells is not as stringent as other commonly used antibiotics; we recommend not to use Puromycin selection in 96-well liquid media culture.
3. LB contains 10 g of tryptone, 5 g of yeast extract and 5 g NaCl per litre.

2.5. Restriction Enzymes

*I-Sce*I and *I-Ceu*I are available from New England Biolabs. *I-Ppo*I is from Promega.

2.6. Electroporation of E. coli

Electroporation is performed under the following condition: 1.75 kV, 25 μF with the pulse controller set at 200 Ω. After electroporation, 1.0 mL LB is added to the cuvette which is subsequently incubated at 32°C for at least an hour prior to plating.

3. Methods

3.1. Construction of a lacZ Reporter Conditional Null Targeting Vector

To generate a cko allele, two *lox*P sites flanking exon 4 of Bcl11a gene are introduced into the genome. Upon expression of tissue-specific or inducible Cre recombinase, recombination occurs between the two *lox*P sites, resulting in deletion of the intervening genomic DNA sequence, thus creating a null allele. We describe here steps to make a cko targeting vector for a multi-purpose allele that can serve as a conventional KO, a conditional KO and a reporter allele. For high-throughput operation in 96-well plates, please refer to our recent publication *(27)*. The overall strategy is illustrated in **Fig. 2.1A–D**.

3.2. Conferring Recombineering Competence to BAC Cells by Transformation with pSim18

1. Inoculate BAC cells (containing region of interest) into 1.0 mL of LB with Chloramphenicol (in a 15-mL polypropylene tube) for overnight growth at 37°C with shaking at 200 rpm.
2. Transfer cells into a 1.5-mL eppendorf tube and pellet cells by spinning the tube at maximum speed using a benchtop centrifuge for 25 s.
3. Decant supernatant and wash three times with ice-cold water. Cells are collected by spinning at maximum speed using a benchtop centrifuge for 25 s at each wash step.
4. Resuspend cells in 50 μL of ice-cold water with 1 ng of pSim18 and perform electroporation.
5. Add 1.0 mL of LB to the cuvette and incubate transformation mixture at 32°C for 1 h.
6. Plate out cells onto a LB-Hygro plate and incubate the plate at 32°C overnight.

3.3. Targeting Selection Cassettes to BACs

1. Pick one Hygro^R BAC colony and inoculate it into 1.0 mL of LB with Chloramphenicol and Hygromycin (in a 15-mL polypropylene tube) and incubate overnight at 32°C with shaking at 200 rpm (**Fig. 2.1A, B**).
2. Inoculate 25, 35, 45 and 55 μL of the overnight culture into four 15-mL polypropylene tubes, each containing 1.0 mL of fresh LB, and incubate at 32°C with shaking at 200 rpm for 2 h.
3. Without measuring OD, transfer the cultures to separate wells in a 42°C heat block (Grant Instrument, Cambridge, UK) and incubate for 15 min.
4. Put the heat block on ice and incubate for 5 min.

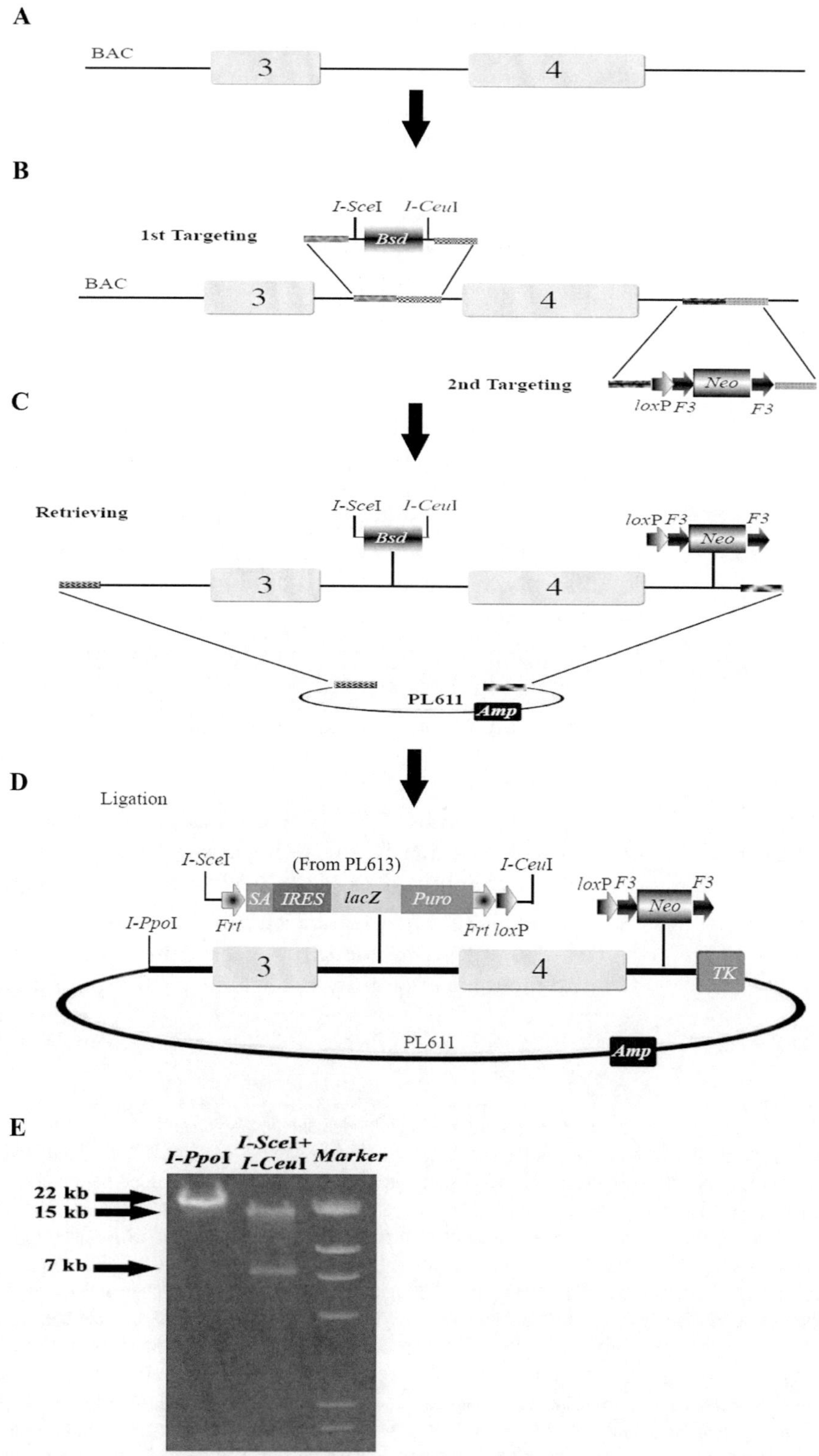

Fig. 2.1. The workflow of generating a reporter conditional null targeting vector. (**A**) The BAC clone containing the region of interest is made recombineering-competent by transforming with pSim18 plasmid. Exon 4, which encodes the main

5. Transfer cells into four 1.5-mL eppendorf tubes and centrifuge at maximum speed using a benchtop centrifuge for 25 s. Decant supernatant and wash three times with ice-cold water. Cells are collected by spinning at maximum speed using a benchtop centrifuge for 25 s at each wash step.
6. Combine cells from the four tubes and resuspend them in 50 μL of ice-cold water with about 300 ng–1.0 μg PCR product (*Bsd* cassette) and perform electroporation.
7. Add 1.0 mL of LB to the cuvette and incubate the transformation mixture at 32°C for 1 h.
8. Plate out cells onto a LB-Bsd plate and incubate the plate at 32°C overnight.
9. Pick 10 Bsd^R colonies. Streak the colonies onto two plates, LB-Amp and LB-Hygro, and incubate the plates at 32°C for overnight. The desired colonies should be Amp^S and $Hygro^R$. Sensitivity to Amp indicates that these Bsd^R colonies are true targeted ones and not the contamination from the original Bsd plasmid. Bsd^R-$Hygro^R$ colonies still retain pSim18 and can be directly used for the next round of recombineering.
10. The Amp^S-Bsd^R-$Hygro^R$ cells are used for targeting the PCR-amplified *Neo* cassette to the BAC (repeating steps 1–9).
11. The Amp^S-Bsd^R-Kan^R-$Hygro^R$ cells are now ready for the retrieval step.

3.4. Retrieving Genomic DNA to a Plasmid Backbone

1. Inoculate one Bsd^R-Kan^R-$Hygro^R$ BAC colony into 1.0 mL of LB with Kan in a 15-mL polypropylene tube and incubate the tube overnight at 32°C with shaking at 200 rpm (**Fig. 2.1C**).
2. Inoculate 25, 35, 45 and 55 μL of overnight culture into four 15-mL polypropylene tubes, each containing 1.0 mL of LB and incubate at 32°C with shaking at 200 rpm for 2 h.

Fig. 2.1 (continued) functional domains of Bcl11a protein, is the intended deletion region and would be flanked by *loxP* sites in the conditional allele. (**B**) The *Bsd* cassette flanked by two rare cutter sites, *I-Sce*I and *I-Ceu*I, is targeted to intron 3. Subsequently, the *loxP-F3-PGK-EM7-Neo-F3* (*Neo*) cassette is targeted into intron 4. Shaded bars represent the short homology arms (50–80 nts) used for recombineering. In a typical cko vector, we select between 4 and 5 kb genomic DNA as the left homology arm (5′), and 2–3 kb as the right homology arm (3′). The genomic DNA region to be deleted is generally between 1 and 7 kb. (**C**) The doubly targeted genomic DNA on the BAC is retrieved to PL611 (Amp^R). (**D**) The *Bsd* cassette is replaced by the *lacZ* reporter cassette (from PL613) in a simple restriction digestion and ligation process. The final targeting vector has the *lacZ* reporter flanked by two *FRT* sites with one *loxP* site, and has the *F3*-flanked *Neo* cassette with another *loxP* site. Finally, the negative selection marker *MC1TK* is added to the vector backbone by recombineering. The final targeting vector is linearized with *I-Ppo*I for ES cell transfection. The true conditional knockout allele is obtained by excising the *lacZ* and the *Neo* cassettes with Flpe in the targeted ES cells or in the mouse germline. We have shown that *FRT* and *F3* sites do not recombine in the mouse germline with the constitutive presence of Flpe. (**E**) Restriction digestion patterns of the final targeting vector. Restriction digestion with *I-Ppo*I linearizes the targeting vector (22 kb). Digestion with both *I-Sce*I and *I-Ceu*I excises the 7-kb *lacZ* reporter cassette.

3. Without measuring OD, transfer the cultures to individual wells in a 42°C heat block and incubate for 15 min.
4. Put the heat block on ice and incubate for 5 min.
5. Transfer cells into 1.5-mL eppendorf tubes and centrifuge at maximum speed using a benchtop centrifuge for 25 s. Decant supernatant and wash three times with ice-cold water. Cells are collected by spinning at maximum speed using a benchtop centrifuge for 25 s at each wash step.
6. Combine cells from the four tubes and resuspend them in 50 μL of ice-cold water with about 300 ng–1.0 μg PCR product (PL611 retrieval cassette) and perform electroporation.
7. Add 1.0 mL of LB to the cuvette and incubate transformation mixture at 32°C for 1 h.
8. Plate out cells onto a LB-Amp plate and incubate the plate at 32°C overnight.
9. There are usually hundreds or even thousands of Amp^R colonies on the plate. In some cases, most of the colonies are background from either self-ligation or intra-molecular recombination of the retrieval plasmid vector. The true retrieval plasmid, however, can be easily identified through re-transformation from the background Amp^R cells because only the true retrieved plasmid carries the *Bsd* and *Neo* Cassettes.
10. Add 2 mL of LB to the plate, and swirl the plate to collect cells.
11. Isolate plasmids from the cell mixture using Qiaprep Spin Miniprep kit (Qiagen) and electroporate into DH10B electro-competent cells with 1.0 μL of the plasmid preparation. After 1-h incubation at 37°C, plate the transformants onto a LB-Kan plate.
12. Inoculate three Kan^R colonies for plasmid preparation and restriction digestion. These Kan^R colonies are the correctly retrieved plasmid.

3.4.1. Replacement of Bsd Cassette with the lacZ Reporter

1. Set up restriction digestion reactions of the retrieved plasmid (10 μL, about 1.5 μg) or PL613 (10 μL, about 1.5 μg): 2 μL of *I-Sce*I, 1 μL of *I-Ceu*I (New England Biolabs, NEB), 3 μL of NEB buffer 4, 0.3 μL of BSA and 13.7 μL of water and incubate at 37°C for 2 h (*see* **Note 4**) (**Fig. 2.1D**).
2. Purify the restriction-digested retrieved plasmid using Qiaprep Mini-preparation columns (Qiagen) and elute DNA in 30 μL of PCR-grade water (*see* **Note 5**).
3. Run the restriction digestion reaction of PL613 through a 1.0% agarose gel, and purify the *lacZ* reporter cassette (the 7 kb band) using QIAquick Gel Extraction kit (Qiagen).

4. Set up ligation reaction of purified digestion products using 12 μL of *lacZ* reporter cassette (600 ng), 10 μL of purified retrieved plasmid (50 ng), 2.5 μL of T4 DNA ligase buffer with 1.0 μL of T4 DNA ligase (New England Biolabs) and incubate at room temperature for 2 h (*see* **Note 6**).
5. Add 5 μL of ligation products to chemical-competent TOP10 cells (Invitrogen) and incubate on ice for 30 min (can use electroporation for the transformation).
6. Heat shock at 42°C without shaking for 30 s.
7. Add 250 μL of SOC and incubate transformation mixture at 32°C for 1 h.
8. Plate out cells onto a LB-Puro-Kan plate and incubate the plate at 37°C overnight. Colonies are typically observed after 16–24 h.
9. Inoculate four $Puro^R$-Kan^R colonies into 3 mL of LB (in 15-mL polypropylene tubes) with Kanamycin, and culture at 37°C overnight with shaking at 200 rpm.
10. Isolate plasmids using Qiaprep Spin Miniprep kit (Qiagen) and set up restriction digestion to confirm the identity of the plasmid.

3.4.2. Targeting the Negative Selection Cassette to the Vector Plasmid Backbone

This step is to add the *Cm-MC1TK* cassette to the plasmid backbone.

1. Inoculate 25, 35, 45 and 55 μL of overnight culture of EL350 cells (recombineering-competent) into four 15-mL polypropylene tubes, each containing 1.0 mL of fresh LB and incubate at 32°C with shaking at 200 rpm for 2 h.
2. Without measuring OD, transfer the cultures to separate wells in a 42°C heat block and incubate for 15 min.
3. Transfer the heat block on ice and incubate for 5 min.
4. Transfer cells into individual 1.5-mL eppendorf tubes and centrifuge at maximum speed using a benchtop centrifuge for 25 s. Decant supernatant and wash three times with ice-cold water. Cells are collected by spinning at maximum speed using a benchtop centrifuge for 25 s at each wash step.
5. Combine cells from the four tubes and resuspend them in 50 μL of ice-cold water with about 10–100 ng purified *Cm-MC1TK* cassette and perform electroporation.
6. Add 1.0 mL of LB to the cuvette and incubate transformation mixture at 32°C for 1 h.
7. Plate out cells onto a LB-Kan-Cm plate and incubate the plate at 32°C overnight.

8. Inoculate a few Puro^{R}-Kan^{R}-Cm^{R} colonies into 3 mL of LB with Kanamycin in 15-mL polypropylene tubes and grow at 37°C overnight with shaking at 200 rpm.
9. Isolate plasmid DNA using Qiaprep Spin Miniprep kit (Qiagen). Dilute the plasmid DNA 1:100. Use 1.0 μL of the diluted DNA to electroporate into DH10B cells and plate onto a LB-Kan plate. This re-transformation step eliminates plasmid multimers formed during recombineering.
10. Isolate plasmid DNA using Qiaprep Spin Miniprep kit (Qiagen) to obtain the final targeting vector.

3.4.3. Verification of the Final Targeting Construct

1. Set up restriction digestions using 1.0 μL of the final targeting vectors (150 ng) with either 2 μL of *I-Ppo*I or 2 μL of *I-SceI* plus 1.0 μL of *I-Ceu*I in a total reaction volume of 30 μL and incubate at 37°C for 2 h.
2. Run the restriction digestion products through a 1% gel to check for expected digestion patterns (**Fig. 2.1E**).
3. Sequence the final targeting vectors to verify the key DNA junctions, including the two junctions between the plasmid backbone and the two ends of the retrieved genomic DNA fragment, and the *lox*P, *FRT* and *F3* sites.

4. Notes

1. pSim18 is prepared in the conventional plasmid preparation way. Since it is a low-copy plasmid, the yields are usually lower than regular high-copy plasmids used in the laboratories such as pBluescript.
2. The 50–80 nucleotides homology in the long oligos should avoid genomic regions that have known repeats and stretches of Gs (over three). The two homology arms flanking a selection cassette should not have more than five nucleotides identical between them. We find that HPLC-purified long oligos perform better than desalted-only oligos. HPLC presumably eliminates most of the incorrect oligos that are present in significant amounts in the crude desalted oligos.
3. The selection cassettes need to be gel-purified prior to PCR amplification. We usually use 500 ng plasmid DNA for digestion in a 30 μL volume with 20 units of each restriction enzyme. The reaction is usually for 2 h at 37°C. In case small amount of background is still present after gel purification, PCR products are digested with *Dpn*I restriction enzyme. *Dpn*I only cleaves DNA from dam+ strains. Plasmid

DNA prepared from commonly used *E. coli* strains, such as DH5α and DH10B, are sensitive to *Dpn*I digestion. We test the PCR-amplified cassette DNA to make sure that no background colony is obtained after electroporating the DNA into commercially purchased electro-competent DH10B cells.

4. The activity of *I-Sce*I in buffer 4 is 50%; therefore, we used twice as much of *I-Sce*I in the double digestion reaction with *I-Ceu*I.
5. Briefly, add water to the digestion mixture or PCR tube to make volume to 100 μL. Add 500 μL PB from Qiaprep kit and mix well. Load the mixture to a Qiaprep Mini-preparation column. Wash the column and elute DNA according to the plasmid mini-preparation protocol.
6. We found that increasing the molar ratio of the insert (*lacZ* reporter cassette) vs. the vector (retrieved backbone) to 10:1 results in a drastic increase in the number of $Puro^R/Kan^R$ colonies. This is probably due to the fact that the *lacZ* reporter (7 kb) has to compete with the smaller *Bsd* cassette (0.6 kb) in the retrieved plasmid (**Fig. 2.1D**). Gel purification of the targeting vector after *I-Sce*I/*I-Ceu*I digestion to eliminate the *Bsd* fragment further improves the ligation efficiency.

Acknowledgement

This work is supported by The Wellcome Trust.

References

1. Evans MJ, Kaufman MH. Establishment in culture of pluripotential cells from mouse embryos. Nature 1981;292:154–6.
2. Martin GR. Isolation of a pluripotent cell line from early mouse embryos cultured in medium conditioned by teratocarcinoma stem cells. Proc. Natl. Acad. Sci. USA 1981;78:7634–8.
3. Bradley A, Evans M, Kaufman MH, Robertson E. Formation of germ-line chimaeras from embryo-derived teratocarcinoma cell lines. Nature 1984;309:255–6.
4. Thomas KR, Capecchi MR. Site-directed mutagenesis by gene targeting in mouse embryo-derived stem cells. Cell 1987;51:503–12.
5. Smithies O. Forty years with homologous recombination. Nat. Med. 2001;7:1083–6.
6. Mansour SL, Thomas KR, Capecchi MR. Disruption of the proto-oncogene int-2 in mouse embryo-derived stem cells: a general strategy for targeting mutations to non-selectable genes. Nature 1988;336:348–52.
7. van der Weyden L, Adams DJ, Bradley A. Tools for targeted manipulation of the mouse genome. Physiol. Genomics 2002;11:133–64.
8. Collins FS, Rossant J, Wurst W. A mouse for all reasons. Cell 2007;128:9–13.
9. Bradley A, Hasty P, Davis A, Ramirez-Solis R. Modifying the mouse: design and desire. Biotechnology (NY) 1992;10:534–9.
10. Ferrara N, Carver-Moore K, Chen H, et al. Heterozygous embryonic lethality induced by targeted inactivation of the VEGF gene. Nature 1996;380:439–42.

11. Gu H, Marth JD, Orban PC, Mossmann H, Rajewsky K. Deletion of a DNA polymerase beta gene segment in T cells using cell type-specific gene targeting. Science 1994; 265:103–6.
12. Zhang Y, Buchholz F, Muyrers JP, Stewart AF. A new logic for DNA engineering using recombination in *Escherichia coli*. Nat. Genet. 1998;20:123–8.
13. Copeland NG, Jenkins NA, Court DL. Mouse genomic technologies recombineering: a powerful new tool for mouse functional genomics. Nat. Rev. Genet. 2001;2:769–79.
14. Court DL, Sawitzke JA, Thomason LC. Genetic engineering using homologous recombination. Annu. Rev. Genet. 2002;36:361–88.
15. Sawitzke JA, Thomason LC, Costantino N, Bubunenko M, Datta S, Court DL. Recombineering: in vivo genetic engineering in *E. coli*, *S. enterica*, and beyond. Methods Enzymol. 2007;421:171–99.
16. Mythili E, Kumar KA, Muniyappa K. Characterization of the DNA-binding domain of beta protein, a component of phage lambda red-pathway, by UV catalyzed cross-linking. Gene 1996;182:81–7.
17. Yu D, Ellis HM, Lee EC, Jenkins NA, Copeland NG, Court DL. An efficient recombination system for chromosome engineering in *Escherichia coli*. Proc. Natl. Acad. Sci. USA 2000;97:5978–83.
18. Cassuto E, Lash T, Sriprakash KS, Radding CM. Role of exonuclease and protein of phage lambda in genetic recombination. V. Recombination of lambda DNA in vitro. Proc. Natl. Acad. Sci. USA 1971;68:1639–43.
19. Zhang P, Li MZ, Elledge SJ. Towards genetic genome projects: genomic library screening and gene-targeting vector construction in a single step. Nat. Genet. 2002;30:31–9.
20. Datsenko KA, Wanner BL. One-step inactivation of chromosomal genes in *Escherichia coli* K-12 using PCR products. Proc. Natl. Acad. Sci. USA 2000;97:6640–5.
21. Murphy KC. Use of bacteriophage lambda recombination functions to promote gene replacement in *Escherichia coli*. J. Bacteriol. 1998;180:2063–71.
22. Lee EC, Yu D, Martinez de Velasco J, et al. A highly efficient *Escherichia coli*-based chromosome engineering system adapted for recombinogenic targeting and subcloning of BAC DNA. Genomics 2001; 73:56–65.
23. Liu P, Jenkins NA, Copeland NG. A highly efficient recombineering-based method for generating conditional knockout mutations. Genome Res. 2003;13:476–84.
24. Angrand PO, Daigle N, van der Hoeven F, Scholer HR, Stewart AF. Simplified generation of targeting constructs using ET recombination. Nucleic Acids Res. 1999;27:e16.
25. Valenzuela DM, Murphy AJ, Frendewey D, et al. High-throughput engineering of the mouse genome coupled with high-resolution expression analysis. Nat. Biotechnol. 2003;21:652–9.
26. Cotta-de-Almeida V, Schonhoff S, Shibata T, Leiter A, Snapper SB. A new method for rapidly generating gene-targeting vectors by engineering BACs through homologous recombination in bacteria. Genome Res. 2003;13:2190–4.
27. Chan W, Costantino N, Li R, et al. A recombineering based approach for high-throughput conditional knockout targeting vector construction. Nucleic Acids Res. 2007;35:e64.

Chapter 3

Gene-Trap Vectors and Mutagenesis

Silke De-Zolt, Joachim Altschmied, Patricia Ruiz, Harald von Melchner, and Frank Schnütgen

Abstract

Gene trapping can be used to introduce insertional mutations into the genome of mouse embryonic stem cells (ESCs). The method has been adapted for high-throughput use, in an effort to inactivate all genes in the mouse genome. Gene trapping is performed with vectors that simultaneously inactivate and report the expression of the trapped gene and provide a molecular tag for its rapid identification. Gene-trap approaches have been used successfully in the past by both academic and commercial organizations to create libraries of ESC lines harboring mutations in single genes that can be used for making mice. Presently, approximately 70% of the protein-coding genes in the mouse genome have been disrupted by gene-trap insertions. Here we describe the basic methodology used to induce and characterize gene-trap mutations in ESCs.

Key words: High-throughput mutagenesis, gene trapping, ES cells.

1. Introduction

With the complete sequencing of the human and mouse genomes, attention has shifted towards the comprehensive functional annotation of mammalian genes *(1, 2)*. Among the various approaches for addressing gene function, the most relevant for extrapolation to human genetic diseases is mutagenesis in the mouse. Although several model organisms have been used in a variety of mutagenesis approaches, the mouse offers particular advantages including a genome structure and organization closely related to the human genome, a manageable size and a relatively short breeding time. Most importantly, the mouse is one of the few mammals for which

Ralf Kühn, Wolfgang Wurst (eds.), *Gene Knockout Protocols: Second Edition, vol. 530*

DOI 10.1007/978-1-59745-471-1_3 Springerprotocols.com

embryonic stem cells (ESCs) are available. ESCs grow indefinitely in tissue culture without losing their pluripotency and can therefore be converted into mice after extensive in vitro manipulation.

The original strategy developed for gene inactivation in ESCs is gene targeting by homologous recombination, which requires a detailed knowledge of gene structure and organization as well as its physical isolation in a targeting vector (*see* **Chapters 1** and **2**). While the availability of the mouse genome sequence greatly assists the gene-targeting approach and high-throughput methodologies for targeting vector construction are being successfully pursued (Bill Skarnes and Francis Stewart, personal communication), the generation of mutant alleles by homologous recombination remains laborious and expensive as it requires individual vector designs for every single gene.

A second approach for gene inactivation in ESCs is gene trapping *(3)*. Unlike gene targeting by homologous recombination, gene trapping is not restricted by regional determinants of homologous recombination efficiency and can be used to modify a large number of genes in the mouse genome in a single experiment. Gene trapping is performed with gene-trap vectors which insert DNA elements that interfere with gene expression into a large collection of mostly random chromosomal sites. The most widely used vectors contain a promoterless reporter and/or selectable marker gene flanked by an upstream splice acceptor (SA) site and a downstream polyadenylation sequence (pA) (**Fig. 3.1A**). When inserted into an intron of an expressed gene, the cassette is transcribed from the endogenous promoter yielding a fusion transcript in which the upstream exons are spliced to the reporter/selectable marker gene. Since transcription is terminated prematurely at the inserted polyadenylation site, the processed fusion transcript encodes a truncated and nonfunctional version of the cellular protein plus the reporter/selectable marker. Thus, gene traps simultaneously mutate and report the transcriptional activity of the trapped gene and provide a molecular tag for the rapid identification of the disrupted gene.

Gene-trapping cassettes lacking a SA have also been used to target the exons rather than the introns of expressed ESC genes *(4–6)*. However, a more recent analysis of a large number of insertion sites revealed that only about 15% of these gene traps disrupted exons (Geoff Hicks, personal communication). The vast majority of productive gene trap events resulted from intron insertions by activation of cryptic SA sites located upstream of the insertions *(7)*. Since these cryptic SA sites are generally weaker than the endogenous sites, gene-trap cassettes are frequently spliced out of the processed transcripts thereby precluding a null mutation. Conventional gene-trap vectors can target most functional classes of genes expressed in ESCs with the notable exception of secretory pathway genes. This is because the capture of a

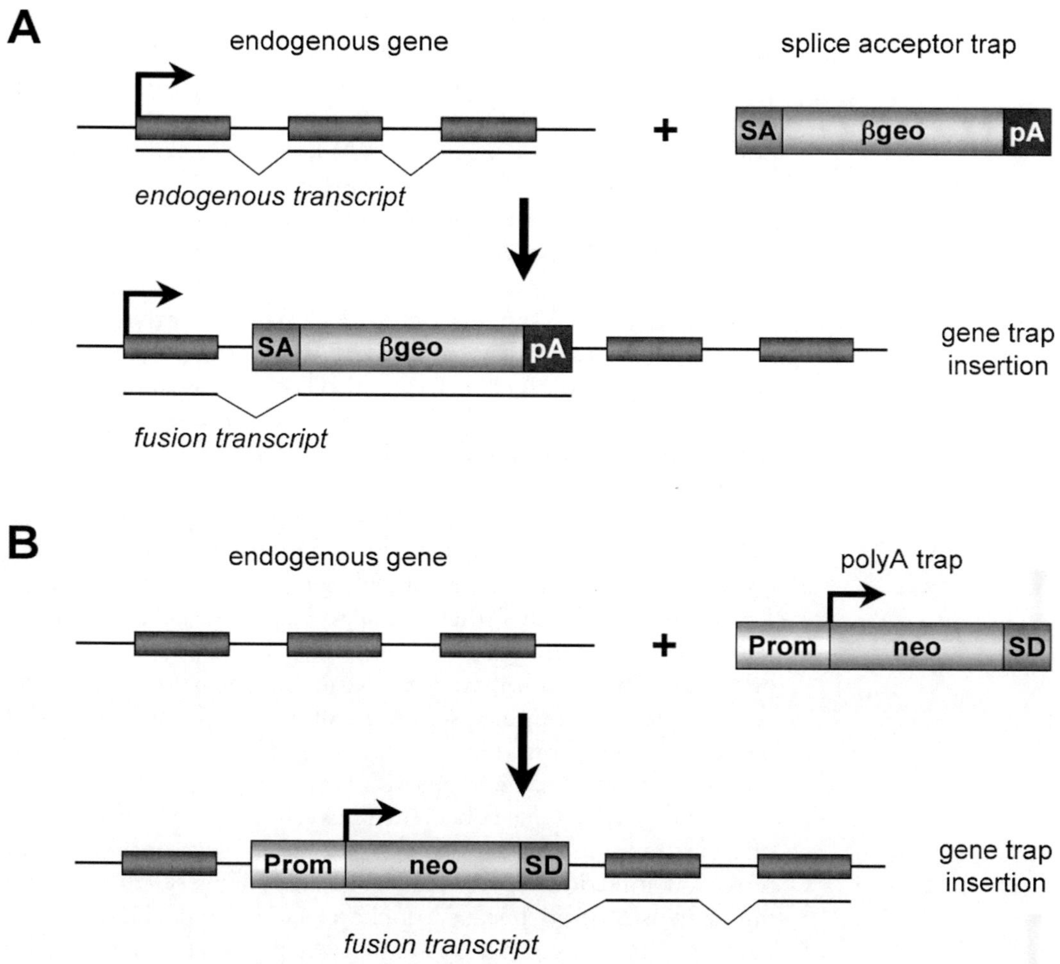

Fig. 3.1. Types of gene-trap vectors. (A) Conventional splice acceptor (SA) gene trap consisting of a SA, a promoterless β-galactosidase-neomycin-phosphotransferase fusion gene (βgeo) and a polyadenylation sequence. Splicing to upstream exons results in a gene-trap fusion transcript from which βgeo is translated. SA, splice acceptor; pA, polyadenylation sequence. (B) PolyA-trap containing a neomycin-phosphotransferase gene (neo) as a selectable marker gene flanked by a constitutive promoter and a splice donor sequence. Shown is only the 3′ segment of a polyA trap. Inserted into introns, the neo gene expressed from the constitutive promoter is spliced to the downstream exons resulting in a fusion transcript terminating in the endogenous polyadenylation site. Prom, constitutively active promoter; SD, splice donor.

signal sequence leads to the secretion of the trapped protein leaving the cells vulnerable to selection. To prevent this, we and others have inserted a transmembrane domain (TM) into the gene-trap vectors, which effectively prevented the secretion of trapped secretory proteins by anchoring them to the cell membrane *(8–10)*.

Gene trapping has been used extensively by both academic and private organizations to assemble libraries of ESC lines harboring mutation in single genes *(11–13)*. Collectively, the existing gene-trap resources cover about 70% of all protein-coding genes in the mouse genome *(14)*. However, recent estimates indicate that approximately 30% of all genes are not expressed in ESCs and therefore untrappable by conventional gene-trap vectors. To mutate non-expressed genes, vectors have been designed that contain in addition to a conventional gene-trapping cassette an independently expressed selectable marker gene coupled to a downstream splice donor sequence. When inserted into an intron, the selectable marker is spliced to the downstream exons yielding a fusion transcript that terminates in the endogenous polyadenylation site (**Fig. 3.1B**). Since the fusion transcript is constitutively expressed the vectors can trap genes independently of their expression. Because polyA sequence capture is essential for this approach, the vectors are referred to as "polyA" traps *(15)*. While the older polyA traps preferentially inserted into last introns, newer generations successfully avoid this bias by using vector designs protecting against the nonsense-mediated decay of fusion transcripts initiating upstream of the last intron *(16)*.

In a further development of the gene-trapping approach, vectors generating conditional ready mutations have been recently introduced *(17)*. Equipped with two directional site-specific recombination systems, these vectors enable temporally and spatially restricted mutagenesis in somatic cells. Moreover, by introducing specific recombinase target sequences into the genome, the vectors create multipurpose alleles allowing virtually any DNA sequence to be inserted into the gene-trap locus by recombinase-mediated cassette exchange (RMCE) *(17)*.

Two methods have been extensively used for the delivery of gene-trapping cassettes to ESCs, that is, electroporation and retroviral infection. Currently, gene-trap delivery by retroviral infection is favored because unlike electroporated DNA, retroviral insertions cause little if any postinsertional damage to the target and/or proviral DNA. Since avoidance of postinsertional damage is a prerequisite for the precise mapping of insertion sites, most ongoing high-throughput screens employ retroviral protocols *(18)*.

Trapped gene identification was traditionally performed using 5′-RACE to amplify the cellular sequences contained within the gene-trap fusion transcripts (gene-trap sequence tags, GTSTs) expressed at the insertion sites. During a period in which genomes were poorly annotated, when genes were largely defined by cDNAs, these RACE tags provided confirmation of a successfully trapped gene and were useful for identifying novel genes. However, RACE tags have two major disadvantages. First, they cannot reveal the exact position of insertion sites, which are usually at some considerable distance from the RACE tags and therefore cannot be used directly for mouse genotyping. Second, because 5′-RACE is entirely dependent on gene expression, the major determinant for identifying a trapped gene in ESCs is the level of its expression. As most high-throughput trapping screens employ highly sensitive G418 selection, the levels of trapped gene expression are often below the gene identification thresholds imposed by 5′-RACE. As a result, a variable proportion of ESC lines are not amenable to GTST identification. In general, only about half of the trapped ESC lines isolated from previous large-scale screens provided a RACE tag. To achieve a more efficient characterization of gene-trap events, high-throughput PCR protocols have been recently developed for the amplification of genomic sequences flanking the gene-trap insertion sites *(18)*.

Here we provide protocols for gene-trap vector delivery to ESCs and describe the methods used for the molecular characterization of gene-trap insertion sites.

2. Materials

1. 10x PCR buffer: 20 mM Tris-HCl, pH 8.4, 500 mM KCl.
2. dNTP mix: 10 mM each of dATP, dCTP, dGTP and dTTP in 10 mM Tris-HCl, pH 8.0.
3. Ethanol/NaCl solution: 75 mM NaCl in ethanol; this solution is obtained by adding 15 μL of 5 M NaCl per mL of ethanol.
4. Gelatine solution: 0.1 % (w/v) gelatine in water, autoclave, store at room temperature.
5. Lysis buffer: 10 mM Tris-HCl, pH 7.5, 10 mM EDTA, 10 mM NaCl, 0.5 % (w/v) sodium dodecyl sulfate (SDS), 1 mg/mL proteinase K.
6. MEF medium: DMEM with 4.5 g/mL glucose supplemented with 2 mM glutamine, 1 mM sodium pyruvate, 1x nonessential amino acids (Invitrogen), 10% (v/v) pretested fetal bovine serum (Hyclone) and 50 μM β-mercaptoethanol.
7. Phoenix-Eco medium: DMEM with 4.5 g/mL glucose (Invitrogen) supplemented with 2 mM glutamine and 10% (v/v) fetal bovine serum.
8. Polybrene stock solution: 5 mg/mL polybrene (hexadimethrine bromide, SIGMA) in water, store in small aliquots at –20°C.

3. Methods

3.1. Gene-Trap Vector Delivery to ESCs

Gene-trap cassettes are delivered to the ESCs either by electroporation (transfection) of a plasmid vector or by infection with a retroviral vector. Besides a standard tissue culture facility, both methods do not require any special equipment, except for an electroporation apparatus.

3.1.1. Electroporation of ESCs with Plasmid Vectors

The preferred method for transfecting ESCs with plasmid-based gene-trap vectors is electroporation because conditions can be adjusted to favor single-copy integrations. Plasmids are linearized before electroporation to trigger illegitimate recombination, which requires double-strand DNA breaks *(19)*. Since linear DNA is rapidly degraded by exonucleases leading to strand recession at both ends, restriction sites are generally chosen at a distance from the gene-trap cassette for protecting its integrity. A single electroporation typically requires 10^8 cells, which is equivalent to at least six semiconfluent 90-mm (P90) tissue culture dishes.

1. Add 3 mL of trypsin to each dish and incubate for 3 min at 37°C.
2. Gently pipette the ESCs up and down to obtain a single-cell suspension. Depending on the type of ESCs used, it might be advisable to resuspend the cells after blocking the trypsin reaction (e.g., E14TG2a.4 ESCs).
3. Block trypsin by adding 3 mL of medium supplemented with 10% (v/v) fetal bovine serum and transfer cells into a conical tube.
4. Centrifuge cells for 3 min at 270*g*.
5. Resuspend cells in 10 mL of PBS and count.
6. Transfer 1×10^8 cells to a 15-mL conical tube.
7. Centrifuge for 5 min at 250*g*.
8. Resuspend cell pellet in 500 μL of ice-cold PBS and add 120 μg of the linearized plasmid.
9. Adjust the final volume to 900 μL by adding ice-cold PBS.
10. Transfer the suspension to a 4-mm electroporation cuvette and carefully pipette the electroporation mix up and down.
11. Zap the cells at 0.8 kV, 3 μF (settings for BioRad GenePulser).
12. Remove the cuvette from its holder, flick immediately to stabilize pH and leave on ice for 10–20 min.
13. Carefully transfer the cell suspension to a 15-mL conical tube containing 12 mL of ESC medium supplemented with 1000–1500 U/mL leukemia inhibitory factor (LIF) depending on the ESC line (*see* **Note 1**).
14. Distribute 1-mL aliquots of the suspension into 12 freshly gelatinized P90 dishes (*see* **Note 2**) each containing 9 mL ESC medium (*see* **Note 1**) and LIF. Use mouse embryonic fibroblasts (MEFs) as feeder layer if ESCs have not been adapted to feeder-free growth (*see* **Note 3**).

15. Incubate overnight.
16. On the next day start selection (*see* **Section 3.2**).

3.1.2. Infection of ESCs with Retroviral Gene-Trap Vectors

Infectious, replication-incompetent retroviral gene-trap particles are obtained by transfecting the retrovirus gene-trap plasmid into a retrovirus packaging cell line expressing the proteins required for virus assembly, reverse transcription and integration (gag, pol, env). Supernatants from these cultures containing the retroviral particles are used to infect ESCs. To ensure single-copy integrations, the ratio between the number of target cells and infectious gene-trap particles (multiplicity of infection, MOI) should be kept below 0.25.

To calculate the MOI, virus titers of individual supernatants need to be determined prior to ESC infection. As there is no independently expressed selectable marker in the conventional gene-trap vectors, titers are based on the number of gene-trap events obtained with a particular reporter/selectable marker combination. With a β-galactosidase-neomycin-phosphotransferase (βgeo) combination, for example, only 10–20% of the viruses have been shown to induce a gene-trap event *(20)*. To determine the virus titer in a supernatant collected from the gene-trap virus producer cells, infect ESCs with serial dilutions of supernatant as described below and count colonies after selecting in the appropriate antibiotic (e.g., G418 for βgeo vectors). For βgeo vectors multiply the number of colonies by 5 to obtain the approximate virus titer.

3.1.3. Production of Retroviral Supernatants

We produce gene-trap retrovirus by collecting supernatants from Phoenix-Eco cells (*see* **Note 4**) after transfecting the gene-trap plasmid. Phoenix-Eco cells carry stable integrations of expression plasmids encoding the gag, pol and env proteins of Moloney murine leukemia virus (MMLV). The protocol we describe here is adapted to P90 culture dishes but can be scaled up or down according to specific needs. Each P90 dish yields 7 mL of virus supernatant, which is usually sufficient for infecting 4.5×10^5 ESCs distributed equally between three P90 dishes (i.e., 1.5×10^5 cells/dish). Virus supernatant can be either used directly or stored frozen at –80°C. Note that one freezing/thawing cycle reduces virus titers by about 50%. Therefore, try to avoid repeated freezing/thawing cycles.

1. Seed 3×10^6 Phoenix-Eco cells into one P90 cell culture dish; make sure that cells are evenly distributed over the entire surface and incubate overnight.
2. Prepare the following transfection mixes: (a) add 10 μg of plasmid to 700 μL of DMEM (Invitrogen) without supplements and vortex briefly, (b) add 20 μL Lipofectamine 2000 (Invitrogen) to 700 μL of DMEM without supplements and vortex briefly. Incubate (a) and (b) for 5 min at room temperature.
3. Combine (a) and (b), vortex briefly and incubate at room temperature for 20 min.

4. In the meantime carefully wash the cells once in 7 mL of PBS; bear in mind that Phoenix-Eco cells detach very easily from the culture dish so make sure they are still there before you proceed.
5. Overlay the cells with 7 mL of regular growth medium for Phoenix-Eco cells.
6. Add the transfection mix dropwise to the cells and try to spread it evenly across the dish.
7. Incubate overnight.
8. On the next morning carefully wash the cells once with 7 mL of PBS.
9. Add 7 mL of ESC medium containing LIF (*see* **Note 1**). It is important to switch to ESC medium at this stage, as the virus supernatant will be used for ESC infection without further purification.
10. Incubate overnight.
11. On the next morning collect the supernatant. If you need more virus you can repeat Steps 9–11 2–3 times. However, since the medium acidifies over multiple collection rounds, supernatants from repeated collection cycles should be diluted in ESC medium at least threefold prior to infection.
12. Pass the supernatant carefully through a 0.45-μM cellulose acetate syringe filter; do not use nitrocellulose filters as nitrocellulose binds retroviral membrane proteins and decreases virus titers.
13. Add 1 μL of polybrene stock solution per milliliter of virus suspension to obtain a final concentration of 5 μg/mL. Polybrene increases virus infectivity by changing the charge of the cell membrane.
14. Use virus supernatant directly for infections or store in small aliquots at –80°C. DO NOT flash freeze!

3.1.4. Infection of ESCs with Gene-Trap Retroviruses

1. One day prior to infection seed 1.5×10^5 ESCs into a P90 cell culture dish freshly coated with gelatine (*see* **Note 2**). If ESCs require feeder layers, prepare plates with feeders the day before seeding the ESCs (*see* **Note 3**).
2. Incubate overnight.
3. Next morning wash the dishes with 7 mL of PBS.
4. Add 2 mL of retrovirus supernatant (*see* **Section 3.1.3**) and 2 mL of fresh ESC medium supplemented with 5 μg/mL polybrene. The exact composition of the medium depends on the ESC line used (*see* **Note 1**). Gently tilt the dishes in your hand to spread the supernatant evenly over the ESCs.
5. Incubate for 4–4.5 h.

6. Add 6 mL of ESC medium and incubate for another 24 h.
7. Wash with 7 mL of PBS and start appropriate selection.

3.2. Selection and Isolation of Trapped ESC Clones

To trap as many genes as possible use the lowest concentration of antibiotic that still kills 100% of the employed ESCs. Keep in mind that with the promoterless gene-trap vectors (*see* **Fig. 3.1A**) the higher the concentration of antibiotic the higher the bias towards trapping highly expressed genes.

1. Incubate electroporated or infected cells in fresh ESC medium for 24 h.
2. Replace the ESC medium with selection medium. If you use G418, make sure to use the appropriate concentration of "active" antibiotic. We routinely use 130 μg/mL of active G418 (Invitrogen) for E14TG2a.4 and 150 μg/mL for TBV-2 ESCs.
3. Incubate for at least 10 days with daily medium changes until colonies become visible.
4. For feeder-dependent ESCs seed MEFs into freshly gelatinized 96-well plates one day before picking the colonies (*see* **Note 3**); for feeder-independent ESCs use gelatinized 96-well plates (*see* **Note 2**).
5. For picking of the clones remove the medium from the P90 dishes containing the drug-resistant colonies and cover the cells with a thin layer of prewarmed PBS. Under a sterile hood, displace the colonies from the dish using a Pipetman (P20) with sterile tips and transfer the cells to round-bottom 96-well plates containing 30 μL of PBS per well. Do not pick more than 48 colonies at a time (beginners should start with 12 colonies) before you proceed to the next step.
6. Add 30 μL of 2x trypsin solution (Invitrogen) to each well and incubate for 2–5 min until cells can be easily dispersed by tapping the plate.
7. Add 200 μL of ESC medium to each well, resuspend the cells by pipetting up and down, transfer the cells to the pretreated 96-well plates (*see* Step 4) and incubate immediately. Repeat Steps 4–6 until all colonies are picked.
8. Incubate overnight.
9. Change medium the next morning and keep on changing every day until cultures are about 80% confluent.
10. Trypsinize cells and expand for freezing and/or further analysis.

3.3. Trapped Gene Identification

Here we describe protocols for 5′-RACE, inverse PCR and splinkerette PCR (SPLK-PCR) routinely used for the identification of genes trapped with the FlipRosaβgeo gene-trap vector *(17)*. The sequences of all primers are listed in **Table 3.1**.

Table 3.1
PCR and sequencing primers

Primer	Sequence (5′-3′)
3′SEQ	GCTAGCTTGCCAAACCTACAGGTGG
3′SPLKrev1	AGTCATAGACACTAGACAATCGG
3′SPLKrev2	CAGTCAATCGGAGGACTGGCG
5′SEQ	TTGTGGTCTCGCTGTTCCTTGGG
5′SPLKrev1	CGACCAGCTGTGCGCATAGTG
5′SPLKrev2	TTTGGCAAGCTAGCACAACC
I09	GGTGCAGGATATCCTGCTGATG
I10	TCAGGTCATGGATGAGCAGACG
I11	AGACGATTCATTGGCACCATGC
I12	GGTAGCCAGCGCGGATCATC
I13	TGTTTTGACCGCTGGGATCTGC
I14	ACTATCCCGACCGCCTTACTGC
I15	GTCTCAGAAGCCATAGAGCCC
I16	CGAGCCCCAGCTGGTTCTTTC
IPCRU3	CCTCCGATTGACTGAGTCGCCC
IPCRU4	TACCCGTGTATCCAATAAACCC
Lac2	CAAGGCGATTAAGTTGGGTAACG
Lac3	CCGTGCATCTGCCAGTTTGAGGGG
RACE-1	GACCACGCGTATCGATGTCGACTTTTTTTTTTTTTTTTG
RACE-3	GACCACGCGTATCGATGTCGAC
Splirev1	GCTAGCTTGCCAAACCTACAGGTGG
Splirev2	GCCAAACCTACAGGTGGGGTCTTT
SPLK-A	CGAAGAGTAACCGTTGCTAGGAGAGACCGTGGCTGAATGAGACTGGTGTCGACACTAGTGG
SPLK-B	GATCCCACTAGTGTCGACACCAGTCTCTAATTTTTTTTTTCAAAAAAA
SPLKfwd1	CGAAGAGTAACCGTTGCTAGGAGAGACC
SPLKfwd2	GTGGCTGAATGAGACTGGTGTCGAC

Note that inverse PCR as well as SPLK-PCR are not well suited for plasmid gene-trap vectors due to the unpredictable pre- and/or postinsertional modifications precluding the recovery of unambiguous GTSTs.

3.3.1. 5′-RACE

5′-RACE is used to amplify the cellular sequences (GTSTs) contained within the gene-trap fusion transcripts expressed at the insertion sites (**Fig. 3.2**).

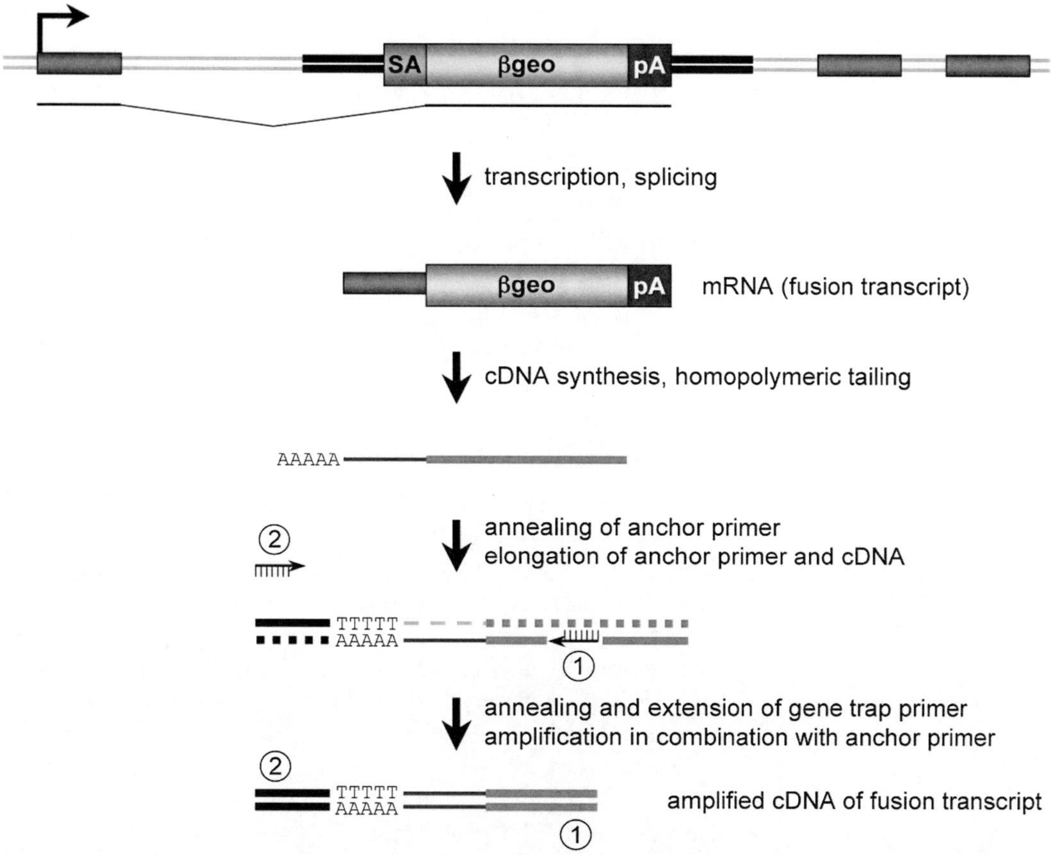

Fig. 3.2. **5′-RACE strategy for the amplification of gene-trap fusion transcripts.** The cell-provirus fusion transcript expressed at the insertion site is first reverse-transcribed into cDNA. After homopolymeric tailing and annealing of an anchor primer the exon sequences appended to the fusion transcripts are amplified by PCR using anchor (2) and gene-trap-specific (1) primers. SA, splice acceptor; *β*geo, *β*-galactosidase-neomycin-phosphotransferase fusion gene; pA, polyadenylation sequence.

1. Grow cells on 48-well plates until confluent.
2. Purify total RNA from each well using the RNeasy Mini Kit (Qiagen) or equivalent. Dissolve the purified RNA in 35 μL of DEPC-water.
3. Mix 5 μL of the RNA with 1 μL random hexamers (pdN_6, 100 ng/μL) and 5 μL of DEPC-water.
4. Keep at 65°C for 5 min and immediately place on ice.
5. Add 4 μL of 5x first strand buffer (Invitrogen), 1 μL of dNTP mix, 2 μL of 0.1 M DTT and 1 μL of RNAsin (Promega, 20–40 U/μL).
6. Incubate for 10 min at 25°C and 1 min at 42°C.
7. Immediately add 1 μL of Super Script II reverse transcriptase (200 U/μL, Invitrogen) and incubate for 1 h at 42°C.

8. Heat-inactivate reverse transcriptase by incubating for 15 min at 70°C.
9. Add 1 μL of RNAseH (2 U/μL, Invitrogen) and incubate for 20 min at 37°C.
10. Purify the cDNA using a PCR purification kit (e.g., Qiagen) and elute in 35 μL of water.
11. Transfer 23 μL of cDNA solution to a new tube, incubate at 70°C for 5 min and place on ice immediately.
12. Add 4 μL of 10x restriction buffer 4 (New England Biolabs), 4 μL of 25 mM $CoCl_2$, 4 μL of 2.5 mM dATP, 0.5 μL of terminal transferase (20 U/μL, New England Biolabs), 4.5 μL of water and incubate for 10 min at 37°C.
13. Stop the reaction by incubating for 10 min at 70°C.
14. Set up the PCR reaction in a 0.2-mL thin-walled PCR tube in a total volume of 50 μL by mixing 5 μL of 10x PCR Buffer, 1.5 μL of 50 mM $MgCl_2$, 1 μL of dNTP mix, 1 μL of a 10 μM stock solution of the anchor primer RACE-1, 1 μL of a 10-μM stock solution of the Lac3 primer, 0.3 μL of Taq polymerase (5 U/μL, Invitrogen), 5 μL of the tailed cDNA and 35.2 μL of water. Run the following PCR program: 94°C 2 min, 10 cycles of 94°C 30 s, 60°C 30 s, 72°C 40 s followed by 25 cycles of 94°C 30 s, 58°C 30 s, 72°C 40 s with a gradual increase of elongation times by 10 s per cycle and a final elongation at 72°C for 7 min.
15. Add 1 μL of the first PCR reaction directly to PCR tubes containing the same reaction mix except for the primers. Add 1 μL of each RACE-3 and Lac2 primers from 10-μM stock solutions. Proceed with the PCR reaction for 35 cycles of 94°C 30 s, 58°C 30 s, 72°C 1 min and initial and final denaturation/elongation steps of 94°C, 2 min and 72°C 7 min, respectively.
16. Purify the amplification products using a PCR purification kit and sequence directly using the Lac2 primer. Note that even if amplification products produce a smear on agarose gels, sequencing will still be successful in most cases.

3.3.2. Inverse PCR

Inverse PCR uses circularized genomic DNA restriction fragments as templates for amplification. These are obtained by cleaving the genomic DNA with a class II restriction enzyme cutting near the ends of the inserted gene-trap provirus and in the flanking DNA. Amplification of the flanking cellular DNA is achieved with proviral sequence primers that point in opposite directions within the provirus/flanking DNA sequence circle. The linear amplification product obtained from this reaction includes the genomic DNA directly adjacent to the inserted provirus (**Fig. 3.3**).

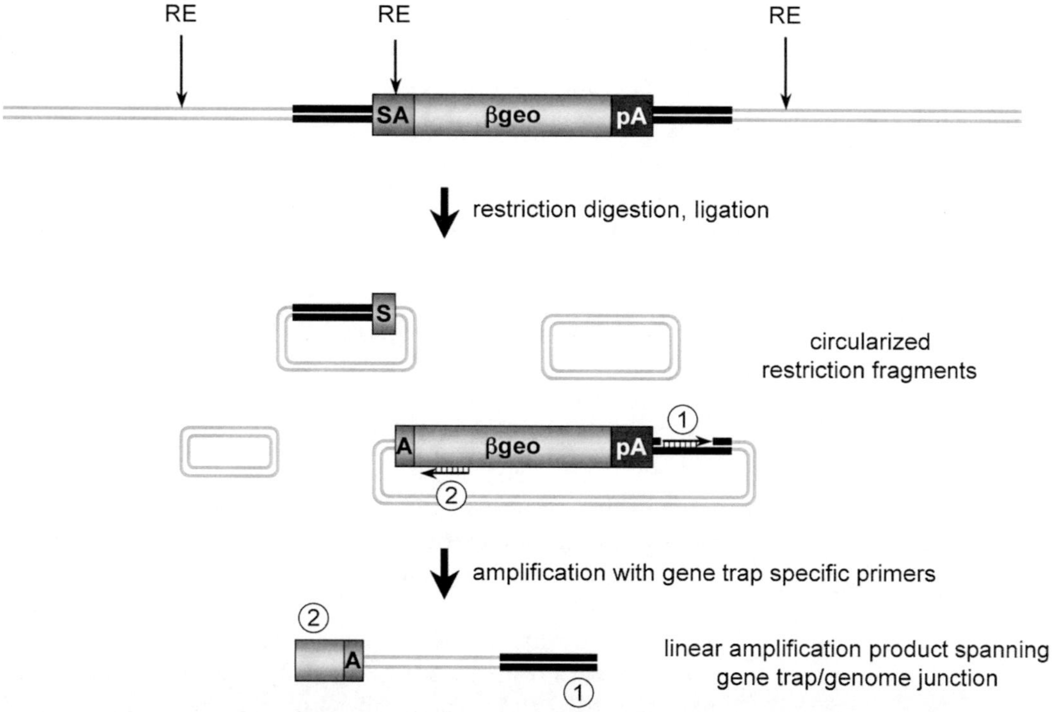

Fig. 3.3. **Inverse PCR strategy for the amplification of genomic sequences flanking the gene-trap insertion sites.** Genomic DNA from trapped ESC lines is digested with a class II restriction enzyme with cleavage sites in the gene-trap vector and the flanking genomic DNA. The resulting restriction fragments are ligated under conditions favoring intramolecular circularization. DNA circles are then subjected to PCR using primers complementary to vector sequences (1, 2) pointing in opposite directions. Using appropriate primer combinations the cellular sequences 5′ and 3′ to the inserted gene trap can be amplified separately. RE, restriction enzyme; SA, splice acceptor; *β*geo, *β*-galactosidase-neomycin-phosphotransferase fusion gene; pA, polyadenylation sequence.

The choice of restriction enzymes and PCR primers depends on the sequence of the integrated gene-trap provirus. Select enzymes with high religation efficiency to ensure optimal circularization of the restriction fragments. It is also advisable to use at least two different enzymes in parallel to increase sequence retrieval efficiency.

1. Grow trapped ESCs on 48-well plates until confluent.
2. Purify genomic DNA from single wells using standard methods (e.g., DNeasy Blood & Tissue Kit, Qiagen) and elute the DNA in 150 μL of water.
3. Digest 1–5 μg of genomic DNA with *Nsp*I or *Ssp*I in a final volume of 100 μL and incubate at 37°C overnight.
4. Purify the digested DNA using a PCR purification kit (e.g., Qiagen) and elute in 50 μL of water.
5. Incubate 45 μL of the eluate with 800 U of T4-DNA ligase (cohesive end ligation units; New England Biolabs) in a final volume of 300 μL of 1x ligase buffer at 16°C overnight. The relatively large volume is chosen to favor intramolecular religation.

6. Purify the DNA by using the PCR purification kit and elute in 50 μL of water.
7. For PCR amplification mix in a 0.2-mL thin-walled PCR tube 5 μL of the ligation products with 5 μL of 10x PCR buffer, 1.5 μL of 50 mM $MgCl_2$, 1 μL of dNTP mix, 1 μL of each of the appropriate primers (*see* **Table 3.2**) from 10 μM stock solutions, 0.3 μL of Taq polymerase (5 U/μL, Invitrogen) and 35.2 μL of water to obtain a final volume of 50 μL.
8. Place the tubes in a thermal cycler and start cycling using the following touchdown program: 1 cycle at 94°C for 4 min, 10 cycles of 94°C 30 s, 57°C 30 s with a 0.2°C annealing temperature decrease per cycle, 72°C, 2 min. This is followed by 30 cycles of 94°C 30 s, 55°C 30 s, 72°C 2 min and a final elongation step at 72°C for 7 min.
9. Check the resulting PCR products on an agarose gel. Depending on the amount of product usually appearing as a faint smear on the gel, use 1–5 μL of the first PCR reaction without further purification for a nested PCR reaction using the reaction mix identical to that described in Step 7 except for the nested primers (*see* **Table 3.2**) of which you add 1 μL from 10-μM stocks. Place tubes into the cycler, denature for 4 min at 94°C and allow the reaction to proceed for 35 cycles of 94°C 30 s, 58°C 30 s, 72°C 1 min followed by final elongation step at 72°C for 7 min.
10. Purify the PCR products from an agarose gel and sequence with either of the two primers used for the nested amplification (*see* **Table 3.2**).

3.3.3. Splinkerette PCR

Like inverse PCR, SPLK-PCR is used for the amplification of DNA fragments positioned between a region of known sequence and a nearby restriction site. SPLK-PCR also includes a ligation step in which a generic, double-stranded oligonucleotide (adaptor) is attached to the genomic restriction fragments obtained

Table 3.2
Primer combinations for inverse PCRs on ESC clones harboring the gene-trap FlipRosa*β*geo

Enzyme	Junction	Primers first PCR	Primers nested PCR
*Nsp*I	5′-junction	IPCRU3/I14	IPCRU4/I13
	3′-junction	Splirev1/I16	Splirev2/I15
*Ssp*I	5′-junction	IPCRU3/I10	IPCRU4/I09
	3′-junction	Splirev1/I12	Splirev2/I11

after cleaving with a class II restriction enzyme. The trick for specific PCR amplification is the hairpin structure of the adaptor's bottom strand and a generic primer corresponding to the adaptor's upper strand *(21)*. Both insure that only a fully elongated bottom strand enters the exponential phase of the PCR reaction (**Fig. 3.4**).

For the gene-trap FlipRosaβgeo, we digest the genomic DNA with *Bst*YI and amplify the 5′ and 3′ gene-trap flanking sequences in two separate reactions. The primers for both PCRs are listed in **Table 3.3**.

For other vectors other enzymes might be more suitable. When you choose an enzyme make sure it can be heat-inactivated because there is no purification step before the splinkerette adaptor ligation. In addition, modify the splinkerette adaptor sequence according to the expected restriction fragment ends.

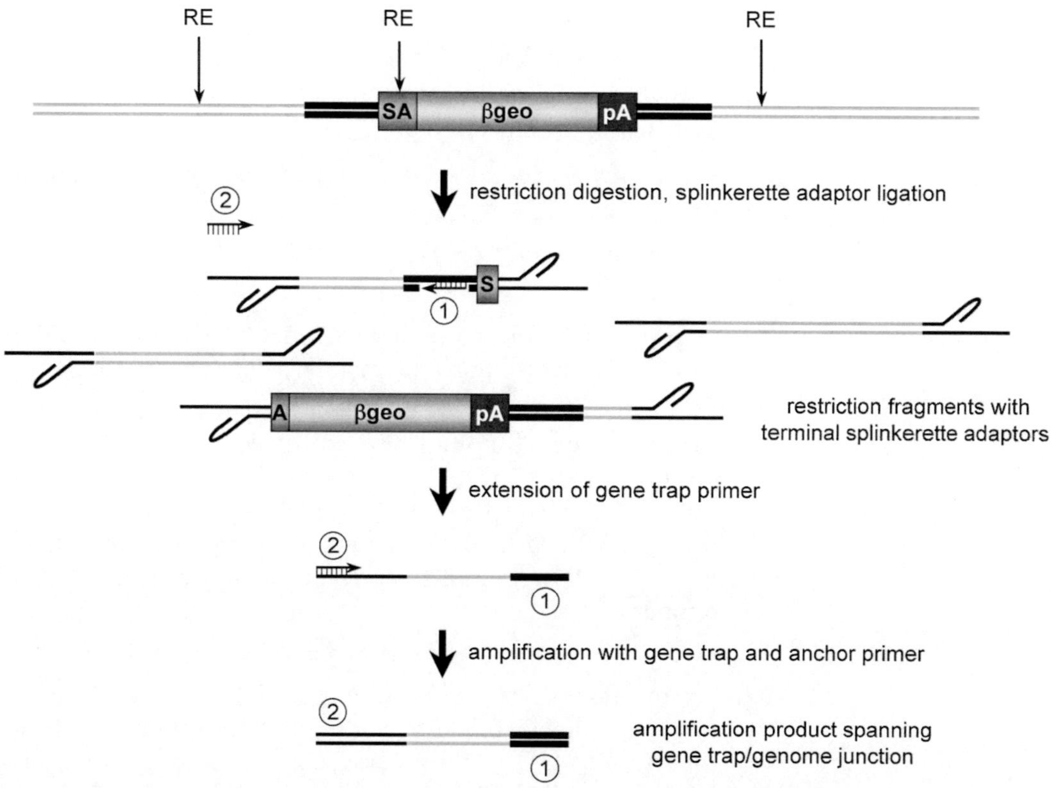

Fig. 3.4. **Splinkerette PCR strategy for the amplification of genomic sequences flanking the gene-trap insertion sites.** Genomic DNA from trapped ESC lines is digested with a class II restriction enzyme that cleaves within the inserted gene-trap vector and in the flanking genomic DNA. The cleaved DNA is ligated to a double-stranded splinkerette adaptor whose bottom strand folds back to form a hairpin loop. The ligated DNA is subjected to a PCR reaction using primers corresponding to the bottom strand of the gene-trap vector (1) and to the 5′ – non-base-paired – end of the adaptor (2). This ensures that only a fully elongated bottom strand enters the exponential phase of the PCR reaction. Like inverse PCR this method allows the separate amplification of the vector–genome junctions at both ends of the gene trap. RE, restriction enzyme; SA, splice acceptor; βgeo, β-galactosidase-neomycin-phosphotransferase fusion gene; pA, polyadenylation sequence.

Table 3.3
Primer combinations for splinkerette PCRs on ESC clones harboring the gene-trap FlipRosaβgeo

Junction	Primers first PCR	Primers nested PCR	Sequencing primer
5′-junction	SPLKfwd1/5′SPLKrev1	SPLKfwd2/5′SPLKrev2	5′SEQ
3′-junction	SPLKfwd1/3′SPLKrev1	SPLKfwd2/3′SPLKrev2	3′SEQ

1. Grow trapped ESCs on 96-well plates until confluent.
2. Add 40 μL of lysis buffer to the cells.
3. Incubate for 2 h at 60°C and allow to cool to room temperature.
4. Add 15 μL of ethanol/NaCl solution.
5. Precipitate DNA for 1 h at room temperature.
6. Centrifuge for 15 min at 2800*g*. Wash the DNA three times with 150 μL of 70% (v/v) ethanol.
7. Air-dry pellet. You can also use a speed vac, but make sure not to overdry.
8. Add 40 μL of 5 mM Tris-HCl, pH 8.0.
9. Allow DNA to dissolve 12–16 h at room temperature.
10. Digest 9 μL of the genomic DNA in a final volume of 20 μL with *Bst*YI according to the manufacturer's instructions.
11. In the meantime prepare double-stranded splinkerette adaptor by mixing 400 pmol of each SPLK-A and SPLK-B oligonucleotides (**Table 3.1**) in a total volume of 40 μL of 5 mM Tris-HCl, pH 7.5, 25 mM NaCl, 0.5 mM DTE to obtain a 10 μM final concentration. Denature for 3 min at 97°C and allow strands to anneal by slowly cooling to room temperature.
12. Heat-inactivate restriction enzyme for 20 min at 80°C.
13. Set up ligation in a final volume of 30 μL using 20 μL of restriction digest, 0.33 μM of the annealed splinkerette adaptor and 400 U T4-DNA ligase (cohesive end ligation units; New England Biolabs) according to the manufacturer's instructions.
14. Incubate 12–16 h at 16°C.
15. Purify ligation products using a PCR purification kit and elute the DNA into 45 μL of 5 mM Tris-HCl, pH 8.0.
16. Set up a PCR reaction in a 0.2-mL thin-walled PCR tube in a volume of 15 μL containing 1x PCR buffer (Eppendorf) supplemented with additional 0.25 mM $MgCl_2$, 0.2 mM dNTPs,

0.2 μM of each primer (*see* **Table 3.3**), 0.4 U Taq Polymerase (Eppendorf) and 5 μL of the adaptor ligated DNA. Be aware that the PCR reaction buffer supplied by Eppendorf already contains an unspecified concentration of $MgCl_2$, the additionally supplied Mg^{++} is on top of that. Run the following PCR program: 94°C for 75 s as initiation step, 2 cycles of 94°C 20 s, 64°C 15 s and 30 cycles of 94°C 20 s, 60°C 15 s, 72°C 2 min and a final elongation step at 72°C for 7 min.

17. Use 1 μL of the primary amplification reaction without further purification in a nested PCR using the primers listed in **Table 3.3** in a final volume of 10 μL. Run the following protocol: 94°C 60 s initial denaturation step followed by 35 cycles of 94°C 20 s, 60°C 15 s, 72°C 60 s and final elongation step at 72°C for 7 min
18. Use 1 μL of the final amplification product for direct sequencing using primers 5′ SEQ or 3′ SEQ (**Table 3.1**).

4. Notes

1. Each ESC line requires a specific medium formulation, which should be obtained from the source of the ESC line. All ESC media are supplemented with 1000–1500 U/mL LIF; the concentration depends on the ESC line (e.g., use 1000 U/mL for E14TG2a.4 and 1500 U/mL for TBV-2).
2. Gelatinized tissue culture surfaces are required for culturing MEFs and ESCs. To gelatinize tissue culture dishes or multiwell plates cover the bottom of the culture vessels with gelatine solution for at least 10 min at room temperature. Before seeding MEFs or ESCs aspirate the gelatine solution and wash once with PBS.
3. Tissue culture dishes or multiwell plates with MEF feeder cells have to be prepared the day before the ESCs are seeded onto the feeder layer. To prepare feeder layer plates grow early passage MEFs (≤passage 4) on gelatine-coated culture dishes (*see* **Note 2**) until cells are about 80% confluent. Incubate MEFs for 2 h at 37°C, 5% CO_2 in MEF medium containing 10 ng/mL Mitomycin C. Wash the cells four times in 15 mL of PBS and trypsinize by adding 7 mL 1x trypsin solution (Invitrogen). Stop reaction after 3–5 min by adding 7 mL of MEF medium, harvest cells in a 15-mL Falcon tube and spin for 5 min at 250*g*. Resuspend cells in MEF medium, count and seed immediately at a density of 1×10^4 cells/cm^{2}onto freshly gelatinized dishes or multiwell plates.

4. The Phoenix-Eco cell line is a 293T cell-derived, second-generation retrovirus packaging line widely used for the production of helper-free ecotropic retroviruses (for more details see www.stanford.edu/group/nolan). Phoenix-Eco cells are much less adherent than many other cells you are used to and therefore require gentle handling during washings and medium change. To obtain optimal transfection efficiencies and good virus titers make sure the cells never reach confluence in culture. Phoenix-Eco cells carry stable integrations of gag/pol and env expression cassettes. Selectable markers coexpressed with the viral proteins confer resistance to hygromycin and diphtheria toxin. To ensure highest packaging efficiency, Phoenix-Eco cell should undergo reselection every 2–3 months by growing in selection medium containing 300 μg/mL hygromycin B and 1 μg/mL diphtheria toxin for 7 days.

Acknowledgements

We thank all members of the von Melchner laboratory for helpful discussions and suggestions. This work was supported by grants from the Bundesministerium für Bildung und Forschung (BMBF) to the German Gene Trap Consortium and from the Deutsche Forschungsgemeinschaft (DFG) to HvM.

References

1. Austin CP, Battey JF, Bradley A, et al. The knockout mouse project. Nat. Genet. 2004;36:921–4.
2. Auwerx J, Avner P, Baldock R, et al. The European dimension for the mouse genome mutagenesis program. Nat. Genet. 2004;36:925–7.
3. Stanford WL, Epp T, Reid T, Rossant J. Gene trapping in embryonic stem cells. Methods Enzymol. 2006;420:136–62.
4. von Melchner H, Ruley HE. Identification of cellular promoters by using a retrovirus promoter trap. J. Virol. 1989;63:3227–33.
5. von Melchner H, Reddy S, Ruley HE. Isolation of cellular promoters by using a retrovirus promoter trap. Proc. Natl. Acad. Sci. USA 1990;87:3733–7.
6. Hicks GG, Shi EG, Li XM, Li CH, Pawlak M, Ruley HE. Functional genomics in mice by tagged sequence mutagenesis. Nat. Genet. 1997;16:338–44.
7. Osipovich AB, White-Grindley EK, Hicks GG, et al. Activation of cryptic 3′ splice sites within introns of cellular genes following gene entrapment. Nucleic Acids Res. 2004;32:2912–24.
8. Skarnes WC, Moss JE, Hurtley SM, Beddington RS. Capturing genes encoding membrane and secreted proteins important for mouse development. Proc. Natl. Acad. Sci. USA 1995;92:6592–6.
9. Mitchell KJ, Pinson KI, Kelly OG, et al. Functional analysis of secreted and transmembrane proteins critical to mouse development. Nat. Genet. 2001;28:241–9.
10. De-Zolt S, Schnütgen F, Seisenberger C, et al. High-throughput trapping of secretory pathway genes in mouse embryonic stem cells. Nucleic Acids Res. 2006;34:e25.
11. Wiles MV, Vauti F, Otte J, et al. Establishment of a gene-trap sequence tag library to

generate mutant mice from embryonic stem cells. Nat. Genet. 2000;24:13–4.

12. Hansen J, Floss T, Van Sloun P, et al. A large-scale, gene-driven mutagenesis approach for the functional analysis of the mouse genome. Proc. Natl. Acad. Sci. USA 2003;100:9918–22.
13. Zambrowicz BP, Abuin A, Ramirez-Solis R, et al. Wnk1 kinase deficiency lowers blood pressure in mice: a gene-trap screen to identify potential targets for therapeutic intervention. Proc. Natl. Acad. Sci. USA 2003;100:14109–14.
14. Skarnes WC, von Melchner H, Wurst W, et al. A public gene trap resource for mouse functional genomics. Nat. Genet. 2004;36:543–4.
15. Niwa H, Araki K, Kimura S, Taniguchi S, Wakasugi S, Yamamura K. An efficient gene-trap method using polyA trap vectors and characterization of gene-trap events. J. Biochem. (Tokyo) 1993;113: 343–9.
16. Shigeoka T, Kawaichi M, Ishida Y. Suppression of nonsense-mediated mRNA decay permits unbiased gene trapping in mouse embryonic stem cells. Nucleic Acids Res. 2005;33:e20.
17. Schnütgen F, De-Zolt S, Van Sloun P, et al. Genomewide production of multipurpose alleles for the functional analysis of the mouse genome. Proc. Natl. Acad. Sci. USA 2005;102:7221–6.
18. Horn C, Hansen J, Schnütgen F, et al. Splinkerette PCR for the more efficient characterization of gene trap event. Nat. Genet. 2007;39:807–8.
19. Merrihew RV, Marburger K, Pennington SL, Roth DB, Wilson JH. High-frequency illegitimate integration of transfected DNA at preintegrated target sites in a mammalian genome. Mol. Cell Biol. 1996;16:10–8.
20. Friedrich G, Soriano P. Promoter traps in embryonic stem cells: a genetic screen to identify and mutate developmental genes in mice. Genes Dev. 1991;5:1513–23.
21. Devon RS, Porteous DJ, Brookes AJ. Splinkerettes – improved vectorettes for greater efficiency in PCR walking. Nucleic Acids Res. 1995;23:1644–5.

Chapter 4

Chromosome Engineering in ES Cells

Louise van der Weyden, Charles Shaw-Smith, and Allan Bradley

Abstract

Chromosomal rearrangements, such as deletions, duplications, inversions and translocations, occur frequently in humans and can be disease-associated or phenotypically neutral. To understand the genetic consequences of such genomic changes, these mutations need to be modelled in experimentally tractable systems. The mouse is an excellent organism for this analysis because of its biological and genetic similarity to humans, the ease with which its genome can be manipulated and the similarity of observed affects. Through chromosome engineering, defined rearrangements can be introduced into the mouse genome. The resulting mouse models are leading to a better understanding of the molecular and cellular basis of dosage alterations in human disease phenotypes, in turn opening new diagnostic and therapeutic opportunities.

Key words: Chromosomal rearrangements, Cre recombinase, loxP, mouse genome, embryonic stem cell.

1. Introduction

1.1. Chromosomal Rearrangements in Humans

Chromosomal rearrangements, such as deletions, duplications, inversions and translocations, occur frequently in humans. Deletions may be associated with developmental syndromes such as DiGeorge and Williams syndrome (germline deletions) as well as tumourigenesis (somatic deletions). Duplications probably occur more frequently because the biological consequences are often less severe than those of deletions. Three separate whole-chromosome trisomies (those for chromosomes 13, 18 and 21) are compatible with postnatal life, while no whole-autosome deletions are similarly recognized. Translocations are associated with an increased risk of infertility, miscarriage and congenital malformations; they

Ralf Kühn, Wolfgang Wurst (eds.), *Gene Knockout Protocols: Second Edition, vol. 530,*

DOI 10.1007/978-1-59745-471-1_4 Springerprotocols.com

are also important in certain types of cancer, including leukaemia. They are usually associated with little or no net loss or gain of genetic material (balanced), but genes that are normally tightly regulated may be taken out of their normal context and novel fusion products may be generated, which exert pathological effects. Translocations may be inherited in unbalanced form leading to significant loss and/or gain of genetic material, with corresponding pathological consequences. Inversions are also usually balanced, though, as with translocations, there may be loss or gain of material at the breakpoints, and they also may be inherited in unbalanced form, with similar consequences.

1.2. Inducing Chromosomal Rearrangements in Mice

Where human genetic analysis has reached its natural limits, genetic manipulation of mice has begun to make significant progress in unravelling the mechanistic basis of human chromosomal diseases. Chromosomal rearrangements can occur spontaneously in mice (albeit at a low rate) or be induced by exposure to chemical or physical mutagens. However, although these mutagens have generated some valuable mouse models for human diseases, such as the segmental trisomy 16 (Ts65Dn) mouse (a model for Down syndrome (DS)) *(1, 2)*, their usefulness is limited by the fact that both the endpoints and type of rearrangements are random and cannot be predetermined.

These limitations have been overcome by the development of "chromosome engineering" techniques, based on gene targeting in embryonic stem (ES) cells and the Cre/*loxP* site-specific recombination (*see* **Note 1**) *(3)*. With this system, the type of chromosomal rearrangement depends upon the orientation of the *loxP* sites and the relative position of selection cassettes used to recover cells that have undergone the long-range *Cre/loxP* recombination event (*see* **Fig. 4.1**). This strategy relies on the conservation of the arrangement of genes on the chromosomes in large blocks with similar gene order (synteny between the mouse and human chromosomes).

1.3. Chromosome Engineering

The principle behind chromosome engineering is simple: by two consecutive gene-targeting events one can deliver specific recombination recognition sites (*loxP* sites) to two predefined loci in the genome, and following expression of Cre recombinase the desired rearrangement is generated through site-specific recombination between the two *loxP* sites. An important aspect of this technology is the ability of Cre to catalyze recombination between sites that are located far apart in the genome, even when located on different chromosomes. Chromosome engineering strategies have been used by numerous different groups to generate mouse models that accurately recapitulate human chromosomal rearrangements (reviewed in *(4)*). These strategies differ mainly in their methods for identification of the recombinants: using positive selection

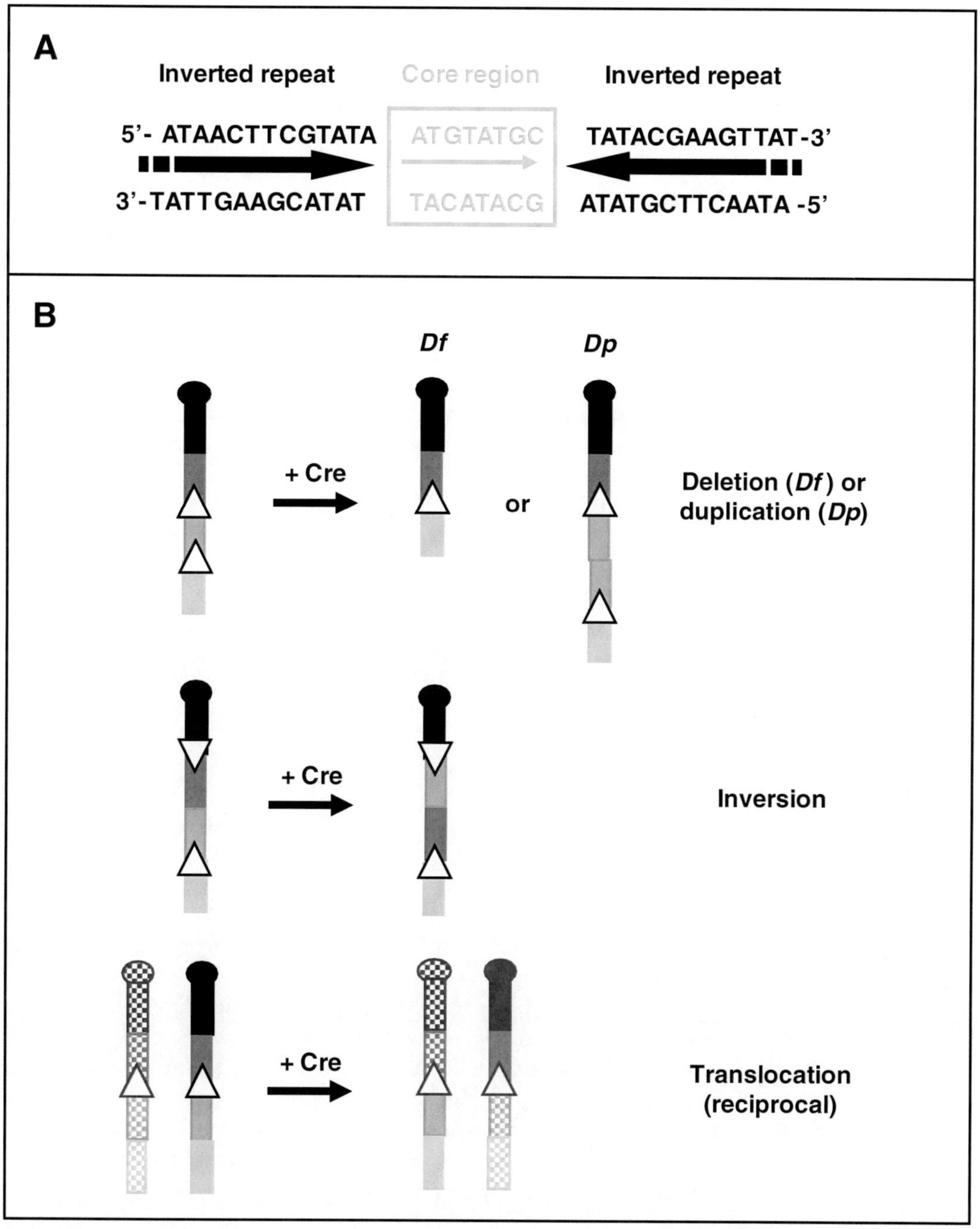

Fig. 4.1. **Recombinase reactions catalyzed by Cre.** (**A**) The target site recognized by the P1 bacteriophage recombinase, Cre, is called *loxP* (locus of cross over (*x*) in *P1*). The *grey arrow* in the core region of the *loxP* site indicates the direction (orientation) of the *loxP* site. (**B**) The orientation of the loxP sites determines the type of recombination event that occurs: a *cis* recombination event between two *loxP* sites (*white triangles*) in the same orientation will lead to deletion (*Df*) or duplication (*Dp*) of the flanked DNA sequence; recombination between *loxP* sites orientated in opposite directions results in an inversion of the flanked DNA sequence; recombination between two *loxP* sites in *trans* will lead to the reciprocal exchange (translocation) of the regions that flank the *loxP* sites.

(5, 6), using negative selection *(7)* or screening without selection *(8)*. Selection is useful when the rearrangement occurs at low efficiency, such as when generating large inversions. The use of positive selection is preferable to negative selection as it ensures specificity of the recombined products by allowing the survival of clones that have undergone a precise rearrangement. By contrast, negative selection gives a much higher background as a variety of other genetic changes in the cell that affect the selection cassette could allow the clones to survive. In addition, positive selection allows for rearrangements without a net loss of genetic material to be readily engineered (such as duplications, inversions and translocations) while negative selection is only applicable to deletions. Therefore, due to the obvious advantages of positive selection, we will focus on the experimental design and analytical tools for the positive-selection-based chromosome engineering strategy. Although the two groups that independently developed the positive selection strategy used a virtually identical scheme *(5, 6)*, we base our discussions on the vector system developed in our laboratory.

2. General Strategy

2.1. Selection of Endpoints

Chromosome engineering cassettes are generally targeted to a specific chromosomal location. Genes of known chromosomal location may be used as endpoints, thanks to sequencing and annotation of the mouse genome (*see* the MGSC Ensembl Mouse Genome Server at http://www.ensembl.org/Mus_musculus/). Numerous genes have been used as endpoints for generating chromosomal rearrangements, including *p63 (9)*, *Es2/Dgsi* and *Ufd1l (10)*, *Csn3* and *Zfp179 (11)* and the *HoxB* cluster *(12)* to name a few. Alternatively, SSLP microsatellite markers make useful endpoints as they have also been genetically mapped (*see* the Whitehead Institute STS Physical Map of the Mouse at http://www-genome.wi.mit.edu/cgi-bin/mouse/index). Numerous SSLP markers have been successfully used as the endpoints for engineered chromosomal deletions on mouse chromosome 11 *(13, 14)*.

2.2. Generating the Chromosomal Rearrangement

Once the endpoints have been selected, the first step involves the targeting of an insertion (or replacement) vector containing a *loxP* site, a positive selection cassette (e.g. neomycin) and one of two complementary, but non-functional, fragments of the *Hprt* gene into the desired locus (selected endpoint) of the ES cell genome (*see* **Fig. 4.2**) *(3)* (*see* **Note 2**). ES cell clones with a *loxP* site targeted to the first endpoint can be identified by positive

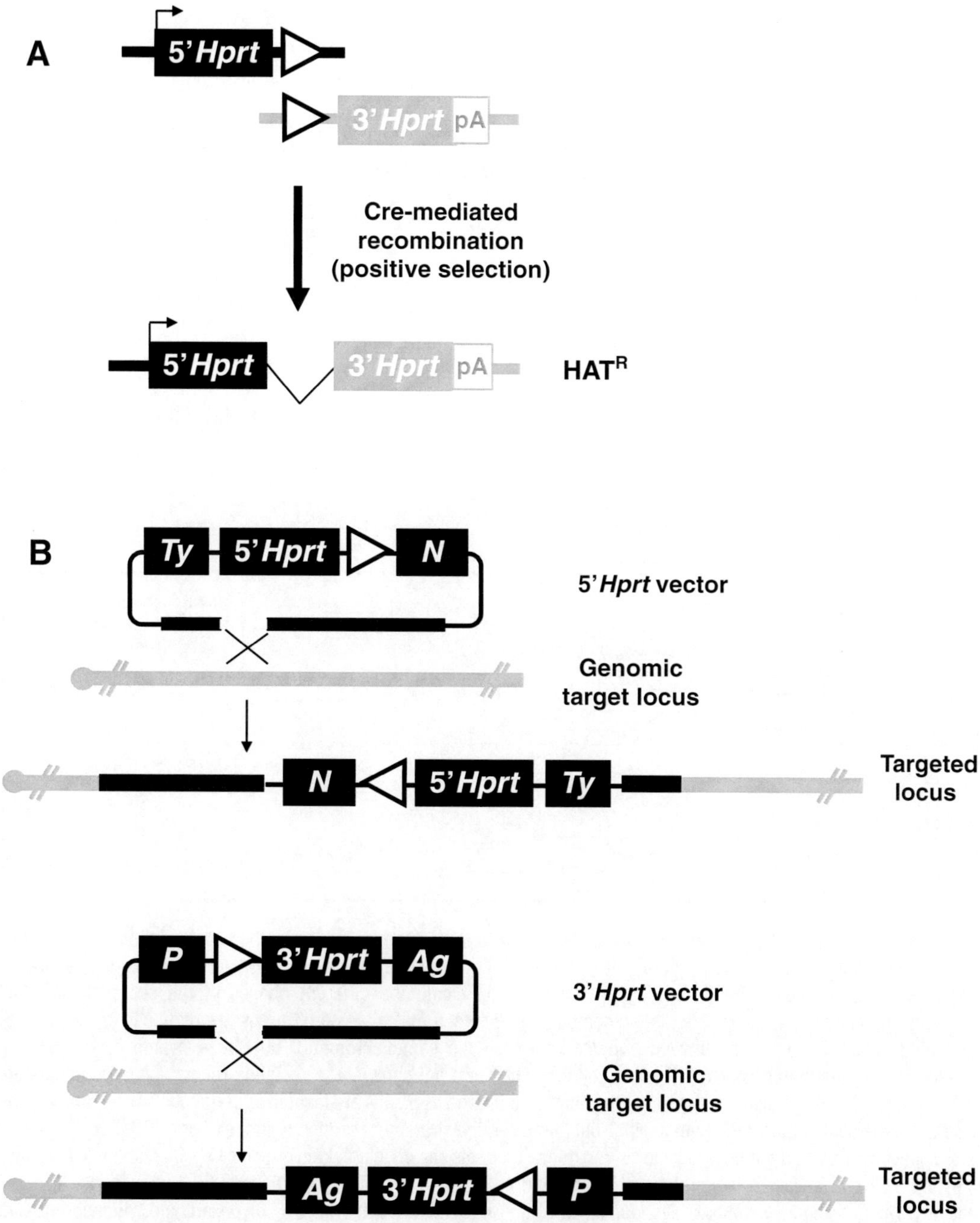

Fig. 4.2. **Gene targeting in ES cells.** (**A**) The two-part *Hprt* minigene system used to select for Cre-mediated recombination consists of the 5′*Hprt* (exons 1–3; *black box*) and 3′*Hprt* (exons 4–9 and polyA (pA); *grey box*) cassettes. Recombination between loxP sites (*white triangles*) reconstitutes a functional *Hprt* minigene, rendering cells resistant to HAT (aminopterin, hypoxanthine, thymidine). (**B**) Insertional targeting vectors can be used to insert *loxP* sites, positive selectable markers, hypoxanthine phosphoribosyltransferase (*Hprt*) gene fragments and coat-colour markers to pre-determined loci in the ES cell genome *(5)*. The vector (*thin black line*) is linearized in the region of homology (*thick black line*) to stimulate targeted insertion into the locus (*thick grey line*). The *5-Hprt* vector contains the neomycin selectable marker (*N*), the 5′ end of the *Hprt* minigene, a *loxP* site (*white triangle*) and the *Tyrosinase* minigene (*Ty*) coat-colour marker. The *3′-Hprt* vector contains the puromycin selectable marker (*P*), the 3′ end of the *Hprt* minigene, a *loxP* site (*white triangle*) and the

selection and followed by genotyping (Southern blot or PCR). The second step involves the targeting of a second vector containing a *loxP* site, a different positive selection cassette (e.g. puromycin) and the complementary fragment of the *Hprt* gene into the second endpoint (*see* **Fig. 4.2**). The generation of these targeting vectors is detailed below (**Section 2.3**). To induce *loxP*

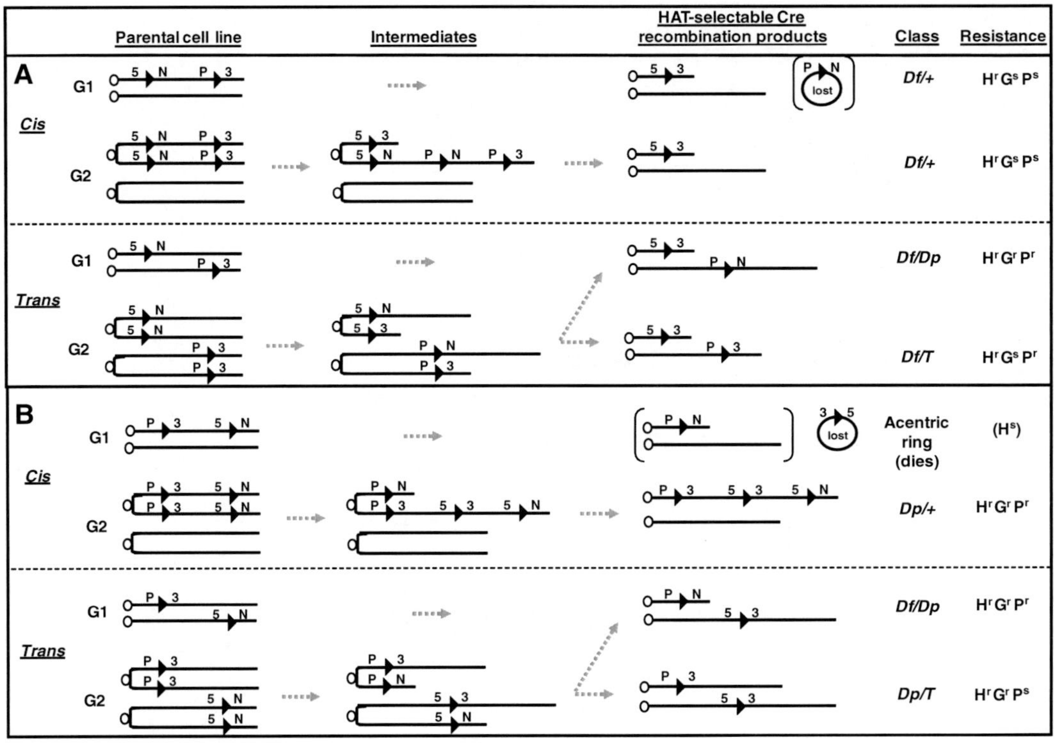

Fig. 4.3. **Engineering a deletion and/or duplication in embryonic stem cells.** When the two *loxP* sites are targeted in the same orientation with respect to the centromere, the half *hprt* cassettes can lie in two orientations, (**A**) outside the floxed region or (**B**) inside the floxed region. G1 and G2 indicate the different phases of the cell cycle in which recombination occurs. For G2 events, only recombination between *loxP* sites on different chromatids is considered: G2 recombination can also occur between two *loxP* sites on the same chromatid, but these events have the same consequence as the corresponding G1 events and are therefore not shown. The resulting recombination products and hence drug sensitivity ("resistance") will depend upon whether the *loxP* sites are located in *cis* (on the same chromosome) or *trans* (on the two chromosome homologs). Note that using the strategy described in the text, only viable HAT-resistant recombination products are recovered and scored (those products shown in brackets are not HAT selectable). A *loxP* site is indicated by a *black triangle*, a centromere is indicated as an *open circle*. Abbreviations: 5, *5′ hprt* cassette; 3, *3′ hprt* cassette; *Df*, deficiency (deletion); *Dp*, duplication; G, G418 (neomycin); H, HAT (hypoxanthine, aminopterin and thymidine); N, neomycin selection cassette (conferring resistance to G418); P, puromycin selection cassette (conferring resistance to puromycin); r, resistant; s, sensitive; *T*, targeted.

(Fig. 4.2. continued) *K14-Agouti* transgene (*Ag*) coat-colour marker. Cre-mediated recombination unites the 5′ and 3′ cassettes and restores *Hprt* activity (which is required for purine biosynthesis), thereby allowing the desired recombination events to be selected for in HAT.

recombination, a Cre-expression vector is electroporated into the double-targeted clones and the recombinants are selected in HAT (hypoxanthine, aminopterin and thymidine). Commonly used examples of such vectors include pOG231 *(15)*, pTurboCre (GenBank accession no. AF334827), pCrePAC *(16)* and pBS185 *(17)*.

The type of chromosomal rearrangement generated and selected from the double-targeted ES cells is determined by the *loxP* configuration, the order of the *Hprt* fragment and whether they have been targeted to the same or homologous chromosomes. *LoxP* sites in the same orientation on the same chromosome generate a chromosomal deletion or duplication event (*see* **Fig. 4.3**), whereas *loxP* sites in the opposite orientation on the same chromosome generate a chromosomal inversion (*see* **Fig. 4.4**) (discussed in more detail in **Section 7.2**). Translocations are generated when *loxP* sites, orientated in the same direction relative to their respective centromeres, are targeted to non-homologous chromosomes (*see* **Fig. 4.5**). The use of chromosomal engineering to study human chromosomal disorders such as deletions and duplications relies on the conservation of the order of genes in the two genomes. The

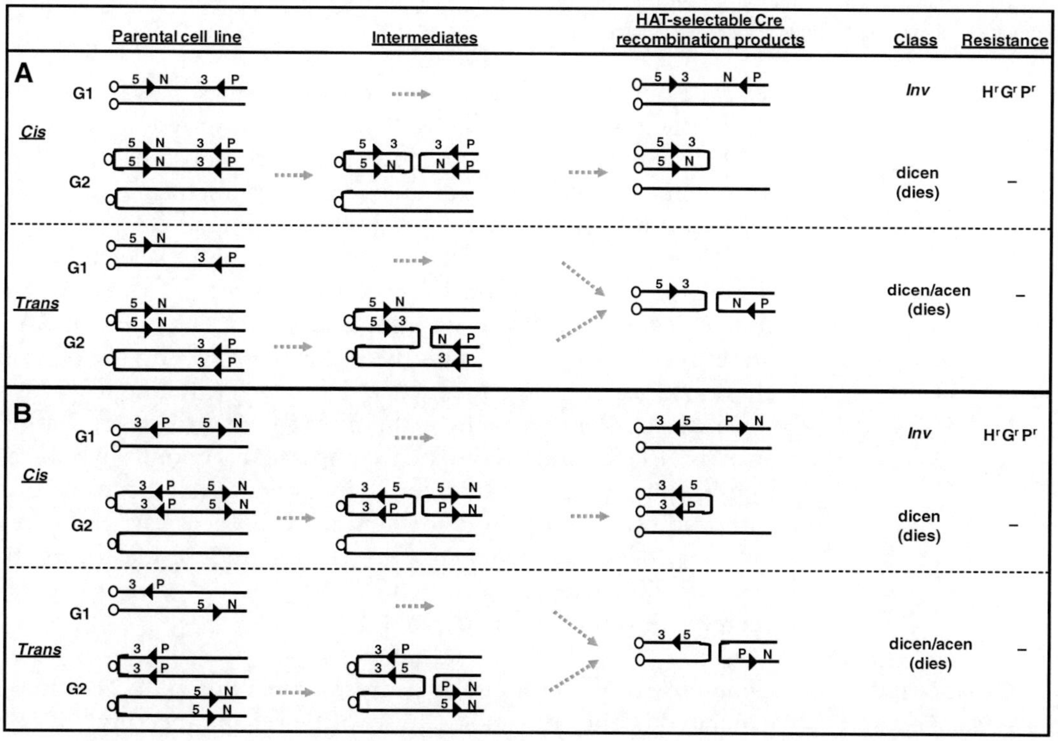

Fig. 4.4. **Engineering an inversion in embryonic stem cells.** When the two *loxP* sites are targeted in the opposite orientation with respect to the centromere, either **(A)** facing towards or **(B)** away from each other, a chromosomal inversion will only occur if the *loxP* sites are in *cis* (on the same chromosome). Note that using the strategy described in the text, only viable HAT-resistant recombination products are recovered and scored. Details of the abbreviations and symbols are the same as for **Fig. 4.3** (with additional abbreviations: acen, acentric; dicen, dicentric).

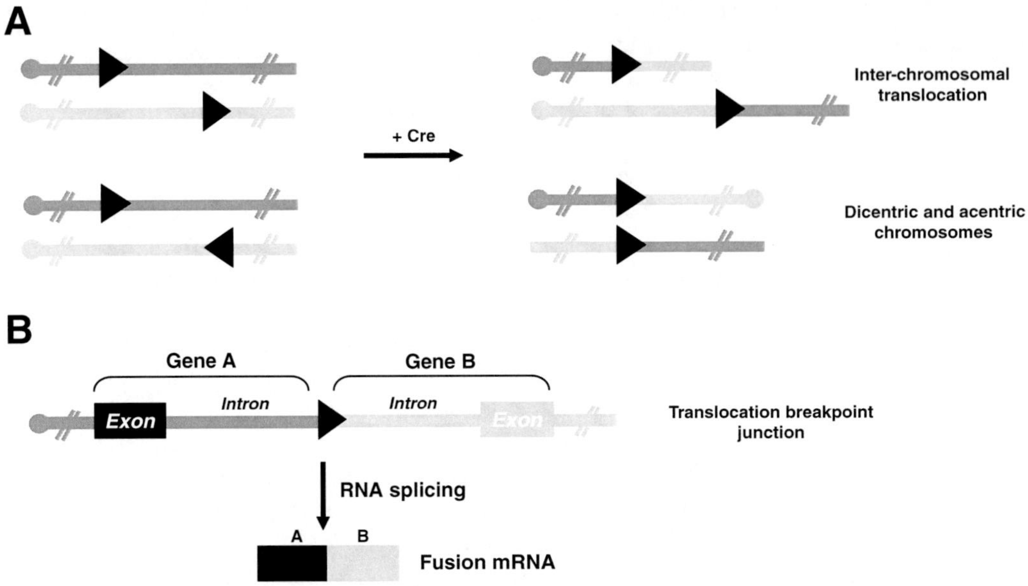

Fig. 4.5. **Engineering chromosomal translocations.** (**A**) Cre-mediated recombination leads to chromosomal translocation or dicentric and acentric chromosomes, depending on the relative orientations of the *loxP* sites (*black triangles*) on two non-homologous chromosomes. (**B**) An in-frame fusion mRNA and protein can be generated by engineering an appropriate junction at the translocation breakpoint.

relative orientation of the genes in the two species is also important when generating translocations. If one gene is inverted (with respect to the centromere) in one species, interchromosomal *Cre/loxP* recombination will generate an acentric fragment and dicentric chromosome instead of a balanced translocation.

The Cre-mediated recombination event can be selected for in culture because a functional *Hprt* cassette is reconstituted, which confers resistance to the drug using selection in HAT (**Fig. 4.2A**). Other methods for identifying ES cell clones carrying the desired chromosomal rearrangement include junction fragment analysis (by PCR or Southern blot), fluorescence in situ hybridization analysis (FISH) and array-based comparative genome hybridization (CGH), which are all discussed in more detail in **Section 7**. As with traditional gene-targeting strategies, ES cells carrying chromosomal rearrangements are injected into mouse blastocysts to generate chimaeras, from which the progeny that carry the engineered chromosomes are derived.

2.3. Generation of Gene-Targeting Vectors for Chromosome Engineering

Gene-targeting vectors, such as those shown in **Fig. 4.2B**, can be generated in the conventional way or by recombineering *(5, 18)*. However, the indexed "two-library" system, composed of two complementary libraries of pre-made gene-targeting insertion vectors *(3, 19)*, greatly reduces the number of cloning steps required for generating gene-targeting vectors, and is discussed in more detail in **Section 2.3.1**.

2.3.1. Libraries of Targeting Vectors

A rate-limiting step of engineering chromosome rearrangements in mice, particularly in large-scale experiments, is the molecular cloning required for building the individual targeting vectors. To streamline this process, a "two-library" system for large-scale chromosome engineering has been developed *(3)*. This system is composed of two complementary libraries of pre-made gene-targeting insertion vectors generated by cloning a 129 Sv genomic DNA genomic library into one of two vector backbones, containing either (1) the *5′Hprt* cassette, a *loxP* site and a positive selection marker (neomycin; *PGK-neo-bpA*) or (2) the *3′Hprt* cassette, a *loxP* site and an alternative positive selection marker (puromycin; *PGK-puro-bpA*). In addition, these libraries are equipped with visible coat-colour markers (either the *tyrosinase* minigene (*Ty*) or *K14-Agouti* transgene (*Ag*)), which are discussed in more detail in **Section 2.3.2**.

The unique feature of these libraries is that each clone is a ready-made (insertional) targeting vector, requiring only linearization prior to electroporation into ES cells. Only one contiguous genomic insert can be cloned into a vector when constructing a library; thus, these libraries are used for insertional targeting rather than the more widely used replacement vectors. With insertional targeting, linearization of the targeting vector must be made within the homologous sequence (the genomic insert), and the entire insertion vector is integrated into the target site, including the homologous sequences of the vector, such that the recombinant allele becomes a duplication of the target homology separated by the heterologous sequences in the vector backbone (reviewed in *(20)*). Targeting can be assessed by Southern blot analysis using an external probe; however, it is often more convenient to prepare a vector with a gap in the homologous sequence. In this case, this DNA fragment can be used as an "external probe" or a PCR assay can be designed using primers specific for the gap (which is repaired during targeting) and vector. An additional feature of this gene-targeting library system is that clones isolated from these libraries can also be used for analyzing single gene function via the knockout (null allele) approach, as demonstrated by targeted disruption of the *p63* locus *(9)*.

Recently, these libraries have been made more accessible by indexing. More than 150,000 clones from these libraries have been end-sequenced and mapped against the genome. These are distributed as a public resource through the Mutagenic Insertion and Chromosome Engineering Resource (MICER) website (http://www.sanger.ac.uk/PostGenomics/mousegenomics/) *(19)*. All mapped clones are displayed in the Ensembl mouse genome browser *(21)* under the DAS source "MICER". Vectors from these libraries are randomly distributed across the genome, but on average one vector is available for every 39 kb of genomic sequence. If the vector backbone is not in the correct orientation, or if an alternative vector backbone is required, the genomic insert can be inverted or shuttled between backbones using the *Asc*I restriction enzyme sites which flank the insert. In addition, alternative genomic inserts can be cloned into the vector backbone by recombineering fragments of genomic DNA *(22)*, such as from the arrayed 129S7 BAC library derived from AB2.2 cells *(23)* or the 129/Ola BAC library derived from E14.1 cells *(24)*, two of the most widely used ES cell lines.

2.3.2. Marking Chromosomal Rearrangements with Coat-Colour Genes

A useful feature of chromosomal rearrangements in *Drosophila* is that many of them are marked with a dominant visible marker enabling a chromosome to be followed without selection or genotyping. In mice, such a feature is potentially very useful for stock maintenance and for genetic screens. The tyrosinase minigene (*Ty*) has been used to mark transgenes with a visible pigment marker (causing a greyish coat on an otherwise albino background) *(25, 26)*, and the *K14-Agouti* transgene (*Ag*), in which the *Keratin-14* promoter drives expression of the *Agouti* gene *(27)* and results in a "butterscotch" coat in black agouti or non-agouti mice *(3)* (*See* **Note 3**).

These two coat-colour minigenes have been used as visible markers for targeted alleles in ES cell-derived mice and are incorporated into the "two–library" system of targeting vectors for chromosome engineering (the *5′Hprt* vector carries *Ty* and the *3′Hprt* vector carries *Ag*; *see* **Fig. 4.2B**) *(3)*.

3. Chromosomal Deletions

3.1. Uses for Engineering Chromosomal Deletions

Chromosome engineering can be used to generate mouse models of human microdeletion syndromes. For example, mice heterozygous for a 1.2-Mb deletion between *Es2* and *Ufd1l* on mouse chromosome 16 show cardiovascular abnormalities resembling those found in DiGeorge syndrome patients *(10)*. When these mice were crossed to a strain harbouring a duplication of the same region, the mutant phenotype was functionally rescued, indicating that the defects seen in DiGeorge patients can be explained by haploinsufficiency of one or more genes in that region.

The chromosomal engineering of deletions can also be used to identify tumour suppressor genes (TSGs) without prior knowledge of the gene function. Mice possessing a deletion encompassing a putative TSG should exhibit increased tumourigenesis, since they only possess a single copy of a TSG, and tumour-specific loss or inactivation of the remaining allele can be used to clone the causative gene. For example, to functionally identify novel TSGs mapping to human *1p36*, chromosome engineering was used to generate mouse strains with deletions (*df*) or duplications (*dp*) of this region and these models were used to identify a 4.3-Mb region that encodes a potent regulator of proliferation, cellular senescence, apoptosis and tumourigenesis (namely *Chd5*, which functions as a tumour suppressor) *(28)*.

3.2. Radiation-Induced Chromosomal Deletions

Radiation-induced mutagenesis has also been successfully performed to generate deletion complexes in ES cells that have been examined as alleles in mice *(29–31)*. Radiation-induced deletions can be selected in a specific genomic region by targeting a negative selection cassette such as HSVtk to the desired site in the genome, followed by irradiation. Clones that have deleted the *tk* cassette can be identified by culturing them in a medium containing FIAU (1,2′ deoxy-2′-fluoro-β-D-arabinofuranosyl-5-iodouracil), which kills cells that express HSVtk *(7, 32–34)*. The key to this technology is the use of ES cells whose developmental potential is not compromised by the radiation dosages required to induce deletions (300–500 rads). F_1 hybrid ES cells (such as 129/Sv^{Jae} × BALB/cJ or 129/Sv^{Jae} × C57BL/6 J cells) retain the ability to create germline chimaeras following dosages of up to 400 rads *(32)*. An additional benefit of the F_1 hybrid lines is that the allelic polymorphism between parental chromosomes enables rapid characterization of deletion sizes by PCR typing of microsatellite markers.

3.3. In Vivo-Generated Chromosomal Deletions (TAMERE and STRING)

Chromosomal deletions (and duplications) can be generated in vivo using Cre-mediated targeted meiotic recombination (TAMERE) technology. In this application, *loxP* sites are bred together in the same mouse in *trans*. Following chromosome pairing in meiotic prophase and Cre expression through the *Sycp1/Cre* transgene (which produces Cre in male spermatocytes during the stages when chromosome pairing occurs), recombination between *loxP* sites occurs which can generate recombinant chromosomes with deletions and duplications *(35)* (*see* **Fig. 4.6**). For example, TAMERE was used to generate a systematic series of *HoxD* cluster deletions which have allowed investigation of the regulation of the *HoxD* locus as well as the spontaneous semidominant mutation (*Ironside*) which causes severe hind-limb paralysis *(36)*. More recently, five distinct *loxP*-containing knockout alleles in the protocadherin (*Pcdh*) gene clusters and a strong Cre driver (*Hprt*-Cre; a strong CAG promoter-driven Cre expression cassette inserted into the X-linked *Hprt* locus *(37)*) was used to generate large deletions (and duplications) by trans-allelic recombination between homologous chromosomes in somatic cells and the germline, including deletion of the α cluster (228 kb) and deletion of the α to β cluster (730 kb) with very high efficiencies (~9.3% in the latter case) *(38)*.

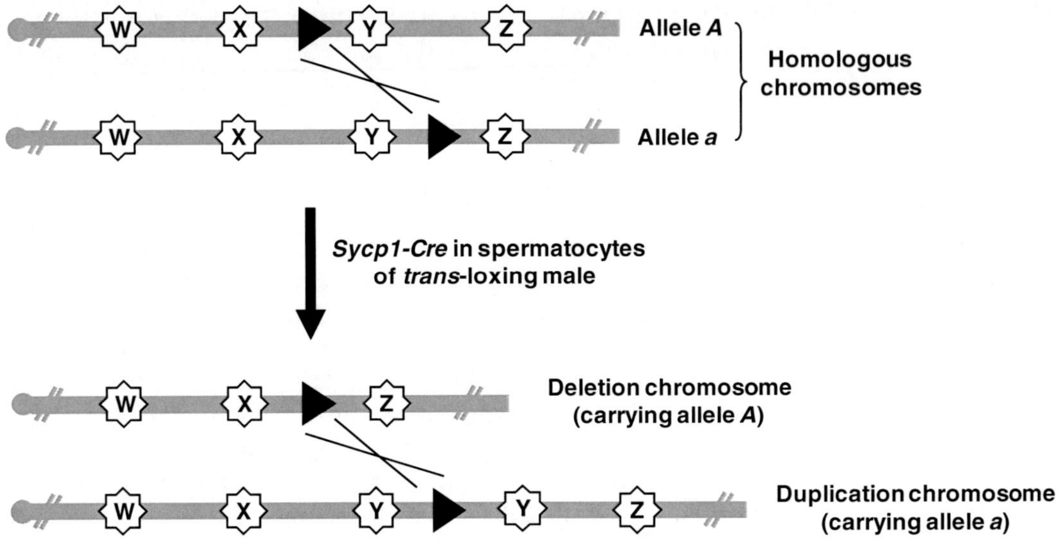

Fig. 4.6. **Chromosome engineering through Cre-mediated targeted meiotic recombination (TAMERE).** Allele *A* has a *loxP* site (*black triangle*) between genetic loci (*stars*) X and Y and allele *A* has a *loxP* site between genetic loci Y and Z. Following chromosome pairing in meiotic prophase and Cre expression through the *Sycp1-Cre* transgene, *trans*-allelic recombination occurs, generating deletion and duplication chromosomes.

Thus the advantage of this technology is that pre-existing chromosomes with targeted *loxP* sites can be bred together. However, it was not found to be effective in generating either a deletion or translocation in vivo when the *loxP* sites were 3.9 Mb apart on the same or both copies of mouse chromosome 16 *(39)*, suggesting this method may be of limited use for large genomic rearrangements. By taking advantage of naturally occurring crossovers to bring two *loxP*-containing alleles onto the same chromosome, an alternative method (designated STRING) has been described that uses Cre-mediated recombination to generate deletions (and inversions) of several megabases *(40)*. This sequential targeted recombination-induced genomic (STRING) approach, in which animals already containing an appropriate *loxP* site are bred and the offspring selected for recombination events in vivo, has been shown to generate deletions with fair efficiency (10–30% for *Itga6* and *HoxD* and ~1% for *HoxD* and *Cd44*) *(40)*. However, because STRING is dependent on using naturally occurring homologous recombination-induced crossovers to bring the *loxP* site to the same chromosome, this methodology is limited to *loxP* sites that are separated on homologous chromosomes by several megabases to ensure reasonable crossover efficiency.

3.4. Generation of Nested Chromosomal Deletions

One extension of the chromosome engineering strategy is the application to generate nested chromosomal deletions, a series of variably sized, overlapping deletions at a predetermined genomic locus. This strategy was used to show that haploinsufficiency of the *Tbx1* gene contained within the 1.2-Mb deletion interval of DiGeorge mice was responsible for the aortic arch defects seen in these patients *(41)*. Furthermore, if the genomic locations of the endpoints are known, then nested deletions can be extremely useful for mapping novel recessive mutations *(42)*.

To avoid having to generate individual targeting vectors for the nested endpoints, retroviral integration of a second *loxP* site and selection cassette can be used *(43)*. Deletion complexes can be anchored to a predetermined location in the genome by targeting of the *5-Hprt-loxP* cassette. The *3-Hprt-loxP* cassette is then randomly inserted into the ES cell genome by retrovirus-mediated integration, generating a library of ES cell clones with the same targeted endpoint and a collection of random endpoints (*see* **Fig. 4.7**). This method has been used to generate nested deletions from a few kilobases to several megabases (Mb) at the *Hprt* locus

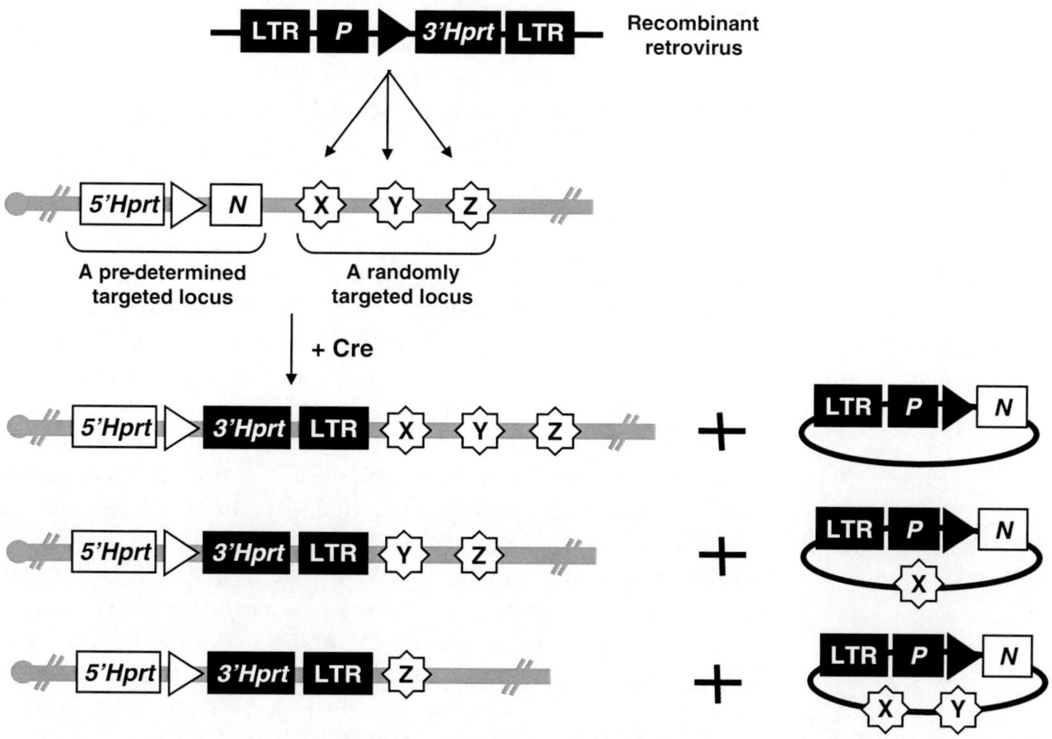

Fig. 4.7. **Nested chromosomal deletions induced with a retroviral vector.** The first deletion endpoint is fixed by targeting the *5′-Hprt* cassette, the neomycin resistance gene (*N*) and a *loxP* site (*white triangle*) to a predetermined locus. The second step involves the integration of the *3′-Hprt* cassette, the puromycin resistance cassette (*P*) and a second *loxP* site (*black triangle*) into a random site in the ES cell genome, using a recombinant retroviral vector. Here are depicted G1 recombination events from retroviral orientations that result in nested deletions of the randomly targeted gene (the stars depict genetic markers within this chromosome). Cre catalyses recombination between the *loxP* sites, resulting in the expression of the *Hprt* gene and hence survival in HAT medium of ES cell clones carrying the recombinant chromosomes. The nested deletions can be identified from a pool of HAT-resistant clones on the basis of their sensitivity to G418 and puromycin. LTR, long terminal repeat.

(43). It has also been used to generate ES cells carrying subdeletions within the ~2-Mb deletion of the syntenic region of the human Smith–Magenis syndrome (SMS) common deletion *(44)*; mice carrying these subdeletions showed the same obesity and craniofacial abnormalities seen in mice carrying the ~2-Mb deletion (*Df(17)*) *(11)*, suggesting that the genes responsible for this syndrome are located within the small deletions *(44)*.

Electroporation has also been used to introduce a plasmid vector carrying the second *loxP* site and selection cassette randomly into the ES cell genome *(45)*. However, this method increases the risk of additional genomic rearrangements occurring at the insertion site, and tandem repeats of a vector may be introduced at the insertion site (although these can be reduced to a single locus by the activity of Cre on a head-to-tail concentrate) *(42)*. Endpoints generated by random insertion can be characterized by cloning the genomic DNA that flanks the deletion endpoints and by mapping these junction fragments onto a physical map of the region. More recently, the *piggyBac* transposon was shown to be an effective tool for introducing *loxP* sites randomly throughout the mouse genome, and allowing for Cre-*loxP*-mediated generation of deletions and/or duplications *(38)*.

Nested deletions can also be efficiently generated by irradiation *(29, 30, 32, 33, 46, 47)*, although they require additional extensive characterization to define each deletion interval *(42)*.

3.5. Important Considerations in the Generation of a Chromosomal Deletion

The major factor limiting the generation of deletions in ES cells is the size of the rearranged interval. Larger deletions lead to ES cell lethality or may cause a severe growth disadvantage of the cells in culture *(13)*. This may affect the germline potential of the ES cells, although cells carrying deletion alleles of several centimorgans have been successfully transmitted through the germline *(5)*. However, even relatively small deletions of only a few centimorgans may still cause embryonic lethality in heterozygotes if the deleted region contains one or more happloinsufficient genes (i.e., the lethality is due to a dosage effect of that gene(s)) *(48)*. In addition, even when the deletion is heterozygous viable, it may be associated with other developmental defects *(49)*. Thus, it is best to produce chimaeras by injecting deletion/duplication ES cells which carry a deletion on one chromosome and a duplication on the other chromosome (resulting from Cre recombination between *loxP* sites on homologous chromosomes (*in trans*)) since cells containing both the deleted and duplicated regions are genetically balanced. This strategy also allows for the subsequent production of both deletion and duplication mice (necessary when the deletion is heterozygous lethal and as such can only be produced and maintained in a duplication background).

Another cause for deletion-associated heterozygous lethality is imprinting, although this is only an issue when the gene(s) involved is maternally imprinted (i.e., paternally expressed). This is because ES cells are typically XY in origin and as such a maternally expressed gene inherited from a male chimaera is not expressed. However, when a gene is paternally expressed, the deletion progeny will be a functional null for this gene, such that if the imprinted gene is essential for normal development, the deletion progeny will not survive. This problem can be circumvented through the use of XX or XO ES cells, allowing transmission through the female germline. Or if the deletion size is small, it can be generated by breeding mice targeted at both endpoints to a universal deletor line *(50, 51)*.

4. Chromosomal Duplications

In autosomal duplications, the copy number increases from two to three. Duplication of a whole chromosome results in trisomy, the most well-known of which involves chromosome 21 and results in DS. Segmental duplications describe a situation where a portion of a chromosome is duplicated and can arise in the offspring of balanced translocation carriers or can occur during meiosis if there is crossing over between homologous chromosome pairs that are out of alignment (resulting in one with duplicated genes and the other having the corresponding deletion). The first segmentally trisomic DS mouse model (Ts65Dn) was generated by a radiation-induced reciprocal translocation *(52)*. Ts65Dn mice contain an extra copy of the region of mouse chromosome 16 corresponding to part of human chromosome 21 (including the DS critical region). These mice exhibit some characteristics of DS such as developmental delay, craniofacial dysmorphology, impaired learning and behavioural abnormalities *(2, 53, 54)*.

Using chromosomal engineering, duplications can be generated when two *loxP* sites are placed in the same orientation on the same chromosome (*see* **Fig. 4.3**). Such a situation can also result in the generation of a deletion event; however, duplications can be specifically generated with an appropriate order and orientation of the selection cassette (this is explained in detail in **Section 7**). Reeves and colleagues used chromosome engineering to create mice that were trisomic or monosomic for only the mouse chromosome segment orthologous to the DS critical region (Ts1Rhr mice) and found that these genes alone were not sufficient to recapitulate the facial phenotype (the characteristic craniofacial dysmorphologies seen in DS individuals) *(55)*. More recently, to generate a more complete trisomic mouse model of DS, a 22.9 Mb duplication spanning the entire human chromosome 21 syntenic region on mouse chromosome 16 was generated using chromosome engineering, and these mice showed a cardiovascular and gastrointestinal phenotype similar to those of patients with DS *(56)*.

5. Chromosomal Inversions

Chromosomal inversions occur when a single chromosome breaks in two places and the material between the breaks is inverted. Chromosomal inversions are useful genetic tools for recovering and maintaining mutations in model organisms such as *Drosophila* and mice, due to the suppression of recombination that occurs in the inverted region (as recombination products between an inverted and a wild-type chromosome do not produce viable gametes) (reviewed in *(57)*). Thus inversions are useful for recovering and/or maintaining recessive mutations that are not easily recovered or maintained as homozygotes.

Inversions in the mouse have traditionally been generated by chemical or radiation mutagenesis *(58, 59)*; however, cumulatively these are only a small fraction of the mouse genome, and most do not carry a phenotypic marker and/or are too large to effectively suppress recombination. Moreover, the few that are available are on a variety of different genetic backgrounds. Inversions can be generated using chromosomal engineering technology (as detailed in **Fig. 4.4**). The inversion can also be marked with a dominant marker (such as the dominant *K14-Agouti* coat-colour gene) so that progeny carrying the marked inversion chromosome can be readily identified. Chromosomes carrying a phenotypically marked engineered inversion are known as "balancer chromosomes". Because they suppress crossing over during mitotic recombination, balancer chromosomes can be used for stock maintenance (to maintain the integrity of mutagenized chromosomes).

Balancer chromosomes can also be used for large-scale mutagenesis screens, such as ENU screens (reviewed in *(20)*). For example, in intercrosses between siblings inheriting the balancer chromosome and a mutagenized chromosome, absence of non-balancer-carrying progeny indicates the presence of one or more recessive lethal mutations on the mutagenized chromosome *(42, 60)*. The first specifically engineered mouse balancer chromosome was constructed on chromosome 11, to facilitate the isolation of ENU-induced recessive mutations on this chromosome. This balancer chromosome contains a 24-cM inversion between *Trp53* and *Wnt3* that is recessive lethal and dominantly marked with a K14-Agouti transgene *(3)*. The inversion functions as a balancer chromosome because it can be used to maintain a lethal mutation in the inversion interval as a self-sustaining *trans*-heterozygous stock.

Since this first report, balancer chromosomes have been generated for segments on mouse chromosomes 15 *(61)*, 11 *(62)* and 4 *(14)*. Thus balancer chromosome strains provide key advances for mutagenesis screens, stock maintenance and tracking quantitative traits.

6. Chromosomal Translocations

Translocations are rearrangements involving two different chromosomes and consist of the exchange of chromosomal material between chromosomes. When these occur spontaneously there is often a small overall loss or gain of genetic material in a "balanced" translocation. Chromosomal translocations are common occurrences in cancer, particularly lymphoid malignancies. The principle consequences of translocations that result in cancer are that (1) the translocation brings a gene under the influence of different regulatory elements, resulting in altered expression due to a position effect, or (2) a fusion gene encoding a chimaeric protein is generated *(63, 64)*. Although mouse models for several human leukaemias have been established by tissue-specific expression of fusion proteins transgenes *(65)* and knockin constructs expressing a fusion protein under the control of the appropriate endogenous promoter *(66)*, they fail to recapitulate the situation found in humans as these mice carry the fusion gene from conception and it is expressed in large numbers of cells, whereas in humans the chromosomal translocation is believed to occur at later stages and is believed to arise in a single cell *(67, 68)*. In addition, these mice models possess only one fusion protein, whereas in balanced chromosomal translocations in humans two fusion proteins are generated *(69)*. Thus to accurately recapitulate chromosomal translocations found in human tumours, chromosome engineering needs to be deployed in vivo.

The feasibility of generating chromosome translocations in ES cells was first demonstrated by two different groups in 1995. Smith and co-workers demonstrated the generation of a chromosomal translocation commonly found in mouse plasmocytomas *(6)*, while Van Deursen and colleagues generated a translocation to mimic the human chromosomal translocation t(6;9) associated with acute myeloid leukaemia *(8)*. Chromosomal translocations in mice have now been engineered to model several translocations found in human leukaemias, including t(8;21) *(70)* and t(9;11) *(71)* (reviewed in *(4)*).

Translocations can be generated by chromosomal engineering when the *loxP* sites are targeted to the relevant chromosomes provided that the *loxP* sites are orientated in the same direction relative to their respective centromeres (*see* **Fig. 4.5A**). If the *loxP* sites are oriented in opposite directions then recombination will result in acentric chromosomes (without a centromere) and dicentric chromosomes (containing two centromeres). Cells which carry these products will not be viable (*see* **Fig. 4.5A**). The efficiency of Cre/*loxP* recombination between non-homologous chromosomes is several orders of magnitude lower than that of the recombination between *loxP* sites on the same chromosomes. The frequency of Cre/*loxP*-mediated recombination between non-homologous chromosomes is also lower than that obtained when *loxP* sites are inserted within a few megabases of each other on homologous chromosomes *(42)*.

To generate a fusion protein from a chromosomal translocation, the targeting vectors need to allow the two genes to be linked through their introns (with the *loxP* site embedded in the junction region of the breakpoint), such that RNA splicing will generate an in-frame fusion mRNA and protein (*see* **Fig. 4.5B**). Only pairs of genes with the same transcriptional orientations relative to their centromeres can be engineered to produce fusion proteins *(42)*. This technique has been utilized to generate mice carrying a *loxP* site in the *Mll* gene and the *Enl* gene (*Mll-loxP;Enl-loxP* mice). Recombination in vivo resulted in mice carrying the fusion gene equivalent to the *MLL–ENL* fusion found in human leukaemias with t(11;19) and the mice developed myeloid tumours with rapid onset and penetrance *(72)*.

To induce the translocation in vivo, mice carrying both targeted endpoints can be crossed with transgenic mice expressing a tissue and/or temporal-specific Cre. The choice of Cre strain is an important consideration, as it impacts on the specific cell type which expresses the fusion gene and thus the phenotype of the mouse; this is discussed further in **Section 8** (and reviewed in *(4)*). For example, *Mll-loxP;Enl-loxP* mice bred to mice in which Cre was expressed under the control of the haematopoietic *Lmo2* gene, which allows the translocations to be generated in multipotent stem cells, resulted in mice developing myeloid tumours *(72)*. However, when the same mice were bred to mice specifically expressing Cre in T cells and their progenitors (under the control of the *Lck-Cre* transgene), they developed lymphoid tumours (*Mll–Enl* fusions have been found in both the myeloid and lymphoid lineage in human leukaemias) *(73)*. The use of tissue-specific Cre mice circumvents the problem of transmitting the translocation through the male germline, as the presence of chromosomal translocations in male germ cells can cause infertility *(74)*.

7. Analysis of HAT-Resistant Cre-loxP Recombination Products

There are 5 complementary techniques to analyze the HAT-resistant Cre recombination products: Cre recombination efficiency, selectable marker retention, junction fragment analysis, FISH and CGH *(75)*.

7.1. Cre Recombination Efficiency

The efficiency with which Cre catalyzes site-specific recombination can be used to identify if the two *loxP* sites have inserted on the same chromosome or not. The "recombination efficiency" can be assessed in an in vitro transient transfection assay, where the efficiency can be expressed as the ratio of the number of HAT-resistant colonies relative to the number of cells electroporated, after electroporation with a Cre expression plasmid. The relative Cre recombination efficiency is very helpful in distinguishing among different recombination products. Using this assay on several genetic intervals from 2 cM to 60 cM, the general conclusion is that Cre recombination efficiency on the same chromosome decreases over increasing genetic distance *(13)*.

When the physical distance is small, *cis* deletions, *cis* duplications and *cis* inversions occur at similarly high Cre recombination efficiency, such that only small quantities of Cre expression plasmid needs be transfected or the electroporated cells need to be diluted before plating, in order to be able to pick single colonies. In contrast, *trans* recombination to generate deletion/duplications has a moderately low efficiency (~2–3 orders of magnitude lower) for similar distances. However, if the *loxP* sites are in opposite orientations on different chromosomal homologs, then no viable recombination products will be generated, regardless of the distance between the *loxP* sites. Therefore, for short distances (1) the generation of HAT-resistant colonies at high efficiency indicates recombination has occurred in *cis* to generate deletions, duplications or inversions, (2) the absence of HAT-resistance colonies indicates that the l*oxP* sites are in opposite orientations in *trans* and (3) the number of HAT-resistant colonies for *trans* deletion/duplications falls somewhere in between these values.

When the physical distance is large (10–20 Mb), the situation is different. Large deletions can cause lethality or severe growth disadvantages *(13)*, thus the observed Cre recombination efficiency can be very low since survival will depend on compensating genetic changes. By contrast, inversions are relatively benign since they do not involve gains or losses of genetic material. Thus the number of HAT-resistant colonies tends to relate to the distance between the *loxP* sites without the impact of selection for particular genotypes. If the inverted *loxP* sites are not on the same chromosome then recombination between them produces acentric and dicentric chromosomes and cells with these products are unviable. Thus the observed frequency of recombination is very low or zero. Colonies would still be unviable if the *loxP* sites are in opposite orientations in *trans*.

7.2. Selectable Marker Test

This test takes advantage of the design of chromosome engineering cassettes shown in **Fig. 4.2**, as the selectable marker for gene targeting (neomycin or puromycin) and the half *hprt* gene (*5′hprt* or *3′hprt*) are on opposite sides of the *loxP* site so they are always separated from each other after Cre recombination; this forms the basis for the selectable marker test (detailed in 75) (*see* **Note 4**). There are four possible combinations, with respect to the orientation of the *loxP* sites relative to each other and the half *hprt* cassettes (*see* **Figs. 4.3** and **4.4**):

1. *LoxP* sites in the same orientation with respect to the centromere and both the half *hprt* cassettes are outside of the *loxP*-flanked (floxed) region (*see* **Fig. 4.3A**). If the *loxP* sites are in *cis*, recombination leads to a simple deletion chromosome that is resistant to HAT, yet sensitive to puromycin and G418. If the *loxP* sites are in *trans*, recombination results in a deletion and duplication. If these chromosomal rearrangements segregate, the cells will be HAT- and puromycin-resistant but G418-sensitive; if they do not segregate, the cells will be resistant to HAT, puromycin and G418.
2. The two *loxP* sites in the same orientation with respect to the centromere and both the half *hprt* cassettes are inside the floxed region (*see* **Fig. 4.3B**). If the *loxP* sites are in *cis*, recombination can yield a deletion that is HAT-sensitive so will not be recovered. However, it can also yield a duplication which will be HAT-resistant (as well as puromycin- and G418-resistant). *trans* recombination results in a situation analogous to that described in (1) above (segregation of the deletion/duplication chromosomes results in cells that will be HAT- and G418-resistant but puromycin-sensitive, and non-segregation results in cells resistant to all three drugs).
3. The two *loxP* sites in opposite orientation with respect to the centromere with the *5′hprt* cassette outside of the floxed region (*see* **Fig. 4.4A**). *cis* recombinations lead to inversions, that are HAT-, puromycin- and G418-resistant. *trans* recombinations result in acentric and dicentric chromosomes and the cells do not survive.
4. The two *loxP* sites in opposite orientation with respect to the centromere with the *3′hprt* cassette outside of the floxed region (*see* **Fig. 4.4B**). The outcomes for this situation are as described in (3) above.

7.3. Junction Fragment Analysis

When the recombination occurs it generates a junction fragment that is unique to the rearranged allele, due to the process of gap repair that occurs during the targeting event. This junction fragment can be analysed either by PCR (*see* **Fig. 4.8A**) or Southern blotting (*see* **Fig. 4.8B**). If using PCR, the PCR products generated can be sequenced to determine the precision of the rearrangement at the nucleotide level, though this is more of a precautionary step and not strictly necessary as all rearrangements analysed by PCR and sequencing to date have proved to be precise rearrangements *(75)*. It should be noted, however, that each junction fragment has both a unique as well as a

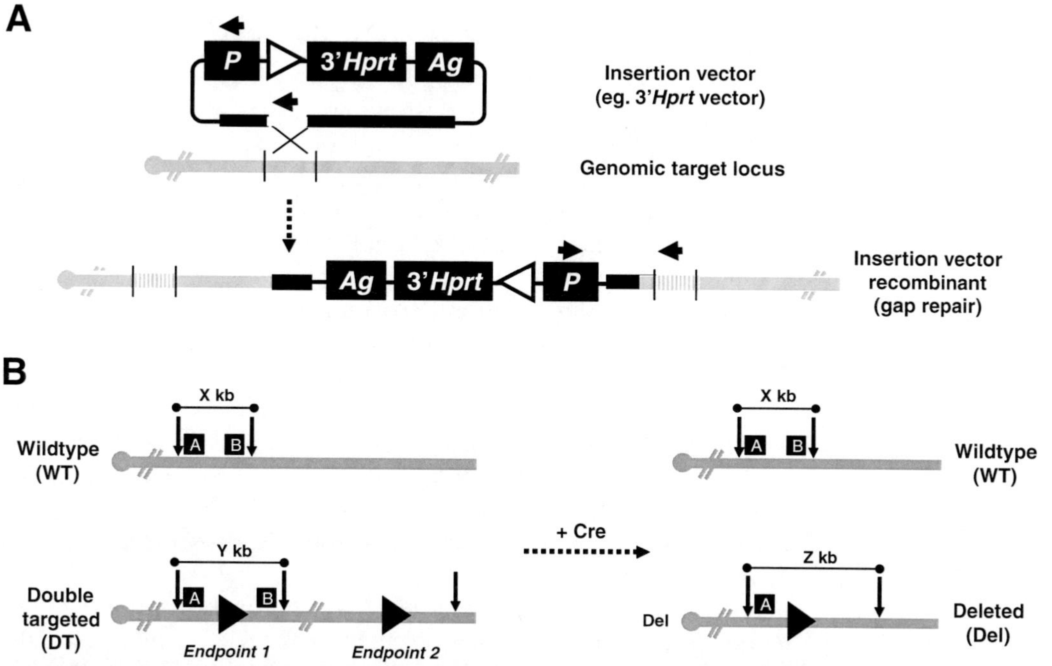

Fig. 4.8. **Schematic of PCR and Southern blotting analysis of chromosomal rearrangements.** (**A**) PCR screening for insertion events can take advantage of the process of gap repair. The insertion vector contains a gap at the linearization point (demarcated by two *vertical lines* in the genomic target locus and the insertion vector recombinant). The *arrow heads* represent PCR primers: one specific for the positive selection marker and the other for the sequences which correspond to the gap that is repaired during the target event. (**B**) Southern blotting analysis. WT is the wild-type locus, DT is the double-targeted locus and Del is the locus for the deletion allele. Probe A lies outside of the deletion at endpoint 1 and detects restriction fragments of different sizes (X kb for WT, Y kb for DT and Z kb for Del alleles). Probe B lines inside of the deletion at endpoint 1 and detects the WT and DT alleles, but not the Del allele. In the case of deletion/duplication, inversion or translocation, probe B can detect a unique fragment for the rearranged allele. In addition to probes at endpoint A, a similar strategy can be used to develop probes at endpoint B.

universal region because the same selection markers are used for every rearrangement. Thus the sequences flanking the *loxP* site are universal, consisting of the HPRT minigene and the neo and puro selection cassettes. Thus PCR of this junction over a relatively small distance is a useful universal genotyping assay but it will not distinguish among different rearrangements. To determine more conclusively if the rearrangement has occurred, Southern blotting can be used to analyse the recombination products (**Fig. 4.8B** illustrates the scheme for such an analysis). Southern blotting has the advantage over PCR as a larger portion of the junction region can be accessed. Since the breakpoints will have been confirmed as being targeted in these experiments, a probe internal to the homology region can be used, as well as the original external probes used for gene-targeting analysis. It is good practice to analyse junctions at both breakpoints.

7.4. Fluorescence In Situ Hybridization (FISH)

FISH is an extremely effective method to analyse the rearrangements generated by chromosomal engineering, particularly for larger rearrangements (examples of FISH analysis on several chromosomal rearrangements is shown in *(75)*). To detect a chromosomal deletion and duplication, two probes are needed: one probe internal to the floxed interval (termed "test probe") and another one external to the floxed internal (termed "control probe"); each probe is labelled with a different fluorochrome. A deletion is indicated by the absence of a signal for the test probe on one chromosome. Duplications are indicated by duplicated signals along the chromosome, independent from the pair of signals for each probe arising from the two chromatids of each metaphase chromosome (i.e., four signals, in two pairs, are expected for a probe within the duplicated region). Although large duplications ($\geq$10 cM) can be easily detected by FISH, smaller deletions (less than a few centimorgans) may have the duplicated signals only visible as a single dot on metaphase spreads; thus interphase nuclei must be examined.

To detect a translocation between non-homologous chromosomes, at least two probes need to be used, one proximal to the breakpoint on one chromosome and the other distal to the breakpoint on the second chromosome. To detect an inversion, two probes internal to the inversion are needed; a change in the order of the two probes relative to the centromere indicates an inversion. It is useful to use an additional probe that lies external to the inversion as a control (to aid in determining the relative order of the two internal probes, though in many cases the unlabelled centromere is adequate for this purpose).

7.5. Array CGH

The array CGH technique has been developed to detect chromosomal copy number changes on a genome-wide and/or high-resolution scale. Initially, although BAC microarrays were used for mouse CGH studies, the resolving power of these analyses was limited because high-density whole-genome mouse BAC microarrays were not available. However, a mouse BAC microarray containing 2,803 unique BAC clones from mouse genomic libraries at 1 Mb intervals has been developed and reproducibly identified DNA copy number alterations including single-copy deletions and regional amplifications, such as found in ES cell lines carrying genetically engineered chromosomes (e.g. a 2 Mb heterozygous deletion between *D4Mit117* and *D4Mit246* on chromosome 4, and a 6.9 Mb heterozygous deletion from *Mpo2* to *Chad2* on chromosome 11) *(76)*. Thus array CGH is a useful technique for identifying chromosomal deletions and duplications (but not inversions or translocations).

8. Tissue- and Temporal-Specific Chromosome Engineering

An important aspect of the Cre/*loxP* system is that it allows a chromosomal rearrangement to be generated in a temporal-, spatial- and/or tissue-specific manner (discussed in more detail in *(20)*). Such tight regulation over the expression of Cre and thus the expression of the chromosomal rearrangement(s) allows for the generation of more accurate mouse models of human disease. For example, chromosomal rearrangements can be generated somatically by transmitting targeted chromosome(s) carrying unrecombined *loxP* sites through the germline. Thus the desired rearrangement can be generated somatically by breeding this targeted chromosome(s) into a background where Cre is expressed under the control of a tissue-specific promoter and/or Cre can be activated with a small molecule. This is important if the chromosomal rearrangement would otherwise cause an embryonic lethality or a developmental anomaly (as can occur with large deletions). The fusion genes created by translocations are also problematic both because of their affect on normal development as well as infertility in translocation heterozygotes. Moreover, somatic activation of a translocation is a better mimic of the mutational mechanism occurring in cancer. For example, several deletions of a few centimorgans around the *Hsd17b1* locus on chromosome 11 are heterozygous lethal *(48)*; however, a 2-cM *Hsd17b1-D11Mit199* double-targeted mouse line (deletion substrate) crossed with a cardiac-specific Cre deleter mouse line (MyHC-Cre) *(77)* produced progeny that had inherited the MyHC-Cre transgene and 2-cM substrate, and showed Cre-mediated recombination specifically in heart, but not any of the other tissues examined (the deletion efficiency in the heart was about 10%) *(13)*. As discussed in **Section 6**, mice carrying *loxP* sites in both the*Mll* and *Enl* genes developed translocations when bred to the appropriate Cre driver line. For instance, myeloid tumours developed when these two alleles were bred to *Lmo2-Cre* mice *(72)* and lymphoid tumours developed when bred to *Lck-Cre* mice *(73)*.

Thus the main methodology for tissue-specific Cre-mediated excision is the use of established transgenic lines expressing Cre under the control of a promoter with the required specificity. A database of such Cre-expressing mice has been established (*see* http://www.mshri.on.ca/nagy). However, in certain applications, retroviral, adenoviral or even protein delivery systems could also be used to achieve tissue-specific administration of Cre. However, this is more generally applied to targeted alleles, for instance intranasally administered adenoviral Cre-mediated removal of a floxed stop element to allow expression of a mutant *K-ras* in lung tissue *(78)*. In addition, chemically inducible forms of Cre have also been used, namely the tet on/off system (rtTA transactivator) and the mutant oestrogen receptor ligand-binding domain system (Cre-ER^{T}) (reviewed in *(79)*). The most common transcriptional control system used to regulate Cre activity is the tet on/off system, which is based on the use of tetO operator sequences of the tet operon inserted in the promoter controlling the expression of Cre recombinase, and tetracycline-binding transactivators consisting of fusions of the VP16 transactivation domain from herpes simplex virus and the tetracycline repressor protein (TetR) from *Escherichia coli*. In the rtTA transactivator system, the administration of doxycycline to the mouse results in activation of the rtTA. The rtTA transactivator system has been used to achieve regulatable Cre activity in various tissues, such as intestinal epithelium *(80)*, liver *(81)* and lung *(82)*. Post-translational control of Cre activity is enabled by protein fusions with hormone receptor ligand-binding domains, the most commonly used being Cre-ER^{T} and Cre-ER^{T2}, in which the ligand-binding domain of a mutated human oestrogen receptor (ER^{T}), which is bound by tamoxifen or its derivative 4-hydroxytamoxifen (4-OHT), has been fused with Cre; Cre-mediated recombination upon 4-OHT administration by i.p. injection results in Cre-mediated recombination. A number of mouse lines expressing tamoxifen-inducible Cre are available, and include Cre activity being restricted to the liver *(83)*, the intestine *(84, 85)*, the testis *(86)*, melanoblasts/melanocytes *(87)* or astrocytes *(88)*, to name but a few.

9. Concluding Remarks

Chromosome engineering in mice has applications in two major fields in human genetics: developmental syndromes in clinical genetics, and cancer genetics. The human cytogenetics literature contains hundreds of examples of autosomal loci that are sensitive to copy number change. Developmental abnormalities occur when copy number at these sites deviates from the expected value of two. For some of these (termed genomic disorders), the presence of flanking segmentally duplicated regions of DNA means that the size of the chromosomal imbalance is identical, or nearly so, in the majority of cases, giving rise to a distinctive and recognizable phenotype. The synteny between the human and the mouse genomes has enabled chromosome engineering to be applied in the mouse to genetically dissect some of these syndromes, most notably DiGeorge syndrome *(10)*.

However, the great majority of copy number changes reported are too rare, too variable and the phenotypes insufficiently recognizable, to enable the resulting disorder to be classified as a syndrome. Further refinement of these disorders must await the identification of additional patients with smaller deletions, or they can be analysed by mouse chromosome engineering. Once a phenotype has been confirmed in mice, further refinement of the deleted region can be rapidly achieved by constructing nested subsets of deletions leading to the identification of the susceptibility gene(s) at these loci.

Mouse models of human cancers are important for understanding determinants of overt disease and for "preclinical" development of therapeutic strategies. Chromosomal changes such as deletions and translocations underlie many human leukaemias, sarcomas and epithelial tumours. Chromosomal engineering can be used to model a specific known common genetic change or to confirm or refute the role of a specific genetic change in carcinogenesis.

10. Notes

(1) *Site-specific recombination.* The simplest site-specific recombination systems are those composed of a recombinase enzyme and its target sequence. These systems allow for the deletion, insertion, inversion or translocation of specific regions of DNA. Two such recombinase systems are widely used in mouse genetics. The Cre-*loxP* system from the bacteriophage P1 and the Flp-*FRT* system from the budding yeast *Saccharomyces cerevisiae*. Both Cre and Flp cleave DNA at a distinct target sequence (*see* **Fig. 4.1**) and ligate it to the cleaved DNA of a second identical site, to generate a contiguous strand. The orientation of these target sites relative to each other directs the type of modification catalyzed by the recombinase (detailed in **Fig. 4.1**). Currently, the most widely used site- specific DNA recombinase system in both ES cells and mice is the Cre-*loxP* system.

(2) *Choice of ES cell line for chromosome engineering.* Only ES cell with an inactivated *Hprt* gene (such as the AB2.2 line *(5)*) can be used with the chromosome engineering libraries because the Cre-*loxP*-mediated recombination event generates a functional *Hprt* minigene, which is used to select the ES cell clones that contain the desired rearrangement.

(3) *Coat-colour genes*. *K14-Agouti* is an effective marker for visual genotyping of targeted alleles and engineered chromosomes while the *Tyrosinase* minigene is more position-dependent as a genetic marker (reviewed in *(75)*).

(4) *Selectable marker tests.* It must be noted that selectable marker tests do not always give definite results. It is possible that cells can sometimes exhibit partial resistance/sensitivity due to the presence of sister cells with reciprocal recombination products that are rescued by cross-feeding. In general, faster-acting drugs give less ambiguous test results. For example, the puromycin test is generally more reliable than the G418 test. In addition, it is advisable to perform control plating experiments at relatively low cell densities with all three selection drugs to determine whether the resistance to a particular drug is caused by contaminating cells or a property of HAT-resistant clones.

References

1. Davisson MT, Schmidt C, Reeves RH et al. Segmental trisomy as a mouse model for Down syndrome. *Prog Clin Biol Res* 1993;384:117–33.
2. Reeves RH, Irving NG, Moran TH et al. A mouse model for Down syndrome exhibits learning and behaviour deficits. *Nat Genet* 1995;11:177–84.
3. Zheng B, Mills AA, Bradley A. A system for rapid generation of coat color-tagged knockouts and defined chromosomal rearrangements in mice. *Nucleic Acids Res* 1999;27:2354–60.
4. van der Weyden L, Bradley A. Mouse: chromosome engineering for modeling human disease. *Annu Rev Genomics Hum Genet* 2006;7:247–76.
5. Ramirez-Solis R, Liu P, Bradley A. Chromosome engineering in mice. *Nature* 1995;378:720–4.
6. Smith AJ, De Sousa MA, Kwabi-Addo B, Heppell-Parton A, Impey H, Rabbitts P. A site-directed chromosomal translocation induced in embryonic stem cells by Cre-loxP recombination. *Nat Genet* 1995;9:376–85.
7. Li ZW, Stark G, Gotz J et al. Generation of mice with a 200-kb amyloid precursor protein gene deletion by Cre recombinase-mediated site-specific recombination in embryonic stem cells. *Proc Natl Acad Sci USA* 1996;93:6158–62.
8. Van Deursen J, Fornerod M, Van Rees B, Grosveld G. Cre-mediated site-specific translocation between nonhomologous mouse chromosomes. *Proc Natl Acad Sci USA* 1995;92:7376–80.
9. Mills AA, Zheng B, Wang XJ, Vogel H, Roop DR, Bradley A. p63 is a p53 homologue required for limb and epidermal morphogenesis. *Nature* 1999;398:708–13.
10. Lindsay EA, Botta A, Jurecic V et al. Congenital heart disease in mice deficient for the DiGeorge syndrome region. *Nature* 1999;401:379–83.
11. Walz K, Caratini-Rivera S, Bi W, Fonseca P et al. Modeling del(17)(p11.2p11.2) and dup(17)(p11.2p11.2) contiguous gene syndromes by chromosome engineering in mice: phenotypic consequences of gene dosage imbalance. *Mol Cell Biol* 2003;23:3646–55.
12. Medina-Martinez O, Bradley A, Ramirez-Solis R. A large targeted deletion of *Hoxb1-Hoxb9* produces a series of single-segment anterior homeotic transformations. *Dev Biol* 2000;222:71–83.
13. Zheng B, Sage M, Sheppeard EA, Jurecic V, Bradley A. Engineering mouse chromosomes with Cre-loxP: range, efficiency, and somatic applications. *Mol Cell Biol* 2000;20:648–55.
14. Nishijima I, Mills A, Qi Y, Mills M, Bradley A. Two new balancer chromosomes on mouse chromosome 4 to facilitate functional annotation of human chromosome 1p. *Genesis* 2003;36:142–8.
15. O'Gorman S, Dagenais NA, Qian M, Marchuk Y. Protamine-Cre recombinase transgenes efficiently recombine target sequences in the male germ line of mice, but not in embryonic stem cells. *Proc Natl Acad Sci USA* 1997; 94:14602–7.
16. Taniguchi M, Sanbo M, Watanabe S, Naruse I, Mishina M, Yagi T. Efficient production of Cre-mediated site-directed

recombinants through the utilization of the puromycin resistance gene, pac: a transient gene-integration marker for ES cells. *Nucleic Acids Res* 1998;26:679–80.

17. Sauer B, Henderson N. Targeted insertion of exogenous DNA into the eukaryotic genome by the Cre recombinase. *New Biol* 1990;2:441–9.
18. Liu Q, Li MZ, Leibham D, Cortez D, Elledge SJ. The univector plasmid-fusion system, a method for rapid construction of recombinant DNA without restriction enzymes. *Curr Biol* 1998;8:1300–9.
19. Adams DJ, Biggs PJ, Cox T et al. Mutagenic insertion and chromosome engineering resource (MICER). *Nat Genet* 2004;36:867–71.
20. van der Weyden L, Adams DJ, Bradley A. Tools for targeted manipulation of the mouse genome. *Physiol Genomics* 2002; 11:133–64.
21. Hubbard TJ, Aken BL, Beal K et al. Ensembl. *Nucleic Acids Res* 2007; 35:D610–7.
22. Liu P, Jenkins NA, Copeland NG. A highly efficient recombineering-based method for generating conditional knockout mutations. *Genome Res* 13: 476–484.
23. Adams DJ, Quail MA, Cox T et al. A genome-wide, end-sequenced 129 Sv BAC library resource for targeting vector construction. *Genomics* 2005;86:753–8.
24. Ohtsuka M, Ishii K, Kikuti YY et al. Construction of mouse 129/Ola BAC library for targeting experiments using E14 embryonic stem cells. *Genes Genet Syst* 2006;81:143–6.
25. Yokoyama T, Silversides DW, Waymire KG, Kwon BS, Takeuchi T, Overbeek PA. Conserved cysteine to serine mutation in tyrosinase is responsible for the classical albino mutation in laboratory mice. *Nucleic Acids Res* 1990;18:7293–8.
26. Overbeek PA, Aguilar-Cordova E, Hanten G et al. Coinjection strategy for visual identification of transgenic mice. *Transgenic Res* 1991;1:31–7.
27. Kucera GT, Bortner DM, Rosenberg MP. Overexpression of an Agouti cDNA in the skin of transgenic mice recapitulates dominant coat color phenotypes of spontaneous mutants. *Dev Biol* 1996;173:162–73.
28. Bagchi A, Papazoglu C, Wu Y et al. CHD5 is a tumor suppressor at human 1p36. *Cell* 2007;128:459–75.
29. You Y, Bergstrom R, Klemm M et al. Chromosomal deletion complexes in mice by radiation of embryonic stem cells. *Nat Genet* 1997;15:285–8.
30. Schimenti JC, Libby BJ, Bergstrom RA et al. Interdigitated deletion complexes on mouse chromosome 5 induced by irradiation of embryonic stem cells. *Genome Res* 2000;10:1043–50.
31. Chick WS, Mentzer SE, Carpenter DA, Rinchik EM, Johnson D, You Y. X-ray-induced deletion complexes in embryonic stem cells on mouse chromosome 15. *Mamm Genome* 2005;16:661–71.
32. You Y, Browning VL, Schimenti JC. Generation of radiation-induced deletion complexes in the mouse genome using embryonic stem cells. *Methods* 1997;13:409–21.
33. Thomas JW, LaMantia C, Magnuson T. X-ray-induced mutations in mouse embryonic stem cells. *Proc Natl Acad Sci USA* 1998;95:1114–9.
34. Goodwin NC, Ishida Y, Hartford S, Wnek C, Bergstrom RA, Leder P, Schimenti JC. DelBank: a mouse ES-cell resource for generating deletions. *Nat Genet* 2001;28:310–1.
35. Herault Y, Rassoulzadegan M, Cuzin F, Duboule D. Engineering chromosomes in mice through targeted meiotic recombination (TAMERE). *Nat Genet* 1998; 20:381–4.
36. Tarchini B, Huynh TH, Cox GA, Duboule D. HoxD cluster scanning deletions identify multiple defects leading to paralysis in the mouse mutant Ironside. *Genes Dev* 2005;19:2862–76.
37. Tang SH, Silva FJ, Tsark WM, Mann JR. A Cre/loxP-deleter transgenic line in mouse strain 129S1/SvImJ. *Genesis* 2002; 32:199–202.
38. Wu S, Ying G, Wu Q, Capecchi MR. Toward simpler and faster genome-wide mutagenesis in mice. *Nat Genet* 2007; 39:922–30.
39. Olson LE, Tien J, South S, Reeves RH. Long-range chromosomal engineering is more efficient in vitro than in vivo. *Transgenic Res* 2005;4:25–32.
40. Spitz F, Herkenne C, Morris MA, Duboule D. Inversion-induced disruption of the Hoxd cluster leads to the partition of regulatory landscapes. *Nat Genet* 2005;37:889–93.
41. Lindsay EA, Vitelli F, Su H et al. Tbx1 haploinsufficiency in the DiGeorge syndrome region causes aortic arch defects in mice. *Nature* 2001;410:97–101.
42. Yu Y, Bradley A. Mouse genomic technologies: engineering chromosomal rearrangements in mice. *Nat Rev Genet* 2001;2:780–90.

43. Su H, Wang X, Bradley A. Nested chromosomal deletions induced with retroviral vectors in mice. *Nat Genet* 2000;24:92–5.
44. Yan J, Keener VW, Bi W et al. Reduced penetrance of craniofacial anomalies as a function of deletion size and genetic background in a chromosome engineered partial mouse model for Smith–Magenis syndrome. *Hum Mol Genet* 2004;13:2613–24.
45. LePage DF, Church DM, Millie E, Hassold TJ, Conlon RA. Rapid generation of nested chromosomal deletions on mouse chromosome 2. *Proc Natl Acad Sci USA* 2000;97:10471–76.
46. Kushi A, Edamura K, Noguchi M, Akiyama K, Nishi Y, Sasai H. Generation of mutant mice with large chromosomal deletion by use of irradiated ES cells: analysis of large deletion around *hprt* locus of ES cell. *Mamm Genome* 1998;9:269–73.
47. Bergstrom DE, Bergstrom RA, Munroe RJ et al. Overlapping deletions spanning the proximal two-thirds of the mouse t complex. *Mamm Genome* 2003;14:817–29.
48. Liu P, Zhang H, McLellan A, Vogel H, Bradley A. Embryonic lethality and tumorigenesis caused by segmental aneuploidy on mouse chromosome 11. *Genetics* 1998;150:1155–68.
49. Yu YE, Morishima M, Pao A et al. A deficiency in the region homologous to human 17q21.33–q23.2 causes heart defects in mice. *Genetics* 2006;173:297–307.
50. Schwenk F, Baron U, Rajewsky K. A cre-transgenic mouse strain for the ubiquitous deletion of loxP-flanked gene segments including deletion in germ cells. *Nucleic Acids Res* 1995;23:5080–1.
51. Su H, Mills AA, Wang X, Bradley A. A targeted X-linked CMV-Cre line. *Genesis* 2002;32:187–8.
52. Davisson MT, Schmidt C, Akeson EC. Segmental trisomy of murine chromosome 16: a new model system for studying Down syndrome. *Prog Clin Biol Res* 1990;360:263–80.
53. Davisson MT, Schmidt C, Reeves RH et al. Segmental trisomy as a mouse model for Down syndrome. *Prog Clin Biol Res* 1993;384:117–33.
54. Richtsmeier JT, Baxter LL, Reeves RH. Parallels of craniofacial maldevelopment in Down syndrome and Ts65Dn mice. *Dev Dyn* 2000;217:137–45.
55. Olson LE, Richtsmeier JT, Leszl J, Reeves RH. A chromosome 21 critical region does not cause specific Down syndrome phenotypes. *Science* 2004;306:687–90.
56. Li Z, Yu T, Morishima M et al. Duplication of the entire 22.9 Mb human chromosome 21 syntenic region on mouse chromosome 16 causes cardiovascular and gastrointestinal abnormalities. *Hum Mol Genet* 2007;16:1359–66.
57. Hentges KE, Justice MJ. Checks and balancers: balancer chromosomes to facilitate genome annotation. *Trends Genet* 2004;20:252–59.
58. Roderick TH, Hawes NL. Two radiation-induced chromosomal inversions in mice (*Mus musculus*). *Proc Natl Acad Sci USA* 1970;67:961–7.
59. Roderick TH, Hawes NL. Nineteen paracentric chromosomal inversions in mice. *Genetics* 1974;76:109–17.
60. Zheng B, Sage M, Cai WW et al. Engineering a mouse balancer chromosome. *Nat Genet* 1999;22:375–8.
61. Chick WS, Mentzer SE, Carpenter DA, Rinchik EM, You Y. Modification of an existing chromosomal inversion to engineer a balancer for mouse chromosome 15. *Genetics* 2004;167:889–95.
62. Klysik J, Dinh C, Bradley A. Two new mouse chromosome 11 balancers. *Genomics* 2004;83:303–10.
63. Rabbitts TH. Chromosomal translocations in human cancer. *Nature* 1994;372:143–9.
64. Mitelman F, Mertens F, Johansson B. A breakpoint map of recurrent chromosomal rearrangements in human neoplasia. *Nat Genet* 1997;15:417–74.
65. Look AT. Oncogenic transcription factors in the human acute leukemias. *Science* 1997;278:1059–64.
66. Corral J, Lavenir I, Impey H et al. An Mll-AF9 fusion gene made by homologous recombination causes acute leukemia in chimeric mice: a method to create fusion oncogenes. *Cell* 1996;85:853–61.
67. Okuda T, Cai Z, Yang S et al. Expression of a knocked-in AML1-ETO leukemia gene inhibits the establishment of normal definitive hematopoiesis and directly generates dysplastic hematopoietic progenitors. *Blood* 1998;91:3134–43.
68. Yergeau DA, Hetherington CJ, Wang Q et al. Embryonic lethality and impairment of haematopoiesis in mice heterozygous for an AML1-ETO fusion gene. *Nat Genet* 1997;15:303–6.
69. Pollock JL, Westervelt P, Kurichety AK, Pelicci PG, Grisolano JL, Ley TJ. A bcr-3 isoform of RARα-PML potentiates the development of PML-RARα-driven acute

promyelocytic leukemia. *Proc Natl Acad Sci USA* 1999;96:15103–8.

70. Higuchi M, O'Brien D, Kumaravelu P, Lenny N, Yeoh EJ, Downing JR. Expression of a conditional *AML1-ETO* oncogene bypasses embryonic lethality and establishes a murine model of human t(8;21) acute myeloid leukemia. *Cancer Cell* 2002; 163–74.
71. Collins EC, Pannell R, Simpson EM, Forster A, Rabbitts TH. Inter-chromosomal recombination of *Mll* and *Af9* genes mediated by cre-loxP in mouse development. *EMBO Rep* 2000;1:127–32.
72. Forster A, Pannell R, Drynan LF et al. Engineering de novo reciprocal chromosomal translocations associated with *Mll* to replicate primary events of human cancer. *Cancer Cell* 2003;3:449–58.
73. Drynan LF, Pannell R, Forster A et al. *Mll* fusions generated by Cre-loxP-mediated de novo translocations can induce lineage reassignment in tumorigenesis. *EMBO J* 2005; 24:3136–46.
74. Lyon MF, Meredith R. Autosomal translocations causing male sterility and viable aneuploidy in the mouse. *Cytogenetics* 1966;5:335–54.
75. Zheng B, Mills AA, Bradley A. Introducing defined chromosomal rearrangements into the mouse genome. *Methods* 2001; 24:81–94.
76. Chung YJ, Jonkers J, Kitson H et al. A whole-genome mouse BAC microarray with 1-Mb resolution for analysis of DNA copy number changes by array comparative genomic hybridization. *Genome Res* 2004; 14:188–96.
77. Agah R, Frenkel PA, French BA, Michael LH, Overbeek PA, Schneider MD. Gene recombination in postmitotic cells. Targeted expression of Cre recombinase provokes cardiac-restricted, site-specific rearrangement in adult ventricular muscle in vivo. *J Clin Invest* 1997;100:169–79.
78. Johnson L, Mercer K, Greenbaum D et al. Somatic activation of the K-ras oncogene causes early onset lung cancer in mice. *Nature* 2001;410:1111–6.
79. Garcia-Otin AL, Guillou F. Mammalian genome targeting using site-specific recombinases. *Front Biosci* 2006;11:1108–36.
80. Saam JR, Gordon JI. Inducible gene knockouts in the small intestine and colonic epithelium. *J Biol Chem* 1999;274: 38071–82.
81. Schönig K, Schwenk F, Rajewsky K, Bujard H. Stringent doxycycline dependent control of Cre recombinase in vivo. *Nucleic Acids Res* 2002;30 e134.
82. Perl AK, Wert SE, Nagy A, Lobe CG, Whittsett JA. Early restriction of peripheral and proximal cell lineages during formation of the lung. *Proc Natl Acad Sci USA* 2002;99:10482–7.
83. Schuler M, Dierich A, Chambon P, Metzger D. Efficient temporally controlled targeted somatic mutagenesis in hepatocytes of the mouse. *Genesis* 2004;39:167–72.
84. El Marjou F, Janssen KP, Chang BHJ et al. Tissue-specific and inducible Cre-mediated recombination in the gut epithelium. *Genesis* 2004;39:186–93.
85. Kemp R, Ireland H, Clayton E, Houghton C, Howard L, Winton DJ. Elimination of background recombination: somatic induction of Cre by combined transcriptional regulation and hormone binding affinity. *Nucleic Acids Res* 2004;32:e92.
86. Weber P, Schuler M, Gerard C, Mark M, Metzger D, Chambon P. Temporally controlled site-specific mutagenesis in the germ cell lineage of the mouse testis. *Biol Reprod* 2003;68:553–9.
87. Yajima I, Belloir E, Bourgeois Y, Kumasaka M, Delmas V, Larue L. Spatiotemporal gene control by the Cre-ERT2 system in melanocytes. *Genesis* 2006;44:34–43.
88. Hirrlinger PG, Scheller A, Braun C, Hirrlinger J, Kirchhoff F. Temporal control of gene recombination in astrocytes by transgenic expression of the tamoxifen-inducible DNA recombinase variant CreERT2. *Glia* 2006;54:11–20.

Chapter 5

Gene Modification in Embryonic Stem Cells by Single-Stranded DNA Oligonucleotides

Marieke Aarts, Marleen Dekker, Rob Dekker, Sandra de Vries, Anja van der Wal, Eva Wielders, and Hein te Riele

Abstract

Oligonucleotide-mediated gene targeting is an attractive alternative to current procedures to subtly modify the genome of mouse embryonic stem (ES) cells. However, oligonucleotide-directed substitution, insertion or deletion of a single or a few nucleotides was hampered by DNA mismatch repair (MMR). We have developed strategies to circumvent this problem based on findings that the central MMR protein MSH2 acts in two different mismatch recognition complexes: MSH2/MSH6, which mainly recognizes base substitutions; and MSH2/MSH3, which has more affinity for larger loops. We found that oligonucleotide-mediated base substitution could effectively be obtained upon transient suppression of MSH2 protein level, while base insertions were effective in ES cells deficient for MSH3. This method allows substitution of any codon of interest in the genome.

Key words: Subtle gene modification, single-stranded oligonucleotides, DNA mismatch repair, codon substitution, mouse mutant, gene therapy.

1. Introduction

Knockout mice are invaluable for the study of gene function and have provided useful models for human genetic disorders. Thus far, the vast majority of mutant mice have been generated by "gene disruption" through homologous recombination in embryonic stem (ES) cells *(1)*. However, for more accurate mimicking of human disease and in view of the modular nature of many proteins that are composed of separate domains, each responsible for specific interactions and functions, there is an urgent need for "subtle" mutants that affect only one function of a protein but leave others intact.

Ralf Kühn, Wolfgang Wurst (eds.), *Gene Knockout Protocols: Second Edition, vol. 530*

DOI 10.1007/978-1-59745-471-1_5 Springerprotocols.com

Current procedures for subtle modification of the ES cell genome are costly and time-consuming. Briefly, in a two-step protocol, homologous recombination is used to introduce the subtle mutation into the gene of interest concomitantly with a dominantly/negatively selectable marker gene, the latter subsequently being removed via intrachromosomal homologous or Cre/*lox*-mediated site-specific recombination *(2, 3)*.

An alternative approach to introduce subtle gene modifications may be the use of synthetic single-stranded DNA oligonucleotides. Since this procedure has been proven successful in the yeast *Saccharomyces cerevisiae (4)* and in human cells *(5)*, extensive efforts have been made to establish reliable protocols for oligonucleotide-directed gene modification that can eventually be applied in human gene therapy. Thus far, these procedures made use of chemically modified single-stranded oligonucleotides, chimeric RNA/DNA oligonucleotides or triple-helix-forming oligonucleotides *(6–9)*, all containing phosphorothioate linkages or 2′-*O*-methyl-RNA residues to protect oligonucleotides from intracellular nucleolytic degradation *(10)*. Several reports have addressed the relevance of transcription *(11, 12)*, DNA replication *(13)*, homologous recombination *(14)* and DNA damage repair *(15, 16)*; however, the mechanism of transfer of genetic information from the oligonucleotide to the target remains largely elusive.

Several years ago, we have demonstrated the feasibility of subtle gene modification in ES cells by "non-chemically modified" single-stranded DNA oligonucleotides *(17)*. However, we found that "oligo-targeting" was strongly suppressed by DNA mismatch repair (MMR) activity. As a readout for oligo-targeting, we used inactive neomycin resistance (*neo*) genes carrying either a mutation in the start codon (AAG) or a two-base pair frameshift mutation (extra GT) disrupting the open reading frame. Single copies of these reporters have been inserted into the *Rosa26* locus of wild-type and *Msh2*-deficient (*Msh2* $^{-/-}$) ES cells. The latter cells lack the mismatch recognition protein MSH2 and are completely devoid of MMR activity *(18)*. We found that single-stranded deoxyribo-oligonucleotides of ±35 residues substituting, deleting or inserting one or two nucleotides could restore *neo* activity (giving G418-resistance) in *Msh2* $^{-/-}$ ES cells with a frequency of 1–7 per 10^5 cells *(17)*. In wild-type cells, MSH2 activity suppressed single-nucleotide substitution 150-fold and single-nucleotide insertion over 700-fold. Thus, oligo-targeting was only effective in the absence of MMR *(17)*, an observation that has been confirmed in *Escherichia coli (19)*. Unfortunately, constitutive MMR deficiency leads to numerous inadvertent mutations, hampering the general application of oligonucleotide-directed gene modification.

To circumvent this problem we have exploited previous findings that MSH2 acts in two different mismatch recognition complexes: MSH2/MSH6 and MSH2/MSH3 *(20)*. MSH2/MSH6 mainly recognizes base substitutions and small loops of one or two nucleotides, while the MSH2/MSH3 complex has more affinity for larger loops of unpaired bases.

Gene knockout (21). By comparing the efficacy of a large series of insertion oligonucleotides, we found that four-nucleotide insertions are predominantly recognized by the MSH2/MSH3 dimer. Occasionally, a four-nucleotide insertion was suppressed by MSH2/MSH6 activity. We found a simple rule for minimizing recognition by MSH2/MSH6 and hence achieving effective four-nucleotide insertions in *Msh3*$^{-/-}$ cells: the four-nucleotide insertion must not be capable of forming base pairs with the complementary strand and should be placed between G or C residues. As *Msh3*$^{-/-}$ cells do not show an overt mutator phenotype and *Msh3*-deficient mice do not develop cancer, these cells may become the routine target cells for oligonucleotide-mediated "gene disruption". As oligonucleotide-mediated gene modifications are generally not selectable, we have designed a PCR protocol that allows identification and subsequent cloning of modified cells from the large excess of non-modified cells. We have used *Msh3*$^{-/-}$ ES cells to generate TAAA insertions in, among others, the Fanconi anemia *Fancf* gene and the Roberts syndrome gene *Esco2*.

Codon substitution (22). Base substitutions were recognized by both heterodimers. We therefore developed a procedure to transiently suppress the level of MSH2 in wild-type cells by RNA interference. By expression of a short-hairpin RNA sequence against *Msh2* from a transiently transfected pSUPER vector (pS-MSH2), we achieved tenfold reduction of MSH2 level for approximately 3 days after which MSH2 level raised back to wild type. During this period cells were permissive for oligonucleotide-mediated substitution of four nucleotides reaching frequencies of 60–80% of the levels in *Msh2*$^{-/-}$ cells. Simple nucleotide substitutions were still largely suppressed as was spontaneous mutagenesis. These results indicate that upon reduction of MSH2 protein level by RNA interference residual MMR activity persists. This activity largely suppresses spontaneous mutagenesis, but is permissive for four-nucleotide substitutions, allowing the replacement of virtually every codon in the genome. We recently found that also transient suppression of the MMR gene *Mlh1* using vector pS-MLH1 renders cells permissive for oligo-targeting. We have applied the MSH2 and MLH1 knockdown strategies in combination with mutation-specific PCR to substitute codons in *Rb*, *Msh2* and *p53*.

Mouse mutants (22). RB(N750F) ES cells were used to generate the first mutant mouse line that was created via oligonucleotide-mediated gene targeting. Similarly, we generated mouse lines from *Fancf*^{TAAA} and *Esco2*^{TAAA} mutant ES cells, demonstrating the general applicability of oligonucleotide-mediated gene modification for the generation of mutant mouse lines.

2. Materials

ES cells are routinely cultured on irradiated primary mouse embryonic fibroblasts (MEFs). However, during transfections and drug selections, cells are cultured on gelatine-coated plates in buffalo rat liver (BRL)-conditioned medium. This is because MEFs or other feeder layers decrease the transfection efficiency of the ES cells by taking up a large proportion of the transfected DNA. In addition, feeder cells may not tolerate the high concentrations of selective drugs that are used. Protocols for the preparation of MEF feeder layers and culturing of ES cells are given in **Chapter 9** of this volume while the use of BRL-conditioned medium is described in reference *(23)*. This chapter provides protocols for oligonucleotide-mediated gene targeting and subsequent subcloning of mutant ES cells. **Figure 5.1** shows a flow diagram of the procedure.

Fetal bovine serum (FBS) used for ES cell culture must not contain any components that promote differentiation and/or inhibit proliferation, enabling ES cells to retain their totipotent state to generate germline chimeras.

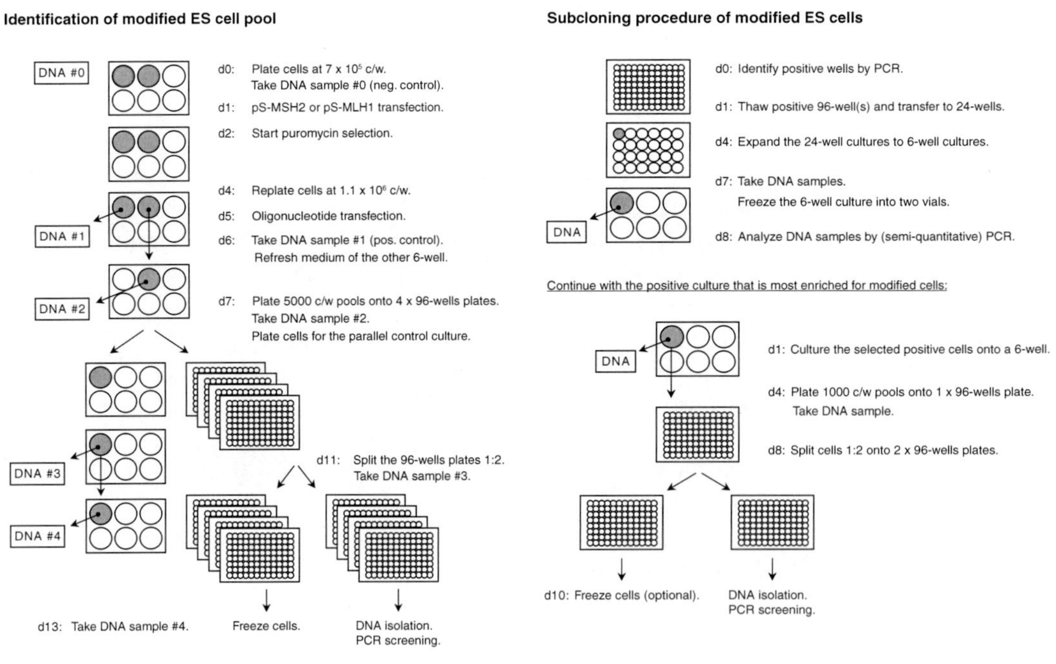

Fig. 5.1. Flow diagram of oligo-targeting procedure and subcloning of modified cells.

2.1. Media and Solutions

1. Complete Medium (CM): 500 mL Glasgow Minimal Essential Medium (GMEM) (GIBCO/Invitrogen, 21710) supplemented with 10% FBS (HyClone), 1 mM sodium pyruvate (100 mM stock, GIBCO/Invitrogen, 11360), 1X non-essential amino acids (100X stock, GIBCO/Invitrogen, 11140).
2. CM+β+LIF (leukaemia inhibitory factor): 100 mL CM, 0.1 mL β-mercaptoethanol (1000X), 1 mL LIF (100X).
3. β-Mercaptoethanol (1000X, 0.1 M): 0.1 mL 2-mercaptoethanol (14.2 M, Merck, 1.15433) and 14.1 mL water. Sterilize by filtration through 0.22-µm Millex-GV filter. Store at 4°C for up to 1 month. Working concentration: 0.1 mM.
4. LIF (100X): Dissolve 1 mL ESGRO® (10^7 Units, Chemicon International, ESG1107) in 99 mL CM. Working concentration: 10^3 U/mL. Store at 4°C.
5. BRL medium: CM conditioned on a monolayer of buffalo rat liver (BRL-3A) cells, filtered through a 0.2-µm filter unit (Nalgene, 450-0020).
6. BRL+β+LIF (60% medium): 100 mL CM, 150 mL BRL medium, 1.5 mL L-glutamine 200 mM (GIBCO/Invitrogen, 25030), 0.25 mL β-mercaptoethanol (1000X), 2.5 mL LIF (100X).
7. Distilled water: 500-mL bottles (GIBCO/Invitrogen, 15230).
8. 1% Gelatin (10X): 5 g (w/v) gelatin (Sigma, G1890) in 500 mL water. Gently heat in microwave oven (do not boil). Sterilize through a 0.2-µm filter unit (Nalgene, 450-0020) while still warm and store aliquots at 4°C. Before use, dilute to 0.1% with water and coat tissue culture dishes for 30 min at RT.
9. Phosphate-buffered saline (PBS): Dulbecco's PBS (1X) without calcium and magnesium, 500-mL bottles (GIBCO/Invitrogen, 14190).
10. Trypsin/EDTA stock (TVP): 500 mL PBS supplemented with 5 mL Trypsin 2.5% (GIBCO/Invitrogen, 15090), 12.5 mL 40 mM ethylenediaminetetraacetic acid (EDTA), 5 mL chicken serum (GIBCO/Invitrogen, 16110). Sterilize by filtration through a 0.2-µm filter unit (Nalgene, 450-0020) and store aliquots at –20°C.
11. 2X TVP: Add 0.1 mL Trypsin 2.5% to 9.9 mL TVP. Store at –20°C.
12. 10X TVP: Add 0.9 mL Trypsin 2.5% to 9.1 mL TVP. Store at –20°C.
13. Puromycin (100X): Dissolve 20 mg (w/v) puromycin (Sigma, P7255) in 10 mL CM. Sterilize through a 0.22-µm filter and store aliquots at –20°C.

14. Targeting oligonucleotide: Unmodified 40-mer synthetic DNA oligonucleotides, deprotected and desalted (Sigma-Aldrich), reconstituted in sterile PBS at 1 μg/μL.
15. TransFast Transfection Reagent (Promega, E2431): Add 400 μL nuclease-free water per vial, vortex vigorously for 10 s and store at –20°C for 16–24 h before use.
16. Lysis Buffer: 100 mM Tris-HCL pH 8.0, 5.0 mM EDTA, 0.2% (w/v) sodium dodecyl sulphate (SDS), 200 mM NaCl, 200 μg/mL (freshly added) Proteinase K (Merck, 24568.0100).
17. DirectPCR Lysis Reagent for cells (Viagen Biotech, 302-C).
18. $T_{10}E_{0.1}$ buffer: 10 mM Tris-HCl (pH 8.0), 0.1 mM EDTA (pH 8.0).
19. PCR primers: 100 μM in nuclease-free water.
20. Taq DNA polymerase: 5 U/μL.
21. 10 mM dNTP mix: Dilute each dNTP (dGTP, dATP, dCTP, dTTP) in nuclease-free water to 10 mM and store aliquots at –20°C.
22. 50X TAE: Dissolve 242 g Tris-base, 16.81 g EDTA, 20.5 g sodium acetate in 1 L distilled water. Adjust to pH 7.9 with acetic acid.
23. Agarose gels: Melt 3% (w/v) multipurpose agarose (Roche) in 0.5X TAE by gently boiling in a microwave oven. Cool until hand warm, add ethidium bromide and pour into prepared gel former. Run gels at 80–100 V.

2.2. General Equipment

1. Tissue culture plates
   a. 96-well microwell plates, flat bottom, 0.3 cm^2 (Costar, 3596).
   b. 24-well microwell plates, 1.8 cm^2 (Costar, 3527).
   c. 6-well microwell plates, 10 cm^2 (Costar, 3506).
   d. 10 cm TC Petri dishes (Falcon, 353003).
2. Sterile 1, 2, 5, 10 and 25 mL plastic pipettes.
3. Sterile 10, 20, 200 and 1,000 μL filter tips.
4. Sterile 15 and 50 mL tubes.
5. 1.8-mL CryoTube vials with external thread (Nunc, 375418).
6. 12-well multichannel pipette.
7. VACUBOY hand operator (INTEGRA Biosciences, 155 500) with eight-channel plastic aspiration adapter for disposable tips with ejector (INTEGRA Biosciences, 155 520).
8. 50-mL disposable polystyrene reagent reservoirs (Costar, 4871).
9. 96-tube freezing boxes (Greiner, 975561) with 1.3-mL polypropylene freezing tubes with attached strip caps, 8.5 × 44 mm (Greiner, 102261).

10. Inverted microscope for routine morphological analysis of ES cells.
11. CASY 1 cell counter (Schärfe System, model TT or DT).
12. Humidified tissue culture incubator maintained at 37°C and 5% CO_2.
13. Tissue culture laminar flow cabinet.
14. Water bath: 37°C.
15. Freezers: –20°C and –80°C.
16. Liquid nitrogen container.
17. DNA thermal cycler (MJ Research, PTC-200).
18. Electrophoresis equipment (Bio-Rad).

3. Methods

3.1. Design of Targeting Oligonucleotide

1. Typical targeting oligonucleotides consist of ~40 nucleotides. The center of the oligonucleotide contains 3–4 bases that comprise the desired genetic alteration (substitution or insertion of nucleotides); the remainder is identical to the gene of interest (*see* examples in **Fig. 5.2B**).
2. The 5′ and 3′ homologous arms are 18 nucleotides in length.
3. Successful gene targeting experiments require optimal sequence identity between the targeting oligonucleotide and the genomic DNA sequence. For oligonucleotide design, ideally use genomic DNA sequence information derived from the same mouse strain as the ES cells (in our case, mouse strain 129Ola). However, this information is often not available.
4. Upon alignment of the targeting oligonucleotide with its complementary genomic sequence, mismatches will be formed. If possible, avoid creating mismatches that are well recognized by the MMR system, such as G/T mismatches.
5. By creating four-nucleotide substitutions, virtually every codon in the mouse genome can be replaced based on the redundancy of the genetic code. However, a different codon usage of the novel codons may affect expression of the target gene. Try to choose codons with high codon usage (*see* www.kazusa.or.jp/codon).
6. Oligonucleotides can be designed in the "sense" or "antisense" orientation, that is, complementary or identical to the transcribed strand of the target gene (*see* **Note 1**).

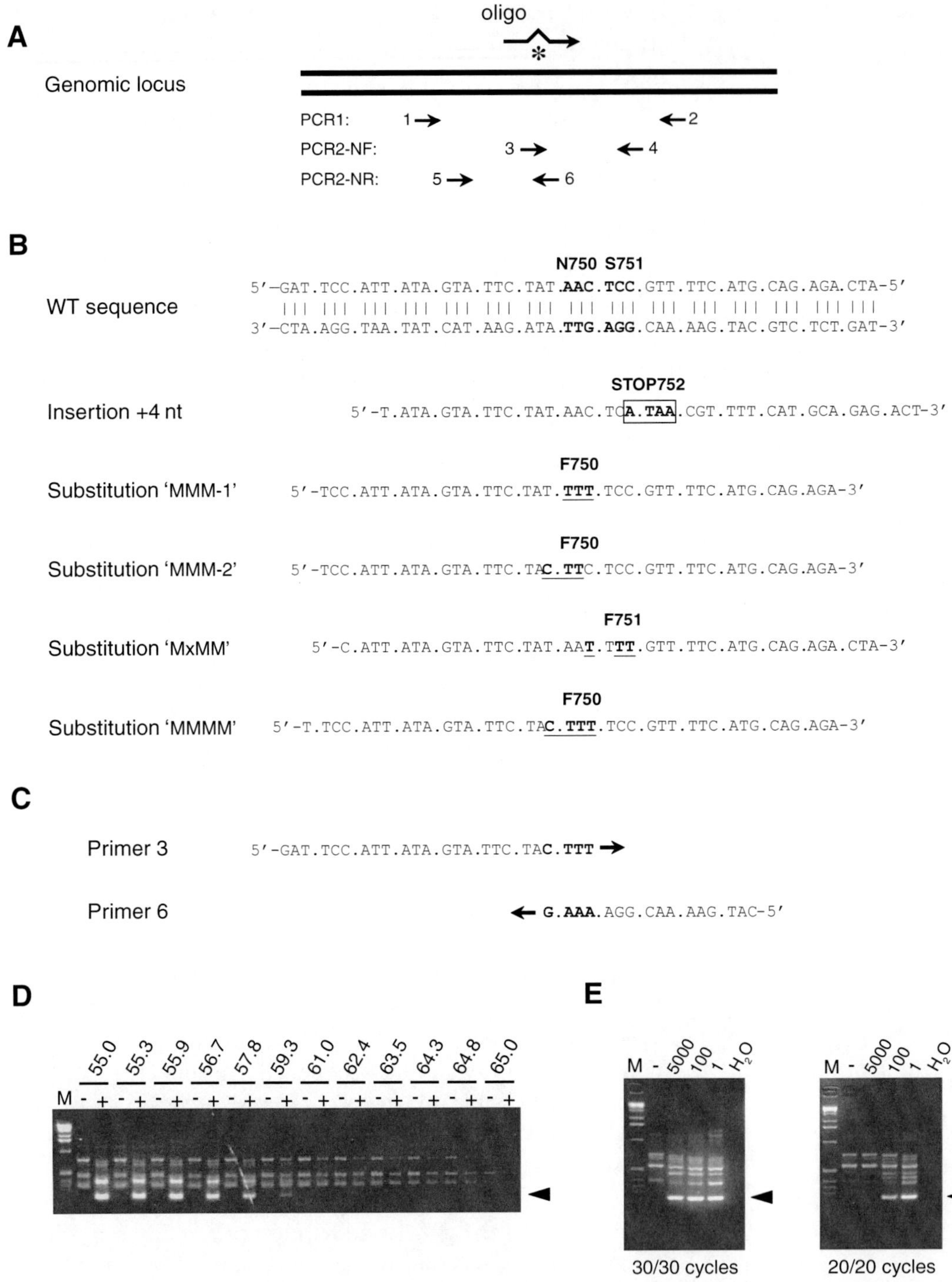

Fig. 5.2. **Targeting oligonucleotide and PCR primer design.** (**A**) Schematic representation of PCR-based detection of modified cells. A single-stranded DNA oligonucleotide is designed to introduce a genomic modification in any given gene. Primer pair 1/2 is used to pre-amplify a DNA fragment surrounding the modification. This fragment is the template in PCR2 using nested primer pairs 3/4 or 5/6 of which primers 3 and 6 are specific for the mutation. *Asterisk* indicates modification. *Arrows* indicate the location and direction of PCR primers. NF, nested forward; NR, nested reverse. (**B**) Genomic DNA

7. Substitution oligonucleotides: Choose an alteration that affects four adjacent bases (MMMM; M indicates mutation), three adjacent bases (MMM), or non-adjacent bases like MxMM (**Fig. 5.2B**).
8. Insertion oligonucleotides: The four-nucleotide insertion must not be capable of forming base pairs with the complementary strand and should be placed between G or C residues (**Fig. 5.2B**).

3.2. Generation of Codon Substitutions in Wild-Type ES Cells

*3.2.1. Knockdown of Msh2 or Mlh1 in Wild-Type ES Cells (see **Note 2**)*

Day -4

1. Gelatinize one well of a 6-well plate.
2. Thaw irradiated feeder MEFs into a 37°C water bath and resuspend in 10 mL CM.
3. Spin for 5 min at 300*g*, remove the supernatant and carefully resuspend the MEFs in CM by gentle pipetting.
4. Plate MEFs onto one 6-well (10 cm^2) plate. Feeder layer cells should be prepared at least 1–2 days before ES culture.

Day -3

1. Thaw ES cells in a 37°C water bath.
2. Spin the cells in 10 mL CM for 5 min at 300*g*, remove the supernatant and resuspend the cells in 3 mL CM+β+LIF.
3. Plate ES cells onto one 6-well plate with MEFs at a density of approximately 2×10^6 cells per well.

Day 0

1. Wash the cells with PBS and trypsinize with 0.5 mL 10X TVP for 5 min at 37°C (*see* **Note 3**). Add 0.5 mL BRL+β+LIF (60% medium) and resuspend by pipetting.
2. Count the number of cells.
3. Plate cells onto two gelatin-coated 6-well plates at a density of 7×10^5 per well in 3 mL BRL+β+LIF (60% medium) (*see* **Note 4**).
4. Take 1×10^6 cells for DNA isolation (DNA #0, negative control for PCR) (*see* **Section 3.4.3**).

sequence and possible targeting oligonucleotides to introduce a stop codon (Insertion +4 nt), a substitution of codon 750 (Substitution MMM-1, MMM-2, MMMM), or codon 751 (Substitution MxMM). M indicates altered nucleotide. (**C**) Sequence of mutation-specific primers 3 and 6 for detection of substitution "MMMM". (**D**) Gradient PCR to optimize the annealing temperature of PCR2. Genomic DNA from untransfected ES cells (DNA #0) is used as negative control (–), whereas sample DNA #1 is used as positive control (+). Temperatures are shown above the lanes. In this example, annealing at 56°C is optimal for screening. *Arrowhead* indicates mutation-specific band. Lane M, molecular mass standards. (**E**) Semi-quantitative PCR analysis of positive cell cultures from subsequent screening rounds to show enrichment for modified cells. PCR amplification of genomic DNA from 5,000, 100 and 1 cells/well cultures using 30 cycles (*left panel*) or 20 cycles (*right panel*) in both PCR1 and PCR2. *Arrowhead* indicates mutation-specific band. Lane M, molecular mass standards. (–), Genomic DNA from untransfected ES cells.

5. Dissolve the pS-MSH2 or pS-MLH1 vector in sterile PBS at 1 μg/μL. Store overnight at 4°C.
6. Thaw one vial of TransFast Transfection Reagent and add 400 μL nuclease-free water. Vortex vigorously and store at –20°C.

Day 1

1. Vortex TransFast Transfection Reagent before use.
2. For each well, add 3 μg of pS-MSH2 or pS-MLH1 vector and 27 μL of TransFast to 1.4 mL CM without FBS. Vortex the mixture and leave 10–15 min at RT.
3. Wash the cells with PBS. Vortex the mixture again and add to the cells. Incubate for 75 min at 37°C/5% CO_2.
4. Add 4 mL BRL+β+LIF (60% medium) and incubate overnight at 37°C/5% CO_2.

Day 2

1. Prepare two gelatin-coated 6-well plates with 3 mL BRL+β+LIF (60% medium) + 40 μL puromycin (100X).
2. Trypsinize transfected ES cells and transfer 2 × 1 mL of cell suspension (i.e., all cells) to two 6-well plates with selective medium (final concentration of puromycin is 20 μg/mL).
3. Incubate for 2 days.

3.2.2. Transfection with Targeting Oligo

Day 4

1. Wash the cells in the two 6-well plates twice with PBS and trypsinize.
2. Count the number of surviving cells (*see* **Note 5**).
3. Replate the ES cells onto two gelatin-coated 6-well plates at 1.1×10^6 cells/well in 3 mL BRL+β+LIF (60% medium) without puromycin (*see* **Note 6**).
4. Dissolve the targeting oligonucleotide in sterile PBS at 1 μg/μL. Store overnight at –20°C.

Day 5

1. To transfect the two 6-well plates with targeting oligonucleotide, add for each well 3 μg of targeting oligonucleotide and 27 μL of TransFast (vortex before use) to 1.4 mL CM without FBS. Vortex the mixture and leave for 10–15 min at RT.
2. Wash the cells with PBS. Vortex the mixture again and add to the cells. Incubate for 75 min at 37°C/5% CO_2.
3. Add 4 mL BRL+β+LIF (60% medium) and incubate overnight at 37°C/5% CO_2.

Day 6

1. Trypsinize and count cells of one 6-well plate (*see* **Note 7**).
2. Take 1×10^6 cells for DNA isolation (*see* **Section 3.4.3**). This sample (= DNA #1) will be used as a positive control for PCR screening (*see* **Note 8**). Discard the remainder of the cells.

3. Refresh the medium of the other 6-well plate with BRL+β+LIF (60% medium).
4. Prepare 12 × 96-well plates with MEFs and 100 μL CM+β+LIF medium. Use a multichannel pipette when handling 96-well plates.

Day 7

1. Trypsinize and count ES cells.
2. Add 2 × 10^6 cells to 40 mL CM+β+LIF medium. Transfer 100 μL portions of cell suspension to 4 × 96-well plates with MEFs and 100 μL CM+β+LIF medium (5,000 cells/well).
3. Plate 1 × 10^6 cells onto one 6-well plate with MEFs. These cells are cultivated in parallel to the 96-well plates in order to monitor the degradation of unincorporated oligonucleotides.
4. Take 1 × 10^6 cells for DNA isolation (= DNA #2) (*see* **Section 3.4.3**).

Day 10

1. Refresh the medium of the 4 × 96-well plates with 5,000 cells/well.
2. Count the number of colonies in each 96-well plate to determine the plating efficiency (should be 500–800 colonies).
3. From the parallel 6-well plate with control cells, take 1 × 10^6 cells for DNA isolation (= DNA #3) (*see* **Section 3.4.3**) and transfer 1 × 10^6 cells to a fresh 6-well plate with MEFs.

Day 11

1. When ES colonies have reached ~90% confluency, split the 5,000 cells/well cultures into duplicates. Trypsinize only two plates at the same time to reduce the stress of trypsinization.
2. Wash the wells with 100 μL PBS and trypsinize with 25 μL 10X TVP for 5 min at 37°C/5% CO_2.
3. Add 175 μL CM+β+LIF and resuspend.
4. Transfer two 100 μL portions of cell suspension to two new 96-well plates with MEFs and 100 μL CM+β+LIF medium.
5. Label the side of each plate so that lids and bases of the plates are both identifiable: four master plates for freezing the cells and four duplicates for DNA isolation. This minimizes confusion during freezing and PCR analysis.

Day 12

1. Refresh the medium of the 8 × 96-well plates with 200 μL CM+β+LIF medium.

3.2.3. Freezing of 96-Well Plates

1. Prepare CM (without LIF) with 11.43% dimethylsulfoxide (DMSO) at 4°C.

Day 13

2. Wash cells of the master plates with PBS and trypsinize with 25 μL 10X TVP for 5 min at 37°C/5% CO_2. When freezing 4 × 96-well plates, trypsinize and process only two plates at the same time to reduce the stress of trypsinization and DMSO.

3. Add 175 µL CM plus 11.43% DMSO and resuspend carefully (final concentration: 10% DMSO).
4. Transfer 200 µL of trypsinized cells to individual tubes of a 96-tube freezing box (Greiner, 975561). Store cells in freezing boxes on ice until all plates have been prepared for freezing (*see* **Note 9**).
5. Freeze the cells by a rate-controlled cooling program and store them in liquid nitrogen. Alternatively, wrap the freezing boxes in paper towels and place them overnight at –80°C. Transfer to liquid nitrogen the next day.
6. Isolate DNA from the four duplicate plates (*see* **Section 3.4.1** or **3.4.2**).
7. From the parallel 6-well plate with control cells, take 1 × 10^6 cells for DNA isolation (= DNA #4) (*see* **Section 3.4.3**). Discard the remainder of the cells.

Day 14

1. Identify wells containing mutated cells in the 4 × 96-well DNA plates by PCR1 and PCR2 (*see* **Section 3.4**).

3.3. Generation of Gene Knockouts in Msh3-Deficient ES Cells

1. Plate thawed *Msh3*-deficient ES cells onto MEFs in 3 mL CM+β+LIF at a density of approximately 2 × 10^6 cells per well. Cultivate the cells to (semi)confluency for 3 days.
2. Wash cells with PBS and trypsinize with 0.5 mL 10X TVP for 5 min at 37°C/5% CO_2. Add 0.5 mL BRL+β+LIF (60% medium) and resuspend by pipetting. Count the number of cells.
3. Replate the ES cells onto two gelatin-coated 6-well plates at 9 × 10^5 cells/well in 3 mL BRL+β+LIF (60% medium).
4. Take 1 × 10^6 cells for DNA isolation (DNA #0, negative control for PCR) (*see* **Section 3.4.3**).
5. Dissolve the targeting oligonucleotide in sterile PBS at 1 µg/µL. Store overnight at –20°C.
6. The following day, proceed with transfection of the targeting oligonucleotide as described above (*see* **Section 3.2.2**, Day 5).

3.4. Identification of Positive Clones by PCR Analysis

3.4.1. Genomic DNA Isolation from 96-Well Plates (Method 1)

1. Remove culture medium by inverting the 96-well plate. Drain the remaining fluid on a tissue. Optional: Cells can be stored dry in a 96-well plate at –20°C.
2. Add 90 µL DirectPCR Lysis Reagent (Viagen Biotech, 302-C) supplemented with 300 µg/mL Proteinase K to wells.
3. Carefully seal the 96-well plates with adhesive covers to prevent evaporation and incubate at least 2 h at 55°C.

4. Heat inactivate for 45 min at 85°C. Lysates are now ready for use in PCR. Mix lysates well by pipetting up and down several times.
5. Store lysates at –20°C (before use in PCR, thaw for 15 min at 85°C).

3.4.2. Genomic DNA Isolation from 96-Well Plates (Method 2)

1. Add 50 μL lysis buffer with 200 μg/mL Proteinase K to dry wells.
2. Carefully seal the 96-well plate and incubate for at least 2 h at 55°C.
3. Add 100 μL of 100% ethanol to the lysates. Seal the plate with silicone sealing film (Bio-Rad, MSA-5001) or adhesive covers that are ethanol-resistant to prevent detachment of the cover and subsequent cross-contamination of the samples. Mix well by carefully inverting the plate at least five times until white precipitates are visible.
4. Centrifuge plates at 3,000*g* for 30 min at RT.
5. Remove supernatant by inverting the plate.
6. Add 100 μL of 70% ethanol. Centrifuge at 3,000*g* for 15 min at RT.
7. Remove supernatant by inverting the plate. Briefly air-dry the DNA precipitates.
8. Add 100 μL $T_{10}E_{0.1}$ and seal plates with adhesive covers to prevent evaporation.
9. Incubate overnight at 55°C.
10. Store lysates at 4°C.

3.4.3. Genomic DNA Isolation from Cell Pellets

1. Spin 1×10^6 cells in an Eppendorf centrifuge at 1,000*g* for 3 min and remove supernatant.
2. Add 100 μL lysis buffer with 200 μg/mL Proteinase K.
3. Incubate for at least 2 h at 55°C.
4. Add 200 μL of 100% ethanol to the lysates and mix well.
5. Centrifuge at 22,000*g* for 30 min at RT.
6. Remove supernatant.
7. Add 200 μL of 70% ethanol. Centrifuge at 22,000*g* for 15 min at RT.
8. Remove supernatant. Briefly air-dry the DNA precipitates.
9. Add 200 μL $T_{10}E_{0.1}$.
10. Incubate overnight at 55°C.
11. Store lysates at 4°C.

3.4.4. PCR Primer Design

For detection of targeted ES cells, primers 1 and 2 are used to amplify a fragment that is subjected to a nested mutation-specific PCR using primers 3/4 or primers 5/6 (**Fig. 5.2A**). Primers 3 and 6 are specific for the planned modification, having 3′ nucleotides that are complementary to the desired alteration (e.g. **Fig. 5.2C** shows primers for identification of the mutation introduced by oligonucleotide MMMM). Due to improper annealing of primer 3 or 6, no product should be formed with unmodified DNA.

1. For PCR primer design, use genomic DNA sequences that are derived from the same mouse strain as the ES cells. However, this is not always possible.
2. PCR1: Design primers that form a 600–800-bp product comprising the altered genomic sequence. The melting temperature (T_m) of the primers should be 60–65°C (*see* www.sigma-genosys.com/calc/DNACalc.asp).
3. PCR2-nested forward (NF): The mutation-specific forward primer (primer 3) should be complementary to the desired genetic alteration. *Important*: The T_m of this primer must be 51–53°C. The T_m of the reverse primer (primer 4) must be 60–65°C. PCR2-NF with primers 3/4 should result in amplification of a 200–300-bp product.
4. PCR2-nested reverse (NR): Similar to PCR2-NF, but now the reverse primer (primer 6) is mutation-specific with a T_m of 51–53°C. The T_m of the forward primer (primer 5) must be 60–65°C. PCR2-NR with primers 5/6 should result in amplification of a 200–300-bp product.

3.4.5. PCR Optimization

1. Use temperature gradients to optimize the annealing temperatures of PCR1 and PCR2 (**Fig. 5.2D**) in order to obtain mutation-specificity. Conditions will depend on the primers and thermal cyclers used.
2. For optimization of PCR1, genomic DNA derived from untransfected ES cells is used as template. Choose the annealing temperature that gives a strong specific signal and no background signal. Annealing is usually performed at 60–65°C.
3. For optimization of PCR2, use sample DNA #1 (*see* **Note 8**) as positive control and genomic DNA derived from untransfected ES cells as negative control. Both control samples are first subjected to PCR1, followed by PCR2 at different annealing temperatures. *Important*: The annealing temperature of PCR2 must be approximately 5°C higher than the T_m of the mutation-specific primer to ensure mutation specificity. Choose the annealing temperature that gives a strong specific signal in the positive control and no signal in the negative control (**Fig. 5.2D**). Annealing is usually performed at 56–58°C.

3.4.6. PCR Reactions

1. Prepare a PCR1 master mix for 440 reactions (4 × 96-well plates) or a smaller quantity as required containing: 7,260 μL nuclease-free water, 1,100 μL 10X PCR buffer containing 15 mM $MgCl_2$, 55 μL 100 μM forward primer, 55 μL 100 μM reverse primer, 220 μL 10 mM dNTPs and 110 μL Taq polymerase (5 U/μL) to a total volume of 8,800 μL.
2. Label 96-well Thermowell PCR plates (Corning, 6511, Model M) and add 20 μL of PCR1 master mix to each well (use a multichannel pipette).
3. In each plate, add 5 μL of sample DNA #1 to well A1 as positive control.
4. Add 5 μL genomic DNA to each well (do not add genomic DNA to well A1).
5. Add two drops of mineral oil (Sigma, M8410) to each well to prevent evaporation.
6. Transfer plates to the PCR machine and amplify for 30 cycles using an initial denaturation step of 94°C for 5 min, followed by 30 cycles of denaturation at 94°C for 30 s, annealing at 60–65°C for 1 min, elongation at 72°C for 1 min 30 s and a final elongation step of 72°C for 10 min.
7. Prepare a PCR2 master mix for 440 reactions (4 × 96-well plates) or a smaller quantity as required containing: 9,020 μL nuclease-free water, 1,100 μL 10X PCR buffer containing 15 mM $MgCl_2$, 55 μL 100 μM NF primer, 55 μL 100 μM NR primer, 220 μL 10 mM dNTPs and 110 μL Taq polymerase (5 U/μL) to a total volume of 10,560 μL.
8. Label new PCR plates and add 24 μL of PCR2 master mix to each well.
9. Add 1.2 μL of PCR1 product to each well and add two drops of mineral oil.
10. Transfer plates to the PCR machine and amplify for 30 cycles using an initial denaturation step of 94°C for 5 min, followed by 30 cycles of denaturation at 94°C for 30 s, annealing at 56–58°C for 1 min, elongation at 72°C for 1 min and a final elongation step of 72°C for 10 min (*see* **Note 10**).
11. Add 5 μL loading buffer to the PCR reactions and separate 15 μL of PCR2 product by electrophoresis on a 3% agarose gel in 0.5X TAE containing ethidium bromide at 80–100 V (*see* **Note 11**).

3.5. Subcloning of Single Mutant ES Cell Clones

3.5.1. Thawing Procedure

1. Cut the tube with the desired clone out of the freezing box and thaw at RT.
2. Transfer the cells to a sterile 15-mL Falcon tube containing 5 mL CM+β+LIF.
3. Centrifuge at 300*g* for 5 min.

4. Resuspend the cells in 1 mL CM+β+LIF and transfer to a 24-well plate with MEFs. Culture the cells to (semi)confluency for 3 days.
5. Expand the 24-well cell culture to a 6-well culture with MEFs. Culture the cells to (semi)confluency for 3 days.

3.5.2. Subcloning Procedure

1. Trypsinize and count ES cells.
2. Take 1 × 10^6 cells for DNA isolation (DNA 5000-2) (*see* **Section 3.4.3**) and freeze the cells from the 6-well culture into four vials.
3. Analyze DNA samples by PCR to assess whether the culture still contains modified cells after freezing (*see* **Section 3.4.6**). Select the well that gives the strongest PCR signal, that is most enriched for the modified ES cells.
4. Thaw and plate cells from the selected well onto one 6-well plate with MEFs. Culture the cells to (semi)confluency for 3 days.
5. Trypsinize and count ES cells.
6. Take 1 × 10^6 cells for DNA isolation (DNA 5000-3) (*see* **Section 3.4.3**) to assess the recovery of the modified cells after expansion.
7. Plate 1 × 10^5 cells onto 1 × 96-well plate with MEFs (1,000 cells/well) for the next screening round. Discard the remainder of the cells.
8. Culture the cells to (semi)confluency for 3 days. Count the number of colonies in each 96-well plate to determine the plating efficiency (should be 100–500 colonies). Subclones may need to be passaged to a fresh 96-well plate before splitting to reach (semi)confluency.
9. Split the 1,000 cells/well cultures into two duplicate 96-well plates.
10. One plate, the master plate, is cultured to semiconfluency (2–3 days) and processed for freezing (*see* **Section 3.2.3** and **Note 12**).
11. The duplicate plate is cultured as confluent as possible for DNA isolation (*see* **Section 3.4.1** or **3.4.2**).
12. Identify wells containing modified cells by PCR.
13. Thaw the master plate and culture all positive wells on 24-wells with MEFs.
14. Expand the 24-well cultures to 6-well cultures on MEFs. Culture to (semi)confluency for 3 days. Take 1 × 10^6 cells for DNA isolation and freeze the remainder of the cells.
15. Analyze the DNA by semi-quantitative PCR (i.e., perform PCR1 and PCR2 both with 20 cycles of amplification instead of 30 cycles) and determine which sample gives the strongest signal (*see also* **Note 13** and **Fig. 5.2E**).

16. Repeat the procedure as described above from Step 4.
    After the 1,000 cells/well subclones, use the following screening rounds (*see* **Note 14**):
    a. Plate 1 × 10^4 cells onto 1 × 96-well plate with MEFs (100 cells/well) (*see* **Note 15**).
    b. Plate 1,000 cells onto 1 × 96-well plate with MEFs (10 cells/well) and, in parallel, plate 6,000 cells in 48 mL of BRL+β+LIF (60% medium) onto six gelatine-coated 10-cm TC Petri dishes for colony picking to obtain single-cell colonies (*see* **Note 16**). Refeed the cells on the 10 cm dishes every 2–4 days. After 6–8 days, individual colonies can be seen. Proceed with **Section 3.5.3**.

3.5.3. Culturing of Single-Cell Colonies (see ***Note 17****)*

1. Prepare a 96-well plate (V-bottom, Costar, 3894) with 15 μL of PBS per well using the multichannel pipette.
2. Use a pipette with sterile 10-μL tips (limit the volume to 2 μL) to scrape off individual colonies and transfer them to the PBS in the 96-well plate. Pick a series of 192–384 colonies (2–4 96-well plates) (*see also* **Note 16**).
3. Add 15 μL of 2X TVP to each well and incubate for 5 min at 37°C/5% CO_2.
4. Add 70 μL of CM+β+LIF medium and resuspend.
5. Transfer trypsinized colonies (100 μL) to a 96-well plate with MEFs and 100 μL CM+β+LIF medium. Refresh the medium the next day.
6. Culture the cells to semiconfluency (2–3 days).
7. Wash the cells with 100 μL PBS and trypsinize with 25 μL 10X TVP for 5 min at 37°C/5% CO_2.
8. Add 175 μL CM+β+LIF and resuspend.
9. Transfer two 100-μL portions of cell suspension to two new 96-well plates with MEFs and 100 μL CM+β+LIF medium.
10. One plate is cultured to semiconfluency (2–3 days) and processed for freezing (*see* **Section 3.2.3**).
11. The duplicate plate is cultured as confluent as possible for DNA isolation (*see* **Section 3.4.1** or **3.4.2** and **Note 12**).

4. Notes

1. If the targeting procedure did not succeed with an oligonucleotide in the "sense" orientation, the targeting is usually effective when repeated with the complementary oligonucleotide, that is, in the "antisense" orientation. PCR primers and conditions for identification of the modified cells remain unchanged.

2. Knockdown of MSH2 generally allows substitution of four nucleotides. Knockdown of MLH1 renders cells permissive to smaller substitutions as well (three nucleotides) and the success rate is generally larger. However, knockdown of MLH1 may lead to a higher spontaneous mutation frequency.

3. 10X TVP is used for trypsinization of ES cells cultivated on MEF feeders, whereas TVP is used for trypsinizing ES cells grown on gelatin-coated dishes.

4. Transfection efficiency of the ES cells should be approximately 25% for plasmids. This can be measured by transfecting GFP or β-galactosidase (LacZ) expression vectors followed by FACS analysis or staining for LacZ activity, respectively. For the ES cell lines we used, the optimal cell density was approximately 1.2×10^6 cells per 6-well plate on the day of transfection which was achieved by plating the cells at 7×10^5 cells per 6-well plate the day before.

5. If the number of cells that are puromycin-resistant after selection for 2 days is less than 7×10^5, transfections were not efficient. Apparently, the cells are in bad shape and it is useless to continue. We usually count between 7×10^5 and 1.5×10^6 cells per 6-well plate.

6. After puromycin selection, the plating efficiency of the ES cells is somewhat decreased and more variable. To deal with this phenomenon, the cells are plated at 1.1×10^6 cells/well rather than 7×10^5 cells/well.

7. The day after transfection, there should approximately be $2–3 \times 10^6$ cells per 6-well plate based on the predicted growth rate of a healthy culture. When cells have been growing much slower or much faster, transfection efficiencies will be sub-optimal. In that case, do not proceed but rather repeat the whole procedure with a fresh batch of ES cells.

8. Targeting oligonucleotides that were transfected into the ES cells may lead to the appearance of false-positives in the PCR analysis. Non-degraded targeting oligonucleotides can participate as PCR primers in PCR1 resulting in amplification of a "false" mutation-specific product. This product is subsequently amplified in PCR2. Usually, targeting oligonucleotides are degraded after 2 weeks of culturing, which we monitor by taking DNA samples (DNA #1, #2, #3 and #4) from a parallel cell culture and subjecting these samples to both PCR1 and PCR2. Sample DNA #1 should always give a mutation-specific product and can actually be used as positive control for optimization of PCR2 (*see* **Section 3.4.5**). Usually, the mutation-specific product is already decreased or gone in sample DNA #2 or DNA #3. This indicates that the

targeting oligonucleotide has sufficiently been degraded and is not likely to amplify false-positive products when screening the 96-well plates.

9. The following procedures can be used as alternative for the freezing boxes. Resuspend trypsinized cells in CM with DMSO and leave the cell suspensions in the 96-well culture plate. Wrap the plate in paper towels. Place it in a small plastic box at –20°C for several hours and then at –80°C (problem: many cells are lost from the culture). Alternatively, keep the ES cells in culture by passaging them to fresh 96-well plates with MEFs.

10. Screen the 96-well plates only with one of the nested PCRs: either use PCR2-NF or PCR2-NR. Select the nested primer pair that gives the most robust signal during PCR optimization. After identification of positive subclones, the other primer pair can be used for confirmation of the screening results.

11. False-positives are rare but can be excluded from the analysis by pre-treating the genomic DNA samples with Exonuclease I (ExoI), which degrades single-stranded DNA. Add 5 μL genomic DNA, 1 μL 10X ExoI buffer, 1 μL ExoI (20 U/μL, NEB, M0293S) and 3 μL of nuclease-free water to a total volume of 10 μL. Incubate for 5 h at 37°C and heat-inactivate for 20 min at 80°C. Use 10 μL of ExoI-treated DNA as template in PCR1 in a 50 μL reaction. For PCR2, use 3 μL of PCR1 product as template in a 25 μL reaction.

12. Alternatively, isolate DNA from the duplicate 96-well plate when the cells in the master 96-well plate are only ~50% confluent. Refresh the medium of the master plate and culture the cells for another day, while performing the PCR analysis on the duplicate DNA plate. Transfer the selected positive cells from the master plate to a fresh 24-well with MEFs and proceed with the subcloning procedure without freezing the cells.

13. Semi-quantitative PCR is used to monitor the enrichment of modified cells during each screening round. Make sure that all DNA samples are dissolved at equal concentrations: the same number of cells into an equal amount of $T_{10}E_{0.1}$. Then, perform both PCR1 and PCR2 with only 20 cycles of amplification. Select the positive cell culture that shows a much stronger signal, that is, is strongly enriched for the modified cells compared to the original positive cell culture from the previous screening round.

14. If ~50% of the 1,000 cells/well plate gives a positive PCR signal, this indicates that the original 5,000 cells/well pool is already strongly enriched for modified cells. Rather than

expanding the positive cells from the 1,000 cells/well plate, continue with the 5,000 cells/well positive culture and plate the cells now at a density of 100 cells/well.

15. When proceeding from the 1,000 cells/well to the 100 cells/well screening round, the number of positive wells usually is 5–20%. However, many of these positive wells are not enriched compared to the 1,000 cells/well positive culture. Therefore, it is necessary to expand a large number of positive wells (24 clones if possible) to identify the most enriched pool by semi-quantitative PCR. If none of the 100 cells/well positives is enriched, reseed the cells from the 1,000 cells/well clone again at 100 cells/well in order to identify an enriched subclone.
16. The 10 cells/well plates are cultured in parallel to the cells for colony picking. The number of positive wells in the 10 cells/well positive culture can be used to estimate the number of colonies that need to be picked. When all picked colonies are negative, plate the 100 cells/well positive culture, pool again for colony picking and pick more colonies based on the results of the 10 cells/well plate. Since most of the 10 cells/well cultures are hardly enriched, it is not useful to continue the subcloning with these cells, rather start from the 100 cells/well culture again.
17. After identification of a positive picked colony, another round of subcloning is necessary because colonies may be derived from multiple cells. Plate cells from the selected colony on 10-cm TC Petri dishes. Pick 24 individual colonies and perform PCR analysis. In case of clonality, all colonies should be positive in the PCR. Use one of these latter colonies for expansion and freeze an appropriate number of vials. After establishment of a pure clonal cell line, verify the presence of the desired modification by sequencing. Isolate total RNA and prepare cDNA by reverse transcription using primers spanning the mutation. Clone the resulting PCR fragment in pGEM®-T Easy vector (Promega, A1360). Sequence multiple clones since both wild-type and modified alleles should be present.

Acknowledgements

We acknowledge financial support for our work on oligonucleotide-mediated gene modification from the Dutch Cancer Society (NKI 2000-2233) and the Netherlands Genomics Initiative (050-71-007 and 050-71-051).

References

1. Capecchi MR. Altering the genome by homologous recombination. Science 1989;244:1288–92.
2. Hasty P, Ramirez-Solis R, Krumlauf R, Bradley A. Introduction of a subtle mutation into the Hox-2.6 locus in embryonic stem cells. Nature 1991;350:243–6.
3. Gu H, Zou YR, Rajewsky K. Independent control of immunoglobulin switch recombination at individual switch regions evidenced through Cre-loxP-mediated gene targeting. Cell 1993;73:1155–64.
4. Moerschell RP, Tsunasawa S, Sherman F. Transformation of yeast with synthetic oligonucleotides. Proc. Natl. Acad. Sci. USA 1988;85:524–8.
5. Campbell CR, Keown W, Lowe L, Kirschling D, Kucherlapati R. Homologous recombination involving small single-stranded oligonucleotides in human cells. New Biol. 1989;1:223–7.
6. Cole-Strauss A, Yoon K, Xiang Y, et al. Correction of the mutation responsible for sickle cell anemia by an RNA–DNA oligonucleotide. Science 1996;273:1386–9.
7. Liu L, Rice MC, Kmiec EB. In vivo gene repair of point and frameshift mutations directed by chimeric RNA/DNA oligonucleotides and modified single-stranded oligonucleotides. Nucleic Acids Res. 2001;29:4238–50.
8. Igoucheva O, Alexeev V, Yoon K. Targeted gene correction by small single-stranded oligonucleotides in mammalian cells. Gene Ther. 2001;8:391–9.
9. Wang G, Seidman MM, Glazer PM. Mutagenesis in mammalian cells induced by triple helix formation and transcription-coupled repair. Science 1996;271:802–5.
10. Oh TJ, May GD. Oligonucleotide-directed plant gene targeting. Curr. Opin. Biotechnol. 2001;12:169–72.
11. Liu L, Rice MC, Drury M, Cheng S, Gamper H, Kmiec EB. Strand bias in targeted gene repair is influenced by transcriptional activity. Mol. Cell Biol. 2002;22:3852–63.
12. Igoucheva O, Alexeev V, Pryce M, Yoon K. Transcription affects formation and processing of intermediates in oligonucleotide-mediated gene alteration. Nucleic Acids Res. 2003;31:2659–70.
13. Li XT, Costantino N, Lu LY, et al. Identification of factors influencing strand bias in oligonucleotide-mediated recombination in *Escherichia coli*. Nucleic Acids Res. 2003;31:6674–87.
14. Ferrara L, Kmiec EB. Camptothecin enhances the frequency of oligonucleotide-directed gene repair in mammalian cells by inducing DNA damage and activating homologous recombination. Nucleic Acids Res. 2004;32:5239–48.
15. Ferrara L, Kmiec EB. Targeted gene repair activates Chk1 and Chk2 and stalls replication in corrected cells. DNA Repair (Amst) 2006;5:422–31.
16. Igoucheva O, Alexeev V, Yoon K. Differential cellular responses to exogenous DNA in mammalian cells and its effect on oligonucleotide-directed gene modification. Gene Ther. 2006;13:266–75.
17. Dekker M, Brouwers C, te Riele H. Targeted gene modification in mismatch-repair-deficient embryonic stem cells by single-stranded DNA oligonucleotides. Nucleic Acids Res. 2003;31:e27.
18. de Wind N, Dekker M, Berns A, Radman M, te Riele H. Inactivation of the mouse Msh2 gene results in mismatch repair deficiency, methylation tolerance, hyperrecombination, and predisposition to cancer. Cell 1995;82:321–30.
19. Costantino N, Court DL. Enhanced levels of lambda Red-mediated recombinants in mismatch repair mutants. Proc. Natl. Acad. Sci. USA 2003;100:15748–53.
20. Kolodner R. Biochemistry and genetics of eukaryotic mismatch repair. Genes Dev. 1996;10:1433–42.
21. Dekker M, Brouwers C, Aarts M, et al. Effective oligonucleotide-mediated gene disruption in ES cells lacking the mismatch repair protein MSH3. Gene Ther. 2006;13:686–94.
22. Aarts M, Dekker M, de Vries S, van der Wal A, te Riele H. Generation of a mouse mutant by oligonucleotide-mediated gene modification in ES cells. Nucleic Acids Res. 2006;34:e147.
23. Smith AG, Hooper ML. Buffalo rat liver cells produce a diffusible activity which inhibits the differentiation of murine embryonal carcinoma and embryonic stem cells. Dev. Biol. 1987;121:1–9.

Chapter 6

Generation of shRNA Transgenic Mice

Christiane Hitz, Patricia Steuber-Buchberger, Sabit Delic, Wolfgang Wurst, and Ralf Kühn

Abstract

RNA interference (RNAi)-mediated gene knockdown has developed into a routine method to assess gene function in cultured mammalian cells in a fast and easy manner. For the use of RNAi in mice, short hairpin (sh) RNAs expressed stably from the genome are a faster alternative to conventional knockout approaches. Here, we describe an advanced strategy for complete or conditional gene knockdown in mice, where the Cre/loxP system is used to activate RNAi in a time- and tissue-dependent manner. Single-copy RNAi constructs are placed into the *Rosa26* locus of ES cells by recombinase-mediated cassette exchange and transmitted through the germline of chimaeric mice. The shRNA transgenic offspring can be either directly used for phenotypic analysis or are further crossed to a Cre transgenic strain to activate conditional shRNA vectors. The site-specific insertion of single-copy shRNA vectors allows the expedite and reproducible production of knockdown mice and provides an easy and fast approach to assess gene function in vivo.

Key words: RNAi, transgenic mice, Rosa26, Cre/loxP, RMCE, shRNA.

1. Introduction

Silencing of gene expression by RNA interference (RNAi) has become a powerful tool for functional genomics in mammalian cells. RNAi is a sequence-specific gene-silencing process that occurs at the messenger RNA (mRNA) level. In invertebrate cells, long double-stranded RNAs (dsRNAs), which are processed into short interfering RNAs (siRNAs) by the ribonuclease Dicer, induce efficient and specific gene silencing. In this sequence-guided process the siRNA antisense strand serves as a template for the RNA-induced silencing complex (RISC). RISC recognises and cleaves the complementary mRNA, which is then rapidly degraded *(1)* (*see* **Fig. 6.1**). In mammalian cells, long dsRNAs (>30 bp) elicit an interferon response resulting in the global

Ralf Kühn, Wolfgang Wurst (eds.), *Gene Knockout Protocols: Second Edition, vol. 530*

DOI 10.1007/978-1-59745-471-1_6 Springerprotocols.com

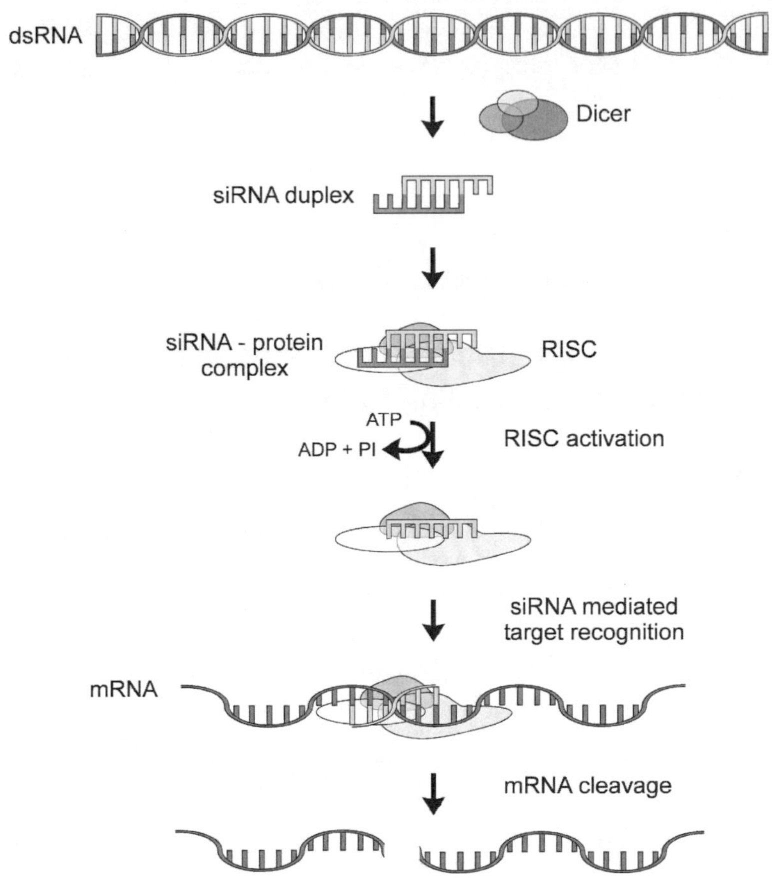

Fig. 6.1. Illustration of the RNAi mechanism.

inhibition of protein synthesis and non-specific mRNA degradation. However, it has been shown that short synthetic dsRNAs trigger the specific knockdown of mRNAs in mammalian cells without interferon activation, if their length is below 30 bp *(1)*. Such synthetic siRNAs can be easily introduced into cultured cells and induce a transient knockdown that enables the study of mammalian gene function within a short time frame. Due to advances in the delivery and design of siRNAs, gene silencing has developed by now into a routine method for in vitro use. Shortly after the establishment of siRNA-mediated transient gene silencing, DNA-based expression vectors were developed that allow the endogenous production of small dsRNAs in mammalian cells *(2–4)*. The vector-derived transcripts are designed to contain a sense and an antisense region that is complementary to a selected mRNA segment. These transcripts can fold back into a stem-loop structure and form short hairpin RNAs (shRNAs) that are processed by Dicer in a similar way as the siRNAs. Since shRNA expression vectors can be stably integrated into the genome, they allow permanent, long-lasting gene silencing in cell lines and organisms.

Soon after these technologies were introduced for use in cultured cells it became an obvious task to explore RNAi-mediated gene silencing in mice also. The number of reports on RNAi in transgenic mice has increased to more than 20, representing all standard methods to generate transgenics *(5)*. shRNA transgenic mice were produced by pronuclear DNA injection *(6)*, infection of zygotes or ES cells with lentiviral vectors *(7)*, random integration into ES cells *(8)* and targeted transgenesis (knockin) of single vector copies into ES cells through recombinase-mediated cassette exchange (RMCE) or homologous recombination *(9, 10)*. The efficiency of gene silencing in transgenic mice reaches, in most cases, knockdown levels in the range of 90% or higher. The phenotypes of knockdown and the corresponding knockout mice were compared in several instances and found to be identical or very similar.

Vector-based transgenic RNAi provides a tool to achieve efficient gene silencing during embryonic development as well as in organs of adult mice *(5)*. As compared to the delivery of siRNA or viral vectors to somatic tissues, for which the uptake or infection rate is a critical issue, transgenic animals harbour the shRNA expression vector in all cells and provide a well-defined experimental setup. However, if transgenic mice are generated by pronuclear injection, random integration into ES cells, or viral infection, a variable number of vector copies integrates into unknown chromosomal locations that could interfere with vector expression. From studies with mice transgenic for RNA polymerase II-driven promoters it is well known that transgene expression can be highly variable among individual lines such that it is necessary to raise several independent lines for each construct and to characterise transgene expression in each line. In case of promoters that are strongly influenced by the chromosomal surrounding, random integration becomes a laborious and time-consuming approach. The problems associated with random vector integrations can be avoided by the targeted insertion of a single vector copy into a predetermined genomic locus and thereby reproducible results can be achieved.

By comparison of heterozygous and homozygous knockout mice, it is well known that 50% reduction in gene expression in heterozygote mutants rarely results in a detectable phenotype. However, little is known about intermediate phenotypes with 70–80% reduced expression of a certain gene. In many cases this could reflect the situation in diseases which are often caused by point mutations in one specific gene. Depending on the efficiency of the used shRNA, RNAi could be used to generate knockdown mice with various preselected knockdown levels. Together with the fast and easy application, this opportunity to gradually regulate gene activity is one of the main advantages of RNAi compared to conventional knockout mice.

In this chapter we provide a collection of working protocols for the streamlined production of single-copy shRNA transgenic mice as described in the publication by Hitz et al. *(9)*. The basic principle shown in **Fig. 6.2** is based on the insertion of shRNA vectors into the *Rosa26* locus of ES cells by RMCE and the subsequent generation of chimaeric mice. Using the same set of tools either constitutive bodywide knockdown mice or conditional Cre/loxP-regulated knockdown mice can be generated.

In **Section 3.1**, the construction of constitutively active shRNA expression vectors is described (*see* **Fig. 6.3**). These vectors consist of an RNA polymerase III promoter-driven shRNA and can be used to generate non-conditional transgenic mice with an all-over knockdown phenotype, similar to conventional knockout mice.

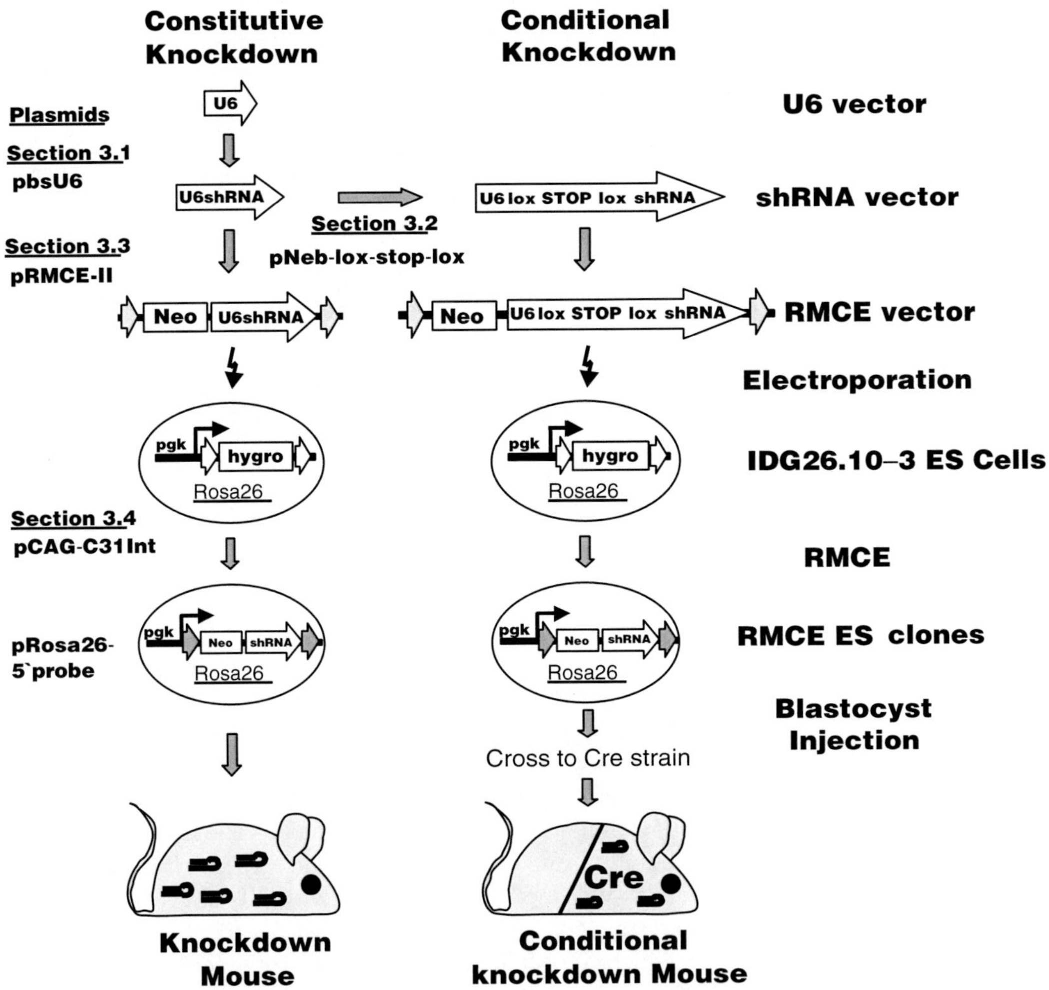

Fig. 6.2. **Production of shRNA transgenic mice by recombinase-mediated cassette exchange (RMCE) in ES cells.** Using the same set of tools either constitutive knockdown mice (left side) or conditional, Cre/loxP-regulated knockdown mice (right side) can be generated. Upon cloning of the shRNA region into a U6 promoter vector the expression unit is converted into an RMCE vector. Constitutive shRNA vectors can be generated by the insertion of loxP-flanked (lox-stop-lox) DNA segment into the shRNA region. Upon electroporation of RMCE acceptor ES cells the shRNA expression unit integrates into the Rosa26 locus. Recombinant ES cell clones are used for the production of germline chimaeras by blastocyst injection. Mice harbouring conditional shRNA vector need to be further crossed with a Cre transgenic strain. The names of required plasmids and the relevant sections of this protocol are indicated.

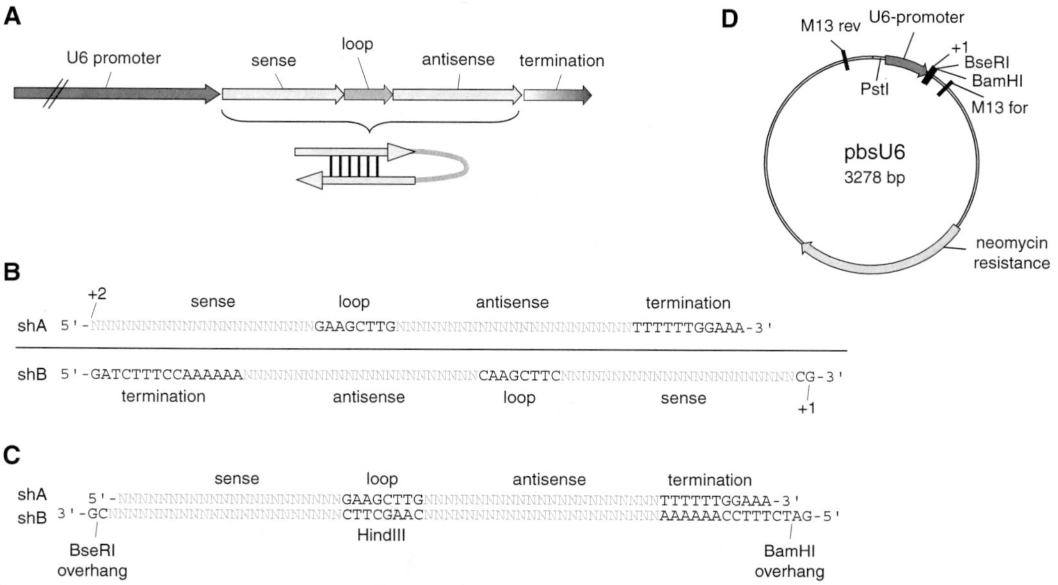

Fig. 6.3. **Generation of shRNA vectors.** ShRNA vectors (**A**) are designed in a way that the U6 promoter drives the expression of the sense, the loop and the antisense sequence of the hairpin. A short termination signal determines the end of the transcript, which folds into a hairpin structure consisting of the sense, loop and antisense part. The oligonucleotides used to clone the shRNA sequence are shown in (**B**). The *grey* sequences (NNN. . .) have to be replaced by your target shRNA sense or antisense sequence, respectively. Upon annealing of shA and shB a dsDNA fragment is formed (**C**) and subcloned into pbsU6 (**D**). +1, transcriptional start; M13rev/M13for, primer binding sites for sequencing; *Bam*HI/*Bse*RI, restriction sites for cloning; *Pst*I/*Hin*dIII, restriction sites for screening (*see* **Section 3.1**).

To overcome embryonic lethality of many mutants and to investigate gene functions in specific tissues or in a time-dependent manner, conditional vectors must be used. A protocol for the conversion of constitutively active shRNA expression vectors into conditional shRNA expression vectors that can be activated by Cre recombinase is given in **Section 3.2**. In these conditional vectors shRNA expression can be regulated by an inserted loxP-flanked stop element that blocks shRNA production (*see* **Fig. 6.4**). After Cre-mediated recombination the stop element is removed and the shRNA will be expressed. This Cre/loxP approach is similar to conditional knockout or knockin strategies, where it is widely used. For the genomic integration of conditional shRNA vectors a single-copy approach is preferable since upon random integration the promoter activity could be negatively influenced by surrounding genomic elements. Furthermore, the integration of multiple copies could lead to unpredictable and non-functional rearrangements upon Cre-mediated recombination. As shown previously *(9)*, one single copy of an shRNA construct in the *Rosa26* locus of the mouse genome is sufficient to achieve strong knockdown. For the insertion of shRNA vectors into the *Rosa26* locus of ES cells we developed an approach for RMCE using the phage ϕC31 integrase (C31Int).

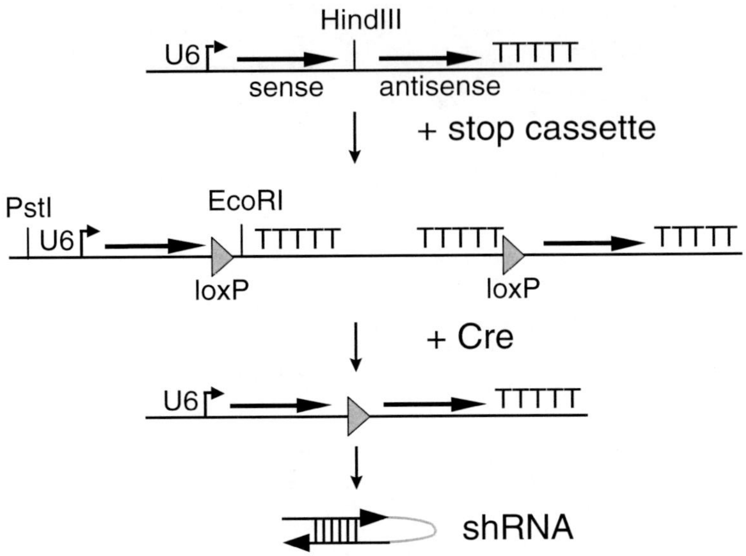

Fig. 6.4. **Conditional shRNA vectors controlled by the Cre/loxP system.** A loxP-flanked stop cassette (from plasmid pNeb-lox-stop-lox) is inserted into the *Hin*dIII site of the non-conditional shRNA vector. This modified conditional shRNA vector is inactive in this state and can be activated by excision of the stop element through Cre-mediated deletion. A single loxP site remains within the shRNA loop of the active construct, but does not affect RNAi efficiency. *Pst*I/*Eco*RI: restriction sites used to test the orientation of the stop cassette (*see* **Section 3.2**).

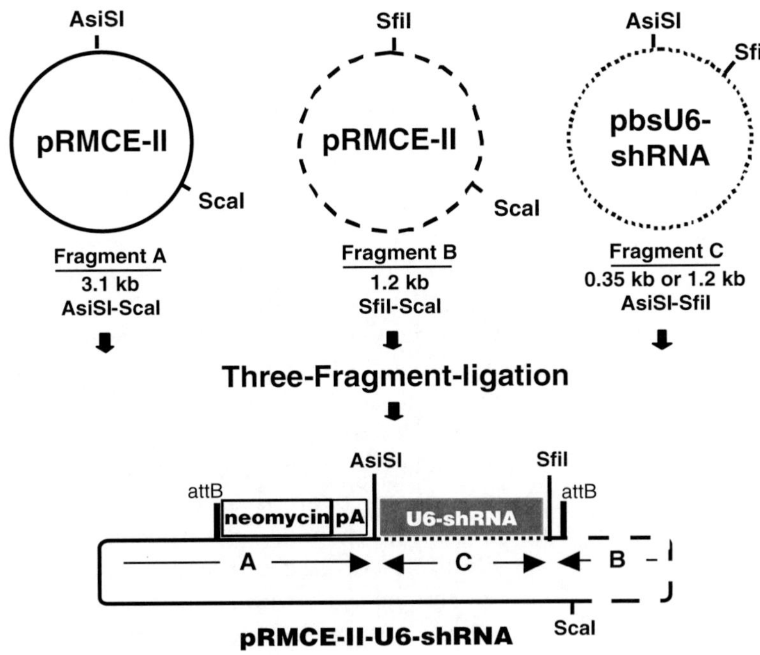

Fig. 6.5. **Generation of pRMCE-II-U6-shRNA by three-fragment ligation.** The fragments A and B are separately isolated from plasmid pRMCE-II and ligated with a third fragment from pbs-U6-shRNA, containing the U6-shRNA expression unit, into the vector pRMCE-II-U6-shRNA for recombinase-mediated cassette exchange (*see* **Section 3.3**).

For RMCE we altered one *Rosa26* allele by introducing a pgk promoter-driven hygromycin resistance gene. The coding and polyA region is flanked by attP recognition sites and will be replaced by an attB-flanked construct from a donor plasmid (pRMCE-II) upon cotransfection of ES cells together with a C31Int expression plasmid (*see* **Fig. 6.6**). The donor vector contains a promoterless neomycin resistance coding and polyA region and the shRNA expression cassette between the attB recognition sites. The subcloning of the shRNA expression cassette into the donor plasmid is described in **Section 3.3** (*see* **Fig. 6.5**). The shift from hygromycin to neomycin resistance in ES cells selects for RMCE events (*see* **Fig. 6.6**). The insertion of the donor plasmid into ES cells by RMCE is described in **Section 3.4**.

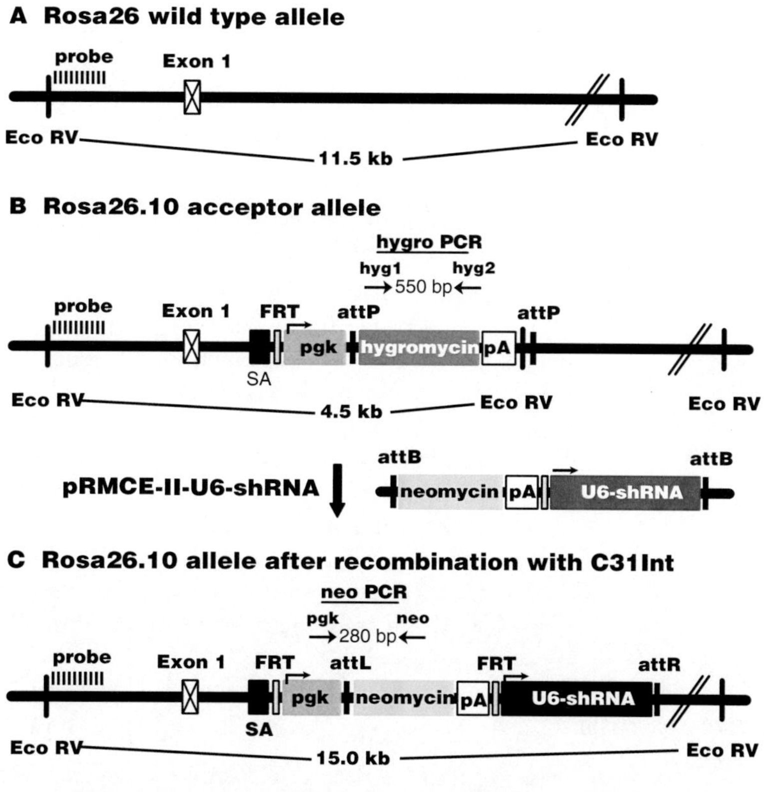

Fig. 6.6. **Insertion of shRNA vectors into the Rosa26 locus by RMCE.** The Rosa26 wild-type locus (**A**) is modified into an acceptor allele for RMCE (**B**) by insertion of a pgk promoter-driven hygromycin resistance gene, precded by a splice acceptor sequence (SA). A pair of attP recognition sites for the integrase of phage ΦC31 (C31Int) flank the coding region and polyA site. During the exchange C31Int replaces the segment between the attP sites with the segment flanked by attB sites on the donor vector containing a promoterless neomycin resistance gene and the U6-shRNA expression cassette (**C**). The attP-flanked hygromycin resistance is lost and the recombination of attP and attB sites forms attL and attR sites that are not further recognised by C31Int. The location of primers used for PCR identification and of the probe and restriction fragments for Southern blot screening of recombinant ES cell clones are indicated (*see* **Section 3.4**).

2. Materials

2.1. General Materials

1. Restriction enzymes:*AsiS*I, *Bam*HI, *Bse*RI, *Eco*RI, *Eco*RV, *Hin*dIII, *Mly*I, *Pst*I, *Sca*I, *Sca*II, *Sfi*I, *XbaI* (NEB).
2. T4 DNA Ligase, 1 U/µL (NEB).
3. *Escherichia coli* DH5α^{TM}(Invitrogen).
4. Luria–Bertani (LB) agar plates: 1% Bacto peptone (BD Biosciences), 0.5% yeast extract (Difco), 0.5% NaCl, 1.5% Bacto agar (Difco). Autoclave, add antibiotics when medium has cooled down to approx. 60°C, aliquot on Petri dishes, let it harden and store the plates at 4°C for up to 4 weeks.
5. LB medium: 1% Bacto peptone (BD Biosciences), 0.5% yeast extract (Difco), 0.5% NaCl. Autoclave and store at 4°C.
6. Ampicillin (Sigma): Dissolve in H_2O at a concentration of 50 mg/mL, aliquot and store at –20°C. Use the stock solution at a dilution of 1:1,000 for LB medium and at 1:500 for LB agar.
7. QIAprep Spin Miniprep Kit, QIAGEN Plasmid Maxi Kit (all Qiagen).
8. Ampuwa® water (Fresenius).
9. Primer for sequencing: M13 forward (5′-GTA AAA CGA CGG CCA GT-3′), M13 reverse (5′-CAG GAA ACA GCT ATG AC-3′), bpA-for (5′-GGG AGG ATT GGG AAG ACA AT-3′).

Plasmids pbsU6, pNeb-lox-stop-lox, pRMCE-II, pCAG-C31Int, pRosa26-5′probe and the cell lines IDG3.2 and IDG26.10-3 are described in ref. *(9)*; vector sequence information can be found on: http://www.rnai.ngfn.de/index_4.htm or http://www.gsf.de/idg/. These plasmids and cell lines are available upon request from R. Kühn (ralf.kuehn@helmholtz-muenchen.de).

2.2. Construction of shRNA Expression Vectors

1. For cell culture materials *see* **Section 2.5**.
2. Plasmid: pbsU6 (*see* **Section 2.1**).
3. ES cell line: IDG3.2 (wild-type ES cells), adapted to feeder-free culture on gelatine-coated dishes (*see* **Section 2.1**).
4. RIPA buffer: 50 mM Tris-HCl (pH 7.4), 1% NP-40, 0.25% sodium deoxycholate, 150 mM NaCl, 1 mM EDTA, Complete® protease inhibitor (Roche, 1 tablet/50 mL), aliquot and store at –20°C.
5. Laemmli buffer 5X: 313 mM Tris-HCl (pH 6.8), 50% glycerol, 10% SDS, 0.05% bromphenolblue, 25% β-mercaptoethanol. Aliquot for longer storage at –20°C; the aliquot in use can be stored at 4°C for several weeks.

6. BCA™ Protein Assay Kit (Pierce).
7. Criterion™ Precast Gel System and Criterion™ Blotter (Biorad).
8. Gels for SDS-PAGE: 10% Criterion™ XT gels (Biorad).
9. SeeBlue® Plus2 Pre-Stained Standard (Invitrogen).
10. MOPS running buffer 1X: 50 mM MOPS, 50 mM Tris-Base, 0.1% SDS, 1 mM EDTA. Adjust pH to 7.7. Prepare 10X running buffer, store at 4°C and dilute prior to use.
11. Tris glycine blotting buffer 1X: 25 mM Tris-Base, 192 mM glycine, 10% methanol. When preparing 10X blotting buffer omit the methanol, store at room temperature and add methanol upon diluting prior to use.
12. Whatman filter paper 3MM (Whatman).
13. PVDF membrane (Pall).
14. TBS-T: 25 mM Tris-HCl (pH 7.6), 137 mM NaCl, 0.05% Tween-20. Prepare 10X stock solution and store at room temperature.
15. Blocking solution: 4% skim milk powder in TBS-T or 5% BSA in TBS-T, depending on the protocol of you antibody. Can be stored at 4°C for up to 2 days.
16. Primary antibody for your gene of interest and for a housekeeping gene as loading control (e.g. β-Actin) and corresponding secondary antibodies (HRP-conjugated).
17. ECL Detection Kit (Amersham).
18. Hyperfilm (Amersham).

2.3. Construction of Conditional shRNA Expression Vectors

1. Plasmid: pNEB-lox-stop-lox (*see* **Section 2.1**).
2. Klenow fragment (NEB)
3. dNTPs: 10 μM each, store at –20°C (NEB).
4. Shrimp alkaline phosphatase (SAP, 1 U/μL, Roche).
5. Cre recombinase (1 U/μL, NEB).

2.4. Construction of RMCE Vectors

1. Plasmid: pRMCE-II (*see* **Section 2.1**).

2.5. Insertion of shRNA Vectors into ES Cells by RMCE

1. Plasmid: pCAG-C31Int (*see* **Section 2.1**).
2. ES cell line: IDG26.10-3 acceptor ES cells (*see* **Section 2.1**).
3. Feeder cells: G418-resistant embryonic fibroblasts.
4. Electroporation cuvette, 0.4 cm (Biorad).
5. Electroporator (Fisher Scientific).
6. Cell culture dishes: 10 cm, 24-well plate, 96-well flat-bottom plate, 96-well "V-bottom" plate (Nunc).

7. F1 ES cell medium: 500 mL Dulbecco's Modified Eagle's Medium (DMEM, Invitrogen), supplemented with 75 mL ES-qualified FCS (PAN Biotech), 9 mL 200 mM L-Glutamine (Invitrogen), 12 mL 1 M HEPES (Invitrogen), 6 mL 100X MEM nonessential amino acids (Invitrogen), 1.2 mL 50 mM β-mercaptoethanol (Invitrogen), 90 µL LIF (10^7 U/mL, Chemicon). Store at 4°C (stable for 4 weeks).
8. G418: Geneticin solution with an active concentration of 50 mg/mL (Invitrogen). Store in single-use aliquots at –20°C.
9. PBS: 1X PBS (Invitrogen) for cell culture, sterile. Store at room temperature.
10. Trypsin: 0.05% Trypsin in Tris saline/EDTA buffer (Invitrogen). Store at 4°C.
11. Gelatine: 0.1% gelatine in water, autoclaved.
12. 2X freezing solution: 50% FCS, 30% F1 ES cell medium, 20% DMSO. Prepare fresh.
13. Wizard® Genomic DNA Purification Kit (Promega).
14. Primers for PCR genotyping: Neo: (5′-GTT GTG CCC AGT CAT AGC CGA ATA G-3′), Pgk (5′-CAC GCT TCA AAA GCG CAC GTC TG-3′), Rosa-5′ (5′-CGT GTT CGT GCA AGT TGA GT-3′), Rosa-3′ (5′-ACT CCC GCC CAT CTT CTA G-3′), Hyg-1 (5′-GAA GAA TCT CGT GCT TTC AGC TTC GAT G–3′), Hyg-2 (5′-AAT GAC CGC TGT TAT GCG GCC ATT G–3′).
15. Probe for Southern blotting: 500-bp *Eco*RI fragment from plasmid pRosa 26-5′probe.
16. Denaturation buffer: 1.5 M NaCl, 0.5 M NaOH. Store at room temperature.
17. Neutralisation buffer: 0.5 M NaCl, 0.1 M Tris-HCl (pH 7.5). Store at room temperature.
18. 20X SSC: 3 M NaCl, 0.3 M sodium citrate. Adjust pH to 7.0 with HCl. Store at room temperature.
19. Whatman filter paper 3MM (Whatman).
20. Hybond N^+ membrane (Amersham).
21. UV-crosslinker, 254 nm (Stratagene).
22. Hybridisation tube (Thermo Scientific).
23. Rapid-hyb buffer (Amersham).
24. Hybaid hybridisation oven (Thermo Scientific).
25. Rediprime™ II labelling kit (Amersham). Store at 4°C.
26. Micro Spin™ S-300 HR columns (Amersham). Store at 4°C.

27. Wash buffer: 2X SSC, 0.1% SDS. Store at room temperature.
28. Exposure cassette with BioMax™ intensifying screen (Sigma).
29. Kodak® BioMax™ MS film (Sigma). Store at 4°C.

3. Methods

3.1. Construction of shRNA Expression Vectors

This protocol describes the design and cloning of shRNAs into the U6 promoter expression vector pbsU6 that is derived from the pShag vector *(4)*. We usually generate 4–5 independent shRNA vectors and compare their gene-silencing efficiencies by transient transfection in ES cells.

3.1.1. Selection and Design of shRNA Sequences

1. Analyse the coding region of the target gene with siRNA prediction programs from, for example, Invitrogen (https://rnaidesigner.invitrogen.com/sirna/) or Sigma-Aldrich (http://www.sigmaaldrich.com/Brands/Sigma_Genosys/siRNA_Oligos/siRNA_Design_Service.html). These programs allow you to search by gene name, gene accession number, or sequence (*see* **Note 1**). All the possible siRNA sequences will be automatically checked in a BLAST search for targeting uniquely your gene of interest.
2. Select five high-ranked target sequences with a length of 21–25 nt (preferably 23 nt) that start with a G. We design shRNAs such that the sense region is transcribed first and the antisense region folds back via the 8-nt loop resulting in the order: U6promoter > target sense > loop > target antisense > termination (*see* **Fig. 6.3A**).
3. For cloning of one shRNA, design two oligonucleotides as shown in **Fig. 6.3B**. For oligonucleotide shA start 5′ with the selected sense sequence omitting the G at position 1, add 8-nt loop sequence (GAA GCT TG), add the complete sense sequence in reverse orientation and add 11-nt termination sequence (TTT TTT GGA AA) (*see* **Note 2**). For oligonucleotide shB start 5′ with the 15-nt sequence containing the termination signal in reverse orientation (GAT CTT TCC AAA AAA), add the complete target sense sequence, add the 8-nt loop sequence in reverse orientation (CAA GCT TC), add the target sense sequence in reverse orientation and the final nucleotides CG. In the templates of the oligonucleotides shA and shB in **Fig. 6.3B** the sense and antisense regions are filled by "N" and have to be complementary to each other.
4. Check your selected shRNA targeting sequence for correctness in both oligonucleotides and virtually anneal shA and shB to check for mismatches due to typing errors.

5. Purchase the designed oligonucleotides at the company of your choice. No phosphorylation or other modifications are needed, but HPLC purification enhances the sequence accuracy of your product.

3.1.2. Cloning of pbsU6-shRNA

The two oligonucleotides are designed such that they can hybridise with each other and form a dsDNA fragment with *Bse*RI- and *Bam*HI-compatible overhangs (*see* **Fig. 6.3C**). With these overhangs they are cloned into the vector pbsU6, opened with *Bse*RI and *Bam*HI. The vector ends remain phosphorylated and ligate to the nonphosphorylated oligonucleotides; finally, the ligation reaction is redigested with *Bam*HI that cuts only religated empty vector. This procedure avoids the purchase of costly phosphorylated oligonucleotides.

1. Dissolve the oligonucleotides shA and shB in water at a concentration of 1 μg/μL, incubate for 15 min at 37°C and vortex vigorously to dissolve the DNA completely. Set up 10 μL of dilutions (100 ng/μL) for each oligonucleotide by mixing 1 μL shA and shB, respectively, with 9 μL water.
2. Mix 2 μL of the diluted shA and 2 μL of the diluted shB with 16 μL water (10 ng of each oligo/μL). Heat the tubes to 100°C and let them cool down to room temperature for 20 min. shA and shB are now annealed and form a dsDNA fragment.
3. For preparation of the vector fragment digest 10 μg pbsU6 plasmid DNA (*see* **Fig. 6.3D**) with 20 Units *Bse*RI and 20 Units *Bam*HI in a total volume of 70 μL and incubate at 37°C for 2 h.
4. Load the restriction digest on a 1% agarose gel, isolate the 3.2-kb vector fragment with the QIAGEN Gel Extraction Kit and determine the concentration of the vector DNA by photometry or by estimation on an agarose gel.
5. Set up a 10 μL ligation reaction by mixing 100 ng purified pbsU6 vector fragment (from Step 4) and 2 μL annealed oligos shA/shB (from Step 2) with 1 μL 10X ligation buffer and 1 μL T4 DNA Ligase. Incubate at room temperature for 1 h. After completed ligation inactivate the enzyme by incubation at 65°C for 10 min.
6. To eliminate self-ligated empty vector redigest the ligation with *Bam*HI by adding 2 μL 10X buffer NEB2, 7 μL H_2O and 1 μL *Bam*HI (10 U/μL). Incubate the reaction for 30 min at 37°C and inactivate the enzyme again for 10 min at 65°C.
7. Use 5 μL of the ligation reaction for transformation of 50 μL competent *E. coli* cells, like DH5α™, and plate cells on LB agar plates containing ampicillin (100 μg/mL). Incubate the plates overnight at 37°C.

8. Pick 12 colonies grown on the agar plates and shake each one in 5 mL LB medium with ampicillin (50 μg/mL) overnight at 37°C. Isolate plasmid DNA from 2 mL of the culture using the QIAprep Spin Miniprep Kit. Use 10 μL of each Miniprep plasmid DNA to set up a 30 μL digestion reaction with 5 Units *Pst*I (cuts upstream of the U6 promoter) and 5 Units *Hin*dIII (cuts within the loop region), incubate for 1 h at 37°C and analyse the digestion products on a 1% agarose gel. Correctly ligated clones show two bands of 310 bp and 2.9 kb; plasmids without insert show only the 2.9-kb vector band.
9. Sequence one positive clone with M13 forward and M13 reverse and compare the obtained sequences to the predesigned plasmid file (*see* **Note 3**).
10. Inoculate a confirmed clone in 200 mL LB medium with ampicillin (50 μg/mL) and shake it overnight at 37°C. Mix 500 μL of the bacterial suspension with 500 μL glycerol and store the stock at –80°C. Isolate the plasmid DNA using the QIAGEN Plasmid Maxi Kit.

3.1.3. Testing shRNA Vectors for Efficiency

We usually transiently transfect shRNA plasmids into ES cells to compare the gene-silencing efficiency of a group of shRNA vectors directed to the same endogenous gene on the protein level (*see* **Notes 4** and **5**). Based on this information the most efficient shRNA vectors can be selected for production of shRNA transgenic mice.

3.1.3.1. Transient Transfections

1. To sterilise the plasmid DNA precipitate 50 μg of supercoiled shRNA plasmid DNA (from **Section 3.1.2**) by adding 1/10 volume of 3 M sodium acetate (pH 5.3) and 2.5 volumes of 100% ethanol. Mix by inverting the tube several times and incubate for 2 h to overnight at –20°C (*see* **Note 6**). Pellet the DNA by centrifugation for 10 min at 16,000*g*. Discard the supernatant, wash the DNA pellet with 70% ethanol and centrifuge again for 10 min at 16,000*g*. To keep the DNA sterile, work in a laminar flow under sterile conditions for the following steps. Remove the supernatant carefully by pipetting and air-dry the DNA pellet. Dissolve the DNA in 25 μL sterile Ampuwa® water, incubate for 15 min at room temperature, mix carefully and determine the DNA concentration by photometry. Adjust the volume to a final concentration of 1 μg/μL (*see* **Note 7**). Store the DNA at –20°C or use it directly for transfection. Prior to the transfection mix 20 μL of the sterile plasmid DNA (corresponding to 20 μg DNA) with 800 μL sterile PBS. As control transfect one sample with a scrambled shRNA construct or with no DNA at all.

2. For details about ES cell culture *see* **Section 3.4.1**. IDG3.2 ES cells are kept on gelatine-coated plates without feeder cells. For one transfection 3 × 10^6 cells are required. Make sure to have enough ES cells for all your transfections on the day of the experiment and keep them in culture for at least 2 days (*see* **Note 8**).
3. On the day of transfection, wash the cells with PBS, trypsinise them and pellet 3 × 10^6 cells for each transfection sample by centrifugation (4 min, 160*g*). Discard the supernatant and resuspend the cells in 800 μL PBS containing the plasmid DNA (from Step 2). Transfer the cell suspension into an electroporation cuvette, thumb the cuvette carefully to the table to remove any air bubbles and electroporate at 320 V and 3 ms cut-off time (*see* **Note 9**). Let the cells rest for 5 min before transferring them into 10 mL medium and plating them on one gelatine-coated 10-cm plate.
4. Incubate plates at 37°C and 5% CO_2. The next day, change the medium to remove dead cells.
5. Two days after the transfection harbour the cells. For this trypsinise the cells, pellet them by centrifugation (4 min, 160*g*) and wash the pellet with PBS. Centrifuge, discard the supernatant and store the pelleted cells at –20°C or continue directly with protein extraction.

3.1.3.2. Detection of Protein Levels by Western Blotting

1. Resuspend the pelleted cells in 100 μL RIPA buffer and ensure complete lysis by pipetting up and down several times. Incubate the suspension for 15 min at 4°C with gentle shaking. Pellet insoluble cell debris by centrifugation for 15 min at 16,000*g* and 4°C (*see* **Note 10**). Transfer the supernatant to a fresh tube and store the protein solution at –20°C if you will not continue immediately.
2. Make sure to keep the protein solution cold during the following steps (work on ice).
3. Measure the protein concentration photometrically, for example, with the BCA™ Protein Assay Kit, and dilute all the samples to the same concentration, for example, 5 mg/mL.
4. Add 1/5 volume of 5X Laemmli buffer to the protein solution, mix gently and heat it to 95°C for 5 min to denature the proteins completely.
5. For running the SDS-PAGE use the Criterion™ Precast Gel System (Biorad) with a 10% *Criterion™XT* gel (*see* **Note 11**). Use 1X MOPS running buffer to run the gel. After rinsing the wells with running buffer or water, load an appropriate amount (e.g. 50 μg) of your protein samples to the gel (*see* **Note 12**) and apply 5 μL of the SeeBlue® Plus2 Pre-Stained

Standard or another appropriate protein marker. Run the gel for approx. 75 min with 200 V depending on the size of your protein of interest.

6. Take the gel out of its cassette, equilibrate it in 1X Tris glycine blotting buffer, soak all blotting pads and filter papers in blotting buffer and build up the blot in the Criterion^{TM} Blotter in the following order: blotting pad, Whatman filter paper, gel, PVDF membrane (pre-equilibrated in 100% methanol), Whatman filter paper, blotting pad. Take care to avoid any air bubbles. Place this blotting assembly into the blotting module with the gel nearest to the cathode and the membrane nearest to the anode. Fill up the blotting chamber with 1X Tris glycine blotting buffer and blot for 2 h at 50 V and 4°C (*see* **Note 13**).
7. Take out the membrane and make sure that the protein marker has been completely transferred to the membrane indicating successful blotting. Place the membrane into an appropriate dish with the protein side up, rinse it with 1X TBS-T and block unspecific binding sites with blocking solution (1 h at room temperature or overnight at 4°C).
8. The antibody incubation procedure (*see* **Note 14**) may vary for different antibodies. Stick to the corresponding protocol if available; otherwise, try the following standard procedure: incubate the membrane in the primary antibody dilution (in TBS-T) for 1 h, wash 3 × 10 min in TBS-T, incubate in the secondary antibody dilution (in TBS-T) for 1 h and wash again 3 × 10 min in TBS-T.
9. Detect antibody signals by chemiluminescence using the ECL Detection Kit and chemiluminescence-sensitive films, for example, Hyperfilm.
10. Compare the band intensities of your gene of interest between the different samples and to the sample, which has been transfected with no/scrambled shRNA construct (wild-type control) (*see* **Notes 15** and **16**). Select the shRNA construct, which mediated the best knockdown as revealed by the Western blot. Continue with cloning of the stop cassette for conditional vectors (*see* **Section 3.2**) or subclone the constitutively active constructs into pRMCE-II (*see* **Section 3.3**).

3.2. Construction of Conditional shRNA Expression Vectors

This protocol describes the conversion of constitutively active shRNA vectors (*see* **Section 3.1**) into conditional vectors that can be activated in vitro or in vivo by Cre recombinase (*see* **Fig. 6.4A**). In the first step a loxP-flanked stop element is inserted into the unique *Hin*dIII restriction site within the shRNA loop region; in the second step the stop element is deleted by Cre recombination in vitro to generate the activated vector. The function of the conditional and the activated version of the vectors should be confirmed by transient transfection into ES cells (*see* **Section 3.1.3**).

3.2.1. Insertion of the loxP-Flanked Stop Element

1. For preparation of the vector fragment digest 5 μg pbsU6-shRNA plasmid DNA (from **Section 3.1**) with 10 Units *Hin*dIII in a total volume of 20 μL and incubate at 37°C for 2 h. Add 1 μL dNTPs (10 μM) and 1 μL Klenow fragment, incubate at room temperature for 20 min. Dephosphorylate the 5′ ends of the digested plasmid by incubation with 1 Unit SAP at 37°C for 1 h. Inactivate all enzymes by incubation at 65°C for 10 min.
2. For preparation of the insert containing the stop cassette digest 10 μg of the plasmid pNEB-lox-stop-lox with 10 Units *Mly*I in a total volume of 40 μL and incubate at 37°C for 2 h. Inactivate the enzyme by incubation at 65°C for 10 min.
3. Load both restriction digests on a 1% agarose gel, isolate the fragments with the QIAGEN Gel Extraction Kit. The vector fragment is 3.3 kb in size and the loxP-stop-loxP insert fragment is 863 bp in size (other fragments of this digest: 263 bp, 486 bp, 502 bp, 1194 bp). Determine the concentration of the DNA fragments by photometry or by estimation on an agarose gel.
4. Set up a 10 μL ligation reaction by mixing 50 ng purified pbsU6-shRNA vector fragment and 35 ng purified loxP-stop-loxP insert fragment with 1 μL 10X ligation buffer and 1 μL T4 DNA Ligase. Incubate at room temperature for 1 h. After completed ligation inactivate the enzyme by incubation at 65°C for 10 min.
5. To eliminate self-ligated empty vector, redigest the ligation with *Hin*dIII by adding 2 μL 10X buffer NEB2, 7 μL H_2O and 1 μL *Hin*dIII (10 U/μL). Incubate the reaction for 30 min at 37°C and inactivate the enzyme again for 10 min at 65°C.
6. Use 5 μL of the ligation reaction for transformation of 50 μL competent *E. coli* cells, like DH5α^{TM}, and plate cells on LB agar plates containing ampicillin (100 μg/mL). Incubate the plates overnight at 37°C.
7. Pick 12 colonies grown on the agar plates and shake each one in 5 mL LB medium with ampicillin (50 μg/mL) overnight at 37°C. Isolate plasmid DNA from 2 mL of the culture using the QIAprep Spin Miniprep kit following manufacturer's instructions. Use 10 μL of each Miniprep plasmid DNA to set up a 30 μL digestion reaction with 5 Units *Eco*RI and 5 Units *Sac*II, incubate for 1 h at 37°C and analyse the digestion

products on a 1% agarose gel. Correctly ligated clones show two bands of 850 bp and 3.4 kb; plasmids with insert in the false orientation show two bands of 150 bp and 4 kb, and plasmids without insert show only the 3.3-kb vector band.

8. Sequence one positive clone with M13 forward and M13 reverse and compare the obtained sequences to the predesigned plasmid file.
9. Inoculate a confirmed clone in 200 mL LB medium with ampicillin (50 μg/mL) and shake it overnight at 37°C. Mix 500 μL of the bacterial suspension with 500 μL glycerol and store the stock at –80°C. Isolate the plasmid DNA using the QIAGEN Plasmid Maxi Kit.

3.2.2. In Vitro Deletion of the loxP-Flanked Stop Cassette

1. To delete the stop cassette and to activate shRNA expression in vitro incubate approx. 1 μg DNA of the conditional shRNA plasmid (from **Section 3.2.1**) with 4 Units Cre recombinase in a total volume of 50 μL 1X Cre buffer at 37°C for 30 min. Inactivate the recombinase for 10 min at 70°C.
2. To eliminate non-recombined plasmids add 1 μL *Eco*RI (10 U/μL) together with 3.3 μL 1 M NaCl and 0.5 μL BSA (100X, NEB) and incubate the reaction for 30 min at 37°C. This restriction digest will have linearized all molecules, which still harbour the stop cassette. Heat-inactivate the enzyme at 65°C for 20 min.
3. Use 2 μL of the recombination reaction for transformation of 50 μL competent *E. coli* cells, like DH5α^{TM}, and plate cells on LB agar plates containing ampicillin (100 μg/mL). Incubate the plates overnight at 37°C.
4. Pick six colonies grown on the agar plates and shake each one in 5 mL LB medium with ampicillin (50 μg/mL) overnight at 37°C. Isolate plasmid DNA from 2 mL of the culture using the *QIAprep Spin Miniprep kit* following manufacturer's instructions. Use 10 μL of each Miniprep plasmid DNA to set up a 30 μL digestion reaction with 5 Units *XbaI*, incubate for 1 h at 37°C, and analyse the digestion products on a 1% agarose gel. Correctly recombined clones show two bands of 460 bp and 2.9 kb and plasmids which have not been recombined by Cre recombinase show two bands of 1.3 and 2.9 kb.
5. Sequence one positive clone with M13 forward and M13 reverse and compare the obtained sequences to the predesigned plasmid file.
6. Inoculate a confirmed clone in 200 mL LB medium with ampicillin (50 μg/mL) and shake it overnight at 37°C. Mix 500 μL of the bacteria suspension with 500 μL glycerol and store the stock at –80°C. Isolate the plasmid DNA using the QIAGEN Plasmid Maxi Kit.

7. Evaluate the efficiency of the conditional shRNA constructs with the stop cassette and after recombination of the stop cassette by Western blotting or another appropriate method (*see* **Section 3.1.3**).
8. Choose the best construct in the conditional state to transfer it into ES cells by RMCE (*see* **Section 3.3**).

3.3. Construction of RMCE Vectors

This protocol describes the cloning of constitutive (from **Section 3.1**) or conditional (from **Section 3.2**) shRNA vectors into the pRMCE-II vector for RMCE into the *Rosa26* locus of mouse acceptor ES-cells (*see* **Section 3.4**). The single cloning step combines the U6-shRNA expression unit with the elements (neomycin resistance, attB sites) required for RMCE.

1. For the insertion of a U6-shRNA expression cassette into the exchange vector pRMCE-II the restriction sites of *Asi*SI and *SfiI* are used. Since both sites are just 4 bp apart, it is more reliable to use a "three-fragment ligation approach" (*see* **Note 17**, **Fig. 6.5**). For this digest the pRMCE-II vector into two separate reactions: first, digest 3 μg pRMCE-II plasmid DNA with 20 Units *ScaI* and 20 Units *Asi*SI in NEB2 buffer with BSA in a total reaction volume of 40 μL; second, digest 3 μg of the same plasmid DNA with 20 Units *ScaI* and 20 Units *SfiI* in NEB2 buffer with BSA in a total reaction volume of 40 μL. Incubate both reactions at 37°C, and switch the temperature for reaction 2 after 1 h to 50°C for another hour to activate *SfiI* digestion. Load both restriction digests on a 1% agarose gel, isolate the 3.1 kb fragment A (*ScaI–Asi*SI) from reaction 1 and the 1.2 kb fragment B (*SfI–ScaI*) from reaction 2 with the QIAGEN Gel Extraction Kit, and determine the concentration of the fragment DNA by photometry or by estimation on an agarose gel. Use 60 ng of each fragment for ligation.
2. For isolation of the U6-shRNA expression cassette from pbsU6-shRNA (from **Section 3.1.2** or **3.2.1**) digest 3 μg of your pbsU6-shRNA plasmid DNA with 20 Units *Asi*SI and 20 Units *SfiI* in NEB2 buffer with BSA in a total volume of 40 μL and incubate at 37°C for 1 h. After 1 h, heat the reaction to 50°C and incubate it for another hour to activate *SfiI* digestion. Load the restriction digests on a 1% agarose gel, isolate the 350 bp *Asi*SI–*SfiI* (constitutive vectors) or the 1.2 kb *Asi*SI–*SfiI* (conditional vectors) fragment C using the QIAGEN Gel Extraction Kit, and determine the concentration of the insert DNA by photometry or by estimation on an agarose gel. Use 75 ng for ligation.
3. Set up a 15 μL ligation reaction by adding 60 ng of the pRMCE-II vector fragments A and B (from Step 1), 75 ng purified U6-shRNA fragment C (from Step 2), 1.5 μL 10X ligation buffer and 1.5 μL T4 DNA Ligase. Mix and incubate at room temperature for 1 h.

4. Use 5 μL of the ligation reaction for transformation of 50 μL competent *E. coli* cells, like DH5α^{TM}, and plate cells on LB agar plates containing ampicillin (100 μg/mL). Incubate the plates overnight at 37°C.
5. Pick five colonies grown on the agar plates and shake each one in 5 mL LB medium with ampicillin (50 μg/mL) overnight at 37°C. Isolate plasmid DNA from 2 mL of the culture using the QIAprep Spin Miniprep Kit. Use 10 μL of each Miniprep plasmid DNA to set up a 30 μL digestion reaction with 5 Units *Sal*I and 5 Units *Hin*dIII (constitutive vectors) or 5 Units *Eco*RI and 5 Units *Sal*I (conditional vectors), incubate for 1 h at 37°C and analyse the digestion products on a 1% agarose gel. Correctly ligated clones in the right orientation show two bands of 1.6 kb and 3.0 kb (constitutive vectors) or 1.7 kb and 3.9 kb (conditional vectors) and plasmids without insert show only the 4.4-kb vector band. The efficiency should be high with 4–5 positive clones out of 5.
6. Sequence one positive clone of your pRMCE-II-U6-shRNA with M13 forward, M13 reverse and bpA-for and compare the obtained sequences to the predesigned plasmid file (*see* **Note 18**).
7. Inoculate a confirmed clone in 200 mL LB medium with ampicillin (50 μg/mL) and shake it overnight at 37°C. Mix 500 μL of the bacteria suspension with 500 μL glycerol and store the stock at –80°C. Isolate the plasmid DNA using the QIAGEN Plasmid Maxi Kit.

3.4. Insertion of shRNA Vectors into ES Cells by RMCE

The donor vector (pRMCE-II-U6-shRNA) for cassette exchange, as prepared following **Section 3.3**, can be integrated by RMCE into ES cells that contain an acceptor vector within the *Rosa26* locus (acceptor ES cells IDG26.10-3). The donor vector contains a pair of attB recognition sites of the integrase of phage ϕC31 (C31Int), flanking a promoterless neomycin resistance coding region and polyA site followed by the shRNA expression unit. The Rosa26 acceptor allele harbours a pair of attP recognition sites that flank a hygromycin resistance coding region and polyA site. In the case of C31Int-mediated recombination between the pairs of attB and attP sites, the hygro-pA segment is replaced by the neo-pA-shRNA unit flanked by attR and attL sites that are not further recognised by C31Int (*see* **Fig. 6.6**). To induce cassette exchange, IDG26.10-3 ES cells are cotransfected with the donor vector together with an expression vector for C31Int (pCAG-C31Int). Upon cassette exchange the ES cells lose the hygromycin-coding region and express the neomycin resistance from the pgk promoter located upstream of the attR site. Following transfection neomycin-resistant ES cell colonies are selected, expanded and analysed by PCR and/or Southern blotting for the presence of the desired RMCE event. Usually ~40% of the neomycin-resistant clones underwent complete RMCE; ~10% contain a partial recombination event and 50% harbour integrations at unrelated genomic sites. For RMCE we routinely use the acceptor cell line IDG26.10-3 ES cells, based on the F1 ES cell line IDG3.2 (C57Bl/6 J × 129S6/SvEvTac).

3.4.1. Electroporation of ES Cells

The pRMCE-II vector with your U6-shRNA fragment and pCAG-C31Int are cotransfected into the acceptor ES cell line IDG26.10-3 by electroporation. The following section describes the preparation of the plasmid DNA and the ES cells for transfection and the electroporation itself.

1. The plasmid DNA for transfection of ES cells must be supercoiled and sterile. For one transfection you need 20 μg of the pRMCE-II vector with the U6-shRNA expression cassette (pRMCE-II-U6-shRNA from **Section 3.3**) and 20 μg of the plasmid pCAG-C31Int.
2. To sterilise the plasmid DNA, precipitate 50 μg of each plasmid as described in **Section 3.1.3.1**, Step 1.
3. The ES cells must be handled under a laminar flow under sterile conditions all the time. All instruments used should be disinfected and sterile. The solutions must be adequate for cell culture use and should be kept sterile and under the conditions indicated by the supplier (*see* **Note 19**).
4. The IDG3.2 and the IDG26.10-3 acceptor ES cells are usually kept on MMC-treated G418 resistant feeder cells to ensure optimal growing conditions. If not mentioned otherwise the cell culture dishes used are coated with MMC-treated feeder cells in a concentration of 1×10^4 cells/cm^2.
5. The ES cells for transfection must be prepared several days in advance. On day 1 thaw a vial of IDG26.10-3 acceptor ES cells in a water bath at 37°C and transfer the suspension to 5 mL F1 ES cell medium. Pellet the ES cells by centrifugation for 4 min at 160*g*. Remove the supernatant, resolve ES cells carefully in 10 mL F1 ES cell medium and plate them on one 10-cm cell culture dish. Change the medium the next day to remove dead cells.
6. At day 3 split the cells in a ratio 1:4 on four fresh 10-cm dishes. For this, remove the medium of the old 10-cm plate and wash the cells with 4 mL PBS. Remove the PBS with the remaining medium (*see* **Note 20**) and add 4 mL trypsin. Transfer the plate back to the incubator for 5 min. After the incubation the colour of the trypsin changes from pink to an orange or even colourless tone and the detached ES cells float freely in the solution. Inactivate the trypsin by adding an equal amount of medium (4 mL). Resuspend and dissociate

the colonies by pipetting up and down several times and transfer the suspension to a new tube. Adjust the volume to 40 mL with medium and distribute the ES cell suspension to four 10-cm plates.

7. On day 4 you have to judge if the ES cells are ready for transfection. They should not grow confluent, meaning they must always have some space between the colonies. In addition, the colonies should not grow too big, as this can lead to differentiation. Usually the colonies look roundish with sharp borders and a light breaking appearance. If colonies are too small, let them grow for another day before continuing with transfection.
8. For one transfection sample you need 6×10^6 ES cells. After trypsinisation use 10 μL of the ES cell suspension for counting in a Neubauer counting chamber (*see* **Note 21**). The cell number of one quadrant multiplied by 10,000 corresponds to the number of cells in 1-mL cell suspension. Transfer the desired volume of the ES cell suspension to a fresh tube and pellet ES cells by centrifugation for 4 min at 160*g*. Wash the pellet once with 4 mL PBS and centrifuge again.
9. Mix 20 μL of the sterile plasmid DNA of the pRMCE-II-U6-shRNA vector and 20 μL of the pCAG-C31Int vector DNA with PBS to a total volume of 800 μL.
10. Resuspend the ES cell pellet in the DNA/PBS mixture (800 μL) and transfer it into an electroporation cuvette. Thump the cuvette carefully to the table to remove any air bubbles and electroporate at 300 V and 2 ms cut-off time. Some foam may occur indicating successful electroporation. Let the ES cells rest for 5 min.
11. Transfer the ES cells into 30 mL of medium and resuspend them. Distribute the suspension to three 10-cm plates (*see* **Note 22**).
12. On the next day, change the medium to remove ES cells floating dead in the medium (*see* **Note 23**).

3.4.2. Selection of Positive Colonies

The pRMCE-II vector carries a promoterless neomycin resistance gene, which allows the easy selection of ES cell clones with incorporated plasmid.

1. At the second day after transfection, start the selection of positive ES cell clones with F1 ES cell medium containing G418 (140 μg/mL). Select the clones for 6 days. Provide fresh medium with G418 every day.
2. After 6 days approx. 40 G418-resistant ES cell colonies should have grown. The colonies should be round with sharp borders. Normally you can identify them macroscopically as white dots on the plate.

3. For picking colonies you need a binocular or a microscope you can install under a laminar flow. Clean everything carefully with 70% ethanol and turn the flow on in advance.
4. Discard the medium from the first 10-cm plate and wash with 5 mL PBS. Add 8 mL PBS and pick approx. eight colonies with a pipette. Transfer each picked colony into feederless "V-bottom" wells of a 96-well plate pre-filled with 50 μL trypsin. Cells should be kept in the trypsin not longer than 20 min.
5. After 10 min at room temperature in the trypsin, add 50 μL medium to the cells and resuspend them by pipetting. Transfer the cells into a normal flat-bottom 96-well plate containing feeder cells and 100 μL medium and put them back into the incubator.
6. Pick approx. 24 colonies in total from all the three 10-cm plates (*see* **Note 24**).
7. Change the medium the next day.
8. The following days you have to expand the ES cells. In general you can split them every two days. The most important factor is to prevent confluence in order to prevent differentiation.
9. In our experience you can transfer ES cells from 96-well plates directly to 24-well plates. When you have to split them again, transfer half of one well to a well of a new 24-well plate and the other half to a well of a 24-well plate coated with gelatine instead of feeder cells.
10. When the cells in the 24-well feeder plate need splitting you can freeze them, so that you do not have to take care of them until you can identify the positive clones (*see* **Note 25**). For freezing, prepare a 2X freezing solution fresh and place it cool. Trypsinise the cells with 150 μL trypsin. After 5 min in the incubator add 150 μL medium and resuspend the cells by pipetting. Add 300 μL of the 2X freezing solution and mix carefully by pipetting. Seal the top of the 24-well plate with adhesive tape and wrap it into paper towels to prevent fast freezing. Put the package into a cardboard box and place it at –80°C.
11. Keep the cells in the gelatine-coated 24-well plate until the medium turns yellow overnight, indicating that the cells have grown fully confluent. Remove the medium and wash the cells with PBS. Now you can freeze the cells dry at –20°C or continue directly with DNA extraction (*see* **Section 3.4.3**).

3.4.3. Identification of Positive Clones

It is possible to identify the positive clones via PCR or via Southern blotting. Both methods are shown in the following.

1. First, extract the genomic DNA of the ES cells from the gelatine-coated 24-well plate, for example, with the Wizard® Genomic DNA Purification Kit as described here.

2. Add 600 μL Nuclei Lysis Solution to each well and lyse the cells by pipetting until you can pipette the solution easily. Transfer the suspension of one well to one 1.5-mL reaction tube.

3. Follow the Wizard® Genomic DNA Purification protocol for DNA extraction from animal tissue (mouse tail) including the RNA digestion. Finally, resuspend the DNA in 60 μL TE for 1 h at 65°C or overnight at 4°C. Store the DNA at 4°C or –20°C.

4. To identify positive clones by PCR screening use primers located in the pgk promoter and in the neomycin resistance gene (primers Neo and Pgk, *see* **Section 2.5** and **Fig. 6.6**; annealing temperature, 65°C; product size, 280 bp; if you use feeder cells harbouring a pgk–neo resistance gene an additional band of 160 bp appears); ES cell clones that did not undergo a complete recombination event (partial recombination, random integration, mixed clones) contain the hygro gene and can be identified by a second, hygro-specific PCR (primers Hyg-1 and Hyg-2, *see* **Section 2.5** and **Fig. 6.6**; annealing temperature, 65°C; product size, 550 bp; if you use feeder cells also harbouring a hygro resistance gene this PCR cannot be used for screening). Thus, for the typing of ES cell clones two PCR reactions (neo, hygro) should be run; for the typing of mice from tail DNA the pgk–neo PCR is sufficient. The *Rosa26* wild-type allele can be identified with the primers Rosa-5′ and Rosa-3′ (*see* **Section 2.5**; annealing temperature, 57°C; product size, 536 bp).

5. To verify correct clones it is mandatory to perform a Southern blot. Only by this method you can exclude partially recombined clones or chromosomal rearrangements. To digest the genomic DNA, mix 20 μL DNA with 3 μL NEB buffer 3, 1 μL spermidine, 0.3 μL BSA and 30 U *Eco*RV. Adjust with H_2O to a total volume of 30 μL. Incubate at 37°C for at least 30 min or overnight.

6. Separate the DNA on a 0.8% agarose gel until the 1 kb marker band leaves the gel.

7. Rinse the gel with water and shake it gently for 30 min in 0.25 M HCl to break very large DNA fragments for an easier transfer. Remove the buffer.

8. Rinse the gel with water and shake it gently for 1 h in denaturation buffer. Remove the buffer.

9. Rinse the gel with water and shake it gently for 1 h in neutralisation buffer. Remove the buffer.

10. Wash the gel with 2X SSC and build up the blot. All Whatman filter papers, the membrane and the paper towels should have the size of the gel. Fill a dish with 20X SSC (buffer reservoir)

and cover it with a Plexiglas plate. Place a Whatman filter paper soaked in 20X SSC onto the plate with the ends hanging into the buffer reservoir. Remove any air bubbles. Cover the borders with plastic foil and carefully transfer the gel onto the plate and the plastic foil. Position the foil so that it encircles the gel (*see* **Note 26**). Place a piece of nylon membrane, for example, Hybond N^+ (Amersham), onto the gel and remove any air bubbles. Mark the backside with a pencil. Place a Whatman filter paper soaked in 20X SSC on the membrane and ensure that there are no trapped air bubbles. Put a stack of paper towels (approx. 4 cm) on the Whatman filter paper. Put a second plate onto the top and load it with a weight of approx. 500 g, for example, a partially filled 500-mL glass flask. Let the blot stay overnight.

11. Remove the membrane from the blot and rinse it with the DNA side up in 2X SSC. Place it on Whatman filter paper and crosslink the DNA to the membrane using UV light (254 nm). Crosslinking will bind the DNA covalently to the membrane.
12. Roll the membrane with the DNA inside and place it into a hybridisation tube with 2X SSC. Uncoil the membrane so that it attaches to the wall of the tube with as less air bubbles as possible. Remove the 2X SSC and replace it with 10 mL Rapid-hyb buffer (Amersham). Prehybridise the membrane at 65°C in a rotating hybridisation oven (*see* **Note 27**).
13. In the meanwhile label 100 ng of the Rosa26 5′ probe, located upstream of the pgk promoter, with the Redipri-me^{TM} II Labelling Kit (Amersham). After incubating the labelling reaction at 37°C for 15–30 min, purify the probe using the Micro Spin^{TM} S-300 HR columns (Amersham). Shake the column and snap off the bottom closure. Loosen the cap and place the column in a 2.0-mL reaction tube. Centrifuge for 1 min at 700*g*. Place the column in a fresh 1.5 mL reaction tube and pipette the labelling reaction on the column. Centrifuge again for 2 min at 700*g* to purify the sample. Measure 1 μL of the probe in a scintillation counter to determine the amount of incorporated radioactivity.
14. Heat the probe to 95°C for 5 min and snap it cool on ice for 5 min. Remove the hybridisation tube from the oven and pour 5 mL of the buffer in a 15-mL tube. Use approx. 300,000 cpm per 1 mL Rapid-hyb buffer, that is, 3,000,000 cpm in 10 mL. Add the desired amount of probe to the 5 mL Rapid-hyb buffer, mix and pour it back in the hybridisation tube. Hybridise the membrane for a minimum of 4 h up to overnight.

15. Wash the membrane two times with Wash buffer for a total of 45 min at 65°C in a shaking water bath until the Geiger counter shows approx. 40 counts. Wrap the membrane in plastic foil and put it into an exposure cassette with Bio-MAX^{TM} intensifying screen (Kodak). Place a BioMax^{TM} MS film for 48–72 h (*see* **Note 28**).
16. The *Rosa26* locus gives rise to a wild-type band of 11.5 kb. Positive clones show additionally a band of 15.0 kb, depending on the size of your U6-shRNA expression cassette. Partially recombined clones show a band of 7.4 kb and the parental IDG26.10-3 ES cells show a band of 4.5 kb.
17. Select the positive clones and thaw the 24-well plate with the ES cells in a water bath at 37°C. Prepare two fresh 24-well plates with 1 mL medium per well and distribute one clone to two wells. Change the medium the next day.
18. Expand the clones to 10-cm plates and freeze aliquots of the clones in liquid nitrogen.
19. Generate mice from the ES cells using blastocyst injection or tetraploid aggregation.

4. Notes

1. When you paste a sequence for shRNA prediction, be sure to use a cDNA sequence and not a genomic one.
2. The first G of the sense sequence in shA is already provided by the vector sequence. Take care to include the final C, complementary to the G provided by the vector sequence, in the antisense part of shA.
3. Vectors containing shRNA sequences may form a complex secondary structure, which leads to a premature stop during sequencing. If you have severe problems with your sequence under standard sequencing conditions, ask specialised companies for reactions under special conditions (similar as for GC-rich sequences).
4. To use the endogenous gene expression for determination of the knockdown efficiency of the constructs, the targeted gene has to be expressed in ES cells. If this is not the case, you can use another cell line or cotransfect an expression vector together with the shRNA constructs.
5. Instead of analysing the protein level on a Western blot (e.g. if you do not have a good antibody), you can also use RNA and analyse it on a Northern blot or by quantitative RT-PCR. But

be aware that knockdown efficiency is not always the same on RNA and on protein level, but protein reduction will be essential for your expected phenotype.

6. It can be sufficient to incubate the reaction just for 30 min at –20°C but the amount of pelleted DNA will be less. If time is restricted, we suggest precipitating more DNA, for example, 100 μg.
7. Due to possible loss of DNA during the precipitation step it is recommended first to use a smaller volume to dissolve the DNA and to dilute the solution, not before determination of the concentration.
8. Freshly thawed cells are not very robust and therefore will not give reliable results.
9. The parameters for electroporation are optimal for the device from Fisher Scientific and IDG3.2 ES cells such that approx. 90% of the electroporated ES cells take up sufficient DNA. For other types of electroporators different settings may be required. An indicator of good conditions is a cell death rate of about 50% after electroporation.
10. If your protein of interest is part of the insoluble fraction of the cell, further extraction procedures will be necessary.
11. Depending on the size of your protein(s) of interest you can vary the percentage of the gel or even use gradient gels (e.g. 4%–12%).
12. Load a minimum of 15 μL to obtain nice bands.
13. For large proteins (>100 kDa) a longer blotting time may be necessary.
14. Additionally to the antibody against your gene of interest, use an antibody against a housekeeping gene like β-Actin as control to show equal loading on all lanes of your blot. You can either use both antibodies subsequently (stripping) or cut the membrane (if the difference in protein size is big enough) to use both antibodies in parallel.
15. To quantify the expression levels impartially you can scan the film and use computer programs, for example, ImageJ, to calculate the band intensities and normalise them to the level of the loading control for each lane. This method harbours a high error rate because of the small range, in which band intensities appear linear on the film, and number of processing steps (membrane -> film -> digital image). For more precise results you should use a chemiluminescent reader, if available, which digitalises the signals directly from the membrane.

16. The parameters for electroporation are selected in a way that approx. 90% of the electroporated ES cells take up the shRNA plasmid. Thus, the maximum level of gene silencing that can be reached in this assay is 90%. A reduction of gene expression by 80% in the ES cell population indicates an shRNA efficiency of approx. 90%.
17. If two restriction sites like *Asi*SI (A) and *SfiI* (B) are neighboured the double digestion may be incomplete and it is not possible to selectively remove the single-digested molecules from the double-digested vector by electrophoresis. This can cause a high background and low cloning efficiency of the desired construct. To circumvent this problem you can choose a third unique restriction site (*ScaI*, C) in the vector sequence and cut the parental vector in two reactions with enzyme A and C and enzyme B and C. Choose enzyme C in a way that the fragments *Asi*SI+*ScaI* and *ScaI*+*SfiI* can be easily distinguished. So, you can control the digestion of both restriction sites in the gel and isolate both fragments. In the ligation you combine your insert with the two vector fragments. Control ligations without insert fragment will give no background and so you can obtain a high rate of positive colonies from your ligations.
18. With the bpA forward primer you can check the 5′ cloning site and with M13 forward you can check the 3′ cloning site for correct integration. If you have a large stop cassette for a conditional hairpin, it might not be possible to sequence the whole insert. The most important sequences are the short hairpin sequences, the lox sites and the attB sites. If you can ensure that these sites are correct, it is not essential to check the stop cassette completely. The 5′ attB site can be screened with the M13 reverse primer.
19. It might be helpful to aliquot and/or freeze the solutions and chemicals to keep them fresh and sterile.
20. It is essential to remove the remaining medium with this washing step since it can inhibit the dissociation of the ES cells.
21. Be aware that the feeder cells are also in the suspension you use for counting. You can distinguish them by their size since the feeder cells are much larger than the ES cells and are not roundly shaped.
22. There may be some clumps in the cuvette, caused by the electroporation.
23. About 50% of the ES cells will be dead after transfection. This is not alarming and just an indication of successful electroporation of the surviving fraction.

24. As precaution you can also pick more clones and discard the clones which differentiate or do not look nicely during the expansion time.
25. As precaution you can also copy your plate to have some backup in case of contamination. For this, split ES cells again to two 24-well plates and freeze them 2–3 days later.
26. The foil should prevent the 20X SSC from bypassing the gel and diffuses directly into the paper towels. You can place the foil also beneath the gel if you do not need the DNA located at that place (e.g. the space above the wells).
27. Take the rotation direction into consideration! Place the tube so that the membrane cannot coil up itself.
28. You can also use exposure cassettes without intensifying screen and films, but then you have to expose the film much longer, for example, 1 week.

Acknowledgements

We thank S. Michailidou, S. Kareth, C. Birke, R. Kneuttinger and A. Tasdemir for excellent technical help and G. Hannon for the pSHAG vector. This work has been funded by the Volkswagen Foundation and the Federal Ministry of Education and Research (BMBF) in the framework of the National Genome Research Network (FKZ:01GR0404).

References

1. Elbashir SM, Harborth J, Lendeckel W, Yalcin A, Weber K, Tuschl T. Duplexes of 21-nucleotide RNAs mediate RNA interference in cultured mammalian cells. Nature 2001;411:494–8.
2. Brummelkamp TR, Bernards R, Agami R. A system for stable expression of short interfering RNAs in mammalian cells. Science 2002;296:550–3.
3. Lee NS, Dohjima T, Bauer G, et al. Expression of small interfering RNAs targeted against HIV-1 rev transcripts in human cells. Nat Biotechnol 2002;20:500–5.
4. Paddison PJ, Caudy AA, Bernstein E, Hannon GJ, Conklin DS. Short hairpin RNAs (shRNAs) induce sequence-specific silencing in mammalian cells. Genes Dev 2002;16:948–58.
5. Kuhn R, Streif S, Wurst W. RNA interference in mice. Handb Exp Pharmacol 2007:149–76.
6. Hasuwa H, Kaseda K, Einarsdottir T, Okabe M. Small interfering RNA and gene silencing in transgenic mice and rats. FEBS Lett 2002;532:227–30.
7. Rubinson DA, Dillon CP, Kwiatkowski AV, et al. A lentivirus-based system to functionally silence genes in primary mammalian cells, stem cells and transgenic mice by RNA interference. Nat Genet 2003;33:401–6.
8. Lickert H, Takeuchi JK, Von Both I, et al. Baf60c is essential for function of BAF chromatin remodelling complexes in

heart development. Nature 2004;432:107–12.

9. Hitz C, Wurst W, Kuhn R. Conditional brain-specific knockdown of MAPK using Cre/loxP regulated RNA interference. Nucleic Acids Res 2007;35:e90.

10. Oberdoerffer P, Kanellopoulou C, Heissmeyer V, et al. Efficiency of RNA interference in the mouse hematopoietic system varies between cell types and developmental stages. Mol Cell Biol 2005;25:3896–905.

Chapter 7

Mutagenesis of Mouse Embryonic Stem Cells with Ethylmethanesulfonate

Robert Munroe and John Schimenti

Abstract

Unraveling the function of the mammalian genome relies heavily on analyses of the laboratory mouse. Because of its powerful genetics and available technologies to manipulate the genome, plus its developmental and physiological similarities to humans, it has become a goal to generate mutations in all mouse genes and analyze the phenotypic consequences. Gene targeting in embryonic stem (ES) cells is the method of choice for making null mutations in known genes of interest. However, forward genetics approaches, in which mutations are produced randomly throughout the genome, has the advantage of producing alleles of varying severity both within known genes, in non-coding regulatory elements, or in other unannotated functional elements. Such forward genetic mutation screens in mice have typically involved treating male mice with *N*-ethyl-*N*-nitrosourea (ENU), followed by three generations of breeding to render potential recessive mutations homozygous, at which time phenotype screens can be performed. An alternative strategy for randomly mutagenizing the mouse genome is by chemical treatment of ES cells. This enables the use of multiple alternative chemicals with different mutational spectra, can reduce breeding to two generations, and impart a higher mutational load. Furthermore, ES cell mutagenesis can be used to create banks of clones that can be screened for point mutations in genes of interest, and to conduct forward genetic screens in vitro to detect potential phenotypes prior to generation of mice. In this chapter, we provide a detailed protocol for mutagenizing ES cells with the point mutagen ethylmethanesulfonate (EMS).

Key words: Ethylmethanesulfonate (EMS), embryonic stem (ES) cells, mouse genetics, mutants, forward genetics, mutagenesis.

1. Introduction

Mouse mutants can be derived by either reverse or forward genetic approaches. Advancements in recent years have facilitated the application of both methods. On the reverse genetics side, the

Ralf Kühn, Wolfgang Wurst (eds.), *Gene Knockout Protocols: Second Edition, vol. 530*

DOI 10.1007/978-1-59745-471-1_7 Springerprotocols.com

Knockout Mouse Project (KOMP) and various gene-trapping resources will soon combine to represent mutations, in ES cells, of nearly all mouse genes. For forward genetic strategies, comprehensive genomic information and DNA sequencing capacity has greatly facilitated positional cloning of induced mutations. These two approaches are highly complementary. Whereas null mutations generated by reverse genetic strategies systematically inform on the essential function of genes, typical forward genetics approaches also produce hypomorphic mutations within genes or in non-coding regulatory elements. Furthermore, it has become clear, as in the cases of various forms of non-coding regulatory RNAs, that there are functional elements of the genome about which we have little knowledge. Random mutagenesis can reveal such novel elements.

The most common method for randomly mutagenizing the mouse genome is by treating male mice with *N*-ethyl-*N*-nitrosourea (ENU), which is an effective mutagen of spermatogonial stem cells *(1)*. Recently, it has been shown that chemical mutagenesis of ES cells is also an effective means to produce mouse mutants, both in a phenotype-driven *(2, 3)* and gene-driven *(4, 5)* fashion. Data from these experiments indicated that ES cell mutagenesis can produce higher mutational loads than classical whole-animal mutagenesis *(4, 6)*, although creating too high a mutational load can make it impractical to breed mice efficiently *(2)*.

Another important advantage of ES cell mutagenesis is that multiple mutagens can be used, each with a characteristically different mutational spectrum. To date, there are reports of using ENU, EMS, ICR-191, and trimethylpsoralen to mutagenize ES cells *(2, 5, 7)*. Because different mutagens have different DNA sequence and context preferences, the mutation rate for a given gene may differ significantly.

Recently, we reported a noteworthy advantage in terms of mouse breeding for conducting whole-animal screens begun with mutagenized ES cells rather than ENU-mutagenized male mice *(8)*. Whereas three generations of mice are required for standard ENU screens for recessive mutations, the production of chimeras from mutagenized ES cells enables a two-generation breeding scheme. After outcrossing the chimera to wild-type mice, G2 daughters can be backcrossed to the chimera to produce G3s that are potentially homozygous for mutations. Alternatively, G2s can be intercrossed. This process takes less than half the time as ENU germline mutagenesis. A key difference compared to the classical ENU paradigm is that chimeras transmit mutagenized versions of all chromosome homologs, whereas G1 males derived from an ENU-mutagenized male are heterozygous for the mutagenized genome. Thus, the number of mutations that can be recovered from the ES cells scheme is double. Another potential advantage for using ES cell mutagenesis is the possibility of conducting phenotype-driven screens in cell culture, followed by generation of mutant mice *(9)*.

In this review, we detail the procedures for effectively treating ES cells with EMS (although as mentioned above, other chemicals such as ENU can be used), quality controlling and monitoring the process, and selecting ideally mutagenized batches for subsequent manipulations.

2. Materials

(1) Tissue culture hood for sterile manipulations.

(2) Water-jacketed CO_2 incubator set at 37°C with 5% CO_2.

(3) Inverted microscope for viewing ES cell cultures.

(4) Tissue culture-grade plates; six-well plates; 60 mm, 100 mm, and 150 mm plates.

(5) 6 thioguanine (6TG) (Sigma # A4660).

(6) Mitomycin C or access to a Cs irradiator (for mitotic inactivation of feeder cells).

(7) Ethylethanesulfonate (1.17 g/mL) liquid (Sigma cat# M0880). Working concentration is 10 mg/mL. Make this by adding 21.4 μL of EMS to 2.48 mL of ES cell media just prior to treatment.

(8) Newbauer improved hemacytometer.

(9) Leukemia Inhibitory Factor (LIF). Commercial source is ESGRO from Thermo-Fisher (cat# 5056885) or Chemicon International (# ESG1106) (*see* **Note 1**).

(10) Gelatin (from porcine skin, Type A, Sigma cat# G1890-500G). Autoclave stock at 2% and make working dilution of 0.2% by diluting into sterile water.

(11) ES cell ("complete") medium. For 100 mL:

Fetal calf serum (FBS)	15 mL (*see* **Note 2**)
100X Penicillin-Streptomycin	1 mL (Invitrogen* cat# 15140-122)
2-mercaptoethanol dilution	1 mL (dilution is 7 μL in 10 mL of sterile H_2O)
Non-essential amino acids	1 mL (Invitrogen cat# 11140-050)
Sodium pyruvate	1 mL (Invitrogen cat# 11360-070)

(continued)

(continued)

DMEM (+ glutamine)	81 mL (Invitrogen cat# 11965-092)
LIF	10 μL

* Note that the Invitrogen products are from its subsidiary Gibco.

(12) Phosphate-buffered saline (Invitrogen, cat# 70011-044 for 500 mL of 10X concentrate).

(13) Trypsin (0.25%)-EDTA (Invitrogen cat# 25200-072).

(14) Low-speed tabletop centrifuge.

3. Methods

3.1. Mutagenizing Embryonic Stem Cells

We must emphasize that if it is the experimentor's intention to generate mice from the mutagenized ES cells, that the experiments must be initiated with cells that have been proven to have excellent germline-colonizing potential. In the most ideal situation, one would start with newly derived or low-passage cultures of ES cells. The genetic background of the ES cells does not seem to be critical, since we successfully generated germline chimeras from highly mutagenized ES of both F1 hybrid and inbred 129 origins *(2)*. However, the ES cells should be male, in order to allow for eventually measuring the induced mutation rate at the X-linked *Hprt* locus (*see* **Section 3.2**). Furthermore, if one intends to determine mutation rate at *Hprt*, it is important that the starting population contains a functional *Hprt* locus, and that there has been little or no accumulation of spontaneous mutations (*see* **Note 4**). This can be determined as described in **Section 3.2**.

Since there have been excellent detailed protocols published on the generation and culture of ES cells, we will not go into depth on this topic; we will assume the user has some experience in handling ES cells *(10, 11)*. Hence, we will concentrate on issues related to the mutagenesis treatment and assessment.

(1) Gelatinize several 60-mm tissue culture plates and seed them with a layer of ~500,000 mitotically inactivated mouse embryonic fibroblasts (miMEFs), also known as "feeders." The MEFs can be mitotically inactivated by exposing to 5,000 rads of gamma radiation. Alternatively, they can be treated with mitomycin C (10 μg/mL for 2.5 h at 37 °C). Also, *see* **Note 3** for feeder types. Allow cells to adhere overnight.

(2) Take a vial of frozen ES cells (containing approximately 1/2–1/3 the contents of a near-confluent 60-mm plate at the time of freezing), quickly thaw at 37°C, add to a 15-mL conical tube containing a few millilitres of complete medium, and pellet in the centrifuge at 200X *g* for 3 min.

(3) Aspirate media and resuspend in 4 mL of complete medium supplemented with LIF. Be sure the cells are in a single-cell suspension. Aspirate media from a feeder plate and add the ES cells.

(4) Grow ES cells for approximately 2–3 days, changing media daily, until the cell population is approximately 1/2 confluent on the plate. Then trypsinize the cells: aspirate media; wash once with PBS; add 1 mL trypsin and incubate 5 min in 37°C CO_2 incubator; add 3 mL of ES medium to a 15-mL conical tube, and 1 mL to the cells; pipette the media on the plate around until all the cells have been dislodged. Transfer to the conical tube, and pipette up and down 25 times vigorously to make a single-cell suspension. Split contents to 6–35 mm plates (or each of six wells in a 6-well plate) with feeders, each containing a total of 2 mL media plus cells.

(5) Grow ES cells for ~1 day or until ~1/2 confluent on the plate. This should represent about 2×10^6 cells.

(6) Change media (replenish with 2 mL complete media plus LIF) and add different concentrations of 10 mg/mL EMS to each well of the six-well plate containing the ES cells: 0, 200, 400, 500, 600, and 800 μg/mL. This can be achieved by adding 0, 40, 80, 100, 120, and 160 μL of EMS respectively, to a final volume of 2 mL on each 35-mm well.

(7) Grow ES cells under EMS treatment for 16 h.

(8) Aspirate the media, wash ES cells three times with 2 mL of PBS per well. Then trypsinize, as in Step 4, except use half the specified volumes. After making a single cell suspension, pellet ES cells at 200X *g* for 3 min, aspirate media, and resuspend in 2.5 mL of complete media.

(9) Determine the concentration of cells in each sample. Plate 1,000 cells onto a gelatinized, feederless, 60-mm plate, in duplicate, for each treatment level (keep the remainder – *see* next paragraph). These plates are for determining the mortality rates due to mutagen treatment. Grow cells for several days or until colonies appear, then count the number of ES cell colonies on each plate. Expect about 10–25% plating efficiency, as assessed by the untreated cells. Plate the remaining treated ES cells onto fresh 6-well plates (each well = 35 mm) with miMEFs. Grow ES cells for 2–3 days, changing

media daily. (The control cells, which will not have reduced viability from mutagen treatment, should be expanded by plating only one-fourth of the remaining cells).

(10) Determine kill rate and number of survivors representing the mutagenized population: First, count the number of colonies on each plate and get the average of the replicates. The number of colonies in the untreated sample determines the plating efficiency. The percentage kill rate due to mutagen treatment is then determined as 100 × [1–(number colonies in treated sample/number colonies in untreated sample)]. In the past, we have found that kill rates of 90%–95% provide a good balance of mutation rate and maintenance of germline-colonizing ability.

(11) Depending on your application, you may want to know the number of distinct ES cell clones representing the surviving mutagenized population. This is estimated by multiplying the number of cells after treatment (Step 9) by the survival percentage (100 – the percentage kill measured in Step 10).

3.2. Determination of Mutation Rate at the Hprt Locus

The easiest way to gauge the induced mutation rate is to use a selectable marker. We have used the X-linked *Hprt* locus, since one can select against expression of this gene by growing cells in the presence of 6TG. Others have used markers such as a hemizygous, autosomal thymidine kinase gene (cells expressing this can be killed by gancyclovir treatment) *(5)*, but this requires that you have such a line and are comfortable generating mice that may contain this transgene. After obtaining clones that have *Hprt* mutations, it is possible to sequence the locus and determine the mutation spectrum, as we have described *(2)*.

(1) Continue growing and passaging the EMS-treated and control ES cells from **Section 3.1**, Step 9 for a minimum of 8 days. This must be done in order to allow for dissolution of wild-type *Hprt* enzyme after induction of mutations. During this expansion, EMS-treated cells can be split and some portion of each treatment level can be frozen back.

(2) After the 8-day period, seed 1 × 10^6 cells (from each EMS treatment level, plus untreated control) onto a gelatinized 150-mm plate, with no feeders. Use 20 mL of complete medium with LIF plus 6TG (10 μg/mL). Also, for each sample, plate 1,000 cells onto a feederless 60-mm gelatinized plate (everything in duplicate) to determine the plating efficiency. Do not apply 6TG to these 60-mm plates.

(3) Grow the cells under 6TG selection for 3 days (except for the 60-mm plates), changing the medium daily. Afterwards, change the medium as needed. Grow a total of 14 days, or until colonies are obvious.

(4) Count the colonies on the 60-mm plates and determine the plating efficiency for each sample (Average number of colonies on the 60-mm plates/1,000). Multiply this by the number of cells plated on the 150-mm plates (1 million) to determine the number of cells that were effectively screened for *Hprt* mutations.

(5) Count the colonies on the 6TG selection plates. Subtract the spontaneous background (from the control plate) from each of the treated samples. Divide this number by the total number of cells screened (Step 4) to get the mutation rate. It is common to represent this in terms of a fraction in which 1 is in the numerator (for example, 1/2,500 per locus mutation rate).

4. Notes

(1) Because commercial LIF is very expensive, many labs choose to make their own recombinant LIF. If you choose to do so, it is important to conduct empirical tests of the ability of your material to inhibit differentiation of ES cells grown off a feeder layer. Briefly, we recommend plating a defined number of ES cells at low density to form discrete colonies (about 1,000 cells/60-mm gelatinized plate), and treat each plate with an increasing amount of home-made LIF. Also, buy a small amount of commercial LIF and add the recommended amount to one plate. Allow the cells to grow for several days, then determine the number of undifferentiated colonies on each plate. Decide which amount of home-made LIF matches the commercial LIF in the ability to prevent differentiation.

(2) It is well known that the quality of FBS varies from batch to batch, so it is advisable to obtain small batches, and test their ability to support healthy ES cell growth. We do it by plating 1,000 cells/100 mm feederless, gelatinzed plate in complete media containing various sources of FBS. Once colonies are visible and well established (about 10 days), they are compared for plating efficiency and differentiation.

(3) Some people use feeder-free ES cell lines. Obviously, this would decrease the amount of effort overall. Also, some people use some version of STO cells as feeders. One should use the culture conditions under which the ES cells have been propagated. However, it should be noted that the ES cell kill rate due to subsequent mutagen treatment may vary depending on the ES cell growth conditions and feeder types and density.

(4) We have never found this to be a problem, but if it is, then a different ES cell line should be used, or the available line should be selected in HAT medium to ensure only HPRT-positive cells are present prior to EMS treatment.

References

1. Russell W, Kelly E, Hunsicker P, Bangham J, Maddux C, Phipps E. Specific-locus test shows ethylnitrosourea to be the most potent mutagen in the mouse. Proc Natl Acad Sci USA 1979;76:5818–9.
2. Munroe RJ, Bergstrom RA, Zheng QY, et al. Mouse mutants from chemically mutagenized embryonic stem cells. Nat Genet 2000;24:318–21.
3. Browning VL, Chaudhry SS, Planchart A, Dixon MJ, Schimenti JC. Mutations of the mouse Twist and sy (fibrillin 2) genes induced by chemical mutagenesis of ES cells. Genomics 2001;73:291–8.
4. Vivian JL, Chen Y, Yee D, Schneider E, Magnuson T. An allelic series of mutations in Smad2 and Smad4 identified in a genotype-based screen of *N*-ethyl-*N*-nitrosourea-mutagenized mouse embryonic stem cells. Proc Natl Acad Sci USA 2002;99:15542–7.
5. Chen Y, Yee D, Dains K, et al. Genotype-based screen for ENU-induced mutations in mouse embryonic stem cells. Nat Genet 2000;24:314–7.
6. Ward JO, Reinholdt LG, Hartford SA, et al. Toward the genetics of mammalian reproduction: induction and mapping of gametogenesis mutants in mice. Biol Reprod 2003;69:1615–25.
7. Greber B, Lehrach H, Himmelbauer H. Characterization of trimethylpsoralen as a mutagen for mouse embryonic stem cells. Mutation Res 2003;525:67–76.
8. Munroe RJ, Ackerman SL, Schimenti J. Genome-wide two generation screens for recessive mutations by ES cell mutagenesis. Mamm Genome 2004;15:960–5.
9. Chen Y, Schimenti J, Magnuson T. Toward the yeastification of mouse genetics: chemical mutagenesis of embryonic stem cells. Mamm Genome 2000;11:598–602.
10. Robertson E. Embryo-derived stem cell lines. In: Robertson E, ed. Teratocarcinomas and embryonic stem cells a practical approach. Oxford: IRL Press; 1987;71–112.
11. Tessarollo L. Manipulating mouse embryonic stem cells. Methods Mol Biol. (Clifton, NJ) 2001;158:47–63.

Part II

Stem Cell Manipulation

Chapter 8

Gene Targeting in Mouse Embryonic Stem Cells

Lino Tessarollo, Mary Ellen Palko, Keiko Akagi, and Vincenzo Coppola

Abstract

The scientific value of a mouse model with a targeted mutation depends greatly upon how carefully the mutation has been engineered. Until recently, our ability to alter the mouse genome has been limited by both the lack of technologies to conditionally target a locus and by conventional cloning. The "cre/loxP" and "recombineering" technologies have overcome some of these limitations and have greatly enhanced our ability to manipulate the mouse genome in a sophisticated way. However, there are still some practical aspects that need to be considered to successfully target a specific genetic locus. Here, we describe the process to engineer a targeted mutation to generate a mouse model. We include a tutorial using the publicly available informatic tools that can be downloaded for processing the genetic information needed to generate a targeting vector.

Key words: Gene targeting, targeting vector, recombineering, cre/loxP, mouse.

1. Introduction

Gene targeting in the mouse is a powerful technology that allows the study of gene function in mammals *(1–4)*. Its initial application was for the establishment of in vivo genetic models for the phenotypic analysis of genes considered developmentally important. This was achieved by inactivating a specific gene in the mouse germline. Other more complex applications include the introduction of genetic mutations to mimic pathological conditions in humans or the replacement of one gene with another to investigate common genetic functions. Currently, it is fairly common to introduce reporter genes (e.g., LacZ, GFP, *(5–7)*) into a locus to pursue gene expression studies or to insert recombination recognition sites (loxP and Frt sites) that allow inactivation of a gene conditionally or induce chromosomal rearrangements *(8–11)*. These applications

Ralf Kühn, Wolfgang Wurst (eds.), *Gene Knockout Protocols: Second Edition, vol. 530*

DOI 10.1007/978-1-59745-471-1_8 Springerprotocols.com

have been made possible and have been greatly facilitated by the development of the cre/loxP and more recently the "recombineering" technologies (*see* **Chapter 2**, this volume, *(12–16)*). However, some basic concepts must be considered when designing a targeting strategy, irrespective of the type of mutation to be utilized.

A common misconception among investigators that are new to gene targeting in mouse embryonic stem (ES) cells is that a gene targeting strategy merely requires two fragments of DNA flanking the genomic sequence of interest to be modified and a selectable marker to identify correctly targeted ES cells. Indeed, while this constitutes the basic concept of homologous recombination in ES cells, there are a number of criteria that the investigator must consider to successfully target a specific locus. A few examples include, the length of genomic sequence used in the vector; the mouse strain of the genomic DNA; the selectable markers used in the vector and the overall design of a screening strategy to identify correctly targeted ES cells. While this may appear complicated at first, the availability of the DNA sequence of both the mouse and human genomes has greatly facilitated this task *(17–19)*. In this chapter, we describe in detail the steps involved in designing a targeting vector including how to search and obtain sequence information from the publicly available genome database.

2. Methods

In order to design a targeting vector, the researcher must first obtain the cDNA and the genomic DNA sequences of the gene of interest to analyze the characteristics of the locus under investigation. Most of these data are available in computed form and can be obtained from public resources such as the UCSC Genome Browser *(20, 21)*. It should be stressed that the investigator should be knowledgeable of the structure and function of the gene to be targeted. This is important when considering which region of the gene should be deleted to most likely result in the successful disruption of that gene. One consideration is the presence of catalytic sites that are essential for the function of the protein. However, it should be noted that deletion of such sites can still potentially generate a protein that has the ability to bind other accessory proteins and could therefore act in a dominant-negative fashion. Moreover, instead of disrupting a gene, deletion of certain exons can potentially generate a modified cDNA through "alternative splicing" creating a functional protein that lacks only the amino acids coded for in the deleted exon/s. In general, we find targeting the first exon a good strategy for completely disrupting a gene, assuming that the Kozak consensus sequence is removed and/or there are no alternative ATG start sites downstream of the targeted exon. Another strategy includes deletion of an exon that potentially leads to an out-of-frame protein.

For those unfamiliar with the tools available to search the genome databases, we include a case study to describe a step-by-step search and retrieval of genetic information from the public database (*see* **Appendix**).

2.1. Construct Design

The most common type of DNA vector used for gene targeting in ES cells is the so-called "sequence replacement vector" *(22, 23)*. It includes two DNA fragments flanking a marker gene used for the positive selection of ES cells (*see* **Fig. 8.1**, *(24)*). Several aspects of

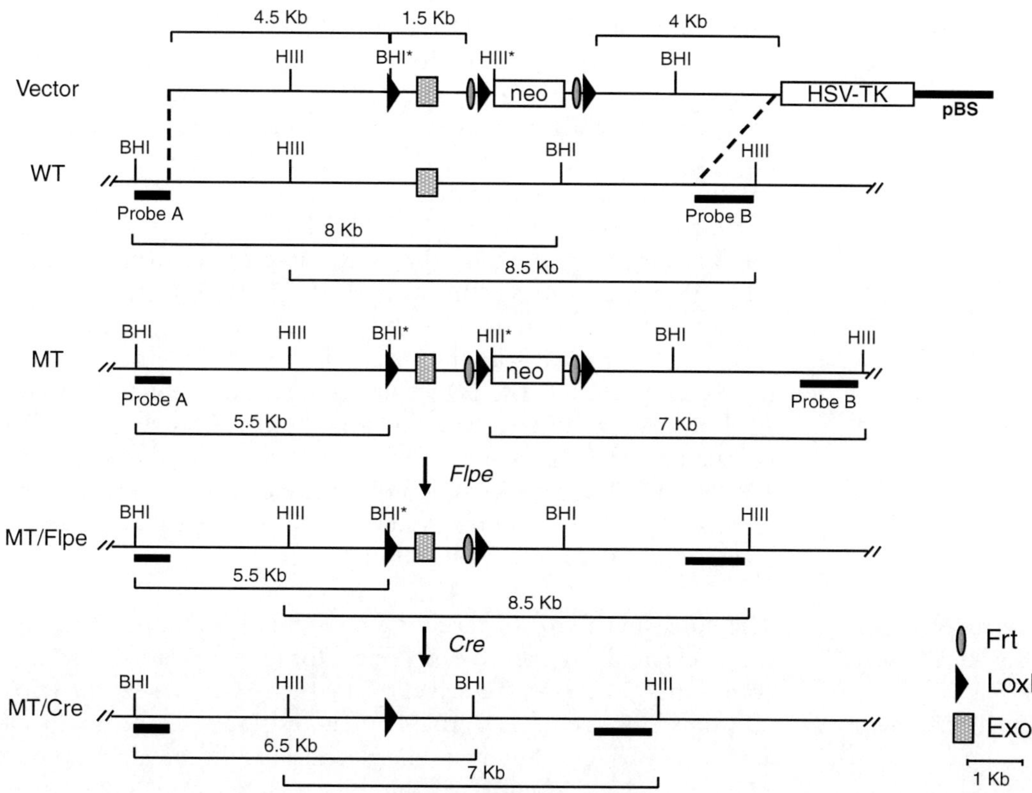

Fig. 8.1. Schematic representation of a representative vector and screening strategy to detect the rearrangements in a specific targeted locus. A targeting vector (*top*) for conditional removal of the exon of interest has been generated using 4.5 kb and 4 kb of upstream and downstream DNA sequence, respectively. An exogenous *Bam*HI site (BHI*) has been added to the upstream loxP site sequence for screening purposes. A *Hin*dIII site (HIII*) present in the neo cassette is also employed for the screening. Note that when the screening strategy has been designed every change in restriction fragment size has been accounted for after each specific recombination event. For example, Probe B detects an endogenous wild-type (WT) band of 8.5 kb that is reduced in size (7 kb) by the addition of the HIII* site present in the neo cassette following homologous recombination. After Flpe-induced recombination to excise the neo cassette the band reverts to the WT 8.5-kb size. However, targeting at the locus is detected by the presence of the 5.5-kb band obtained by *Bam*HI digestion and Probe A. Cre recombination leads to a change of both *Bam*HI and *Hin*dIII restriction fragments that are characteristic of the deletion of the exon (6.5 kb and 7 kb respectively). WT, wild-type allele; MT, targeted allele; MT/Flpe, mutant allele after Flpe recombination; MT/cre, mutant allele after Cre recombination; pBS, Bluescript; HSV-TK, thymidine kinase cassette.

the vector's design should be considered when choosing the genomic region to be targeted. These aspects can affect the frequency of the targeting events and/or affect the targeted locus and normal function of surrounding genes *(25)*. These include the following.

2.1.1. Length of Homology Between the Targeting Vector and the Target Locus

The length of homologous DNA sequence incorporated on either side of the targeted region of interest in a targeting vector significantly influences the frequency of homologous recombination. Targeting of the HPRT locus has shown that the length of homology used in the construct can affect the homologous recombination efficiency 100- to 200-fold *(22, 26)*. Ideally, a total of 8–12 kb should be used with the shortest homologous arm being no less than 2 kb in length.

2.1.2. Source of DNA for Vector Construction

The source of the vector DNA should be isogenic to the ES cells used for the targeting *(22)*. Mouse strain-specific polymorphisms at specific loci can greatly reduce the efficiency of homologous recombination if the vector sequence is somewhat different from the ES cells. Most ES cells are from the 129 Sv background because they are the most efficient at colonizing the mouse germline upon injection into recipient blastocysts *(27, 28)*. However, most of the mouse DNA sequence information that is available comes from "bacterial artificial chromosomes" (BACs) of the C57BL/6 background (see below) that have been used for the public sequencing of the mouse genome. For this reason, in our laboratory we have been using a hybrid 129 Sv–C57Bl/6 ES cell line (v6.4 a gift of John Schimenti; *(29)*) that allows us to use vectors from both 129 Sv and C57BL/6 genetic backgrounds. Most recently, a genome-wide, end-sequenced 129 Sv BAC library has become available for construction of targeting vectors of the 129 Sv background *(30)* (http://www.geneservice.co.uk/products/sanger/bMQ/bMQ.jsp).

2.1.3. Drug Resistance Marker Genes for Selection

2.1.3.1. Positive Selection

The construct used for targeting a specific locus must contain a drug-selectable marker to easily select for recombinants. The most commonly used positive selection marker is the *neomycin* phosphotransferase gene (neo) that confers resistance to the neomycin analog, G418. G418 is an aminoglycoside antibiotic produced by *Micromonospora rhodorangea* that blocks polypeptide synthesis by inhibiting the elongation step in both prokaryotic and eukaryotic cells. Resistance to G418 is conferred by the neo gene that encodes an aminoglycoside 3′-phosphotransferase. Selection in mammalian cells is usually achieved in 3–7 days with concentrations starting from ~300 μg/ml. Rapidly dividing cells are more severely affected; thus, 250–300 μg/ml is effective in killing ES cells that do not contain the neo gene.

Another commonly used selectable marker is the *hygromycin B* phosphotransferase gene. Hygromycin B is an aminoglycoside antibiotic produced by *Streptomyces hygroscopicus*. It inhibits protein synthesis by interfering with translocation and causing mistranslation at the 70S ribosome. Hygromycin B is effective against most bacteria, fungi, and higher eukaryotes. Resistance to hygromycin is conferred by the hph gene from *E. coli*. Hygromycin B is normally used at a concentration of 50–200 μg/ml in mammalian cells and is effective in 2–3 days.

The most widely used selectable marker is the neo gene. Hygromycin is used when a second vector that targets another allele of a gene is required. Other markers for positive selection include genes encoding for resistance to puromycin (inhibits peptidyl transfer on ribosome), Blasticidin S (interferes with the peptide bound formation in the ribosomal machinery), and Zeocin™ (intercalates and cleaves DNA). These markers can be used to introduce multiple targeting events in ES cells. It should be noted that the positive-selectable marker chosen for the vector will influence the type of mouse embryo fibroblast (MEF) feeder layer that is needed to grow the ES cells. For example, if the neo gene is used the feeders must be G418-resistant.

Selectable expression cassettes can easily be obtained from investigators who perform gene-targeting experiments. The main difference between the types of cassettes available is the choice of promoter and the poly-adenylation signal used to control the expression of the selectable marker *(31)*. The phosphoglycerate kinase (PGK) promoter is commonly used with the neo gene. A key element of the promoter used is that it should not be too sensitive to positional effects in the genome to ensure that it will work for the targeted genes that are not expressed in ES cells. The orientation by which the positive selection cassette is inserted in the targeted locus does not matter. After the initial targeting and transmission through the mouse germline the cassette should be removed via cre or flpe recombination to prevent its regulatory elements from affecting the transcription of neighboring genes confounding the phenotypic analysis *(25)*.

2.1.3.2. Negative Selection

Negative selection markers are used to reduce the number of transfected cells containing the exogenous sequence inserted at random loci *(32)*. The most commonly used marker is the thymidine kinase (TK) gene of the herpes simplex virus (HSV). If the TK gene is incorporated into the genome, it confers sensitivity to gancyclovir (2′-nor-2′deoxyguanosine) or FIAU [*1-(2*'-deoxy-2'fluoro-b-D-arabinofuranosyl)-*5*-iodouracil]. These drugs are specific inhibitors of HSV by acting as nucleotide analogues that inhibit cellular growth following incorporation into DNA.

In order for these analogues to be incorporated into DNA, they must be phosphorylated by the TK gene. In the absence of the TK gene, the drug phosphorylation is limited. Low concentrations of gancyclovir or FIAU have no effect on cells. Therefore, the sensitivity of cells to gancyclovir is dependent upon the presence of the HSV tk gene product.

The TK gene is placed at one end of the targeting vector adjacent to one end of the two DNA homology arms. Following a homologous recombination event, the TK is not recombined into the region and the correctly targeted ES cell will be resistant to both G418 (because it has incorporated the neo gene) and gancyclovir or FIAU (because TK is not incorporated during the homologous recombination process) (**Fig. 8.1**).

Another negative selection marker is the diphtheria toxin A-fragment gene (DT-A) that exerts toxicity by inhibiting ADP-ribosylation of elongation factor 2 upon protein synthesis *(33)*. Thus, contrary to the gancyclovir or FIAU it does not have mutagenic potentials and should enhance karyotype stability. The DT-A gene works with the same principle as the TK negative selection gene. The main difference is that if the DT-A gene is randomly incorporated into the genome by non-homologous recombination it will produce diphtheria toxin that kills the cell without the need to add any drug to the cultured cells. A negative selection strategy employing either the TK gene or the DT-A gene can enrich the ratio of homologous recombinant clones by 3–10 fold.

2.1.4. Screening Strategy

The importance of designing a screening strategy to identify homologous recombinant clones cannot be overstated and it should be an integral part of the vector design process. A screening strategy design can include a Southern blotting or PCR methodology for detection of homologous recombination events. Screening by Southern blotting is far superior in assuring that the correctly targeted clones are identified and helps avoid false-positives. We highly recommend that this type of screening be used at least in the initial phases of the analysis. For Southern analysis, specific restriction enzyme digestions should be chosen to recognize homologous recombination events involving the targeting vector (*see* **Note 1**). Moreover, two probes, one 5’ and one 3’, should be identified external to the targeting vector sequence. This means that the probes should not hybridize to the targeting vector. This insures identification of rearrangements that occur only in the targeted locus following the homologous recombination event (**Fig. 8.2E**) without detecting random integrations of the vector in the genome (**Fig. 8.2D**). The restriction enzymes for DNA digestion should be chosen that cut the DNA upstream of the 5’ probe or downstream of the 3’ probe. This allows for detection of rearrangements caused only by homologous recombination events in the targeted locus. A rearrangement can be detected when insertion of the selection cassette (e.g., neo) causes the restriction enzyme fragment to increase in size. However, most times the design of a screening strategy can take advantage of the presence of a restriction enzyme site that is located within the selection cassette itself (*see* in **Fig. 8.1** the *Hin*dIII restriction fragment detected with probe B goes from 8.5 kb in the WT allele to

7 kb in the mutant allele) or a site can be intentionally added when a loxP recombination sequence is included in the targeting vector (see for example the *Bam*HI * site added to the loxP site in **Fig. 8.1** that causes the DNA fragment detected with Probe A to go from 8 kb to 5.5 kb). In general it is advisable to include a restriction enzyme site adjacent to the isolated loxP site to have appropriate means to detect its presence. Some investigators rely on PCR to screen for the presence of a loxP site. However, this can lead to the detection of false-positive clones if the vector has integrated randomly into another site of the genome in addition to the targeted locus. Lastly, when a conditional allele is generated the investigator should include a strategy for detecting rearrangements at both the 5′ and 3′ ends of the construct not only before but also after cre and Flp recombination (**Fig. 8.1**).

2.1.5. Probes

As mentioned above, the probes for the screening strategy should not be part of the targeting vector sequence to allow detection of homologous recombination events at the targeted locus. Moreover, it is critical that the probes are free of repetitive elements whose presence results in a smear on the Southern blot (**Fig. 8.2B,C**). The probes can be synthesized by PCR but even if the sequence from which they are chosen appears repeat-free they should be tested on a blot because computer programs may not recognize all types of repeats. Even a small stretch of repeats in a probe could prevent its use in Southern blot analysis (**Fig. 8.2B**).

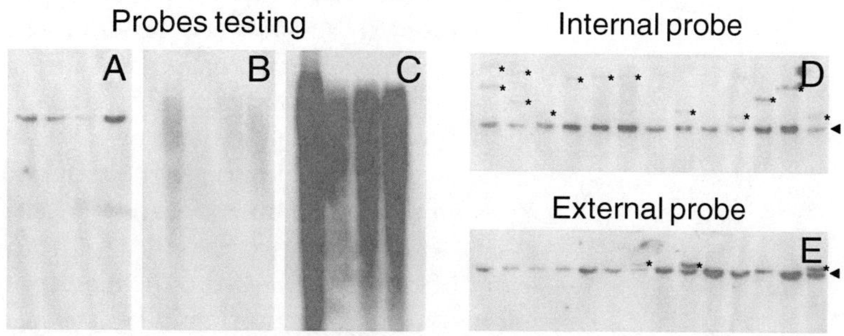

Fig. 8.2. Screening of probes for Southern analysis. Testing of probes for the presence of repetitive sequence (**A–C**). The probe in (**A**) is completely free of repetitive sequence, hence only one clear band is detected by Southern blot hybridization. On the contrary, blots tested in (**B**) and (**C**) have different levels of repetitive sequence as shown by the smear detected in the blot. Note that even the moderate presence of repetitive sequence (**B**) does not allow the use of this probe for analysis. Southern blot analysis of ES cell DNA samples using a probe internal to the targeting vector (**D**) detects bands in almost every lane of the blot (*) in addition to the wild-type band (*arrowhead*). Using a probe that is external to the targeting vector allows detection of a re-arranged band in ES cell clones in which one allele underwent homologous recombination (*, **E**). Note that the band from the targeted allele has intensity similar to that of the wild-type allele (*arrowhead*). Equal molarity of the two bands is a sign of clonality. Only clones with two bands of similar intensity should be pursued for injection into recipient blastocysts.

2.1.6. Vector DNA Sequence

2.1.6.1. Repetitive Elements

Repetitive sequence should be avoided not only when choosing a probe for screening but also when designing the targeting vector per se. Almost 50% of the mouse genome is repetitive *(34)*; thus, it is intuitive that having repetitive sequence in the vector increases the chances that it can integrate throughout the genome by aligning its own repeat sequence to other identical and very abundant sequence in the genome. Nevertheless, because almost half of the mouse genome is repetitive, it is sometimes impossible to avoid the presence of repetitive sequences in a vector. Therefore, during the designing phase, the investigator should choose a region with as few repeats as possible.

2.1.6.2. Sequence Analysis

If the objective of the gene targeting is the simple inactivation of a gene by deletion of a specific genomic region, the task is fairly simple. The vector can include two DNA homology arms that flank the region to be deleted. However, currently it is advisable to generate a vector that targets a gene in a conditional manner by using the cre-loxP/Flpe-Frt technology. This requires the placing of the loxP/Frt sites and the neo selectable marker in an intronic region that does not interfere with the transcription or splicing of the targeted gene. In fact, the objective is to include these elements in such a way as to keep the gene functional until it can be inactivated by the precise spatio-temporal expression of the cre or Flpe recombinases. The most common approach is to include a neo cassette flanked at each end by both Frt and loxP sites and adding an additional isolated loxP site flanking one or more exons (e.g., **Fig. 8.3**). However, intronic regions can contain regulatory elements that control the transcription or splicing of a gene. The best approach in choosing where to insert the loxP and the neo cassette, while limiting the chances of disrupting the gene, is to align and compare the mouse and human genomic DNA sequence (publicly available; see below) in the area to be targeted. This analysis often reveals surprisingly high conservation in intronic areas in addition to the exons. High conservation between species indicates the functional relevance of these regions. Moreover, as a general rule the neo cassette and the loxP site should be placed at least 200–300 bp away from the targeted exon because even if there is no obvious conservation between species these elements could nevertheless interfere with the splicing machinery.

2.1.7. Transfection

The DNA vector should be linearized to facilitate stable integration into the genome by DNA recombination. The preferred method for introducing DNA into ES cells is by electroporation because it allows the introduction of only a few molecules of DNA per cell. This reduces the chances of multiple random integrations of the DNA vector into an ES cell genome that can interfere with the selection process. For example, even if homologous recombination can occur at the targeted locus the random integration of only one other copy of the vector in the genome would cause the ES cell clone to die because of the insertion of the negative selection marker (TK or DT).

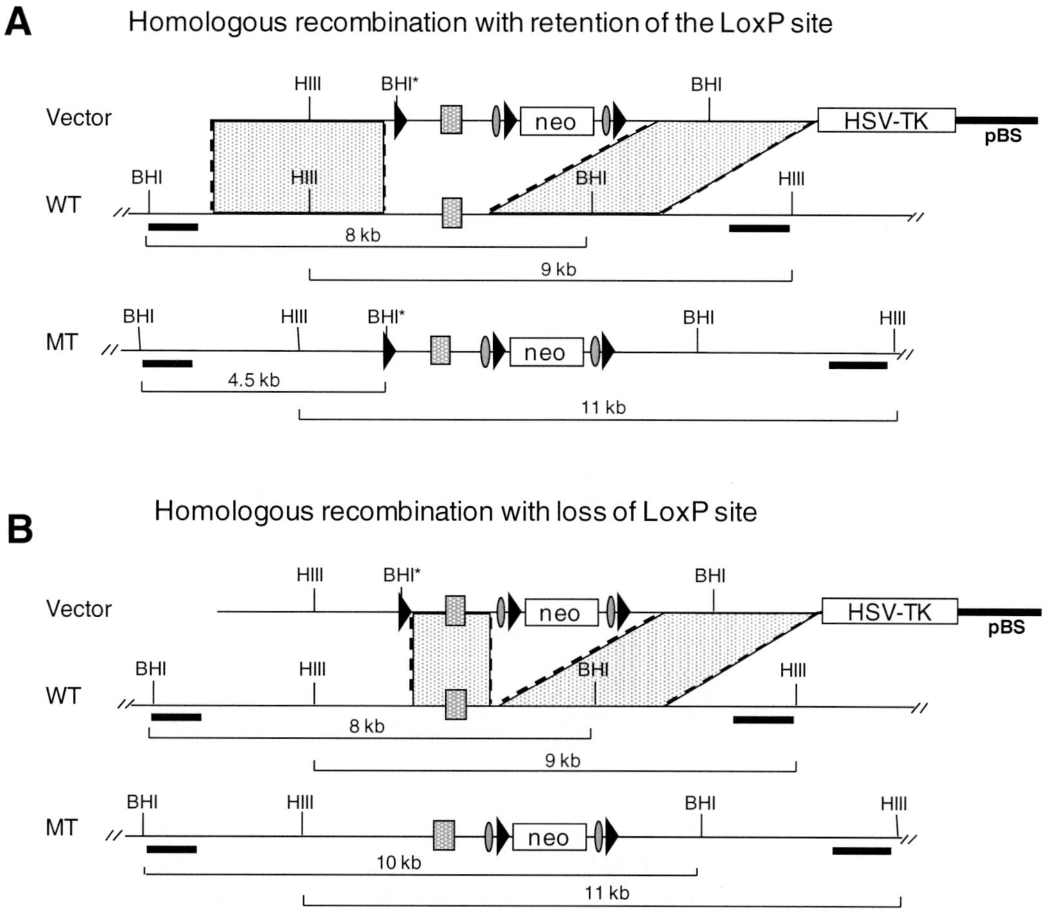

Fig. 8.3. Homologous recombination often leads to the loss of the targeting vector loxP sequence if situated too far away from the neo selectable cassette. (**A**) Example of homologous recombination event leading to the integration of the neo selectable marker and the upstream loxP site with the artificial *Bam*HI site (BHI*). Retention of the loxP site from the targeting vector becomes more difficult during the homologous recombination process if it is situated more than 2 kb away from the selectable marker (**B**). *Shaded areas* indicate the regions between the vector and the targeted locus where homologous recombination occurs. When homologous recombination occurs in the region between the loxP site and the neo cassette the lox P site is not recombined into the locus (**B**). Abbreviations are as in **Fig. 8.1**.

2.2. Summary

An ideal construct should have the following characteristics (**Fig. 8.4**):

(a) DNA sequence used for the vector should be isogenic to the mouse ES cells used for targeting.

(b) Length of DNA sequence homologous to the locus under investigation must range between 7 kb and 12 kb (8–12 kb optimal) with the shortest arm being no less than 2 kb.

A loxP site added to generate a conditional allele should be placed approximately 1–2 kb away from the neo cassette. If placed at a distance greater than 2 kb the chances of "losing" it during the homologous recombination process can be very high because the sequence between the loxP and the neo

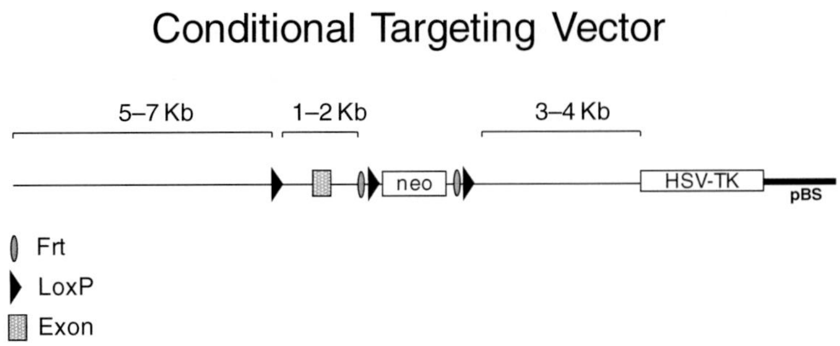

Fig. 8.4. Example of an ideal representative targeting vector for conditional deletion of an exon. Note that the orientation of the construct can be either 5′ to 3′ or 3′ to 5′. The positive and negative selectable markers can be different than the ones indicated (see text).

cassette can undergo homologous recombination with exclusion of the loxP site (*see* **Fig. 8.3**). Greater distances, up to 5 kb, can be acceptable if the homologous recombination frequency at the targeted locus is high (30–50%). The "isolated" loxP site can get "lost" during recombination because there is no selection for the retention of this site.

(c) Selectable markers:

– Neomycin gene (possibily pGKneobpA) for positive selection (selection: 300 μg/mL active G418),

– DT-A (requires no drug for selection) or thymidine kinase gene (possibily pGK TK) for negative selection (selection: 5 μM FIAU).

(d) Restriction enzyme strategy for screening homologous recombination events at both 5′ and 3′ ends of the construct.

(e) Restriction enzyme strategy for screening homologous recombination events at both 5′ and 3′ ends of the construct before and after cre recombination (if applicable).

(f) At least one probe must be external to the genomic DNA sequence used to create the targeting vector.

(g) Unique restriction site for linearization of the vector prior to electroporation.

3. Notes

1. The best restriction enzymes for screening are those that require high salt concentrations. When designing the screening strategy it is desirable to include enzymes such as *Bam*HI, *Bgl*I, *Bgl*II, *Eco*RV, *Hinc*II, *Hind*III, *Nco*I, *Nde*I, *Not*I, *Pst*I, *Spe*I, *Sca*I, and *Stu*I. Other enzymes such as *Eco*RI and *Xba*I are more sensitive to impurities and salt concentration

variations and should be avoided if possible. However, these enzymes can be used if the DNA is purified by the phenol-chloroform procedure (*see* **Chapter 9**).

4. Appendix

How to search and navigate the public databases to obtain and analyze the DNA sequence required to generate a targeting vector.

Case Study: Search and analysis of the genomic sequence of the Cytip/Pscdbp gene.

The task includes retrieving the mouse genomic DNA sequence to design a vector disrupting the exon with the start codon (ATG) of Pscdbp.

Step 1: Retrieve the mRNA of the Pscdbp gene. This will provide the information to identify the ATG initiation codon relative to the genomic sequence. *Tool*: NCBI GeneBank.

Step 2: Identify the genomic structure of the Pscdbp gene. The genomic structure of a gene is important to see the size of the individual exons and introns and whether there are clusters of exons that can be targeted to maximize the amount of cDNA sequence deleted. As indicated above, there are limitations to the amount of genomic sequence that can be flanked by loxP sites (ideally 1–2 kb; **Fig. 8.4**). Thus, the clustering of multiple exons within a few kilobases of genomic sequence may represent a good area to be targeted. *Tool*: UCSC Genome Browser.

Step 3: Obtain 15 kb of genomic DNA sequence upstream and downstream of the Pscdbp's ATG start site with repetitive regions identified within this 30-kb region. Once the exon/s to be targeted are identified (in this case the ATG containing exon), 30 kb of DNA sequence must be downloaded to perform a detailed analysis of restriction enzyme sites (for the screening strategy), to identify repetitive sequence (to be avoided in the generation of probes and possibly to limit in the sequence chosen for the vector) and for genomic sequence comparison among species (to help choose the site for insertion of the loxP sites and the selectable marker cassette). *Tool*: UCSC Genome Browser.

Step 4: Characterize the genomic DNA sequence.

a. Map the restriction enzyme sites. *Tool*: NEBcutter v2.0.

b. Compare the genomic locus among species (mouse vs. human). *Tool*: UCSC Genome Browser or Vista.

Step 1: Get mRNA of Pscdbp gene

1.1. Go to NCBI GenBank (http://www.ncbi.nlm.nih.gov) *(20)*.

1.2. Select "Gene" database, type "Pscdbp" and then click "Go" (*see* **Fig. 8.5**).

Fig. 8.5. Entry window into the NCBI gene database.

1.3. Choose the mouse gene (*Mus Musculus*) Cytip (also known as Pscdbp) from the list. The description for Cytip will appear.

1.4. From the "Genomic regions, transcripts, and products" section click the accession number (i.e., NM_139200). From the mRNA links "GenBank" or "FASTA" format will appear (*see* **Fig. 8.6**).

1.5. Click "GenBank" to obtain the sequence in GenBank format. The output shows annotations of the mRNA such as gene size (i.e., 5735 bases), exons (e.g., exon 1 goes from 1 to 241), and CDS (coding sequences start at nucleotide 68 and end at nucleotide 1147 of the last exon. The rest of the 5735 mRNA nucleotides are untranslated region). At this step, confirm the CDS start position as 68 nt for Pscdbp (*see* **Fig. 8.7**).

1.6. In the "Display" links window, click "FASTA" to obtain the sequence in FASTA format. Save the sequence as a text file, such as Simple Text in Mac and NotePad in Windows, in

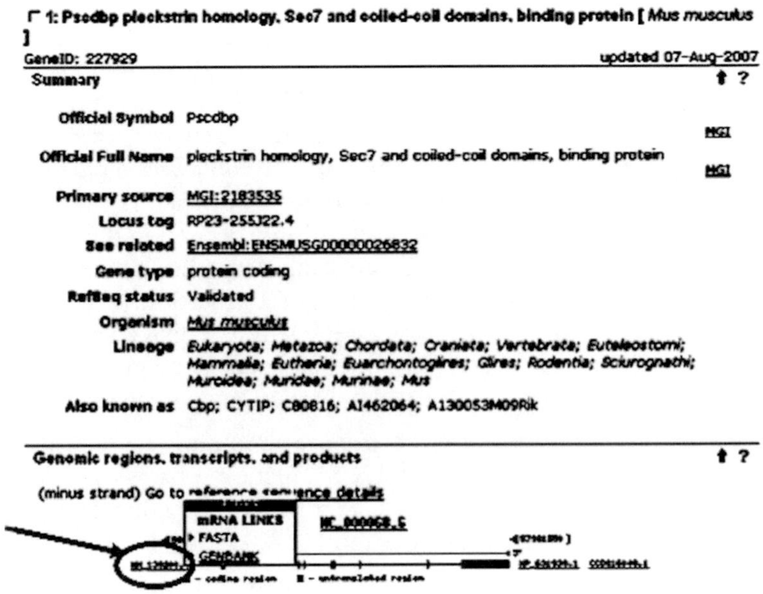

Fig. 8.6. Selection of the Genbank format via the accesssion number (*arrow*).

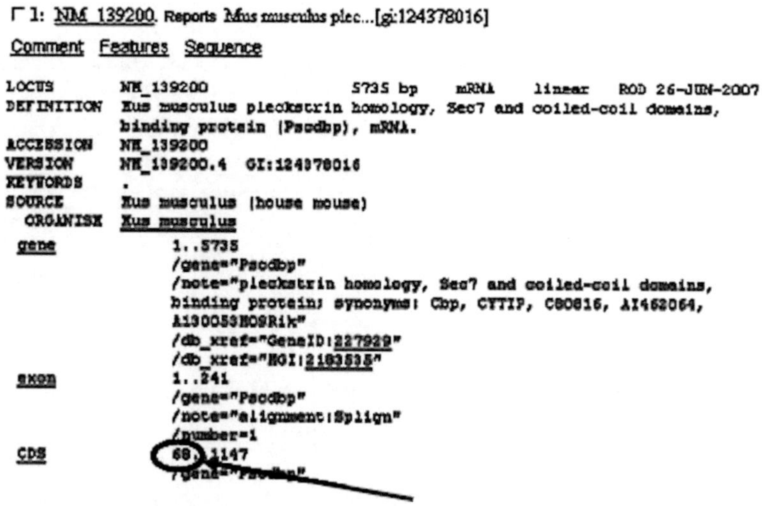

```
1: NM_139200. Reports Mus musculus plec...[gi:124378016]
Comment  Features  Sequence

LOCUS       NM_139200            5735 bp    mRNA    linear   ROD 26-JUN-2007
DEFINITION  Mus musculus pleckstrin homology, Sec7 and coiled-coil domains,
            binding protein (Pscdbp), mRNA.
ACCESSION   NM_139200
VERSION     NM_139200.4  GI:124378016
KEYWORDS    .
SOURCE      Mus musculus (house mouse)
  ORGANISM  Mus musculus
 gene                 1..5735
                      /gene="Pscdbp"
                      /note="pleckstrin homology, Sec7 and coiled-coil domains,
                      binding protein; synonyms: Cbp, CYTIP, C80816, AI462064,
                      A130053H09Rik"
                      /db_xref="GeneID:227929"
                      /db_xref="MGI:2183535"
 exon                 1..241
                      /gene="Pscdbp"
                      /note="alignment:Splign"
                      /number=1
 CDS                  68..1147
                      /gene="Pscdbp"
```

Fig. 8.7. Selection of the start codon of the Pscdbp gene (*arrrow*).

your computer for later usage. If you use MS Word, use "save as Plain text option". FASTA format is a popular sequence format, and many bioinformatics tools accept this format. FASTA format file starts with a ">" followed immediately by a name, and the next line begins the sequence with fixed length (*see* **Fig. 8.8**).

Step 2: Identify genomic structure (exon/intron, ATG site) of Pscdbp gene

2.1. Go to UCSC Genome Browser (http://genome.ucsc.edu) *(21)*.

2.2. Click "Blat" at the top bar to run a sequence alignment program *(35)* against mouse genome (*see* **Fig. 8.9**).

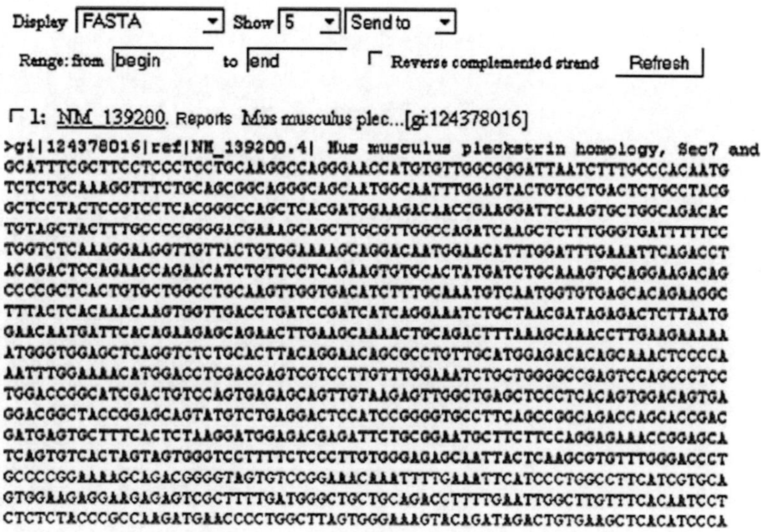

```
Display FASTA   Show 5   Send to
Range: from begin   to end     Reverse complemented strand   Refresh

1: NM_139200. Reports Mus musculus plec...[gi:124378016]
>gi|124378016|ref|NM_139200.4| Mus musculus pleckstrin homology, Sec7 and
GCATTTCGCTTCCTCCCTCCTGCAAGGCCAGGGAACCATGTGTTGGCGGGATTAATCTTTGCCCACAATG
TCTCTGCAAAGGTTTCTGCAGCGGCAGGGCAGCAATGGCAATTTGGAGTACTGTGCTGACTCTGCCTACG
GCTCCTACTCCGTCCTCACGGGCCAGCTCACGATGGAAGACAACCGAAGGATTCAAGTGCTGGCAGACAC
TGTAGCTACTTTGCCCCGGGGACGAAAGCAGCTTGCGTTGGCCAGATCAAGCTCTTTGGGTGATTTTTCC
TGGTCTCAAAGGAAGGTTGTTACTGTGGAAAAGCAGGACAATGGAACATTTGGATTTGAAATTCAGACCT
ACAGACTCCAGAACCAGAACATCTGTTCCTCAGAAGTGTGCACTATGATCTGCAAAGTGCAGGAAGACAG
CCCCGCTCACTGTGCTGGCCTGCAAGTTGGTGACATCTTTGCAAATGTCAATGGTGTGAGCACAGAAGGC
TTTACTCACAAACAAGTGGTTGACCTGATCCGATCATCAGGAAATCTGCTAACGATAGAGACTCTTAATG
GAACAATGATTCACAGAAGAGCAGAACTTGAAGCAAAACTGCAGACTTTAAAGCAAACCTTGAAGAAAAA
ATGGGTGGAGCTCAGGTCTCTGCACTTACAGGAACAGCGCCTGTTGCATGGAGACACAGCAAACTCCCCA
AATTTGGAAAACATGGACCTCGACGAGTCGTCCTTGTTTGGAAATCTGCTGGGGCCGAGTCCAGCCCTCC
TGGACCGGCATCGACTGTCCAGTGAGAGCAGTTGTAAGAGTTGGCTGAGCTCCCTCACAGTGGACAGTGA
GGACGGCTACCGGAGCAGTATGTCTGAGGACTCCATCCGGGGTGCCTTCAGCCGGCAGACCAGCACCGAC
GATGAGTGCTTTCACTCTAAGGATGGAGACGAGATTCTGCGGAATGCTTCTTCCAGGAGAAACCGGAGCA
TCAGTGTCACTAGTAGTGGGTCCTTTTCTCCCTTGTGGGAGAGCAATTACTCAAGCGTGTTTGGGACCCT
GCCCCGGAAAAGCAGACGGGGTAGTGTCCGGAAACAAATTTTGAAATTCATCCCTGGCCTTCATCGTGCA
GTGGAAGAGGAAGAGAGTCGCTTTTGATGGGCTGCTGCAGACCTTTTGAATTGGCTTGTTTCACAATCCT
CTCTCTACCCGCCAAGATGAACCCCTGGCTTAGTGGGAAAGTACAGATAGACTGTGAAGCTCACATCCCA
```

Fig. 8.8. View of the Pscdbp mRNA in the FASTA format.

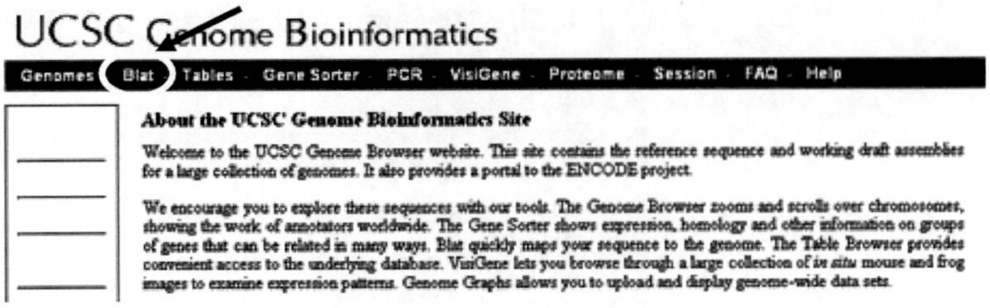

Fig. 8.9. The BLAT search tool of the UCSC genome browser.

2.3. Choose Mouse genome and paste the FASTA format sequence of NM_139200 (from Step 1.6). Click "Submit" (*see* **Fig. 8.10**).

2.4. Choose the best scoring alignment at the top (i.e., 5728). Notice that this gene is on the (−) strand (*see* **Fig. 8.11**).

2.5. Click "details" to view the alignment. The cDNA entered appears at the top in blue upper-case letters. Scroll down to the retrieved genomic sequence where the exons are represented in upper-case blue letters and the intronic regions are lower-case black letters. Highlight, copy, paste, and save the ATG exon (targeted exon) for further analysis. Scroll down further to see the alignment with mouse chromosome 2. You can confirm the ATG initiation position at 68 nt of NM_139200 (Step 1–5) is at chr2:58012466 (*see* **Fig. 8.12**).

2.6. Return to previous page to get to the BLAT search results and click "browser" to view the graphical presentation of gene structure. You can zoom in, zoom out, or move the window by clicking "1.5x" or "<" signs (*see* **Fig. 8.13**).

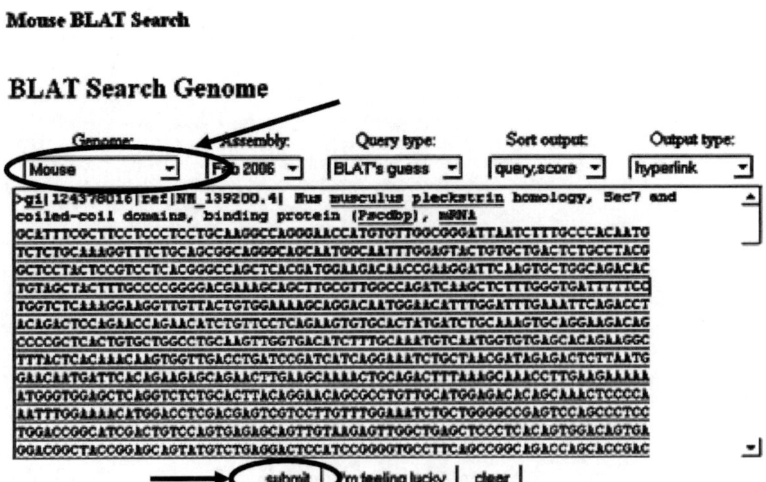

Fig. 8.10. Submission of the Pscdbp FASTA format sequence to the BLAT search.

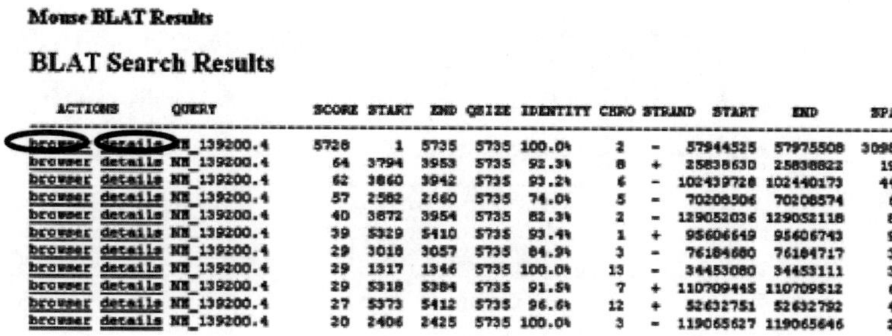

Mouse BLAT Results

BLAT Search Results

ACTIONS		QUERY	SCORE	START	END	QSIZE	IDENTITY	CHRO	STRAND	START	END	SPAN
browser	details	NM_139200.4	5728	1	5735	5735	100.0%	2	-	57944525	57975508	30984
browser	details	NM_139200.4	64	3794	3953	5735	92.3%	8	+	25838630	25838822	193
browser	details	NM_139200.4	62	3860	3942	5735	93.2%	6	-	102439728	102440173	446
browser	details	NM_139200.4	57	2582	2660	5735	74.0%	5	-	70208506	70208574	69
browser	details	NM_139200.4	40	3872	3954	5735	82.3%	2	-	129052036	129052118	83
browser	details	NM_139200.4	39	5329	5410	5735	93.4%	1	+	95606649	95606743	95
browser	details	NM_139200.4	29	3018	3057	5735	84.9%	3	-	76184680	76184717	38
browser	details	NM_139200.4	29	1317	1346	5735	100.0%	13	-	34453080	34453111	32
browser	details	NM_139200.4	29	5318	5384	5735	91.5%	7	+	110709445	110709512	68
browser	details	NM_139200.4	27	5373	5412	5735	96.6%	12	+	52632751	52632792	42
browser	details	NM_139200.4	20	2406	2425	5735	100.0%	3	-	119065627	119065646	20

Fig. 8.11. Selection of the top alignment from the BLAT search results (*circles*).

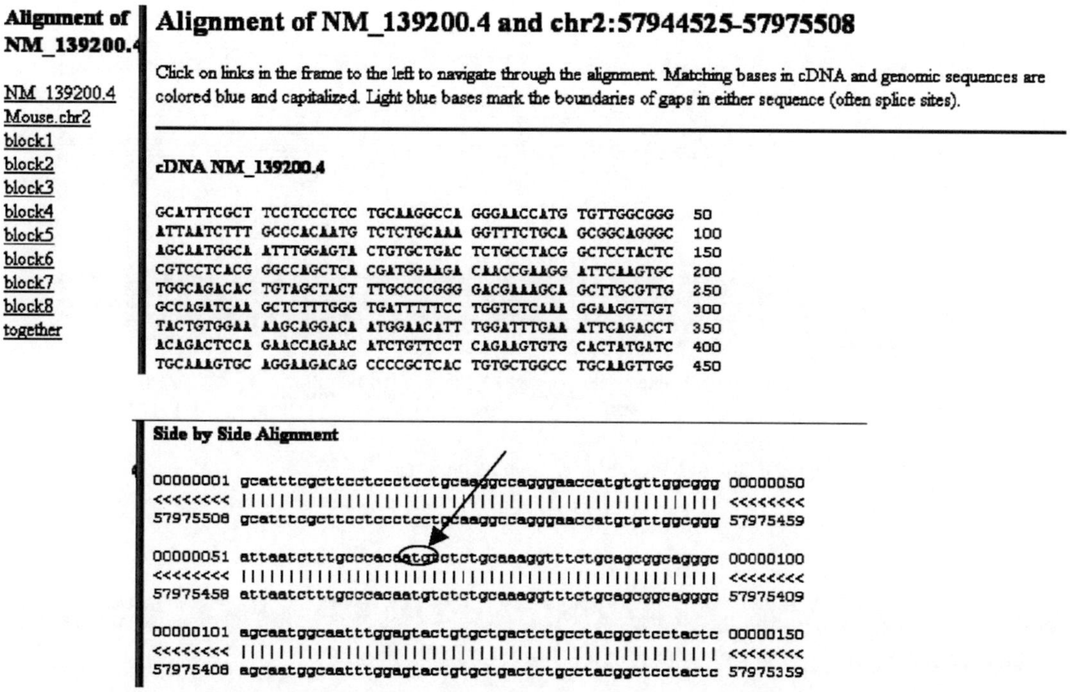

Alignment of NM_139200.4

NM_139200.4
Mouse.chr2
block1
block2
block3
block4
block5
block6
block7
block8
together

Alignment of NM_139200.4 and chr2:57944525-57975508

Click on links in the frame to the left to navigate through the alignment. Matching bases in cDNA and genomic sequences are colored blue and capitalized. Light blue bases mark the boundaries of gaps in either sequence (often splice sites).

cDNA NM_139200.4

```
GCATTTCGCT TCCTCCCTCC TGCAAGGCCA GGGAACCATG TGTTGGCGGG  50
ATTAATCTTT GCCCACAATG TCTCTGCAAA GGTTTCTGCA GCGGCAGGGC  100
AGCAATGGCA ATTTGGAGTA CTGTGCTGAC TCTGCCTACG GCTCCTACTC  150
CGTCCTCACG GGCCAGCTCA CGATGGAAGA CAACCGAAGG ATTCAAGTGC  200
TGGCAGACAC TGTAGCTACT TTGCCCCGGG GACGAAAGCA GCTTGCGTTG  250
GCCAGATCAA GCTCTTTGGG TGATTTTTCC TGGTCTCAAA GGAAGGTTGT  300
TACTGTGGAA AAGCAGGACA ATGGAACATT TGGATTTGAA ATTCAGACCT  350
ACAGACTCCA GAACCAGAAC ATCTGTTCCT CAGAAGTGTG CACTATGATC  400
TGCAAAGTGC AGGAAGACAG CCCCGCTCAC TGTGCTGGCC TGCAAGTTGG  450
```

Side by Side Alignment

```
00000001 gcatttcgcttcctccctcctgcaaggccagggaaccatgtgttggcggg 00000050
<<<<<<<< |||||||||||||||||||||||||||||||||||||||||||||||||| <<<<<<<<
57975508 gcatttcgcttcctccctcctgcaaggccagggaaccatgtgttggcggg 57975459

00000051 attaatctttgcccacaatgtctctgcaaaggtttctgcagcggcagggc 00000100
<<<<<<<< |||||||||||||||||||||||||||||||||||||||||||||||||| <<<<<<<<
57975458 attaatctttgcccacaatgtctctgcaaaggtttctgcagcggcagggc 57975409

00000101 agcaatggcaatttggagtactgtgctgactctgcctacggctcctactc 00000150
<<<<<<<< |||||||||||||||||||||||||||||||||||||||||||||||||| <<<<<<<<
57975408 agcaatggcaatttggagtactgtgctgactctgcctacggctcctactc 57975359
```

Fig. 8.12. Detailed view of the top alignment and identification of the Pscdbp start codon (*arrow*).

Step 3: Obtain +/−15 kb genomic sequence from Pscdbp's ATG site with information about repetitive regions

3.1. Select BLAT and paste the sequence of the exon of interest in the box. Click SUBMIT. This takes you to the BLAT Search Results. Click "browser" to get assembly information about the specific exon of interest as in Step 2.6 (*see* **Fig. 8.14**).

3.2. Click DNA from the top menu. In the boxes indicated type in 15,000 bp upstream and downstream of your exon.
Check "All upper case"; check the "Mask repeats", and "to lower case" boxes.

Check "Reverse complement (get "–" strand)" box (because the exon of interest is on the "–" strand).

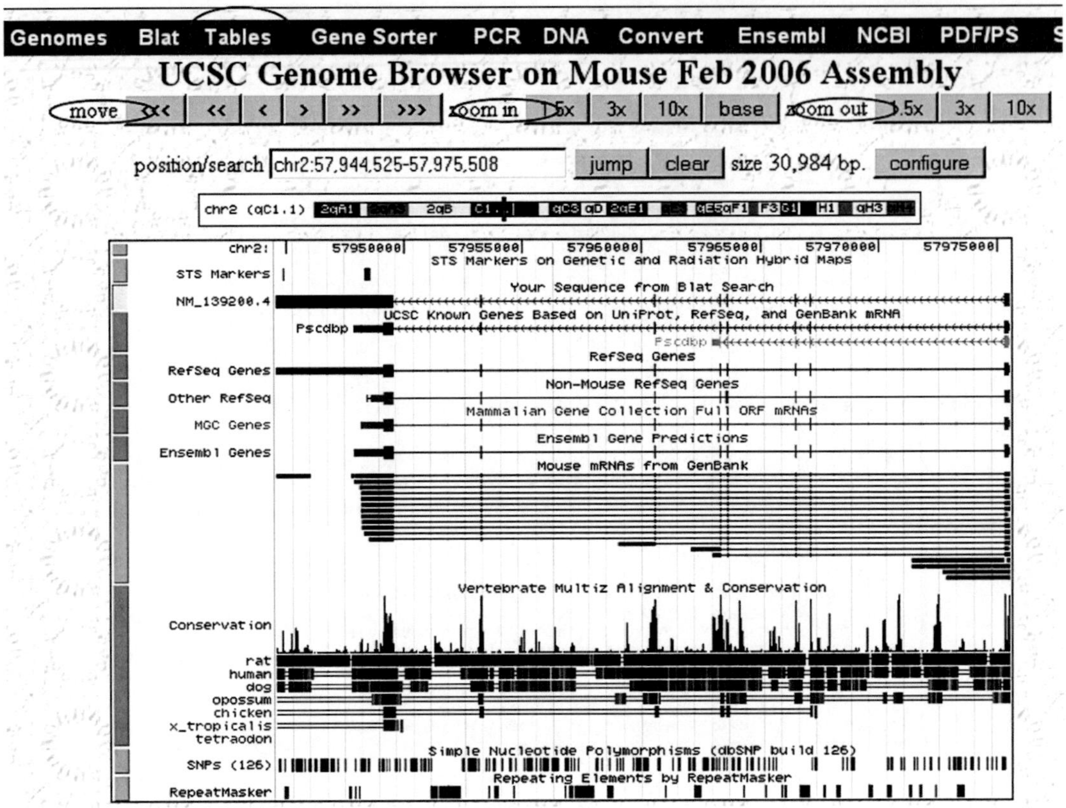

Fig. 8.13. View of the Pscdbp gene structure on the UCSC genome browser.

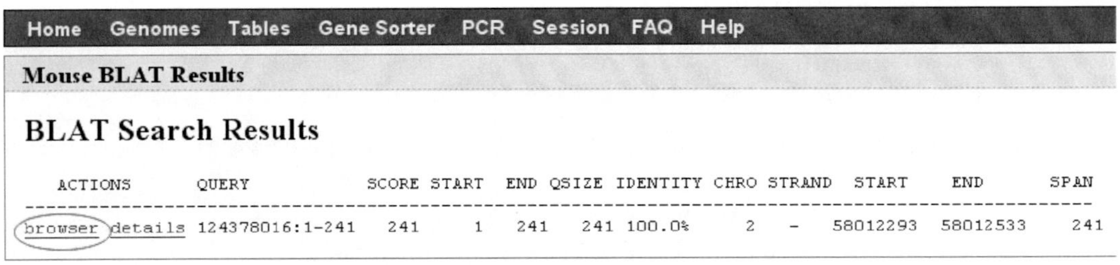

Fig. 8.14. BLAT search for genomic sequence around exon 1 of the Pscdbp gene.

Click "get DNA" (*see* **Fig. 8.15**).

3.3. This sequence is the genomic region 15 kb upstream and downstream of your exon. The lower-case letters represent repeats masked by repeatMasker *(36)*. You should copy and paste this sequence into a word document. You will use it for designing your targeting construct and genotyping strategy. Since there is no numbering you can find the location of the

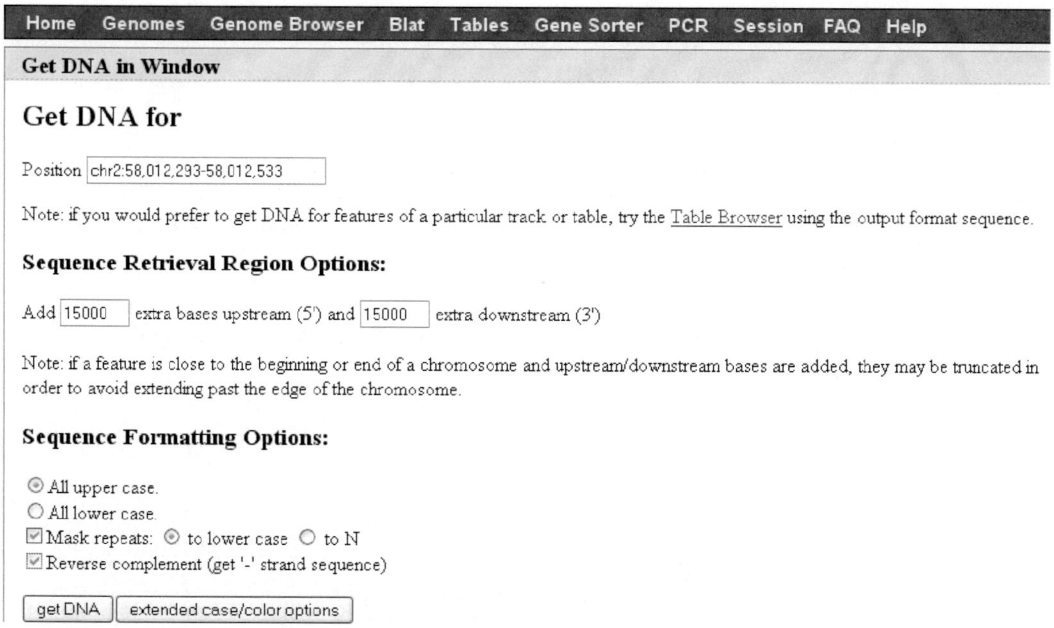

Fig. 8.15. Retrieval of 30 kb genomic sequence around exon 1 of the Pscdbp gene.

ATG-containing exon by doing a search in the word document with a small stretch of sequence from the beginning of the exon. After having located the exon you can see how many repeat sequences (in lower case) are present around the exon to be targeted.

Step 4a: Map restriction enzyme sites

To view cutting sites of major restriction enzymes in the genomic region use NEBcutter v2.0.

4a.1. Go to the New England BioLabs homepage (http://www.neb.com) and choose NEB CUTTER from the menu at the bottom of the page or go to NEBcutter v2.0 directly (http://tools.neb.com/NEBcutter2/index.php).

4a.2. Paste your 30 kb genomic sequence in the box, name your sequence in the box provided (optional), check "All commercially available specificities", and click "Submit" (*see* **Fig. 8.16**).

4a.3. The linear sequence for Pscdbp will appear (*see* **Fig. 8.17**).

Under "Main Options" select "Custom digest". This displays all of the enzymes that will cut your sequence. Choose rare cutter enzymes to simplify the map. Suggested enzymes include *Bam*HI, *Bgl*I, *Bgl*II, *Eco*RI, *Eco*RV, *Hinc*II, *Hin*dIII, *Not*I, *Nco*I, *Pst*I, *Sca*I, *Spe*I, *Stu*I, *Xba*I. Scroll down and check these boxes. Note: Enzymes that do not cut are

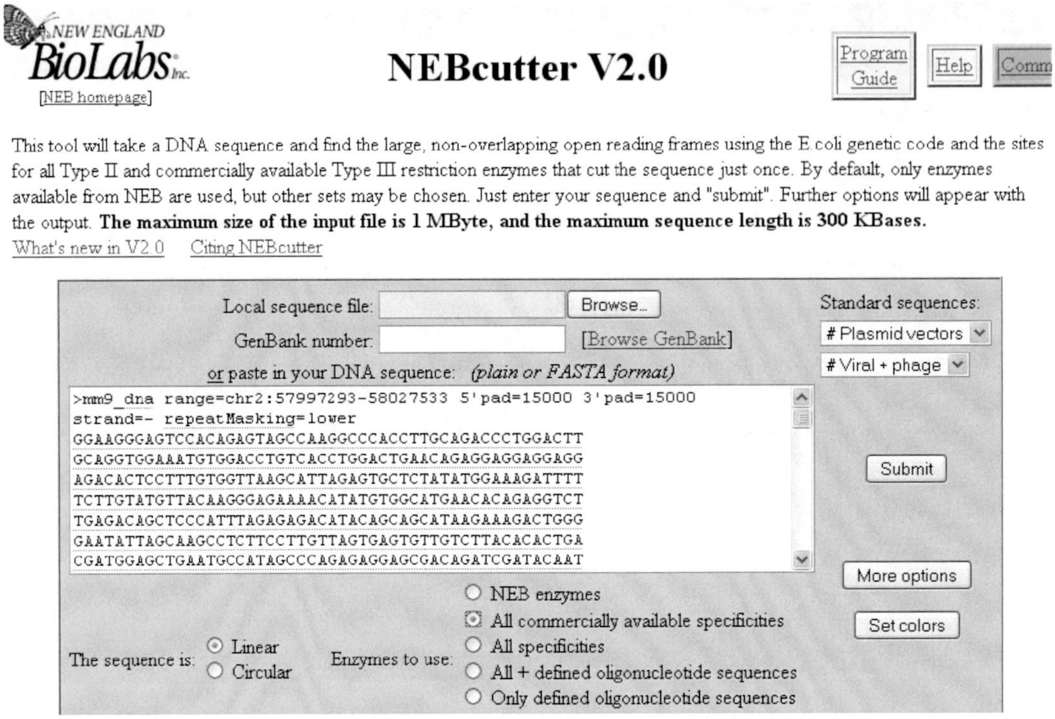

Fig. 8.16. Submission of the 30 kb Pscdbp genomic sequence to the NEBcutter webpage for restriction analysis.

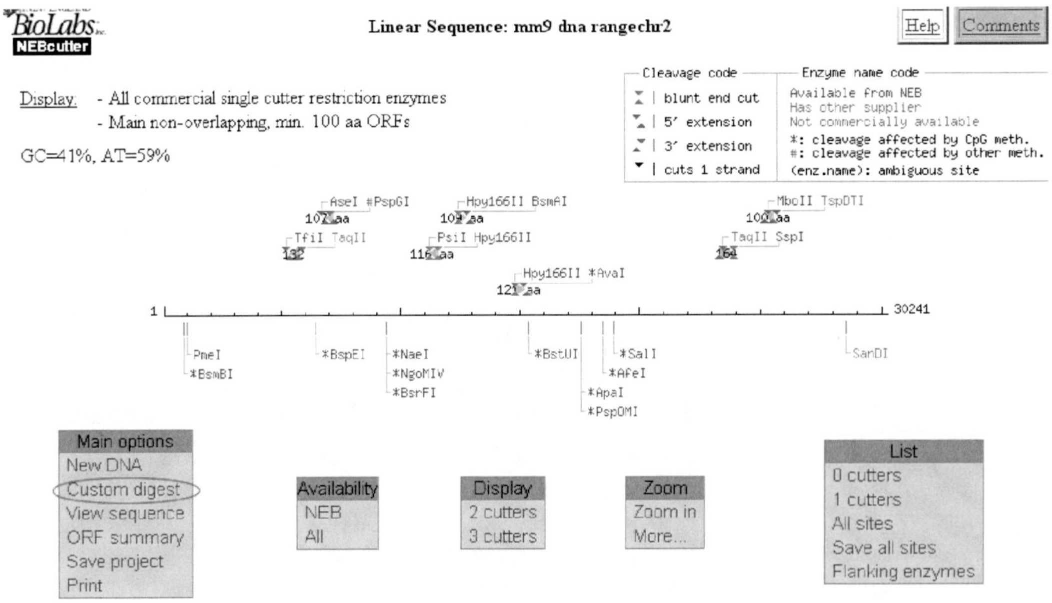

Fig. 8.17. Selection of enzymes for the restriction map of the Pscdbp genomic sequence.

not listed (i.e., *Eco*RV and *Not*I because they do not cut within the 30 kb that is being analyzed). Choose "DIGEST" from the menu at the very bottom of the page.

4a.4. Under "DISPLAY" choose "Alternative" to create a map with one restriction pattern per line. This helps to visually analyze the map and quickly identify restriction enzymes that can be used to generate a screening strategy.

To Print: Under "Main options" choose "Print". Under "Display mode" check "Map only" and then select "Create Image". To see the map click on "Click here to view/download the PDF file" (*see* **Fig. 8.18**). Print the PDF file.

The least frequent cutters are displayed at the bottom. They are most useful in designing a screening strategy. Since the restriction enzyme analysis has been done with 15 kb upstream and 15 kb downstream of the targeted exon, this exon is in the middle of the map. Thus, for example one can envision using *Bgl*I to detect a size increase due to the addition of the selectable marker flanking the exon and using a probe at the 5′ end of the vector, assuming *Bgl*I does not cut within the selectable marker cassette.

Step 4b: Compare sequence among species (mouse vs. human) The mouse and human genomic DNA sequence at the region of insertion of the selectable marker cassette should be aligned and compared to look for areas of sequence conservation. In fact, as mentioned previously the loxP and the neo cassette should be inserted in a non-conserved region to avoid disruption of regulatory elements. Disruption of these regions could cause the generation of a null allele instead of a conditional one. Since the loxP site and the selectable marker cassette (e.g., neo cassette) are placed within 2 kb of each other, 1 kb upstream and 1 kb downstream of the targeted exon should be retrieved from both the mouse and human database to analyze for sequence conservation. Compare the intronic regions directly upstream and downstream of the exon with an alignment or bestfit program.

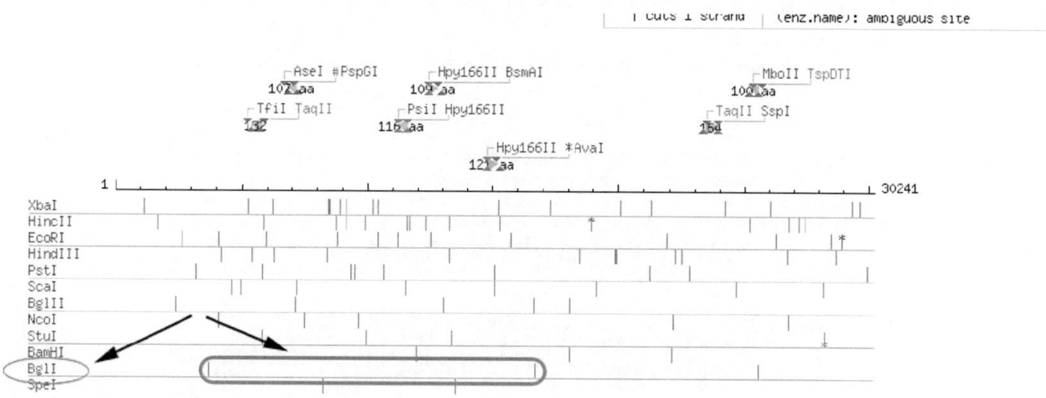

Fig. 8.18. Identification of *Bgl*I as enzyme to screen for targeted ES cell clones by Southern blotting.

4b.1. To retrieve the mouse sequence: Go to UCSC Genome Browser (http://genome.ucsc.edu) and choose BLAT from the top menu.

4b.2. Paste just the ATG exon sequence (targeted exon) in the box and select "SUBMIT". At the "BLAT Search Results" page click "browser". At the top menu, click "DNA" (*see* **Fig. 8.19**).

4b.3. At the "Get DNA for" page, go to the "Sequence Retrieval Region Option" section and type in "1000" in the boxes for add extra bases upstream and downstream. Check the "Mask repeats" and "to lower case" boxes. Check the "Reverse complement" box since the gene is on (–) strand. Click "get DNA". Copy, Paste, and Save this sequence as a TEXT file.

4b.4. Retrieve human Pscdbp sequence by following Step 1. Go to NCBI GenBank (http:/www.ncbi.nlm.nih.gov) and at "search" select "gene", at "for" type in "Pscdbp" and click "Go". Choose a human gene from the list, and obtain the FASTA format human sequence.

4b.5. Obtain the structure of human Pscdbp by following Step 2. Briefly, go to the UCSC Database (http://genome.ucsc.edu). Click BLAT and paste the human Pscdbp sequence into the box. Under "Genomes" choose "Human" and click "SUBMIT". Pick the sequence with the highest score. Click "details" to see the results. Scroll down to "Genomic chr2" and highlight and confirm the ATG exon (in blue upper-case letters). Save the sequence in a TEXT file.

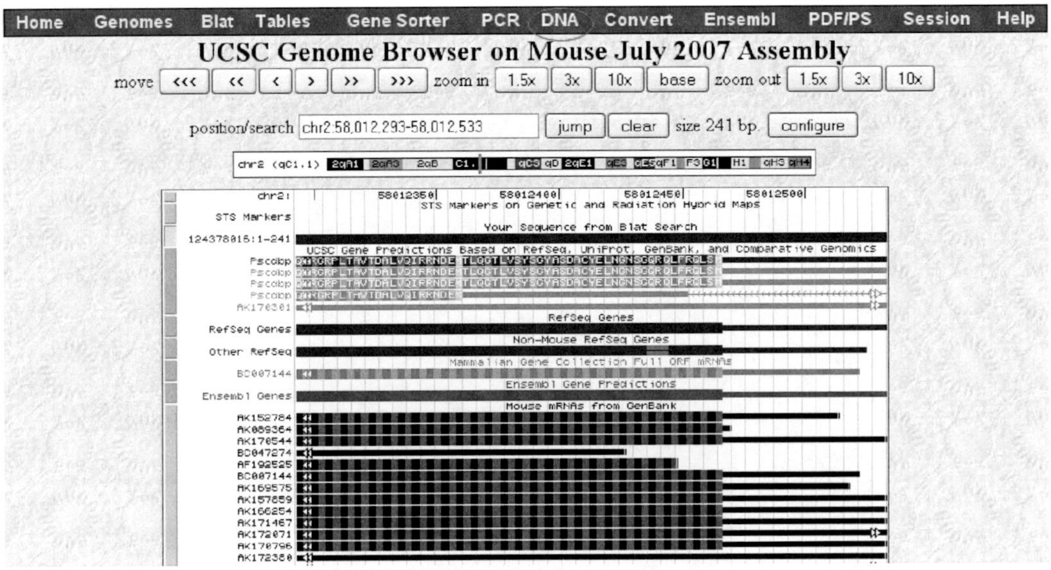

Fig. 8.19. Retrieval of 2 kb of genomic sequence around exon 1 of the mouse and human Pscdbp genes.

4b.6. Retrieve human Pscdbp's first exon's sequence with +/– 1000 flanking sequence by following Steps 4b.2 and 4b.3. Briefly, go to the UCSC Database and use BLAT to align the human ATG exon sequence against human genome. Click "browser" in the result. At the top menu, click "DNA". Go to the "Sequence Retrieval Region Option" section and type in "1000" in the boxes and add extra bases upstream and downstream. Check the "Mask repeats" and "to lower case" boxes. Check the "Reverse complement" box since the gene is on "–" strand. Select "get DNA". Copy, Paste, and Save this sequence as a TEXT file

4b.7. Go to the VISTA website (http://genome.lbl.gov/vista/index.shtml) *(37)* and choose the mVISTA alignment for mouse/human sequence comparison. The program accepts FASTA-formatted sequence files. FASTA format starts with a ">" followed immediately by a name and the next line begins the sequence. Without the ">" sign and a name in the file, VISTA will give you an error message. The sequence must be a TEXT file (*see* **Fig. 8.20**).

4b.8. At the mVista screen, type "2" in the number of species box to compare two species and press submit. Fill in your e-mail address. Select each of the text files for your

```
>human_Pscdbp hg18_dna range=chr2:158007605-158009850
AAAGATTTGCTAACTCTTTCAAATGCTAGTATTGATCTTCAGAGATTTTA
AAAAGCACATTTGACTTTTTGTGCCATATAAAACTTCTCTTTTTGACTTT
TGTTTCGATTCTTCCTTTTACTTGCCTCCCCTACTCCTCCACTGAAACCA
GTTCTCTCAAAACTACCTGCCCCCAACCTCAGCCAAAGGAAAAAAAAAAA
AAGCATCTGTGATCTTGAGTACTTGGTGATGGTGCCCTTGACCCATATGC
ACATGCAGGGACCAACAGGCAATGTTCTTGGTGCCACCTTGGTTTGCCTG
TTACCTTCCTGCCCACTCCTTTCCTCCCCTGGGACCACTTGCTGATGCTG
CCGCAAAGTCTCGTCTAAGAAATGGCAACATTGTGCCCTAAATGTGCTGG
CTTTCCCTTCCCATTGTACTTTGTAAAGATCATGGGTAATTTTTGTAAGG
ACAAAGTAGTAAAAGGACAACCAGAAGAATGCTGAATGGTTACAACCTCA
GTGGCAGATATAATGTCTTTTTGATGAACATTAATGATCTAGGGTATACT
TCAGGGAAAGTACAGTATTACAAGGACTTCGAGAGAGAACATTATGACCG
GCACCTGTAATTCTTGCCCATATTCACAAGAACTGAATGTCCTCGTGTGC
TGACTTCTGCATTGCATTGCTAATGTTAAGTAAGAGAATGGATGGGAAAC
AAATGGTAGCACTTGGGGACATTTTTCTTTCTTCTCTAAAGAAAAAAATG
GCAAATGGTGGTGTGAATGCAACTTGCCACAGAACATAAAAGCGTAAGAA
TTCTGGCTGTTTGAGGTGAAAAAAATAAGTTGATTTTTCTTTAAATTGTAA
AAATTAGCTCCAGGTTCTCTCAGGAGCTTAAAGAAAAAAAGCTTTGAGAA
ATGGGAGTGAATAGCAAGATAGGGTTTGTGTAACAAGTTCCTCAAACCAC
AGAGGTCACATGGGCTCTTTCTGCTTTGCTACTTTTGATTACTTGTCACA
GTTGTACTTTTAGCTTCCCCCATCCTGCAAGGCCACTCAACCATGTGCTA
GCTGGAGTGATCTTTATTCACAATGTCTTTACAAAGGCTCCTGCAACACA
GCAGCAATGGCAATTTGGCGGACTTCTGCGCTGGGCCAGCGTATAGCTCT
TACTCCACACTCACCGGCAGCCTTACGATGGACGATAATAGAAGGATTCA
```

Fig. 8.20. Submission of the mouse and human Pscdbp sequences for alignment at the VISTA webpage.

```
>sequence1 mouse_Pscdbp:1-2241 (+)
>sequence2 human_Pscdbp:1-2213 (+)

000000001  GAACCCACTTCGAGTTCTAGTGCACGCATTCCATTCTAGCTCGGAGCTTTTCTAACTCTT  000000060
>>>>>>>>>   ||                                          | ||| |||||||||  <<<<<<<<<
000000001  AAA------------------------------------------GATTTGCTAACTCTT  000000018

000000061  GTAAACTCTAGTAGTGACCTTCCAAACGACTTTTACGTGCGTGTTTGACTTCT-GTGCTA  000000119
>>>>>>>>>    |||  |||||| ||| ||||  |  || ||| |   ||   |||||||| | |||| |  <<<<<<<<<
000000019  TCAAATGCTAGTATTGATCTTCAGA--GATTTTAAAAAGCACATTTGACTTTTTGTGCCA  000000076

000000120  TTTAAAACTTCTCCTTTTTATTTTTGCTTTGATTCTGCCTTTTACGTACTTTCC-TGCTT  000000178
>>>>>>>>>  | ||||||||||| |||| | ||||| || |||||| |||||||| | | | || | ||   <<<<<<<<<
000000077  TATAAAACTTCTCTTTTTGACTTTTGTTTCGATTCTTCCTTTTACTTGCCTCCCCTACTC  000000136

000000179  CTGGG-TGAAACCCGTTATCTTCAA-CTACTTGAAAAATACAAAATTTA--------AAA  000000228
>>>>>>>>>  ||    ||||||| ||| |||  || |||| ||       | ||  | |        |||  <<<<<<<<<
000000137  CTCCACTGAAACCAGTTCTCTCAAAACTACCTGC----CCCCAACCTCAGCCAAAGGAAA  000000192

000000229  AATAAAAAAAGCAACTGT--TTTCAAGTCCCTCCCCCCACCGTGATGGCACACTG-----  000000281
>>>>>>>>>  || |||||||||| ||||  | |  ||| | |         |||||||  | ||        <<<<<<<<<
000000193  AAAAAAAAAAGCATCTGTGATCTTGAGTACTTG--------GTGATGGTGCCCTTGACCC  000000244

000000282  ------------------CCACAGGAACCATCCT--GTGTTATCTCAGCTTGACTACCCC  000000321
>>>>>>>>>                    | ||||| |   | ||  |||  | ||  | |||  ||    |  <<<<<<<<<
000000245  ATATGCACATGCAGGGACCAACAGGCAATGTTCTTGGTGCCACCTTGGTTTGCCTGTTAC  000000304

000000322  CTTCTTTCCCACTGCTTTCTTCCCCC-----------CTAAAGTTGTAGGTATGAGTAGC  000000370
>>>>>>>>>  |||| | |||||| ||||| |||||            || | | ||  |  | |  | |   <<<<<<<<<
000000305  CTTCCTGCCCACTCCTTTCCTCCCCTGGGACCACTTGCTGATGCTGCCGCAAAGTCTCGT  000000364

000000371  CCCAGATACGAACAATGTTCTGCACTAAATACGCTGCCTTTCCTTCCCCTCCCCGTTATG  000000430
```

Fig. 8.21. Alignment results of mouse and human Pscdbp at the VISTA webpage.

mouse and human sequences as shown and press submit. Your results will be e-mailed to you and will look as shown below (in the box is the end sequence of the exon).

Sequence comparison can be done optimally using VISTA because it is a program that allows you to compare sequences between different species and identify genes with a common ancestor. The sequence comparison resulting from the NCBI Blast does not lead to such a detailed comparison and only highlights a few conserved areas because it is not optimized for sequence comparison between species (*see* **Fig. 8.21**).

Acknowledgments

We thank Eileen Southon for critical reading of the manuscript. "This research was supported by the Intramural Research Program of the NIH, National Cancer Institute."

References

1. Robertson EJ. Using embryonic stem cells to introduce mutations into the mouse germ line. Biol Reprod 1991;44:238–45.
2. Doetschman TC. Gene targeting in embryonic stem cells. Biotechnology 1991;16:89–101.
3. Bradley A. Modifying the mammalian genome by gene targeting. Curr Opin Biotechnol 1991;2:823–9.
4. Capecchi MR. The new mouse genetics: altering the genome by gene targeting. Trends Genet 1989;5:70–6.
5. Godwin AR, Stadler HS, Nakamura K, Capecchi MR. Detection of targeted GFP-Hox gene fusions during mouse embryogenesis. Proc Natl Acad Sci USA 1998;95:13042–7.
6. Gagneten S, Le Y, Miller J, Sauer B. Brief expression of a GFP cre fusion gene in embryonic stem cells allows rapid retrieval of site-specific genomic deletions. Nucleic Acids Res 1997;25:3326–31.
7. Mansour SL, Thomas KR, Deng CX, Capecchi MR. Introduction of a lacZ reporter gene into the mouse int-2 locus by homologous recombination. Proc Natl Acad Sci USA 1990;87:7688–92.
8. Sauer B. Inducible gene targeting in mice using the Cre/lox system. Methods 1998;14:381–92.
9. Dymecki SM. A modular set of Flp, FRT and lacZ fusion vectors for manipulating genes by site-specific recombination. Gene 1996;171:197–201.
10. Dymecki SM. Flp recombinase promotes site-specific DNA recombination in embryonic stem cells and transgenic mice. Proc Natl Acad Sci USA 1996;93:6191–6.
11. Kuhn R, Schwenk F, Aguet M, Rajewsky K. Inducible gene targeting in mice. Science 1995;269:1427–9.
12. Liu P, Jenkins NA, Copeland NG. A highly efficient recombineering-based method for generating conditional knockout mutations. Genome Res 2003;13:476–84.
13. Lee EC, Yu D, Martinez de Velasco J, et al. A highly efficient *Escherichia coli*-based chromosome engineering system adapted for recombinogenic targeting and subcloning of BAC DNA. Genomics 2001;73:56–65.
14. Copeland NG, Jenkins NA, Court DL. Recombineering: a powerful new tool for mouse functional genomics. Nat Rev Genet 2001;2:769–79.
15. Muyrers JP, Zhang Y, Testa G, Stewart AF. Rapid modification of bacterial artificial chromosomes by ET-recombination. Nucleic Acids Res 1999;27:1555–7.
16. Zhang Y, Buchholz F, Muyrers JP, Stewart AF. A new logic for DNA engineering using recombination in *Escherichia coli*. Nat Genet 1998;20:123–8.
17. Mural RJ, Adams MD, Myers EW, et al. A comparison of whole-genome shotgun-derived mouse chromosome 16 and the human genome. Science 2002;296:1661–71.
18. Venter JC, Adams MD, Myers EW, et al. The sequence of the human genome. Science 2001;291:1304–51.
19. Lander ES, Linton LM, Birren B, et al. Initial sequencing and analysis of the human genome. Nature 2001;409:860–921.
20. Maglott D, Ostell J, Pruitt KD, Tatusova T. Entrez gene: gene-centered information at NCBI. Nucleic Acids Res 2007;35:D26–31.
21. Karolchik D, Baertsch R, Diekhans M, et al. The UCSC Genome Browser database. Nucleic Acids Res 2003;31:51–4.
22. Deng C, Capecchi MR. Reexamination of gene targeting frequency as a function of the extent of homology between the targeting vector and the target locus. Mol Cell Biol 1992;12:3365–71.
23. Hasty P, Rivera-Perez J, Chang C, Bradley A. Target frequency and integration pattern for insertion and replacement vectors in embryonic stem cells. Mol Cell Biol 1991;11:4509–17.
24. Capecchi MR. Gene targeting in mice: functional analysis of the mammalian genome for the twenty-first century. Nat Rev Genet 2005;6:507–12.
25. Olson EN, Arnold HH, Rigby PW, Wold BJ. Know your neighbors: three phenotypes in null mutants of the myogenic bHLH gene MRF4. Cell 1996;85:1–4.
26. Hasty P, Rivera-Perez J, Bradley A. The length of homology required for gene targeting in embryonic stem cells. Mol Cell Biol 1991;11:5586–91.
27. Simpson EM, Linder CC, Sargent EE, Davisson MT, Mobraaten LE, Sharp JJ. Genetic variation among 129 substrains and its importance for targeted mutagenesis in mice. Nat Genet 1997;16:19–27.
28. Evans MJ, Kaufman MH. Establishment in culture of pluripotential cells from mouse embryos. Nature 1981;292:154–6.

29. You Y, Bersgtram R, Klemm M, Nelson H, Jaenisch R, Schimenti J. Utility of C57BL/6 J × 129/SvJae embryonic stem cells for generating chromosomal deletions: tolerance to gamma radiation and microsatellite polymorphism. Mamm. Genome 1998;9:232–4.
30. Adams DJ, Quail MA, Cox T, et al. A genome-wide, end-sequenced 129 Sv BAC library resource for targeting vector construction. Genomics 2005;86:753–8.
31. Soriano P, Montgomery C, Geske R, Bradley A. Targeted disruption of the c-src proto-oncogene leads to osteopetrosis in mice. Cell 1991;64:693–702.
32. Capecchi MR. Altering the genome by homologous recombination. Science 1989;244:1288–92.
33. Yagi T, Ikawa Y, Yoshida K, et al. Homologous recombination at c-fyn locus of mouse embryonic stem cells with use of diphtheria toxin A-fragment gene in negative selection. Proc Natl Acad Sci USA 1990;87:9918–22.
34. Martens JH, O'Sullivan RJ, Braunschweig U, et al. The profile of repeat-associated histone lysine methylation states in the mouse epigenome. Embo J 2005;24:800–12.
35. Kent WJ. BLAT – the BLAST-like alignment tool. Genome Res 2002;12:656–64.
36. RepeatMasker. (Accessed at http://www.repeatmasker.org/)
37. Frazer KA, Pachter L, Poliakov A, Rubin EM, Dubchak I. VISTA: computational tools for comparative genomics. Nucleic Acids Res 2004;32:W273–9.

Chapter 9

Manipulating Mouse Embryonic Stem Cells

Eileen Southon and Lino Tessarollo

Abstract

Murine embryonic stem (ES) cells are derived from the inner cell mass of 3.5–day-old embryo and have the ability to colonize the germline and form normal gametes following in vitro genetic manipulations. This remarkable characteristic of ES cells has provided the basis for studying normal gene function in the mouse by targeted mutagenesis. Nevertheless, ES cells are very sensitive and need to be manipulated with care for them to retain totipotency after extensive in vitro manipulations. Here we provide straightforward protocols for proper care of these cells. Special emphasis is placed on aspects that are particularly critical for proper culture of this cell type.

Key words: Embryonic stem cells, totipotency, mouse embryo fibroblasts, in vitro culture.

1. Introduction

Since it was first established, targeted mutagenesis technology has evolved significantly and has been used by a number of laboratories to systematically alter the mouse genome to generate mice with specific targeted gene mutations *(1, 2)*. The improvement in the quality of reagents (ES cells and media) and the sophisticated instrumentation made available to the scientific community has in part been responsible for making this technology of widespread use. However, its successful application relies heavily on the proper techniques to manipulate murine ES cells, the vehicle to introduce a mutation into the mouse genome.

Embryonic stem (ES) cells are derived from the inner cell mass (ICM) of a 3.5-day post coitus (dpc) embryo at the blastula stage of development *(2–4)*. ES cells can be cultured, and still retain their ability to contribute to all cell lineages when reintroduced

Ralf Kühn, Wolfgang Wurst (eds.), *Gene Knockout Protocols: Second Edition, vol. 530*

DOI 10.1007/978-1-59745-471-1_9 Springerprotocols.com

into a host blastocyst *(3, 5–7)*. For historical and practical reasons gene targeting has been performed mainly in ES cells from the 129 mouse strains *(8)*. Strain 129 ES cell lines are easy to derive and can be manipulated in culture while remaining competent to repopulate the mouse germline. However, because 129 strains do not breed well, and are reported to have abnormal anatomy and behavior, it is a widely accepted practice to backcross newly generated mutant mice into another genetic background such as C57BL/6. C57BL/6 mice breed well, are long-lived, and are permissive to expression of most mutations. A few reports have indicated that C57BL/6 ES cell lines can also be used in gene-targeting experiments for the production of mice in a pure C57BL/6 genetic background. However, these cells are not as efficient as 129 cells *(9, 10)*. Hybrid ES cell lines derived, for example, from F1 intercrosses of 129 × C57BL/6 or 129 × BALB/c mice retain the 129 characteristics of easy derivation and ability to be manipulated in culture while remaining competent to repopulate the mouse germline *(11)*. Moreover, they permit one to obtain a specific genetic mutation in a congenic strain of interest. For example, if a mutation is needed in the C57BL/6 background, a hybrid 129 × C57BL/6 ES cell line can still be used for gene targeting with a C57BL/6 DNA vector. This will favor the targeting of the C57BL/6 allele making it easier to obtain a pure congenic line with all genes surrounding the targeted area coming from this strain (*see* **Note 1**). This was not possible when using ES cells from a pure 129 strain since even with repeated backcrosses into the C57BL/6 background, the null-mutation will still be flanked by two 129-derived alleles, while the wild-type locus is flanked by two C57BL/6 alleles *(12, 13)*.

Although 129 ES cells are more resilient than ES cells from other genetic backgrounds, they are nevertheless still sensitive to in vitro manipulations and special care must be taken to preserve their totipotency. For example, good basic tissue culture rules should be followed: ES cells should be passaged as little as possible since prolonged culture periods will affect their ability to contribute to the mouse germline; plated ES cells should never reach confluency or be exposed to exhausted medium for prolonged periods of time; the feeder layer (when required) on which the ES cells are plated should always be healthy and relatively fresh. In most instances, ES cells will show the adverse effects of stress due to poor culture conditions only at the time their totipotency is evaluated by their ability to generate "germline chimeras" (several months or a year after the beginning of the experiments). Here, I will disclose some of the basic principles to establish ES cell technology in a laboratory. These protocols are relatively simple and will be very effective if followed carefully. It is, however, very important to follow additional specifications given by the source of the specific ES cell line in use.

2. Materials

2.1. Equipment

The following equipment can be found in almost every tissue culture area. However, special care should be taken to assure that every piece is decontaminated before use. In particular, the culturing of mycoplasma-positive cell lines should be avoided in areas where ES cells are cultured. We found that while mycoplasmas do not necessarily affect the growth rate of ES cells they can affect their ability to contribute to the mouse germline.

1. Tissue culture incubators: Set at 5% CO_2, 20% O_2 (normal atmospheric concentration), with a saturated aqueous atmosphere. It is extremely important to ensure the proper function of this piece of equipment. In particular, special care should be taken in monitoring precise CO_2 levels every week and ensure that the recovery time after openings is not too extended. If possible, reserve one incubator for ES cell work exclusively, keeping track of its usage and users. Primary mouse embryonic fibroblasts (MEFs) grow particularly well in a low-oxygen condition because it delays senescence. Thus, we use an incubator capable of maintaining a low percentage (3%) of O_2; 5% CO_2 (e.g., Sanyo Scientific #MCO-18M).
2. Inverted Microscope with 5 × and 10 × objectives.
3. Electroporator: Capable of reaching the capacitance of 500 μFarads (e.g., Biorad Gene Pulser)
4. Tissue culture hood with UV light: Before use, make sure that it has been properly decontaminated and the air-flow is working properly. If possible keep it separate from other users. Hood should be wiped down with 70% ethanol before and after use. Also, turn on the UV light when not in use.
5. Table-top centrifuge.
6. Water bath: Set at 37°C for regular use but can also be used at 56°C to heat inactivate serum.

Once a tissue culture area is set up for ES cell use, it is extremely important to periodically check that the instruments are functioning properly.

2.2. Tissue Culture Reagents and Solutions

In the past many of the following reagents were prepared from powder. Now they are purchased pre-made or as stock solutions. This has greatly reduced the risk of variability in quality due to in-house preparation.

2.2.1. Reagents for ES Cell Growth Media

1. Dulbecco's Modified Eagle Medium (DMEM) with high glucose (4500 mg/L; e.g., Invitrogen) without sodium pyruvate (however, some ES cell lines may require sodium pyruvate). We are currently using one ES cell line that grows best in an ES cell-optimized Knockout^{TM} DMEM (e.g., Invitrogen).

2. Fetal calf serum (FCS): FCS can be purchased from several companies that provide ES cell-tested batches. However, an in-house test should be performed for every new lot of FCS purchased to assure that it is optimal for the ES cell line in use in the laboratory (*see* **Note 2**). The two parameters to be tested are (a) plating efficiency and (b) toxicity. Briefly, 10^3 cells are plated onto a 6-cm plate with feeder cells. Media is prepared with 10% or 30% of the new batch of heat-inactivated (30 min at 56°C) FCS. Plate cells in triplicate for every batch of FCS to be tested. Colonies are counted after 5–7 days. The number of colonies is compared between different lots. The presence of more colonies in the 10% FCS media compared to the 30% is an indicator of serum toxicity. These lots should be avoided. Serum can be stored between –20°C and –80°C for up to 3 years.
3. 100X Penicillin-Streptomycin (Pen/Strep): 50 units/mL (final concentration) each of penicillin and streptomycin. 100X solutions are commercially available (e.g., Invitrogen etc.).
4. 100X L-Glutamine: 200 mM (e.g., Invitrogen or GlutaMAX™ (Invitrogen)).
5. 100X mercaptoethanol: 7 µL β-mercaptoethanol in 10 mL of PBS. Store at 4°C and replace weekly.
6. 100X MEM Non-Essential Amino Acids (e.g., Invitrogen).
7. Leukemia inhibitory factor (LIF): Available from Millipore as ESGRO Murine Leukemia Inhibitory Factor 10^7 Units/mL.

   The above components should be added to fresh DMEM to make two types of media.
8. MEF medium: To make 100 mL, combine 83 mL DMEM, 15 mL FBS, 1 mL 100X Pen/Strep, and 1 mL 100X L-Glutamine.
9. ES cell medium: To make 100 mL, combine 81 mL DMEM, 15 mL FBS, 1 mL 100X Pen/Strep, 1 mL 100X L-Glutamine, 1 mL 100X sol β-mercaptoethanol, 1 mL 100X MEM Non-Essential Amino Acids, and 10 µL LIF. ES cell media should be pre-warmed only before use, and it should be kept protected from light. This media should be stored for not more than 2–3 weeks if used with ES cells. Older media can be stored and used for expanding cloned ES cells used for DNA analysis.

2.2.2. Miscellaneous

1. Freezing medium: 60% FCS, 20% D-MEM, 20% DMSO.
2. Phosphate buffered saline (PBS): For example, Invitrogen cat # 14040-133.
3. PBS without Ca^{2+} and Mg^{2+} (PBS w/o): For example, Invitrogen cat # 14190-044.
4. 0.1 % Gelatin. Dissolve 1 g of gelatin (Porcine skin gelatin, cell culture tested; Sigma cat # G 1890) in 1 L of distilled water, aliquot and autoclave; store at 4°C.

5. Dimethyl sulfoxide (DMSO): For example, Sigma cat # D 2650.
6. Trypsin-EDTA: For example, Invitrogen cat # 25300-054.
7. DNase I solution: 100 mg (2000 units/mg) in 7.0 mL 0.154 M NaCl, 50% glycerol (e.g., Roche cat # 104159).
8. Collagenase H solution: 100 mg (>0.5 units/mg) in 7 mL PBS, 50% glycerol (e.g., Roche cat # 1074032).
9. Hyaluronidase: 100 mg (~800 units/mg) in 7 mL PBS, 50% glycerol. (e.g., Sigma cat # H-3884).
10. MEF digestion media: 50 mL DMEM supplemented with 500 μL DNase I, 500 μL collagenase H, and 500 μL hyaluronidase solutions.
11. Mitomycin C (MC). 100X stock: resuspend 2 mg of MC in 2 mL PBS (e.g., Roche cat # 107-409). Filter through a 0.22-μm syringe filter. Aliquot and store at –20°C for up to a month.
12. Geneticin (G418): For example, Invitrogen cat # 10131-035.
13. Gancyclovir : Use Cytovene (Syntex) at 2 μM or FIAU (Oclassen Pharmaceuticals) at 0.5 μM. Different concentrations of Gancyclovir or FIAU should be tested considering the sensitivity of ES cells to these compounds.

2.3. Plasticware and Other Materials

It is recommended that disposable plastic material is used for all tissue culture work.

1. Culture plates:
   a. 10-cm polystyrene dishes (use 10 mL media)
   b. 6-cm polystyrene dishes (use 4 mL media)
   c. 12-well cluster plates tissue culture quality (use 2–2.5 mL media/well)
   d. 24-well cluster plates tissue culture quality (use 1.5–2 mL media/well)
   e. 96-well cluster plates, round bottom, tissue culture quality.
2. 50-mL and 15-mL tubes.
3. 1-mL freezing vials.
4. 1-mL eppendorf tubes.
5. 25 mL, 10 mL, 5 mL, 1 mL plastic pipettes.
6. 200 μL aerosol barrier tips.
7. 12-well multi-channel pipettor.
8. Repeat pipettman.
9. 0.4-cm electrode gap electroporation cuvettes (e.g., Bio-Rad Gene Pulser cuvettes cat # 165-2088).
10. Equipment for dissection: razor blades, scissors, micro dissecting scissors, straight and curved forceps, tweezers.

2.4. Molecular Biology Reagents

For additional information about molecular biology procedures and reagents *see* (*14*).

1. Lysis buffer: 100 mM Tris-HCl pH 8.5, 5 mM EDTA, 0.2% SDS, 200 mM NaCl, 0.2 mg/mL proteinase K. Proteinase K is prepared as a stock solution of 10 mg/mL in water and stored at –20°C. *Note*: Proteinase K should be added fresh to the other components of the lysis buffer just before lysing the cells.
2. 0.1 M spermidine trihydrochloride: For example, Sigma cat # S2501
3. 10X TPE: 1 M Tris-phosphate, 20 mM EDTA. For 1 L use 108 g Tris-base, 15.5 mL 85% phosphoric acid (density 1.679 g/mL), 40 mL 0.5 M EDTA, pH 8.0 and adjust the volume to 1 L.
4. Alkali solution: 1.5 M NaCl, 0.5 M NaCl.
5. 20X SSC: For 1 L dissolve 175.3 g NaCl and 88.2 g sodium citrate in 800 mL water, adjust to pH 7.0 with a few drops of 10 M NaOH and adjust the volume to 1 L.
6. 20X SSCP: For 1 L dissolve 140.2 g NaCl, 88.2 g sodium citrate, 43.7 g Na_2HPO_4, 12.7 g $NaH_2PO_4$and adjust the volume to 1 L.
7. Neutralization solution: For 1 L dissolve 121.1 g Tris Base, 174 g NaCl, and 66.6 mL HCl and adjust the volume to 1 L.
8. Prehybridization solution: 4X SSCP; 1X Denhardt's solution (e.g., Invitrogen cat # 750018; 1% SDS; 80 mg/L boiled sonicated salmon sperm DNA (e.g., Invitrogen cat # 15632-011). Store at 4°C.
9. Hybridization solution: 4X SSCP, 1X Denhardt's solution (*14*), 1% SDS, 100 mg/L boiled sonicated salmon sperm DNA, 10% dextran sulfate. Store at 4°C
10. Loading buffer (6X): 0.25% bromophenol blue; 0.25% xylene cynol; 15% Ficoll type 400.
11. Hybond N(+): Amersham (cat # RPN 203 B)

3. Methods

3.1. Preparation of Feeder Cells

ES cells are usually grown on mitotically inactive primary MEFs or STO fibroblasts (*3*). These mitotically inactive feeder cells provide support to the ES cells and produce growth factors, including LIF, that allow the ES cells to maintain their totipotency. The importance of using good-quality feeders cannot be overstated. Poor-quality feeders cause ES cells to suffer and consequently compromise their totipotency potential (see below).

3.1.1. Isolation of MEFs

MEFs are isolated from 12–14 dpc embryos derived from a mouse strain transgenic for a gene that confers resistance to the antibiotic (e.g., neomycin) used for the selection of ES cell clones subsequent to the electroporation process. The production of good-quality feeder cells, in our experience, is a critical step in obtaining genetically manipulated ES cells that retain their ability to colonize the mouse germline. It is important to note that there is a certain degree of variability between batches of feeder cells produced. However, the growth rate of the primary fibroblasts, from which feeder cells will be derived, can be used as a quality indicator; that is, fast-growing MEFs will generally produce good feeders. As mentioned before (**Section 2.1**) primary MEFs grow particularly well when cultured in low O_2 conditions (3%). We have found that MEFs grown in these conditions generate feeders that are of superior and more consistent quality (*see* **Note 3**). However, most laboratories (including ours until recently) grow MEFs in the standard 5% CO_2, 20% O_2.

1. Sacrifice a 12–14-day pregnant mouse by CO_2 asphyxiation.
2. Aseptically dissect out the uterine horns. With large scissors and forceps cut the skin and then with sterile smaller scissors and tweezers cut the peritoneum. Remove the uterine horns by holding the cervix with sterile tweezers and cutting the connective vaginal side.
3. Rinse uterine horns twice in PBS.
4. Place uterine horns into a 10-cm tissue culture dish with PBS (in tissue culture hood) and dissect out embryos.
5. Place embryos in a fresh dish with 10 mL PBS. Decapitate and eviscerate the embryos.
6. Place carcasses into a clean 10-cm dish and mince them with a sterile razor blade until a gelatinous mass is created.
7. Add 5 mL of DMEM and transfer the mix into a 200-mL flask containing a stir bar.
8. Add 50 mL of MEF digestion media.
9. Stir at ≤37°C on a warm plate for 30–40 min.
10. Pipette off supernatant and save in a 50-mL tube.
11. Add 50 mL of fresh MEF digestion media to remaining chunks in flask and stir again at ≤37°C for 30–40 min.
12. Transfer supernatant to a new 50-mL tube and set aside.

13. Add 20 mL of trypsin-EDTA to remaining particles and stir at 37°C for 15–20 min.
14. Neutralize trypsin with ~30 mL of MEF medium and transfer the supernatant to a 50-mL tube.
15. Spin all three tubes in a centrifuge at 300*g* for 5 min.
16. Pipette off supernatants.
17. Use 5 mL of MEF media per tube to resuspend pellets and pool the three fractions.
18. Bring volume to 10 mL of MEF media per dissected embryo and aliquot 10 mL per 10 cm tissue culture dishes (e.g., for ten embryos, resuspend in 100 mL of media and plate into ten plates).
19. Grow the cells until 80% confluency at which time you may freeze them down (four vials/plate) or trypsinize and replate for feeder production.

3.1.2. Preparation of MEF Feeder Cell Layer

1. Thaw one vial of primary MEFs onto a 10-cm plate with MEF media.
2. Grow cells to confluency and split 1:4. When these four plates are again confluent, split one more time 1:5 (total 20 plates).
3. When MEFs have almost reached confluency, pipette off the medium and add 5 mL of MEF media with MC (50 μL of 100X sol.) to each plate.
4. Incubate for 3–3.5 h at 37°C.
5. After incubation, remove media and wash 3X with PBS.
6. Add 1.5 mL trypsin-EDTA to each plate and incubate 2–3 min (after the three washes with PBS the cells start coming off the plate very quickly) at 37°C.
7. Stop the trypsin reaction by adding 1.5 mL MEF media. Pipette to dissociate the cells. Collect all cells by rinsing the plates with media.
8. Determine the total number of cells.
9. Spin the cells down and resuspend them in an adequate volume of MEF media for immediate plating or for freezing. MEFs can be frozen down at a density of about 7×10^6 per vial for later use. Mitotically inactive MEFs (feeder cells) are sensitive. Therefore, they should be seeded onto gelatin-coated plates to facilitate adhesion.
10. Prepare gelatin-coated plates by covering the dish surface with 0.1% gelatin. After incubating for at least 30 min. at room temperature (overnight is also good) remove the liquid. The plate is ready for seeding the feeder cells. Plates can also be dried for later use.

11. Plate MEFs at the following density onto gelatin-coated dishes in order to obtain a monolayer (**Fig. 9.1A**):

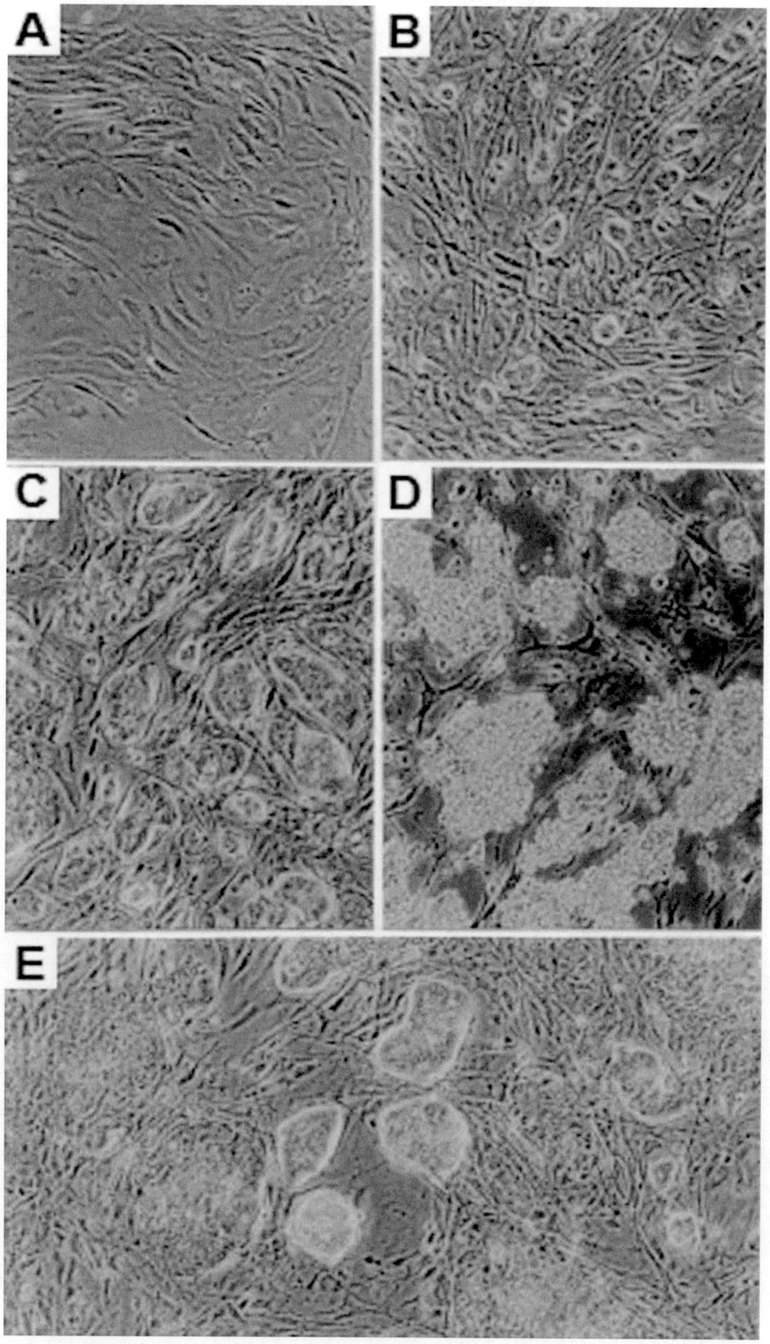

Fig. 9.1. Culturing ES cells. (**A**) Monolayer of feeder cells; (**B**) density at which ES cells should be plated after thawing; (**C**) semi-confluent ES cell culture; (**D**) incubation of ES cell cultures with trypsin for 3–5 min makes ES cell clusters look like grapes; (**E**) highly differentiated ES cell culture (note the high degree of differentiation toward the epithelial lineage). 100 × mag.

Plate size	MEFs
10 cm	3.5×10^6 cells per plate
6 cm	1×10^6 cells per plate
12-well plate	5×10^5 cells per well
24-well plate	2×10^5 cells per well

12. Change the medium every other day. Use the cells within 1 week.

3.2. Culturing ES Cells

ES cells are totipotent cells with a doubling time of approximately 12 h. They should be monitored twice a day to prevent overgrowth. However, they should be seeded at a certain density (about 10%) since they like to grow in clusters (**Fig. 9.1B**). To avoid temperature shock, ES cell media should always be warmed to 37°C before adding to the culture. In addition, since ES cells are grown on feeder cells, special care should be taken in assuring that enough feeder plates are available for splitting. When grown on feeders, ES cells may not require addition of LIF to the media. However, we found that it is a good practice to grow them with this additive to buffer possible feeder deficiency (*see* **Note 4**).

1. Thaw one vial of ES cells (ES cells should be frozen at a concentration of about 2×10^6 cells/vial) by warming quickly at 37°C. Transfer the cells to a tube with 5–6 mL of ES cell medium.
2. Spin down for 10 min at 1000 rpm, remove the supernatant and resuspend the cells into 10 mL fresh ES medium.
3. Seed onto a 10-cm feeder dish.
4. Change medium every day.
5. After about 3 days the cells should be ready for splitting (**Fig. 9.1C**). It is very important to split ES cells before they reach confluency. A confluent culture may result in a high degree of cell differentiation (**Fig. 9.1E**, *see* **Note 5**). It is also very important to culture ES cells only when needed since these cells tend to lose their totipotency after multiple tissue culture passages.
6. Feed the cells 2 h before splitting.
7. Remove the medium and rinse briefly with 5 mL of trypsin-EDTA (37°C).
8. Add 1 mL of fresh trypsin-EDTA and incubate 3–5 min at 37°C.
9. Check under the microscope to see if the colonies look like grapes (**Fig. 9.1D**).

10. Neutralize the trypsin by adding 1 mL MEF medium (during splitting it is not necessary to use ES medium) and dissociate the colonies into a single cell suspension by pipetting at least 15–20 times with a sterile plugged pasteur pipette (the tip of the pasteur pipette should be flame-polished to create a slightly smaller opening).
11. Collect the cells in 5–7 mL MEF medium.
12. Spin down and resuspend in ES cell medium for plating. Usually one subconfluent plate of ES cells is split 1:6 or frozen 1:5.
13. To freeze one plate of ES cells, resuspend the cells in 2.5 mL of DMEM and slowly add 2.5 mL of Freezing medium.
14. Mix gently, aliquot into five cryopreservation vials and store overnight at –80°C in a partially insulated box such that the cells freeze slowly. The next day transfer to liquid N_2.

3.3. Genetic Manipulation of ES Cells

3.3.1. Electroporation

Day 0

1. Split one plate of sub-confluent ES cells onto four dishes such that the next day the cells are ready to be split again (**Fig. 9.2A**) (*see* **Note 6**).

   Day 1
2. Change media in the morning since these cells are plated at higher than usual density (**Fig. 9.2B**).
3. In the afternoon, trypsinize plates as previously described and collect the cells.
4. Spin down cells for 10 min at 300 g and resuspend in 50 mL of room temperature PBS w/o. Count cells. There should be 5–8 $\times$ 10^7 cells.
5. Spin down and resuspend cells to a final concentration of 1.5–2 $\times$ 10^7 cells/0.8 mL PBS w/o (for example, if you have 7 $\times$ 10^7 cells total, resuspend the pellet in 3.2 mL PBS w/o).
6. Transfer 0.8 mL of cell suspension to a cuvette with 20 μL of DNA (1 μg/μL of linearized and purified DNA vector in water). Mix the cells with the DNA by pipetting with a 1-mL plastic pipette.
7. Quickly electroporate the cells. The electroporator is set at 250 V with a capacitance of 500 μFarads. (During the process many cells die making the cell suspension appear somewhat viscous. The viscosity is caused by the DNA released by the lysed cells) (*see* **Note 7**).
8 Transfer the cells from each electroporation cuvette to 30 mL of ES cell medium and plate onto three 10-cm dishes with feeders.
9. Return electroporated ES cells to the incubator.

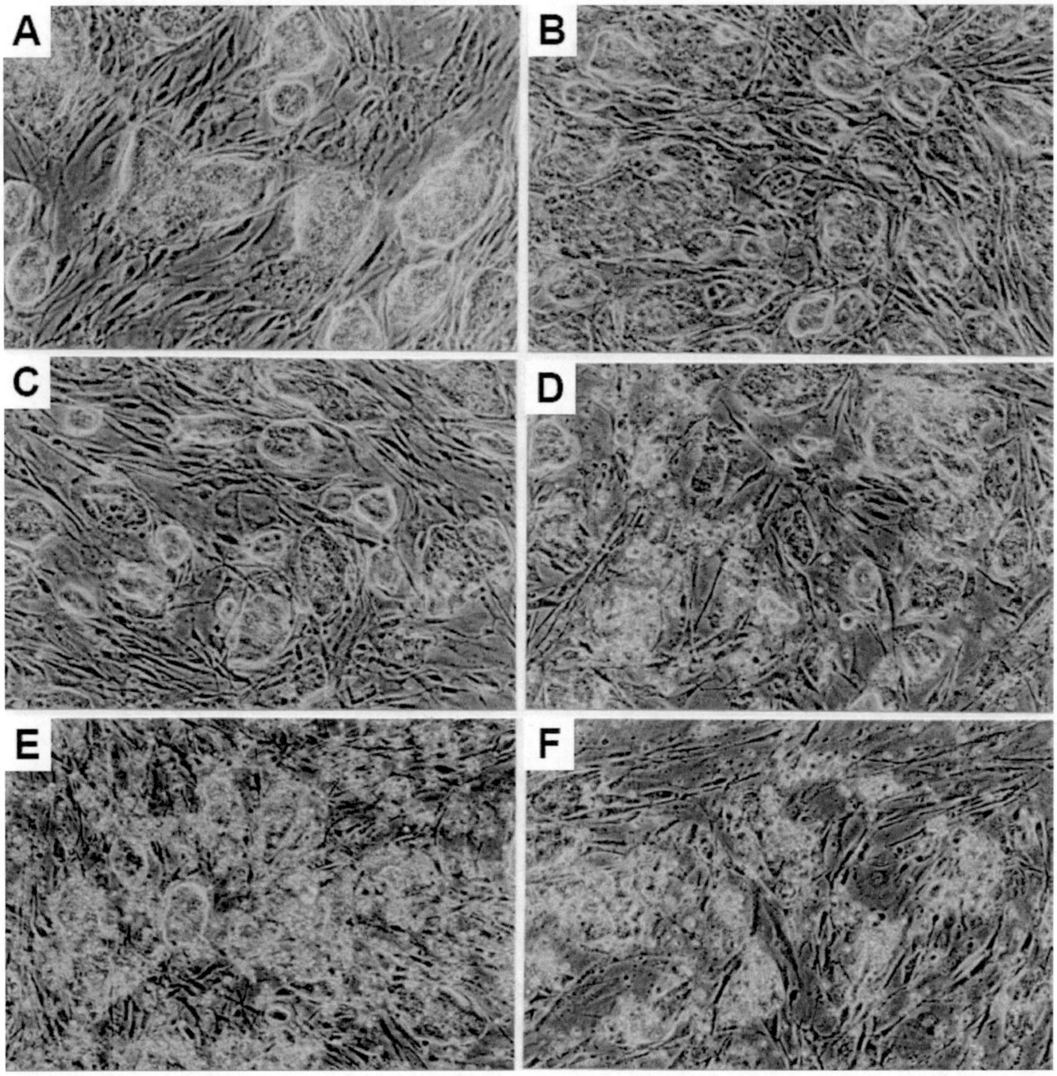

Fig. 9.2. Electroporation and selection. (**A**) Sub-confluent ES cell plate (day 0). (**B**) Appearance of ES cell plate before being used for electroporation (day 1). (**C**) Appearance of cells 24 h after the electroporation (day 2). (**D**) ES cell culture after the first day of selection (day 3). (**E**) Massive cell death is seen after 48 h of selection (day 4). (**F**) Almost all ES cells are dead on the third day of selection; debris is still present (day 5). 100 × mag.

3.3.2. Selection

Day 2

1. Start selection 18–24 h after the electroporation with 250 μg/mL G418 and 0.5 μM FIAU (or 2 μM gancyclovir) for 3–4 days (**Fig. 9.2C**).

Day 3

2. Change media (ES cell medium + G418 + FIAU). Now the cells have stopped growing. Some cells look vacuolated and debris from cell death appears on the plate (**Fig. 9.2D**).

Day 4

3. Massive cells death appears on the plate (**Fig. 9.2E**). Rinse off debris with PBS and change media (ES cell medium + G418 + FIAU).

Day 5

4. Still extensive cell death. Rinse plate with PBS. This is the end of double selection. Change to G418 media (ES cell medium + 250 μg/mL G418). Few ES cells should be present on the plate among the feeders (**Fig. 9.2F**).

Day 6/Day 7

5. Colonies appear on the plate. Rinse plate with PBS if debris is still present and change G418 media every day.

   Prepare 24-well feeder plates for the colonies to be transferred.

6. Resuspend feeder cells to 2×10^5 cells/mL and plate 1 mL/gelatin-coated well.

3.3.3. Isolating ES Cell Clones

Day 8–9

1. Pipette 100 μL/well of trypsin-EDTA to alternate wells of a 96-well plate (**Fig. 9.3**). Do not do more than 12–24 wells at a time.
2. Replace the medium in the dish with the colonies with 5–8 mL PBS. Pipette gently along the side of the plate to avoid detaching of the colonies.
3. Place the open dish under the microscope and score for good-quality colonies (**Fig. 9.4A**). Colonies are chosen based on their morphology and size. Only colonies with irregular tri-dimensional growth should be picked (**Fig. 9.5A, C, E**). Colonies that show signs of differentiation should not be picked (**Fig. 9.5B, D, F**)
4. Pick a colony using a P20 tip mounted on a micropipetman set at 10 μL and transfer to well with trypsin (**Fig. 9.4B**).
5. After picking 12–24 colonies (do not leave colonies for more than 10–15 min in trypsin), incubate the plate for 3–4 min at 37°C.
6. Dissociate colonies by pipetting 10–15 times with a multi-channel P200 pipettor set at 100 μL without making bubbles. 200 μL aerosol barrier tips should be alternated in the multi-channel pipettor such that six colonies at a time can be dissociated and transferred to a 24-well plate (**Fig. 9.3**). This method allows quick colony processing, therefore reducing the stress of trypsinization.

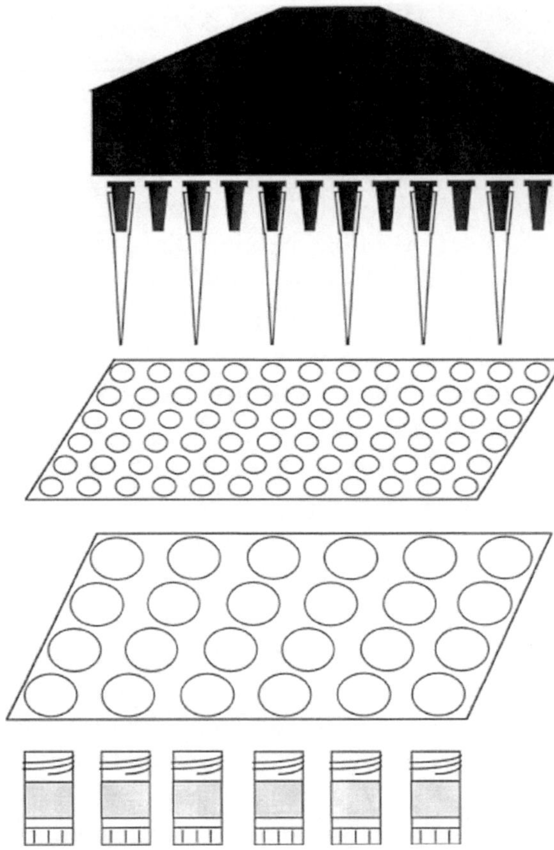

Fig. 9.3. Schematic illustrating how to arrange the tips in the multichannel pipettor in order to perform dissociation (**Section 3.3.3**) and freezing (**Section 3.3.4**) of several ES cell clones simultaneously. After colonies are placed in individual wells of a 96-well plate, cells can be handled with a multichannel pipettor for the dissociation and transfer to a 24-well plate. At the time of freezing the same tip arrangement can be used for dissociation of ES cells and transfer to tubes for cryopreservation.

7. Transfer the cells to a 24-well feeder plate with G418 medium (**Fig. 9.3**). Use 2 mL/well. This medium does not need to be changed the days following colony plating).
8. After picking the colonies, replace the PBS on the 10-cm plate with fresh G418 medium and store for more picking on the following day.

3.3.4. Freezing ES Cell Clones

After plating, individual cells from dissociated colonies will take 4–5 days to produce new clusters of cells. It is important not to let these cells overgrow and differentiate. After a maximum of 6 days every clone should be frozen (*see* **Note 8**).

Day 13–14

1. Remove media from clones in 24-well plates.
2. Rinse each well with 0.5 mL of trypsin.

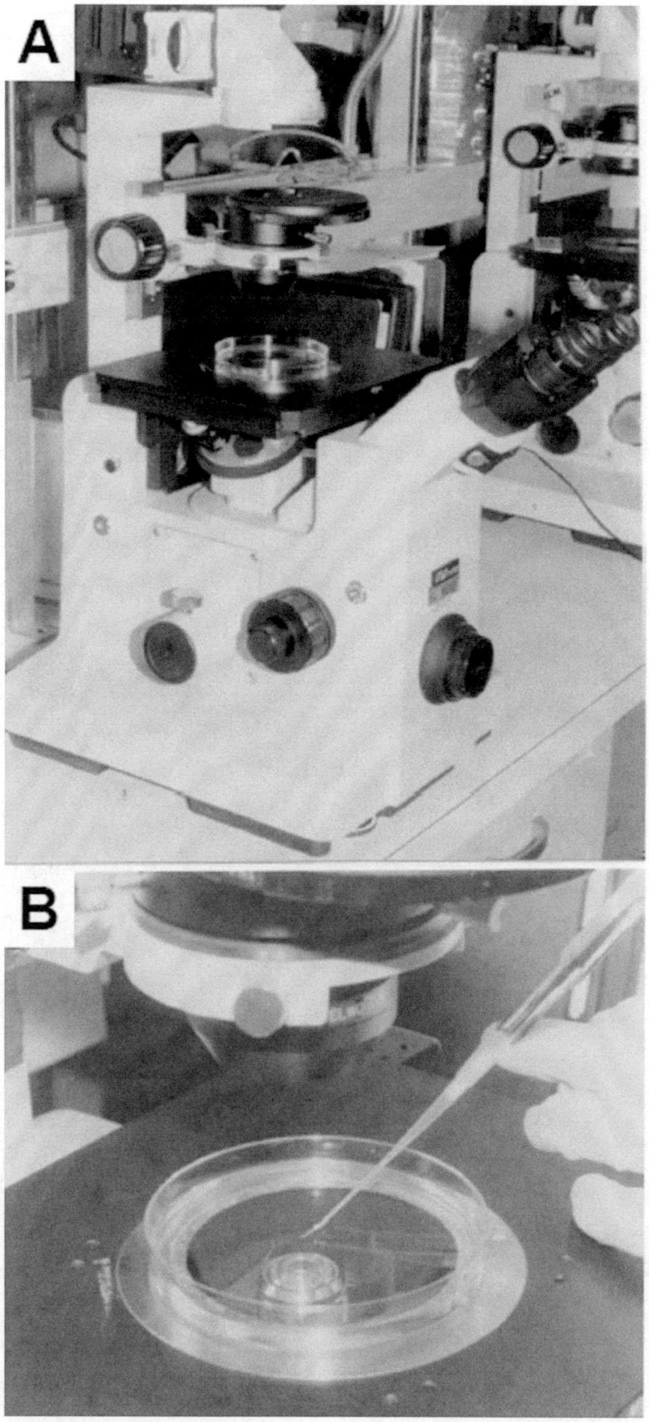

Fig. 9.4. Picking of ES cell clones. (**A**) Typical tissue culture microscope used for viewing and picking of ES cell clones. (**B**) Picking of ES cell clone.

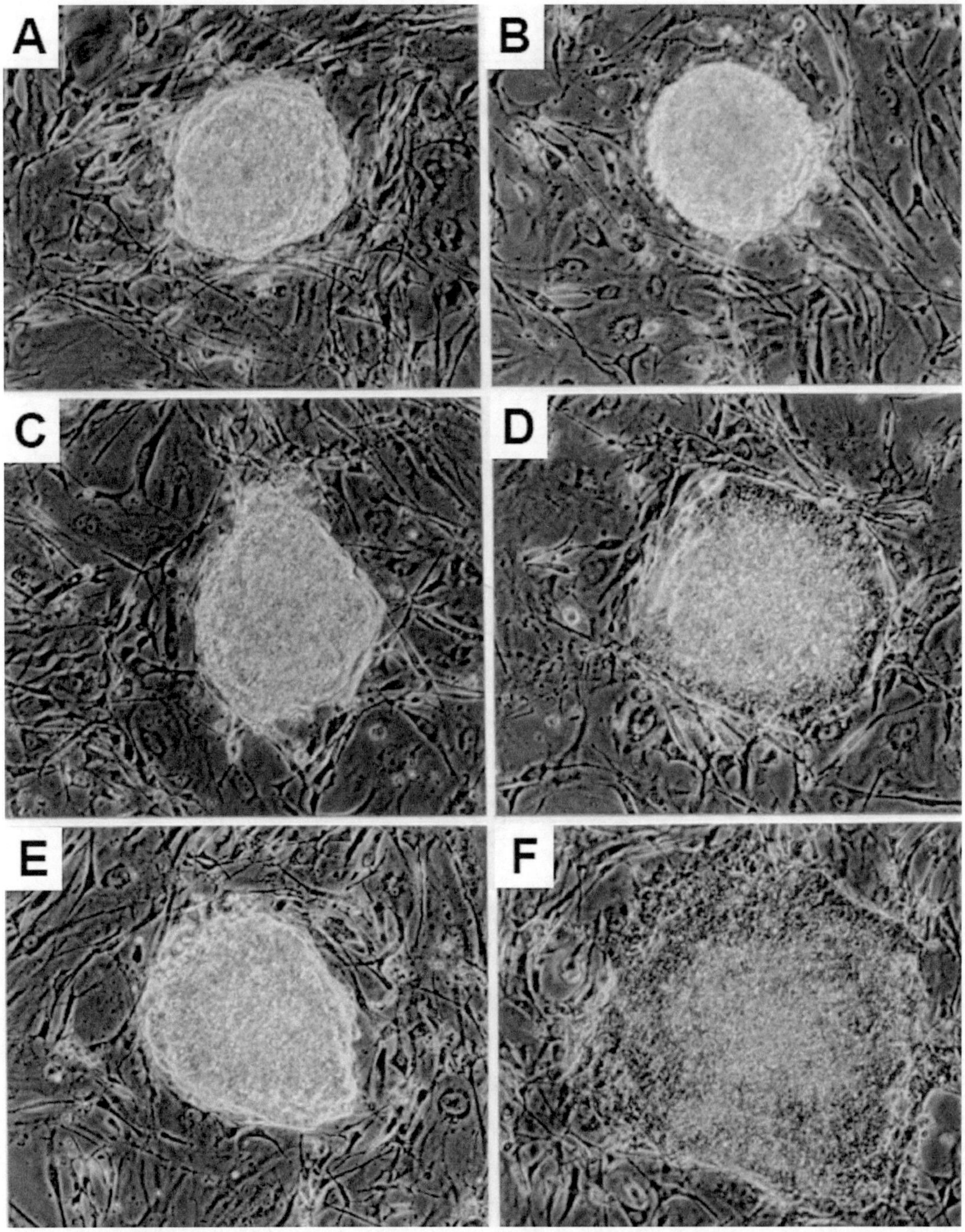

Fig. 9.5. Morphology of undifferentiated (**A**, **C**, **E**) and differentiated (**B**, **D**, **F**) ES cell colonies. Undifferentiated colonies display slightly irregular three-dimensional ES cell growth (**A**, **C**, **E**). Colonies which grow three-dimensionally but appear rounded are differentiating as simple embryoid bodies (**B**; (*3*)) whereas flat colonies (**D**, **F**) are differentiating as epithelial cells. 100 × mag.

3. Add 150 μL of fresh trypsin and incubate the plate at 37°C for 3–5 min (use a repeater pipettman for speed).
4. Neutralize trypsin with 200 μL of MEF media.
5. Dissociate the cells from six wells at a time by pipetting 10–15 times with a multichannel pipettor set at 200 μL with tips arranged as in **Fig. 9.3**.
6. Transfer 200 μL of cells to cryovials containing 200 μL of Freezing media.
7. Mix by pipetting twice with the multichannel pipettor, close the tubes and store in a styrofoam box on top of dry ice (in this way the freezing process will occur slowly).
8. After 2 h transfer the tubes to a –80°C freezer. The clones can be stored until the results from the DNA screening are obtained.
9. The remaining cell suspension from the trypsinized plate can be transferred to a new gelatin-coated 24-well plate (without feeders) with G418 media. These cells will be grown and used for DNA analysis.

3.4. Analysis of DNA from ES Cell Clones

Day 19–21

A key issue at this step is to maximize DNA recovery from these ES cell clones. Therefore, cells can be overgrown and should be harvested at maximum confluency.

3.4.1. Lysis of ES Cell Clones

1. Remove medium from well.
2. Rinse with 0.5 mL of PBS.
3. Remove PBS and add 500 μL of lysis buffer.
4. Let the cells sit for a few minutes until lysis is completed and harvest the lysates to a 1.5-mL eppendorf tube.
5. Incubate overnight in a shaking incubator at 55°C.

3.4.2. DNA Extraction

The following protocol is a very fast, simple, and crude method for DNA extraction and works effectively with DNA restriction endonucleases that require high-salt buffer conditions (e.g., *Bam*HI, *Bgl*I, *Bgl*II, *Hin*dIII, *Eco*RI, *Eco*RV, *Pst*I, *Not*I, *Xba*I, etc.) (modified from *(15)*). However, before proceeding with the extraction of all clones, a test should be performed to ensure that the DNA samples extracted with this method are suitable for analysis with the endonucleases required for the screening for homologous recombination. We recommend the phenol/chloroform method for extraction if cleaner DNA is needed *(14)*.

1. Add 1.0 mL of 100% ethanol to the 0.5 mL of cell lysate.
2. Mix thoroughly by hand until DNA is completely precipitated.
3. Centrifuge for 15 min at high speed in a microfuge.

4. Remove supernatant without dislodging the pellet.
5. Rinse the pellet by adding 0.5 mL of 70% ethanol and vortex.
6. Spin 10 min and remove as much of the supernatant as possible.
7. Allow to air-dry for 5–10 min, but not more otherwise it will be very difficult to resuspend the DNA.
8. Resuspend in 100 μL of sterile water.

3.4.3. Southern Blot Analysis

DNA recovered from one ES cell clone expanded in a 24-well plate is sufficient for at least three restriction digest analyses. The digestions can be performed in a 96-well plate to reduce manipulations and more easily preserve the sequence of the samples.

1. Check the DNA concentration of several clones by spectrophotometric analysis to determine more precisely the average amount of DNA *(14)*. About 10–30 μL contains on average 10 μg of DNA.
2. Prepare a master mix containing 5 μL 10X enzyme buffer, 2 μL 0.1 M spermidine trihydrochloride, 30–50 units of restriction enzyme and water to make a total of 55 μL after addition of 10–20 μL of DNA.
3. Mix the reaction by pipetting in the DNA sample.
4. Incubate for 8–20 h at 37°C.
5. Stop the digestion with 10 μL of Loading buffer.
6. Load the samples on a 0.8% agarose gel in 1X TPE buffer and subject to electrophoresis overnight at ~30 V.
7. Nick the DNA with 245 nm reflected UV light for 5 min or depurinate with 0.25 M HCl for 15 min (with shaking).
8. Soak gel in alkali solution for 45 min (with shaking).
9. Soak the gel in neutralization buffer for 30 min (with shaking).
10. Set up transfer and blot onto nylon membrane (e.g., Hybond N+) overnight in10X SSC. For more complete and detailed protocols for Southern analysis see (*14*) (*see also* **Note 9**).
11. Take apart Southern and soak the membrane in 2X SSC for approximately 30 s.
12. UV crosslink or bake for 1 h.
13. Prehybridize for at least 2 h at 65°C.
14. Hybridize overnight (at least 8 h) at 65°C. Several commercial kits (e.g., DNA labeling kit from Boehringer Mannheim or Stratagene) allow the labeling of probes with radioactively modified deoxyribonucleoside triphosphates. Probes with a specific activity of 1.5–2.0 $\times$ 10^9 d.p.m. (disintegration per minute)/μg should be used to obtain relatively quick results (i.e., with an O/N exposure).

15. Wash blots two times with 0.2X SSCP and 0.1% SDS for 20–30 min at 65°C.
16. If necessary, wash one time with 0.1X SSCP and 0.1% SDS for 20–30 min 65°C.
17. Wrap the filter in Saran Wrap and apply O/N to X-ray film to obtain an autoradiographic image.

3.5. Preparation of ES Cell Clones for Expansion and Injection

1. Quickly thaw the selected clones at 37°C and transfer to 5 mL of MEF media.
2. Spin down for 10 min at 300*g*.
3. Remove the supernatant without dislodging the small pellet.
4. Resuspend in ES media and plate one clone/well on a 12-well feeder plate (from now on the cells should get only ES media).
5. Expand the clone and split into one or two 6-cm feeder plates after 4–5 days.
6. When these cells are ready for splitting freeze down a few vials and split some more cells for a second DNA analysis. With this analysis you want to ensure the identity and the clonality of the positive clones (*see* **Note 10**).
7. Perform injections with cells trypsinized and resuspended in ES medium (*see* **Chapter 14** this volume). Make sure that before a clone is injected it is grown for at least 1 week in ES media without G418.

4. Notes

1. The ability to use a DNA vector obtained from the C57BL/6 genetic background for targeting (*see also* **Chapter 8**, Gene Targeting in Embryonic Stem Cells) allows to take advantage of the publicly available mouse sequence for vector designing and to obtain bacterial artificial chromosomes (BACs) for vector construction without the need to screen a DNA library (BACs can be ordered from http://bacpac.chori.org/).
2. The serum used for growing ES cells is of critical importance, so despite the fact that "ES cell quality" serum is available we strongly recommend to always test the serum before buying a large amount. In our experience, we have found that cheaper, not "ES cell tested", serum can perform at the same level or sometimes even better than the ES cell-certified one.

3. Despite the fact that we have found that MEFs grow better in low-oxygen conditions most laboratories (including ours until recently) grow successfully MEFs in the standard 5% CO_2, 20% O_2. Thus, fast-growing MEFs can be obtained with regular culture conditions but there may be more variability between different MEF batches.
4. Upon receiving an early passage of ES cells it is important to expand them to create a stock of frozen cells to be used for electroporation experiments. A few ES cells from the stock should be tested for the presence of mycoplasma, since contamination may greatly reduce ES cell efficacy. Moreover, a sample of the expanded cells, before genetic manipulations, should be tested for their ability to contribute to the mouse germline upon blastocyst injection. This will give a good indication of the parental ES cell line quality and whether the initial manipulations had an impact on the cells.
5. It is very important to culture ES cells only when needed since these cells tend to lose their totipotency after multiple tissue culture passages, which is why it is critical to expand and make multiple freeze down upon receipt of a vial of a new ES cell line.
6. It is important to split cells 24 h before the electroporation to have a culture where the majority of cells are actively dividing. In fact, it is believed that recombination with the foreign DNA vector occurs during the S phase of the cell cycle.
7. Most of the floating debris and DNA are lost during the pipetting since the DNA tends to stick to the wall of the plastic pipette.
8. For planning purposes, it is important to note that almost all clones picked on a certain day are going to be ready simultaneously. Therefore, materials should be prepared prior to starting (i.e., gelatin-coat 24-well plates, label cryovials, prepare freezing media, arrange tips for multichannel pipettor).
9. Although this protocol works well in our hands we recommend to follow closely the instructions of the manufacturer of the membrane used for Southern analysis.
10. If a clone represents a mixed population of normal ES cells and G418-resistant cells, it can be grown in G418 media until clonality is achieved. However, we have found that in most cases a mixed clone is the result of a correctly targeted clone and a non-targeted, but G418-resistant clone, in which case selection continuation will not help. In that situation we suggest to plate at low density the ES cells and subclone.

Acknowledgments

We thank Mary Ellen Palko for critical reading of the manuscript. This research was supported by the Intramural Research Program of the NIH, National Cancer Institute. The contents of this publication do not necessarily reflect the views or policies of the Department of Health and Human Services, nor does mention of trade names, commercial products, or organizations imply endorsement by the U.S. Government. The Center for Cancer Research, NCI-Frederick has filed an Animal Welfare Assurance with the Office for Protection from Research Risks (OPRR). The protocols herein described have been approved by the NCI-Frederick Institutional Animal Care and Use Committee.

References

1. Evans MJ, Bradley A, Kuehn MR, Robertson EJ. The ability of EK cells to form chimeras after selection of clones in G418 and some observations on the integration of retroviral vector proviral DNA into EK cells. Cold Spring Harb Symp Quant Biol 1985;50:685–9.
2. Evans MJ, Kaufman MH. Establishment in culture of pluripotential cells from mouse embryos. Nature 1981;292:154–6.
3. Robertson EJE. Teratocarcinomas and Embryonic Stem Cells: A Practical Approach. IRL Press, Washington, DC 1987.
4. Martin GR. Isolation of a pluripotent cell line from early mouse embryos cultured in medium conditioned by teratocarcinoma stem cells. Proc Natl Acad Sci USA 1981;78:7634–8.
5. Wassarman PM, DePamphilis ML. Guide to techniques in mouse development. Methods Enzymol 1993;225.
6. Joyner AL. Gene Targeting: A Practical Approach. Oxford University Press, New York 1993.
7. Hogan B, Costantini F, Lacy E. Manipulating the Mouse Embryo: A Laboratory Manual. Cold Spring Harbor Laboratory Press, New York 1986.
8. Nagy A, Vintersten K. Murine embryonic stem cells. Methods Enzymol 2006;418:3–21.
9. Seong E, Saunders TL, Stewart CL, Burmeister M. To knockout in 129 or in C57BL/6: that is the question. Trends Genet 2004;20:59–62.
10. Auerbach W, Dunmore JH, Fairchild-Huntress V, et al. Establishment and chimera analysis of 129/SvEv- and C57BL/6-derived mouse embryonic stem cell lines. Biotechniques 2000;29:1024–8, 30, 32.
11. You Y, Bersgtram R, Klemm M, Nelson H, Jaenisch R, Schimenti J. Utility of C57BL/6 J × 129/SvJae embryonic stem cells for generating chromosomal deletions: tolerance to gamma radiation and microsatellite polymorphism. Mamm Genome 1998;9:232–4.
12. Wolfer DP, Crusio WE, Lipp HP. Knockout mice: simple solutions to the problems of genetic background and flanking genes. Trends Neurosci 2002;25:336–40.
13. McVicar DW, Winkler-Pickett R, Taylor LS, et al. Aberrant DAP12 signaling in the 129 strain of mice: implications for the analysis of gene-targeted mice. J Immunol 2002; 169:1721–8.
14. Sambrook J, Fritsch EF, Maniatis T. Molecular Cloning: A Laboratory Manual, 2nd Ed. Cold Spring Harbor Laboratory Press, New York 1989.
15. Laird PW, Zijderveld A, Linders K, Rudnicki MA, Jaenisch R, Berns A. Simplified mammalian DNA isolation procedure. Nucleic Acids Res 1991;19:4293.

Chapter 10

ES Cell Line Establishment

Heidrun Kern and Branko Zevnik

Abstract

A method is described to establish mouse embryonic stem cell (ESC) lines from hybrid and inbred strains of mice. Attention is paid not only to the methodology for isolation and culture but also to the validation of freshly derived lines, in order to be maintained for prolonged time without significant differentiation or karyotype instability, and to provide reproducible germline transmission in chimaeric mice.

Key words: Embryonic stem cells, ES, establishment, derivation, culture, hybrid, inbred, validation, karyotype, germline transmission.

1. Introduction

Mouse embryonic stem cells (ESCs) were first isolated in 1981 by in vitro culture of cells isolated from the inner cell mass (ICM) of early embryos *(1, 2)*. Under appropriate culture conditions, ESC can proliferate indefinitely while retaining the ability to differentiate into all types of somatic cells. The ability to introduce specific genetic changes in ESCs by homologous recombination *(3, 4)* and to transmit these to the germline of chimaeric mice *(5, 6)* made ESCs an indispensable tool for vertebrate forward genetics.

Since their first description, many different lines have been derived, predominantly from 129 (for an overview see *(7)*), C57BL/6 *(8–14)* and F1 hybrid strains of mice *(15–17)*. Other lines, isolated from strains regarded as somewhat refractory for ESC isolation, include Balb/c *(18)*, DBA-1 and -2, or CBA *(19–22)*, C3H/He *(23)* or NOD *(24)*. The majority of these lines have been successfully used in gene targeting experiments

Ralf Kühn, Wolfgang Wurst (eds.), *Gene Knockout Protocols: Second Edition, vol. 530*

DOI 10.1007/978-1-59745-471-1_10 Springerprotocols.com

and many of them are available either commercially or through the scientific community. It may not be therefore instantly clear, why a laboratory should derive its own ESC line(s).

However, there are a number of reasons, for example, the defined health status of freshly derived ESC lines. Cell lines are frequently contaminated by mycoplasma infections. Such lines are generally no longer competent for germline transmission, and may severely affect the status of a pathogen-monitored mouse colony *(25)*. Equally, ESCs with high passage numbers, undefined history of culture and storage conditions, or freeze–thaw cycles, transmit their genome very poorly in chimaeras, and tend to loose germline competence completely. Moreover, if the karyotype is not tightly monitored, ESCs accumulate chromosome aberrations during culture, most often trisomy 8 or 11 or loss of Y chromosome, which is again refractory to germline transmission *(26)*. Freshly derived lines, on the other hand, generally perform well in germline transmission upon genetic manipulation.

Research may also require the analysis of mutations on specific genetic backgrounds, as phenotypes may be affected by strain differences. Instead of time-consuming and tedious backcrossing of mutated strains to derive congenic lines, it may be worth to derive and manipulate ESC lines directly from the appropriate, mostly inbred, strain or substrain. There are differences though as to how easily ESCs can be derived from different backgrounds.

Undoubtedly the biggest need for ESC derivation stems from the ever-increasing complexity of genetic analysis. For example, it may be desired to study the effects of mutations on a complex disease background such as cancer. The introduction of further alleles into existing models, often conditional alterations requiring the presence of additional inducible or tissue-specific recombinase genes, requires complicated breeding schedules involving multiple generations of mice. The derivation and manipulation of ESCs from the respective disease backgrounds circumvents these efforts. Moreover, such "custom-designed" ESC lines may subsequently serve as flexible systems for a multitude of genetic alterations.

There are numerous protocols available for ESC line derivation from all stages of early embryonic development, namely isolation and culture of blastomeres from two-, four-, or eight-cell-stage embryos *(27–29)*, morulae *(30, 31)* or blastocysts *(1, 2, 9, 12, 32–36)*. ESC lines have even been derived from somatic cell nuclei via nuclear transfer *(37, 38)*.

Some protocols require the use of sophisticated techniques, such as isolation of ICM via immunosurgery of intact blastocysts *(2, 39)*, isolation of epiblast cells from implanted egg cylinder-stage embryos *(19)*, or selective ablation of differentiated cells *(22, 40, 41)*. Other variations include derivation on special feeder layers *(8)*, or in the absence of supporting feeders *(39, 42)*, or require the use of conditioned media *(36)*.

More recent protocol refinements are often direct consequences of the increasing knowledge on stem cell biology and signalling cascades within early embryos. Factors such as FGF, generally present in sera used for ESC culture, seem to inhibit stem cell isolation by advancing differentiation into neuronal lineages *(43)*. Other factors such as BMP4, also present in serum, seem to act in concert with the LIF/Stat3 signalling pathway responsible for ESC maintenance *(44)*. Lastly, pharmacological small-molecule inhibitors have been identified, which inhibit stem cell differentiation pathways, thereby supporting ESC maintenance *(45–47)* or ES cell line isolation *(48)*.

Most of these promising attempts for derivation of ESC lines are now being validated in practice by experimentators. However, in order to maximise success, we will focus on the reliable, commonly used method for derivation of ESCs from intact blastocysts. Embryos at the expanded blastocyst stage are plated intact onto a "feeder" layer of mitotically inactivated murine embryonic fibroblasts (MEFs). ESC colonies of undifferentiated morphology are isolated from disaggregated ICM outgrowths. With this protocol at hand, all laboratories with access to a mouse facility, an embryo manipulation suite and a rigorous ESC culture regime, should be able to derive their own ESC lines.

2. Materials

2.1. Mice

Prerequisite for ESC line derivation is access to a mouse facility, providing mating pairs for blastocyst isolation. For natural matings, mice are best used at 3–6 months of age. If superovulation is employed, fewer females will be required but should be used at pre-puberty age (3–6 weeks old), to optimise embryo yield.

2.2. Equipment for Cell Culture

All plasticware, chemicals, reagents and solutions should be cell culture tested or of analytical grade.

1. Tissue culture incubator (settings: 37°C, 7.5% CO_2).
2. Inverted microscope.
3. Dissecting microscope.
4. Clean bench (Horizontal laminar flow).
5. Tissue culture hood (Class II microbiological safety cabinet).
6. Tabletop centrifuge.
7. Water bath (settings: 37°C).

2.3. Plasticware and Other Materials

1. Embryo handling pipette: 100 μL borosilicate glass micropipettes (Brand 708745).

2. Aspirator tube assembly for microcapillary pipettes (Sigma-Aldrich A-5177).
3. Gas burner.
4. Sterile pipette tips, forceps, scissors, 1-mL tuberculin syringes, 27 gauge × 3/4″ hypodermic needles.
5. High-quality tissue culture (TC) plates and dishes (e.g. Nunc, BD Biosciences): 96-well round-bottom, 48-, 24-, 12- and 6-well flat-bottom plates; 5-cm and 10-cm TC dishes.
6. Sterile filter with 0.22-μm membrane with low protein binding (e.g. Millipore Stericups).

2.4. Reagents and Solutions

2.4.1. Embryo Reagents

1. KSOM (potassium simplex optimised medium; Powdered Media Kit, Millipore, Europe MR-020P-5F).
2. hCZB (HEPES-buffered Chatot–Ziomek–Bavister medium) *(49)* (*see* **Section 2.4.3**, Step 2).
3. Light mineral oil (e.g. Fluka 76235).

2.4.2. ESC Reagents

1. Fetal bovine serum (FBS) (e.g. PAN Systems or other suppliers) (*see* **Note 1**).
2. DMEM: Dulbecco's modified medium (1X), liquid – with L-glutamine, 4,500 mg/L D-glucose, without sodium pyruvate (e.g. Invitrogen 41965-039).
3. Advanced DMEM (1X), liquid (e.g. Invitrogen 12491-015).
4. 200 mM L-Glutamine (100X), liquid (e.g. Invitrogen 25030-024).
5. MEM Non-Essential Amino Acids (100X), liquid (e.g. Invitrogen 11140-035).
6. 100 mM Sodium Pyruvate MEM, liquid (e.g. Invitrogen 11360-039).
7. 1 M HEPES Buffer solution, liquid (e.g. Invitrogen 15630-056).
8. 50 mM 2-Mercaptoethanol in PBS (e.g. Invitrogen 31350-010).
9. Leukaemia Inhibitory Factor (LIF) (e.g. Millipore ESG1107).
10. 0.25% Trypsin with EDTA 4Na (e.g. Invitrogen 25200-056).
11. 2.5% Trypsin (10X), liquid (e.g. Invitrogen 15090-046).
12. Chicken serum (e.g. Invitrogen, 16110-082).
13. DMSO (Sigma, D2650).
14. Gelatine (from porcine skin, cell culture tested) (e.g. Sigma G1890).
15. Penicillin/Streptomycin (Invitrogen, 15140122), use 1:100 (*see* **Note 2**).

16. Mitomycin C (e.g. Sigma M-0503). Exercise caution as mitomycin C is a toxic agent. Minimise exposure. Discard into hazardous waste container.
17. H_2O: embryo tested (e.g. Sigma W1503).
18. Phosphate-buffered saline (PBS) 1X, liquid (e.g. Invitrogen 20012-019).

2.4.3. Embryo Media

1. KSOM Powdered Media Kit: Add 10 μL/mL MEM Amino Acids (50X), 5 μL/mL MEM Non-Essential Amino Acid Solution (100X). Sterile filter and aliquot. Keep at 4°C for up to 1 week.
2. HEPES-buffered CZB media (hCZB; *(49)*): 81.62 mM NaCl, 4.83 mM KCl, 1.18 mM $MgSO_4$ ($7H_2O$), 1.18 mM KH_2PO_4, 0.11 mM EDTA-2Na, 31.3 mM Na-Lactate, 5.55 mM D-Glucose, 100 IU (each) Pen/Strep. Stock salt can be kept at 4°C up to 3 months. For use add to a final concentration: 20 mM HEPES, 5 mM $NaHCO_3$, 0.27 mM pyruvic acid, 1.28 mM $CaCl_2$-$2H_2O$. Adjust pH to 7.5 with 1 M HCL, sterile filter; keep at 4°C up to 2 weeks.
3. Blastocyst isolation (flushing) media. Final concentration: 10% FBS, 1 mM sodium pyruvate, 2 mM glutamine, 25 mM HEPES, 100 μg/mL Penicillin, 100 μg/mL Streptomycin to DMEM (high glucose). Freeze aliquots after filter sterilization.

2.4.4. ESC Media (see **Note 1**)

1. C57BL/6 ESC culture medium. Final concentration: 20% FBS, 2 mM glutamine, 1X non-essential amino acids, 1 mM sodium pyruvate, 0.1 mM 2-mercaptoethanol, 1,200 units/mL LIF in DMEM (high glucose). Filter sterilization is recommended. Store at 4°C and discard after 7 days.
2. Hybrid ESC culture medium. Final concentration: 15% FBS, 2 mM glutamine, 1X non-essential amino acids, 1 mM sodium pyruvate, 20 mM HEPES, 0.1 mM 2-mercaptoethanol, 900 units/mL LIF in DMEM (high glucose). Filter sterilization is recommended. Store at 4°C and discard after 7 days.
3. MEF-Medium: Advanced DMEM + 5% FBS + 2 mM glutamine. Store at 4°C and discard after 7 days.
4. Freezing medium: ES medium + 10% DMSO, prepare fresh.
5. 0.25% Trypsin-EDTA. Trypsin can be used up to 1 week, when stored at 4°C. Add 2% chicken serum to 0.25% trypsin-EDTA.
6. Gelatine (0.1%): Dissolve 0.4 g gelatine in 400 mL embryo-tested water, autoclave immediately and store at room temperature.

2.4.5. Solutions for Isolation of Genomic DNA

1. Lysis buffer: 10 mM Tris-HCl, pH 7.5, 10 mM EDTA, 10 mM NaCl, 5% *N*-Laurylsarcosinate, add 0.5 mg/mL Proteinase K prior to use.
2. Proteinase K, recombinant (e.g. Roche 03115844001).
3. 100% Ethanol.
4. 70% Ethanol.
5. 100 mM Tris-EDTA, pH 7.4.

2.4.6. Karyotype Solutions

1. 0.075 M KCl (prepare careful, hypotonic solution).
2. 5% Acetic acid in ethanol (for washing slides).
3. Fixative: Methanol/acetic acid = 3:1 (always prepare fresh, precool to –20°C).
4. 2% Giemsa in H_2O (Riedel de Haen, cat. no. 32884).
5. Colcemid (e.g. Karyomax® Colcemid®, Invitrogen 15210-057) Exercise caution, as colcemid is a toxic agent. Discard into hazardous waste container.

3. Methods

Protocols for mice handling, the preparation of MEFs from mouse embryos and ESC/embryo manipulations are beyond the scope of this chapter and given elsewhere in this book. **Figure 10.1** provides an overview about the timing of individual steps during the course of ESC generation.

3.1. Preparation of Mitotically Inactivated MEF Feeder Layers

1. Thaw one vial of MEFs and split cells once. When cells reach confluency, replace medium with medium containing 10 µg/mL mitomycin C and incubate for 2–3 h. Remove supernatant and wash three times with PBS. Trypsinise cells as usual and spin down. Resuspend pellet in MEF-medium at a concentration of 1×10^6 cells per mL (count cells with haemocytometer). Seed MEFs on gelatinised dishes (approximately 5×10^4 cells per cm^2 culture area) (*see* **Note 2**).

3.2. Preparation of Gelatine Plates

1. For gelatinising, cover the surface of empty tissue culture plates with 0.1% gelatine. Incubate for 5 min at room temperature. Aspirate gelatine. Plates can be directly used for seeding ESCs or inactivated MEFs.

3.3. Timed Mating

Day 3: Mating

1. Mate 20 females for F1 hybrid ESC derivation (e.g. B6S6F1), and up to 40 females for inbred ESC derivation (e.g. C57BL/6) (*see* **Note 3** and **Fig. 10.2**).

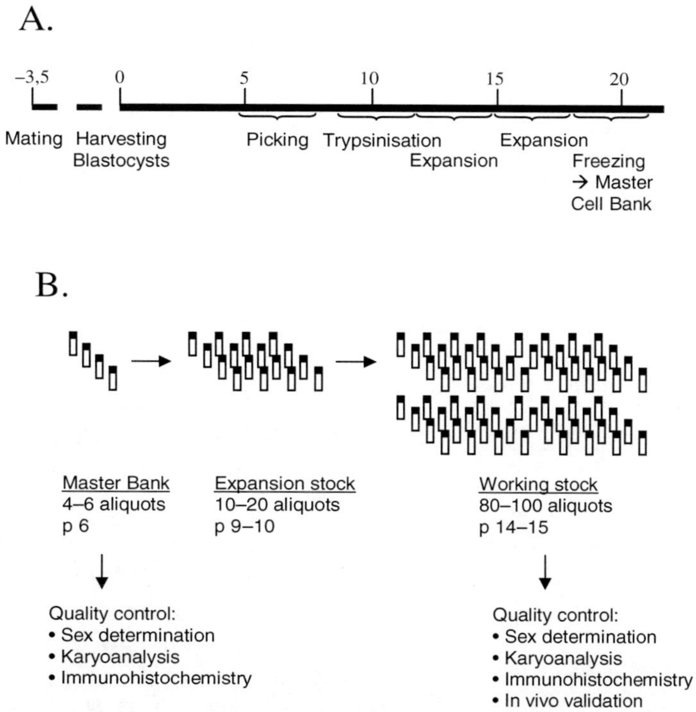

Fig. 10.1. Scheme for creating a validated ES cell line: (**A**) Timetable for derivation (days); (**B**) Establishment of working stock including quality control.

Day 2:

2. Check females for presence of copulation plug (*see* **Note 4**). Plug-positive females (0.5 days postcoitus (dpc)) are separated into individual cages. Leave plug-negative females in cage and add new females for next round of mating. Mating/plug checks are continued until 30–40 blastocysts are obtained.

3.4. Isolation of Blastocysts

Day 1:

1. Inactivate MEFs and seed out on 12-well TC plate. Number of plates depends on the number of blastocysts anticipated.
2. Draw out disposable micropipettes in the flame of a gas burner, break by hand with a diameter of ~200 μm for handling blastocysts with the aspirator tube assembly. Store capillaries in sterile dishes.

   Day 0: Blastocysts are collected from pregnant mice and plated on MEF feeder layer

3. Prepare a 5-cm dish with approximately 30-μL microdrops of KSOM covered with light mineral oil. Equilibrate in incubator set at 37°C and 5% CO_2 (*see* **Note 5**).

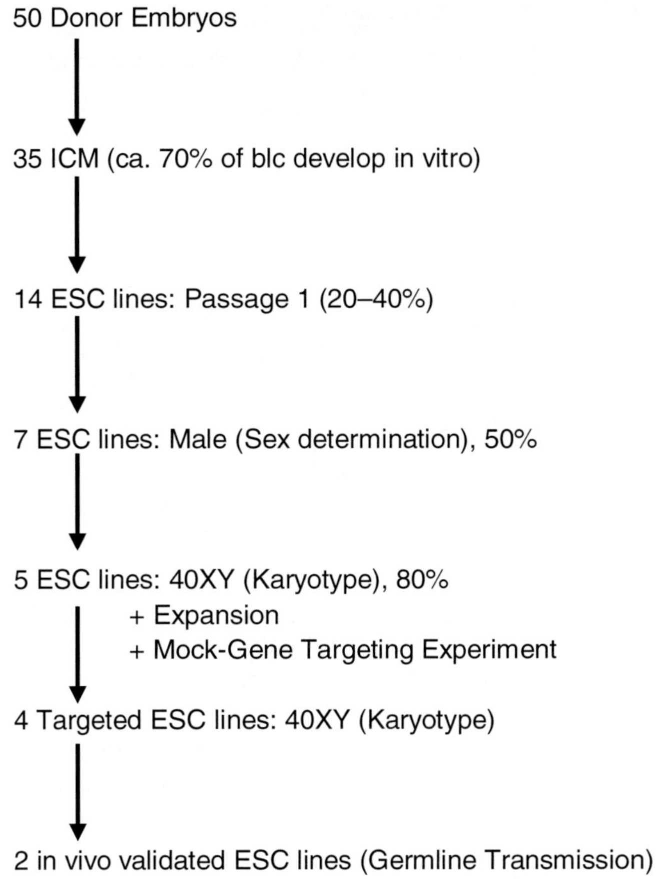

Fig. 10.2. Outline of resources needed to obtain validated ES cell lines from C57BL/6 inbred mice.

4. Prepare a 5-cm dish with approximately 30-μL microdrops of hCZB covered with light mineral oil.
5. Dissect uteri from females at 3.5 dpc and immediately collect in a dish with PBS.
6. Under a clean bench, wash uteri in 10-cm dishes with PBS and remove remnants of fat. Handle with disinfected forceps and scissors.
7. Transfer uteri in clean 10-cm dish.
8. Under a dissection scope, flush blastocysts out of each uterine horn with 0.5–1 mL flushing media using a 1-mL syringe fitted with a 27 gauge × 3/4″ hypodermic needle.
9. Collect blastocysts with an embryo-handling pipette (mouth controlled) in one of the prepared 30-μL microdrops of hCZB and wash through several 30-μL microdrops of hCZB to remove debris.

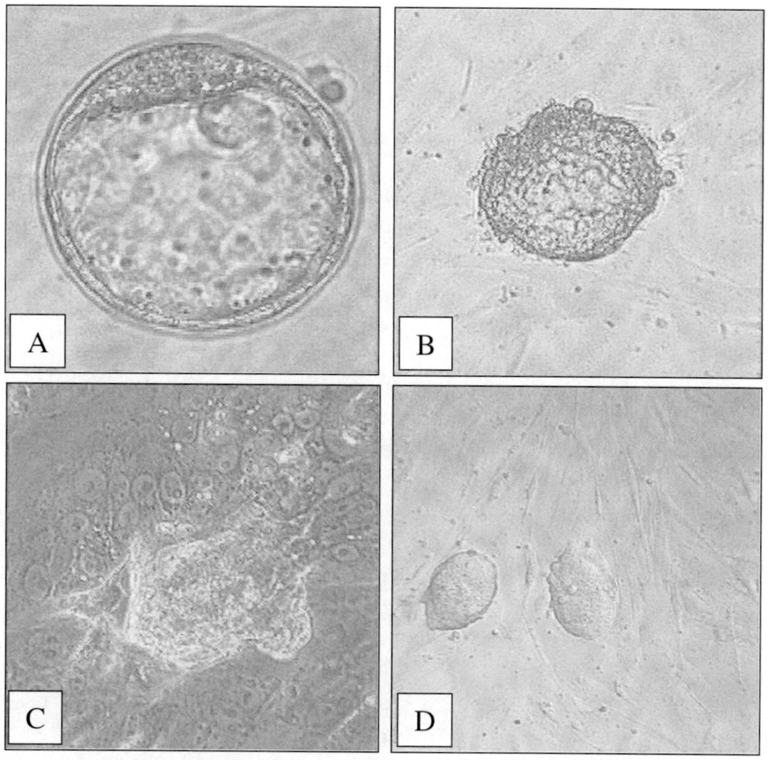

Fig. 10.3. Progressive changes during mouse embryonic stem cell line derivation: (**A**) dpc 3.5 mouse blastocyst; (**B**) day 1 culture, attached to the feeder layer; (**C**) day 4 culture ICM outgrowth; (**D**) passage 1, 3 days upon trypsinisation.

10. Collect blastocysts in a microdrop of KSOM.
11. Incubate microdrop culture for 2–4 h in incubator set at 37°C and 5% CO_2 until blastocysts are well developed. Remaining morulae can be further incubated overnight.
12. Place every single embryo into one well of a 12-well TC plate each containing MEFs (prepared the day before) and pre-warmed ES medium (**Fig. 10.3A**; *see* **Notes 6** and 7).
13. Allow blastocysts to attach to the supporting MEFs (**Fig. 10.3B**) and leave the culture undisturbed for 5–7 days in 37°C and 7.5% CO_2.
14. Monitor blastocysts closely every day and track viability as well as hatching and attachment to the feeder. Usually 2 days after hatching the ICM is apparent inside the blastocyst, later on the ICM appears distinctive and is growing upward (**Fig. 10.3C**).

3.5. ICM Isolation and Dissociation

Day 5–7: ICM outgrowths are selectively picked (see **Note 8**)*, trypsinised and replated onto mitotically inactivated MEFs.*

1. Prepare 96-well flat-bottom TC plates with MMC-treated MEFs the day before picking the ICMs.
2. Draw out bend capillaries (~45°) from glass micropipettes in a gas flame (two for each ICM), break them by hand: capillaries with a tip size of ~200 μm are used to pick and transfer the ICM outgrowth, a tip size of <50 μm (half ICM diameter) is suitable for ICM dissociation; store capillaries in sterile dishes.
3. Prepare a 96-well round-bottom TC plate containing 30 μL/well 0.25% Trypsin-EDTA for each ICM harvested.
4. Aspirate medium from 12-well TC plate containing blastocyst outgrowths, wash once with trypsin, aspirate trypsin immediately.
5. Place 12-well plate under a dissecting microscope placed within a clean bench. Using the 200-μm capillaries fitted with the mouth-controlled aspirator tube assembly, detach ICM from feeder layer and transfer each into an individual well of the trypsin containing 96-well round-bottom TC plate. Try to work fast; otherwise, the wells will dry out.
6. Incubate 5–10 min at 37°C, 7.5% CO_2.
7. Dissociate the ICM under the dissecting microscope using a mouth-controlled 50-μm capillary, by repeated pipetting in and out until single cells are obtained (*see* **Note 9**).
8. Upon dissociation of ICMs transfer suspended cells with the same capillary into a well of a 96-well TC feeder plate, prefilled with 180 μL ES medium.
9. Let cells grow for 2–4 days. Replace medium after 2 days with fresh ES medium.
10. Carefully inspect the wells at day 2 after ICM dissociation under low (50X) and high power (200X) magnification for the presence of colonies that exhibit ESC morphology (group of small cells in close contact forming a colony with clear borderline). Colonies may be marked with a pen at the bottom of the plate and reinspected the next day. A positive well usually exhibits several (2–10) ES colonies (*see* **Note 10**).
    Passage 1: Label positive wells with ES line name and passage no. 1; prepare MEFs for splitting and expansion of the cultures.

3.6. ES Line Expansion

Primary goal is to obtain a master cell bank of a few aliquots from each cell line, frozen in liquid nitrogen. This master cell bank will be used for further expansions. It is advisable to quality control the master cell bank carefully. A fully characterised master cell bank should have the following features: have an acceptable health profile (MAP test or PCR analysis), have a 40XY karyotype, be germline-tested by chimaeric (inbred ESC) or tetraploid analysis (F1 ESC) (*see* **Note 11** and **Fig. 10.1**).

Ensure the availability of freshly prepared MMC-treated plates for all splitting steps.

1. *Passage 2*: If ES cell colonies are visibly expanding (2–4 days after picking of the blastocyst outgrowth) wash the wells once with PBS and add 30 μL trypsin. Incubate 5–10 min at 37°C, then add 75 μL ES medium and resuspend with a pipette. Transfer the contents to a new well containing MEFs and ES medium. The size of the new well depends on the number of ES colonies observed in the trypsinised well (usually again on 96-well TC plate or, if many ES cell colonies are growing, on a 48-well TC plate). Label wells with ES line name, date and passage no. 2 (**Fig. 10.3D**).
2. Change medium the next day.
3. *Passage 3*: Expand ES cells to 48-well or 24-well TC plates containing MEFs and ES medium. The culture area depends on number of ES colonies observed: trypsinise as in step 1 (increase the volume of trypsin and ES medium according to bigger culture area). Label wells with ES line, name and passage no. 3.
4. Change medium the next day.
5. *Passages 4–6*: Proceed with expansion of cell lines until 3–4 semiconfluent wells of a 6-well TC plate are obtained. Change medium daily.
6. Freeze cells from at least two wells of a 6-well TC plate in ES medium + 10% DMSO, label tubes properly with name of cell line, date and passage number. From one well of a 6-well plate usually two aliquots can be frozen. These vials build the master cell bank of the freshly prepared ES cell line.
7. Split one confluent well 1:5–1:6 on gelatinised plates (high split ratio to get rid of the feeders). One well serves for the preparation of genomic DNA for Southern or PCR analysis of sex, 1–3 wells can be used for karyotyping.

3.7. From Master Cell Bank to Working Stock

3.7.1. Generation of an Expansion Stock

Depending on the results of the quality assessment and the amount of experiments planned, an expansion stock should be generated, from which large numbers of working stock aliquots can be produced (*see* **Fig. 10.1**).

1. Thaw one aliquot of frozen cells (from master cell bank) on one well of a 6-well TC plate.
2. Change medium the next day.
3. Split cells 1:3–1:5.
4. Change medium the next day.
5. Split again 1:3–1:5.
6. Freeze aliquots (10–20 aliquots). These aliquots represent the expansion stock with a passage number of usually 9–10.

3.7.2. Generation of Working Stock

To allow reproducible performance in many experiments, it is advisable to generate a well-defined working stock from your freshly derived cell line. Characterisation should again include health monitoring, karyotype analysis and in vivo testing for germline transmission.

1. Thaw one aliquot of the expansion stock on one well of a 6-well TC plate.
2. Change medium the next day.
3. Split on 1 × 10 cm plate.
4. Change medium the next day.
5. Split 1:3–1:5.
6. Proceed splitting steps until 20–25 semiconfluent 10-cm dishes are obtained.
7. Freeze 80–100 aliquots with a passage number of approximately 14–15.

3.8. DNA Analysis

3.8.1. Isolation of Genomic DNA

1. Let cells grow on a 6-well gelatine plate until confluency, change medium every second day without antibiotics. Before adding lysis buffer for the preparation of genomic DNA, take an aliquot of the supernatant to assay potential mycoplasma contamination by PCR. Supernatant can be stored frozen at –20°C.
2. Lyse cells with lysis buffer + 0.5 mg/mL Proteinase K (800 μL per 6-well). Incubate overnight at 56°C in a humidified chamber.
3. Harvest lysed cells in 2-mL Eppendorf tubes, add 1.6 mL 100% ethanol. Invert tube until you see precipitating DNA.
4. Centrifuge for 2 min at high speed in a microcentrifuge.
5. Wash pellet two times with 70% ethanol.
6. Air-dry DNA pellet and resuspend in 100–200 μL 10 mM Tris-EDTA.

3.8.2. Sex Determination

40XY male ESC lines are generally more stable than XX ES lines, which tend to loose one X chromosome during ESC propagation. The XY genotype confers two advantages to chimaeras. Firstly, male chimaeras can produce more offspring than females. Secondly, XY ESCs can sex-convert female genital ridges in male gonads thereby giving rise to 100% ESC-derived sperm *(5)*.

Sex determination by Southern blotting is preferred over PCR analysis. The presence of residual MEF cell DNA, containing male Y chromosomes, may generate false-positive PCR results. Detailed protocols to Southern blotting or PCR are given in molecular biology handbooks.

Southern blotting: ESC genomic DNA are hybridised with a 1.5-kb *Eco*RI fragment of a Y-chromosome probe (pY353) *(50)*.

1. PCR using primers to the Y-specific gene SRY *(51)*: 5′-CAG CCC TAC AGC CAC ATG ATA CTT-3′ and 5′-ACT CCA GTC TTG CCT GTA TGT GAT G-3′; Conditions: 30 s at 94°C, 90 s at 60°C for 40 cycles plus 10 min at 72°C. PCR products are 431-bp fragments.

3.9. Quality Controls

3.9.1. Health Monitoring

Health monitoring of newly derived ESCs is advisable for two reasons. Firstly, ESCs contaminated with mycoplasma may not be transmitted through the germline of chimaeras. Secondly, your animal facility procedures may require screening for and absence of unwanted pathogens. For mycoplasma testing, several kits are commercially available, predominantly assaying for the presence of mycoplasma-specific DNA by PCR reaction. Commercial providers also offer a more extensive health profile of your cells by MAP testing or Virus PCR assay panels *(52)*.

3.9.2. Karyotype

The karyotype of a master ESC bank as well as from a working stock should ideally be determined at a resolution of individual chromosomes (*see* **Note 12**). This can be done by Giemsa staining and banding or fluorescence in situ hybridisation (FISH) and labelling of individual chromosomes. The minimum analysis should encompass counting of chromosomes of at least 30 metaphase spreads, to determine the amount (percentage) of 40XY (if male) ESC genomes. The protocol described here allows counting of chromosome spreads. Cultures on gelatine plates are used to eliminate MEF cells. Start the preparation of the chromosomes one day after the last splitting step. It is important that the cultures are not overgrown (as cells slow down divisions).

1. Replace ES medium from cells growing on gelatine with medium containing 0.2 μg/mL colcemid and incubate for 1–2 h at 37°C/ 7.5% CO_2. Colcemid allows enriched harvest of cells in metaphase, as cells are arrested at this stage.
2. Trypsinise cells as usual and spin down at 250*g*.
3. Wash cells with 10 mL PBS and spin down at 250*g*.
4. Add 1 mL hypotonic solution (0.075 M KCl) very slowly, mix carefully by pipetting and add another 9 mL hypotonic solution dropwise.
5. Incubate 8 min at room temperature.
6. Spin down at 250*g*, aspirate KCl.
7. Add 1 mL ice-cold fixative (methanol:acetic Acid = 3:1) dropwise, mix very carefully by pipetting, add another 9 mL fixative.
8. Spin down at 1,200*g*, aspirate fixative; repeat two times.
9. Resuspend pellet in 100–500 μL fixative (depending on size of pellet). Preparation can be frozen in –20°C.

10. For staining wash high-quality microscope slides in ice-cold ethanol/5% acetic acid, dry slide quickly with lint-free tissue paper and add 3–5 small drops of cell suspension in a row using a 1-mL syringe fitted with a 27-gauge × 3/4″ needle.
11. Air-dry the drops and check the concentration of the cells with a microscope at low-power magnification. If necessary dilute cell concentration by addition of fixative.
12. Stain slides with 2% Giemsa solution for 10 min.
13. Wash slides two times with water. Air-dry slides.
14. Count chromosomes of at least 30 spreads and determine the percentage of 40XY cells.
15. If more than 80% of spreads show a karyotype of 40XY the culture is determined as a good-quality culture.

3.9.3. In Vivo Validation

It is strongly recommended to assay the germline transmission potential of freshly derived ESC lines. Ideally, a mock-gene targeting experiment is performed and clones are injected which are derived under standard conditions (passage numbers/freeze–thaw cycles), to allow direct performance comparison between different cell lines.

1. F1-Hybrid ESCs are most stringently tested in the tetraploid complementation assay. Performance varies by ES clone; however, on average every second ES clone should result in a minimum of 5–10% completely ES cell-derived offspring, upon tetraploid blastocyst injection *(53)*.
2. Inbred ESC lines are tested by standard diploid blastocyst injection or aggregation *(35)*. In our hands, the injection of C57BL/6 ES cell lines into Balb/c hosts gives rise to an average of 76% highly chimaeric males, judged by coat colour contribution. Germline transmission should be obtained from >75% of all male chimaeras tested.

4. Notes

1. We routinely use FBS as a major component of our ES medium; however, replacements for animal sera are available (e.g. ESGRO Complete Clonal Grade Medium, Chemicon; Knockout™ SR, Invitrogen) and have been successfully employed for ESC derivation (*(13, 54*); Tielens, 2006 #2664; Wakayama, 2007 #2586).

   FBS suitability for ESC culture medium varies significantly among suppliers and lot numbers. Suitability is assessed by measuring cloning efficiency and differentiation of ESCs cultured on gelatinised plates in ES medium without addition of LIF. For standard ESC applications, heat inactivation of FBS (to destroy heat-labile complement proteins) is not required.

2. MEFs are primary cells and therefore have a limited life span in culture. They can be trypsinised and passaged 2–3 times. Calculate to thaw one aliquot per week to have always confluent dishes with MEFs for inactivation. MEFs from day-12.5 embryos are preferred, as these may deliver the best support for ESC growth. Dishes and plates with MMC-treated MEFs should always be prepared freshly the day before seeding ESCs.
3. Check up to 12 weeks in advance, if males/females of the desired genotype are available for mating, to allow expansion of your mouse colony. If you would like to stick to a working-week regime, mate animals from Thursday until Monday, to allow isolation of 3.5 dpc blastocysts from Monday until Friday.
4. Plug-check has to occur early in the morning, since females tend to lose their plugs within 12 h upon mating.
5. In our hands, KSOM *(55)* is most efficient in supporting embryo development to the expanded blastocyst stage. Other culture media such as M16 *(56)* or mCZB *(49)* may be used instead. Equally, hCZB for embryo handling at room temperatures may be replaced by M2 medium.
6. To avoid potential bacterial contamination it may be helpful to initially culture isolated blastocysts in ES medium supplemented with Penicillin/Streptomycin (Pen/Strep). However, aim to omit Pen/Strep at later passages as its presence interferes with mycoplasma detection, and is not necessary if aseptic cell culture techniques are followed properly.
7. The efficiency of ES line establishment (in particular for inbred genotypes) is generally enhanced by treatment of isolated blastocysts with the MEK1 inhibitor PD98059 (Cell Signalling Technology, #9900), an inhibitor of the extracellular-signal-regulated kinase (ERK) pathway *(20, 48)*. PD98059 is added to the ES-medium during the blastocyst culture and the dissociation step. Prepare a 50 mM stock of PD98059 in DMSO (5 mg in 374 mL DMSO) and store at –20°C. Dilute into ES medium to a working concentration of 50 μM. It is advisable to monitor MAPK inhibitor-derived ESCs carefully for germline transmission, as the risk for derivation of non-productive ESC lines may be increased.

8. The timing for picking the ICMs is sensitive. For inbred strains we pick the ICMs at day 5–6, for picking of F1 outgrowths we prefer day 6–7. The protocol requires experience in the technique of mouth pipetting; alternatively, a 10-μL pipette may be used.
9. Trypsinisation of up to three blastocyst outgrowths can be handled in parallel. Trypsinisation of each individual outgrowth should be monitored carefully and stopped, when cells are dissociated. Do not let dissociated ICM sit in the trypsin solution for a prolonged period of time. If cell clumps withstand the trypsinisation, refresh the trypsin by adding a drop of 2.5% trypsin and incubate another 2 min. Try to resuspend again. When using the higher-concentrated trypsin, it is advisable to wash suspended cells in ES medium before transferring them into 96-well TC-plates containing MEFs to stop trypsin action completely.
10. The first passage is frequently followed by terminal differentiation mainly to trophectodermal cells. On average, 20–30% of inbred C57BL/6 and 30–60% of hybrid dissociated ICMs should give rise to ES colonies (*see* **Fig. 10.2**).
11. Additional quality controls may include RT-PCR or immunostaining for the expression of the pluripotency genes Oct3/4 *(57)* and Nanog *(58, 59)*, or immunostaining for SSEA-1 *(60)* or alkaline phosphatase.
12. The predominant chromosome aberrations are loss of the Y chromosome, or trisomies of chromosome 8 or 11 *(26)*.

References

1. Evans, M. J. and Kaufman, M. H. Establishment in culture of pluripotential cells from mouse embryos. Nature 1981;292:154–6.
2. Martin, G. R. Isolation of a pluripotent cell line from early mouse embryos cultured in medium conditioned by teratocarcinoma stem cells. Proc Natl Acad Sci USA 1981;78:7634–8.
3. Doetschman, T., Gregg, R. G., Maeda, N., et al. Targetted correction of a mutant HPRT gene in mouse embryonic stem cells. Nature 1987;330:576–8.
4. Thomas, K. R. and Capecchi, M. R. Site-directed mutagenesis by gene targeting in mouse embryo-derived stem cells. Cell 1987;51:503–12.
5. Bradley, A., Evans, M., Kaufmann, M. H. and Robertson, E. Formation of germ-line chimaeras from embryo-derived teratocarcinoma cell lines. Nature 1984;309:255–6.
6. Robertson, E., Bradley, A., Kuehn, M. and Evans, M. Germ-line transmission of genes introduced into cultured pluripotential cells by retroviral vector. Nature 1986;323:445–8.
7. Simpson, E. M., Linder, C. C., Sargent, E. E., Davisson, M. T., et al. Genetic variation among 129 substrains and its importance for targeted mutagenesis in mice. Nat Genet 1997;16:19–27.
8. Ledermann, B. and Burki, K. Establishment of a germ-line competent C57BL/6 embryonic stem cell line. Exp Cell Res 1991;197:254–8.
9. Auerbach, W., Dunmore, J. H., Fairchild-Huntress, V., et al. Establishment and chimera analysis of 129/SvEv- and C57BL/6-derived mouse embryonic stem cell lines. Biotechniques 2000;29:1024–8, 1030, 1032.

10. Kawase, E., Suemori, H., Takahashi, N., et al. Strain difference in establishment of mouse embryonic stem (ES) cell lines. Int J Dev Biol 1994;38:385–90.
11. Kontgen, F., Suss, G., Stewart, C., Steinmetz, M. and Bluethmann, H. Targeted disruption of the MHC class II Aa gene in C57BL/6 mice. Int Immunol 1993;5:957–64.
12. Schuster-Gossler, K., Lee, A. W., Lerner, C. P., et al. Use of coisogenic host blastocysts for efficient establishment of germline chimeras with C57BL/6 J ES cell lines. Biotechniques 2001;31:1022–24, 1026.
13. Cheng, J., Dutra, A., Takesono, A., et al. Improved generation of C57BL/6 J mouse embryonic stem cells in a defined serum-free media. Genesis 2004;39: 100–4.
14. Schoonjans, L., Albright, G. M., Li, J. L., et al. Pluripotential rabbit embryonic stem (ES) cells are capable of forming overt coat color chimeras following injection into blastocysts. Mol Reprod Dev 1996;45:439–43.
15. Yagi, T., Tokunaga, T., Furuta, Y., et al. A novel ES cell line, TT2, with high germline-differentiating potency. Anal Biochem 1993;214:70–6.
16. Tojo, H., Nishida, M., Matsuoka, K., et al. Establishment of a novel embryonic stem cell line by a modified procedure. Cytotechnology 1995;19:161–5.
17. Tielens, S., Verhasselt, B., Liu, J., et al. Generation of embryonic stem cell lines from mouse blastocysts developed in vivo and in vitro: relation to Oct-4 expression. 10.1530/rep.1.00887. Reproduction 2006;132:59–66.
18. Noben-Trauth, N., Kohler, G., Burki, K. and Ledermann, B. Efficient targeting of the IL-4 gene in a BALB/c embryonic stem cell line. Transgenic Res 1996;5: 487–91.
19. Brook, F. A. and Gardner, R. L. The origin and efficient derivation of embryonic stem cells in the mouse. Proc Natl Acad Sci USA 1997;94:5709–12.
20. Buehr, M. and Smith, A. Genesis of embryonic stem cells. Philos. Trans R Soc Lond B Biol Sci 2003;358:1397–402; discussion 1402.
21. Roach, M. L., Stock, J. L., Byrum, R., et al. A new embryonic stem cell line from DBA/1lacJ mice allows genetic modification in a murine model of human inflammation. Exp Cell Res 1995;221:520–5.
22. Gallagher, E. J., Lodge, P., Ansell, R. and McWhir, J. Isolation of murine embryonic stem and embryonic germ cells by selective ablation. Transgenic Res 2003;12:451–60.
23. Kitani, H., Takagi, N., Atsumi, T., et al. Isolation of a germline-transmissible embryonic stem (ES) cell line from C3H/He mice. Zoolog Sci 1996;13:865–71.
24. Nagafuchi, S., Katsuta, H., Kogawa, K., et al. Establishment of an embryonic stem (ES) cell line derived from a non-obese diabetic (NOD) mouse: in vivo differentiation into lymphocytes and potential for germ line transmission. FEBS Lett 1999;455:101–4.
25. Nicklas, W. and Weiss, J. Survey of embryonic stem cells for murine infective agents. Comp Med 2000;50:410–11.
26. Liu, X., Wu, H., Loring, J., et al. Trisomy eight in ES cells is a common potential problem in gene targeting and interferes with germ line transmission. Dev Dyn 1997;209:85–91.
27. Delhaise, F., Bralion, V., Schuurbiers, N. and Dessy, F. Establishment of an embryonic stem cell line from 8-cell stage mouse embryos. Eur J Morphol 1996;34:237–243.
28. Chung, Y., Klimanskaya, I., Becker, S., et al. Embryonic and extraembryonic stem cell lines derived from single mouse blastomeres. Nature 2006;439:216–9.
29. Wakayama, S., Hikichi, T., Suetsugu, R., et al. Efficient establishment of mouse embryonic stem cell lines from single blastomeres and polar bodies. Stem Cells 2007;25:986–93.
30. Tesar, P. J. Derivation of germ-line-competent embryonic stem cell lines from preblastocyst mouse embryos. Proc Natl Acad Sci USA 2005;102:8239–44.
31. Eistetter, H. R. Pluripotent embryonal cell lines can be established from disaggregated mouse morulae. Dev Growth Diff 1989;31:275–82.
32. Axelrod, H. R. Embryonic stem cell lines derived from blastocysts by a simplified technique. Dev Biol 1984;101:225–8.
33. Hogan, B. L. M., Beddington, R. S. P., Costantini, F. and Lacy, E. (eds.) 1994 Manipulating the Mouse Embryo: A Laboratory Manual. 2nd Ed., Cold Spring Harbour Laboratory Press, New York.
34. Robertson, E. J. 1987 Embryo-derived stem cell lines, in Teratocarcinoma and Embryonic Stem Cell: A Practical Approach (Robertson E. J., ed.), IRL Press, Washington, DC, pp. 71–112.
35. Nagy A., G. M., Vintersten, K. (eds.) 2002 Manipulating the Mouse Embryo: A Laboratory Manual. 3rd Ed., Cold Spring Harbor Laboratory Press, New York.

36. Schoonjans, L., Kreemers, V., Danloy, S., et al. Improved generation of germline-competent embryonic stem cell lines from inbred mouse strains. Stem Cells 2003;21:90–7.
37. Munsie, M. J., Michalska, A. E., O'Brien, C. M., et al. Isolation of pluripotent embryonic stem cells from reprogrammed adult mouse somatic cell nuclei. Curr Biol 2000;10:989–92.
38. Wakayama, T., Tabar, V., Rodriguez, I., et al. Differentiation of embryonic stem cell lines generated from adult somatic cells by nuclear transfer. Science 2001;292:740–3.
39. Pease, S., Braghetta, P., Gearing, D., et al. Isolation of embryonic stem (ES) cells in media supplemented with recombinant leukemia inhibitory factor (LIF). Dev Biol 1990;141:344–52.
40. McWhir, J., Schnieke, A. E., Ansell, R., et al. Selective ablation of differentiated cells permits isolation of embryonic stem cell lines from murine embryos with a non-permissive genetic background. Nat Genet 1996;14:223–6.
41. Mountford, P., Nichols, J., Zevnik, B., et al. Maintenance of pluripotential embryonic stem cells by stem cell selection. Reprod Fertil Dev 1998;10:527–33.
42. Nichols, J., Evans, E. P. and Smith, A. G. Establishment of germ-line-competent embryonic stem (ES) cells using differentiation inhibiting activity. Development 1990;110:1341–8.
43. Ying, Q. L., Stavridis, M., Griffiths, D., et al. Conversion of embryonic stem cells into neuroectodermal precursors in adherent monoculture. Nat Biotechnol 2003;21:183–6.
44. Ying, Q. L., Nichols, J., Chambers, I. and Smith, A. BMP induction of Id proteins suppresses differentiation and sustains embryonic stem cell self-renewal in collaboration with STAT3. Cell 2003;115:281–92.
45. Sato, N., Meijer, L., Skaltsounis, L., et al. Maintenance of pluripotency in human and mouse embryonic stem cells through activation of Wnt signaling by a pharmacological GSK-3-specific inhibitor. Nat Med 2004;10:55–63.
46. Miyabayashi, T., Teo, J.-L., Yamamoto, M., et al. Wnt/{beta}-catenin/CBP signaling maintains long-term murine embryonic stem cell pluripotency. Proc Natl Acad Sci USA 2007;104:5668–73.
47. Chen, S., Do, J. T., Zhang, Q., et al. Self-renewal of embryonic stem cells by a small molecule. Proc Natl Acad Sci USA 2006;103:17266–71.
48. Burdon, T., Stracey, C., Chambers, I., et al. Suppression of SHP-2 and ERK signalling promotes self-renewal of mouse embryonic stem cells. Dev Biol 1999;210:30–43.
49. Chatot, C. L., Lewis, J. L., Torres, I. and Ziomek, C. A. Development of 1-cell embryos from different strains of mice in CZB medium. Biol Reprod 1990;42:432–40.
50. Bishop, C. E. and Hatat, D. Molecular cloning and sequence analysis of a mouse Y chromosome RNA transcript expressed in the testis. Nucleic Acids Res 1987;15:2959–69.
51. Tojo, H., Nishida, M., Matsuoka, K., et al. Establishment of a novel embryonic stem cell line by a modified procedure. Cytotechnology 1995;19:161–5.
52. Blank, W. A., Henderson, K. S. and White, L. A. Virus PCR assay panels: an alternative to the mouse antibody production test. Lab Anim (NY) 2004;33:26–32.
53. Eggan, K., Rode, A., Jentsch, I., et al. Male and female mice derived from the same embryonic stem cell clone by tetraploid embryo complementation. Nat Biotechnol 2002;20:455–9.
54. Bryja, V., Bonilla, S., Cajanek, L., et al. An efficient method for the derivation of mouse embryonic stem cells. Stem Cells. 2006;24:844–9.
55. Erbach, G. T., Lawitts, J. A., Papaioannou, V. E. and Biggers, J. D. Differential growth of the mouse preimplantation embryo in chemically defined media. Biol Reprod 1994;50:1027–33.
56. Whittingham, D. G. Culture of mouse ova. J. Reprod Fertil Suppl 1971;14:7–21.
57. Schöler, H. R., Hatzopoulos, A. K., Balling, R., et al. A family of octamer-specific proteins present during mouse embryogenesis: evidence for germline-specific expression of an Oct factor. EMBO J 1989;8:2543–50.
58. Mitsui, K., Tokuzawa, Y., Itoh, H., et al. The homeoprotein Nanog is required for maintenance of pluripotency in mouse epiblast and ES cells. Cell 2003;113:631–42.
59. Chambers, I., Colby, D., Robertson, M., et al. Functional expression cloning of Nanog, a pluripotency sustaining factor in embryonic stem cells. Cell 2003;113:643–55.
60. Solter, D. and Knowles, B. B. Monoclonal antibody defining a stage-specific mouse embryonic antigen (SSEA-1). Proc Natl Acad Sci USA 1978;75:5565–9.

Chapter 11

Generation of Double-Knockout Embryonic Stem Cells

Eva Wielders, Marleen Dekker, and Hein te Riele

Abstract

Gene inactivation in mouse embryonic stem (ES) cells usually affects a single allele that is subsequently transmitted to the mouse germline. Upon breeding to homozygosity the consequences of complete gene ablation can be studied in the context of the complete organism. In many cases, it can be useful to study the consequences of gene ablation already in ES cells, for example, when a cellular phenotype is expected. This requires both alleles of a gene to be disrupted. Besides consecutive targeting by using different selectable marker genes, homozygosity for gene disruption can also be obtained by selecting cells for duplication of (part of) the chromosome carrying the targeted allele with concomitant loss of the wild-type allele.

Key words: ES cells, double knockout, consecutive gene targeting, high drug selection, in vitro studies.

1. Introduction

The inactivation of selected genes in the prokaryotic and eukaryotic genome is a powerful tool in studying their function both at the level of individual cells and in the context of a complete organism. Traditionally, in mouse embryonic stem (ES) cells, the method of gene inactivation relies on their ability to exchange DNA sequences with a high degree of sequence similarity by homologous recombination. Briefly, the procedure involves the generation of a targeting construct that consists of a selectable marker gene flanked by DNA sequences largely identical to the chromosomal locus to be modified. On entry of the targeting construct into the cell, exchange of the flanking sequences with their chromosomal counterparts will result in the introduction of the marker gene into the chromosome thereby disrupting the gene of

Ralf Kühn, Wolfgang Wurst (eds.), *Gene Knockout Protocols: Second Edition, vol. 530*

DOI 10.1007/978-1-59745-471-1_11 Springerprotocols.com

interest *(1)*. Alternatively, gene inactivation can be achieved by random insertional mutagenesis using retrovirus-based gene-trap vectors *(2)*. Determining integration sites may lead to the identification of a disruption in the gene of interest in a library of mutated ES cells. Genetically modified ES cells can subsequently be injected into wild-type blastocysts giving rise to chimeric mice that can transmit the disrupted allele to their offspring. Inbreeding of heterozygous animals will yield offspring carrying the disruption in both alleles. The phenotypic consequences of a complete gene knockout in the mouse germline can provide pivotal information on the function of the gene in development and maintenance of the organism. Moreover, valuable mouse models have been generated for recessive genetic disorders and cancer predisposition syndromes in man.

For a number of reasons, however, it can be desirable or even necessary to achieve complete gene inactivation in ES cells. For many genes, only limited information is gained from their complete inactivation in the mouse germline. This is the case when a complete loss-of-function mutation does not cause any overt phenotype or, on the other hand, leads to early embryonic lethality, which precludes studying its phenotypic consequences at later stages of development and during adult life. To circumvent the latter problem, both copies of a gene can be inactivated in ES cells, which can be used to generate chimeric mice in which embryonic lethality might be prevented by the presence of wild-type cells. This approach has successfully been used in studying the developmental and tumorigenic consequences of an embryonic-lethal loss-of-function mutation in the retinoblastoma gene *Rb (3)*. The total absence of a knockout phenotype may be indicative for complementation by other genes, thus requiring a combination of loss-of-function mutations in different genes. This can be achieved by crossing the relevant knockout mice *(4)*, a complicated effort, when the number of genes involved increases. The generation of chimeric mice using an ES cell line carrying multiple gene knockouts can then be a valuable alternative. Besides the generation of chimeras, mutant ES cell lines can also be studied for their differentiation capacity in vitro or in vivo after subcutaneous implantation into syngeneic or nude mice *(5)*. Finally, the generation of double-knockout ES cell lines can be highly instrumental when studying genes expected to function in a cell-autonomous way, that is, independent of the context of a complete (developing) organism. This is, for example, the case for genes functioning in DNA repair pathways *(6)*.

Clearly, the generation of double-knockout ES cells is a valuable addendum to the generation of knockout mice. Unfortunately, gene targeting only rarely produces double-knockouts in a single experiment. Thus, in a gene targeting experiment aimed at disrupting the *Rb* gene with a Hygromycin-resistance marker, 80%

of the selected colonies resulted from correct integration of the marker into *Rb*. However, in none of the colonies that were analyzed a double-knockout was observed *(7)*. Although in this example targeted ES cell clones were obtained far more efficiently than cells with random integrations of the targeting construct, homologous recombination in an individual cell was apparently restricted to a single event.

Therefore, alternative protocols have been developed to introduce a disruption in the remaining wild-type allele of a single-knockout ES cell line. These are schematically illustrated in **Fig. 11.1**.

1.1. Method 1: Selection at High Drug Concentration

ES cells carrying a disruption in one allele of a particular gene are subjected to a second round of selection at drug concentrations largely exceeding that used to obtain the single-knockout. Highly

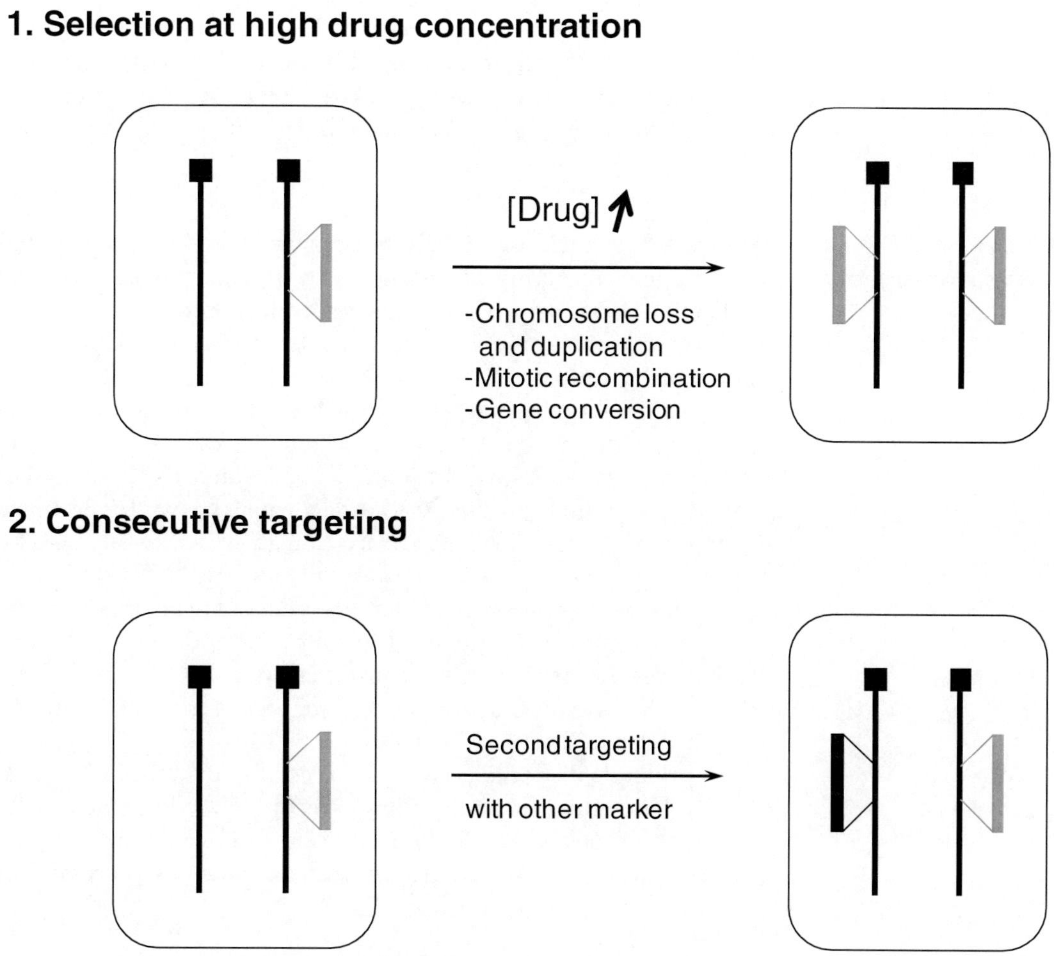

Fig. 11.1. Two methods to generate double-knockout ES cells.

resistant colonies often appear to result from duplication of the targeted allele and concomitant loss of the wild-type allele. This method has originally been described for ES cells carrying a *neo* gene conferring resistance to G418 *(8)*, but also works with Hygromycin B and Puromycin resistance genes. The mechanism by which double-knockout ES cells spontaneously arise in a culture of single-knockout cells likely involves nondisjunction and duplication of the targeted chromosome *(9)*, although mitotic recombination has not been excluded *(10)*. By either mechanism, distant regions of the chromosome become duplicated together with the selectable marker gene. This creates the possibility to render cells homozygous for a particular subtle mutation, for example, introduced by oligonucleotide-directed mutagenesis (*see* **Chapter 5**), provided that a selectable marker is present on the same chromosome. The high-drug selection procedure is relatively simple but the frequency of double-knockout cells among colonies resistant to high drug concentration is highly variable. It can be as low as a few percent; in contrast, for alleles targeted with the poly(A) gene-trap vector GTR1.3, frequencies as high as 85% have been reported *(10)*. The procedure fails when the resistance level of the single-knockout is above achievable drug concentrations or when other mechanisms of gene amplification dominate.

1.2. Method 2: Consecutive Targeting

The wild-type allele in a single-knockout ES cell line is targeted with a new targeting construct carrying another selectable marker as previously used. This procedure requires the construction of a new targeting vector, but is highly reliable *(11)*. Importantly, unlike the first method, a second round of gene targeting with a new marker gene will easily reveal lethality of a complete gene inactivation at the cellular level. In that case, only single-knockouts will be obtained in which the first marker gene is exchanged for the second, leaving the wild-type allele unaltered. Various marker genes are available (**Table 11.1**). The most widely used are *neo*, *hyg*, *pur*, *his*, *bls*, and *Hprt* driven by promoters which are active in ES cells. These include the *3-phosphoglycerate*(*Pgk*) gene promoter *(12)*and the HSV-*Tk*/ PyF441 promoter/enhancer combination from the pMC1*neo* vector *(13)*. Note that the *Hprt* marker gene can only be used in *Hprt*-deficient ES cells.

In principle it should be feasible to combine methods 1 and 2, that is, to introduce into single-knockout ES cells the same targeting construct as was previously used, followed by selection at high drug concentration. In our hands, this method was successful in one case, but failed in another.

Table 11.1
Marker genes for gene targeting in embryonic stem (ES) cells

Marker gene	Selective drug	Concentration[a]	Reference
neo	G418	200 μg/mL	*(13)*
hyg	Hygromycin B	150 μg/mL	*(7, 11)*
pur	Puromycin	1.8 μg/mL	*(14)*
his	L-Histidinol	2 mM	*(15)*
bls	Blasticidin S	5 μg/mL	*(16)*
Hprt[b]	HAT medium:		*(17)*
	Hypoxanthine	100 μM	
	Aminopterin	800 nM	
	Thymidine	20 μM	

[a] For single-knockouts in ES cell line E14 *(18)*.
[b] To be used in *Hprt*-deficient cells.

In case of cellular lethality of a complete gene inactivation, conditional/inducible double-knockouts could be made by using the Cre/lox or Flp/Frt site-specific recombination systems.

Figure 11.2 presents a flow scheme for the generation of double-knockout ES cells. Single-knockout ES cell clones should always be subcloned, in particular when method 1 is applied. This is because ES cell cultures obtained after outgrowth of selected colonies are often bi- or even polyclonal: targeted ES cell clones are contaminated with cells carrying random integrations of the targeting construct. These may preferentially grow out at high drug concentration. Then, the resistance level of single-knockout cells is determined. If this is below tenfold the concentration that is normally used (**Table 11.1**), direct selection for double-knockouts at high drug concentration is likely to be successful. However, sometimes the resistance level of single-knockouts is much higher. We recently found that it is still possible to obtain double-knockouts even when the G418 concentration had to be raised 50-fold. In selected clones the status of the target gene is determined by Southern blot analysis. ES cell clones that have lost the wild-type allele are karyotyped to verify the number of chromosomes. It is highly recommendable to derive subclones of the double-knockout line, which should be checked again for genotype and chromosome number.

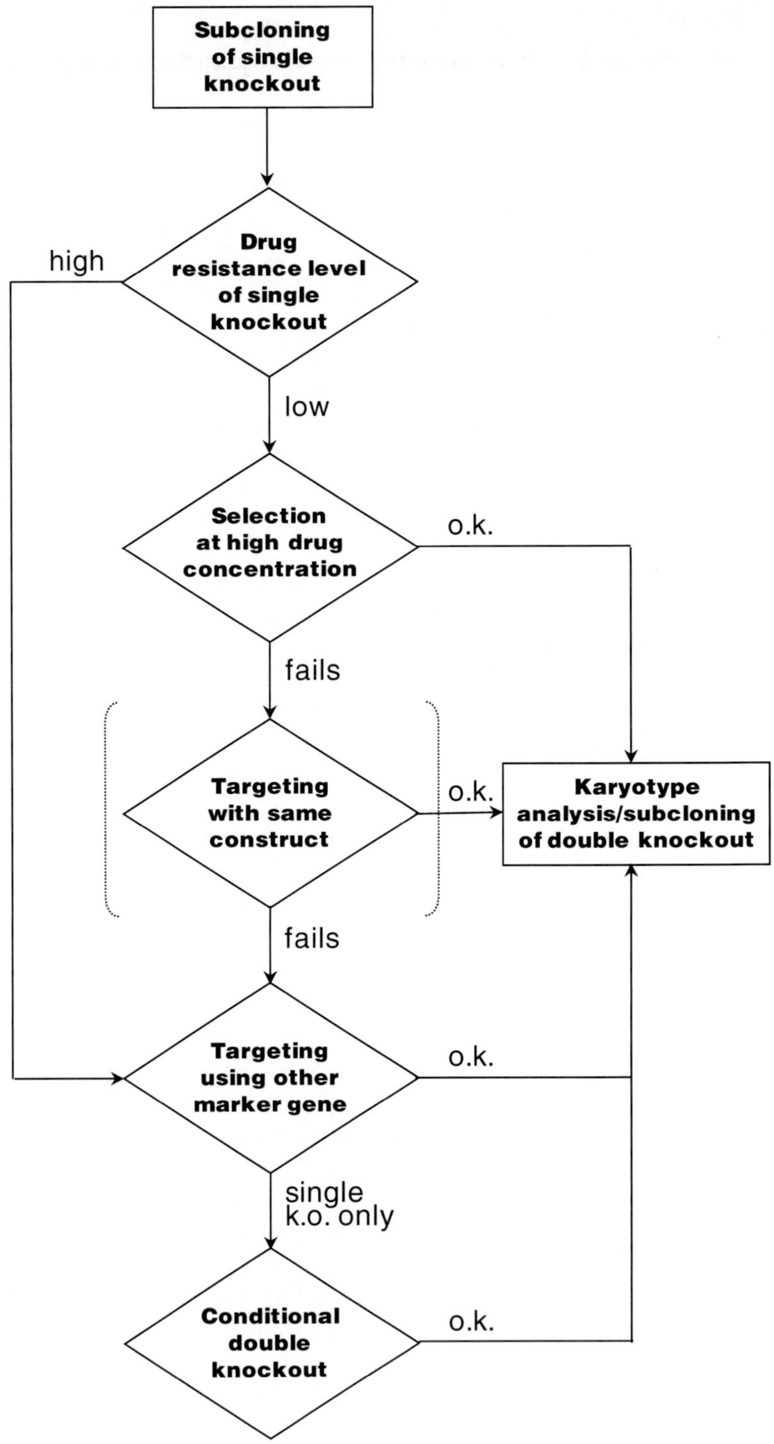

Fig. 11.2. Flow scheme for the generation of double-knockout ES cells.

2. Materials

ES cells are routinely cultured on irradiated primary mouse embryo fibroblasts (MEFs). However, during drug selections, cells are cultured on gelatin-coated plates in buffalo rat liver (BRL)-conditioned medium. This is because MEFs or other feeder layers resistant to the drugs to be used are not always readily available or are not always tolerant to the high concentrations that are required.

Protocols for the preparation of MEF feeder layers, BRL-conditioned medium, and culturing of ES cells are given elsewhere in this volume. This chapter provides protocols for subcloning of ES cell lines and the generation of double-knockout mutants.

2.1. Media and Solutions

1. CM (complete medium): 1X GMEM (GIBCO/Invitrogen, 21710) 500 mL, 100 mM sodium pyruvate (GIBCO/Invitrogen, 11360) 5 mL, 100X non-essential amino acids (GIBCO/Invitrogen, 11140) 5 mL, fetal calf serum 50 mL (FCS is tested for optimal growth of ES cells).
2. CM+β+LIF (leukemia inhibitory factor): 100 mL CM, 0.1 mL β-mercaptoethanol (1000X), 1 mL LIF (100X).
3. BRL medium: CM conditioned on a monolayer of buffalo rat liver (BRL-3A) cells, filtered through a 0.2 µm Nalgene filter unit, 450-0020.
4. BRL+β+LIF (60% medium): 100 mL CM, 150 mL BRL medium, 1.5 mL L-glutamine 200 mM (GIBCO/Invitrogen, 25030), 0.25 mL β-mercaptoethanol (1000X), 2.5 mL LIF (100X).
5. β-Mercaptoethanol (1000X, 0.1 M): 0.1 mL 2-mercaptoethanol (14.2 M, Merck) and 14.1 mL water. Sterilize by filtration through 0.22 µm Millex-GV filter. Store at 4°C for up to 1 month. Working concentration: 0.1 mM.
6. LIF (100X): Dissolve 1 mL ESGRO® (10^7 units, Chemicon International, ESG1107) in 99 mL CM. Working concentration: 10^3 U/mL. Store at 4°C.
7. Distilled water: 500 mL bottles (GIBCO/Invitrogen, 15230).
8. 1% Gelatin (10X): 5 g (w/v) gelatin (Sigma, G1890) in 500 mL distilled water. Gently heat in microwave oven (do not boil). Sterilize through a 0.2 µm Nalgene filter unit, 450-0020, while still warm and store aliquots at 4°C. Before use, dilute to 0.1% with water and coat tissue culture dishes for 30 min at RT.
9. PBS (phosphate-buffered saline): Dulbecco's PBS (1X) without calcium and magnesium, 500 mL bottles (GIBCO/Invitrogen, 14190).

10. Trypsin/EDTA stock (TVP): 500 mL PBS supplemented with 5 mL Trypsin 2.5% (GIBCO/Invitrogen, 15090), 12.5 mL 40 mM ethylene diamine tetraacetic acid (EDTA), 5 mL chicken serum (GIBCO/Invitrogen, 16110). Sterilize by filtration through a 0.2 μm Nalgene filter unit, 450-0020, and store aliquots at –20°C.
11. 2X TVP: Add 0.1 mL Trypsin 2.5% to 9.9 mL TVP. Store at –20°C.
12. 10X TVP: Add 0.9 mL Trypsin 2.5% to 9.1 mL TVP. Store at –20°C.
13. Lysis Buffer: 100 mM Tris pH 8.0, 5 mM EDTA, 0.2% (w/v) sodium dodecyl sulphate (SDS), 200 mM NaCl, 200 μg/mL (freshly added) Proteinase K (Merck 24568.0100).

2.2. Selective Drugs

Solutions are sterilized by filtration through 0.22 μm Millex-GV. Use only disposables and not glassware.

1. G418 (100X): 20 mg G418 (GIBCO/Invitrogen, 11811-031) per mL 60% medium, store at 4°C. Working concentration: 200 μg/mL.
2. Hygromycin B (100X): 15 mg Hygromycin B (Calbiochem, 400050 or 400051) per mL 60% medium, store at 4°C. Working concentration: 150 μg/mL.
3. Puromycin (1000X): 18 mg Puromycin (Sigma, P7255) per 10 mL PBS, store at –20°C. Working concentration: 1.8 μg/mL.
4. Histidinol (100X) 250 mM: 53.5 mg L-Histidinol.2HCl (Sigma, H 6647) per mL 60% medium, store at 4°C. Working concentration: 1.5–2.5 mM.
5. Blasticidin (1000X): 50 mg Blasticidin S (Invitrogen, R21001) per 10 mL PBS, store at –20°C. Working concentration: 5 μg/mL.
6. HAT (100X):

   Hypoxanthine 10 mM in water, neutralized with 10 M NaOH, store at –20°C.

   Aminopterin: 80 μM in water, store at –20°C.

   Thymidine: 2 mM in water, store at –20°C.
7. HAT (50X): HAT Supplement (50X) (GIBCO/Invitrogen, 043-01060H) 100 mL.

2.3. Tissue Culture Plastics

ES cells are grown on standard tissue culture supports. We commonly use:

1. T25: 25 cm^2 tissue culture flask (Costar, 3055)
2. T75: 75 cm^2 tissue culture flask (Costar, 3275)
3. T150: 162 cm^2 tissue culture flask (Costar, 3150)

4. 96-well multidish: 0.3 cm^2 flat bottom (Costar, 3596)
5. 4-well multidish: 1.8 cm^2 (Nunc, 176740)
6. 12-well multidish: 4.0 cm^2 (Costar, 3512)
7. 6-well multidish: 10 cm^2 (Costar, 3506)
8. Tissue culture Petri dish: 100 × 20 mm (Falcon, 3003)
9. 96-tube freezing boxes (Greiner, 975561) with 1.3 mL polypropylene freezing tubes with attached strip caps, 8.5 × 44 mm (Greiner, 102261).

3. Methods

3.1. Subcloning of Single-Knockout ES Cell Clones

3.1.1. Subcloning

1. Prepare a 96-well plate (flat bottom) with a MEF feeder layer. (Use a multichannel pipette when handling 96-well plates).
2. Culture an ES cell line at a small scale (e.g., on a 4-well plate).
3. Trypsinize the ES cell culture and count the number of cells.
4. Add 50 cells to 20 mL of CM+β+LIF.
5. Fill the 96-well plate with 200 µL per well.
6. Grow individual colonies (15–20 will appear after 7 days).
7. Wash the wells with 100 µL PBS and trypsinize with 25 µL 10X TVP for 5 min at 37°C.
8. Add 150 µL CM+β+LIF, resuspend, and transfer the cells to a new 96-well plate with MEFs.
9. Culture the cells overnight and refresh the medium (150 µL).
10. Culture the cells to (semi)confluency (2–3 days).
11. Wash the cells with 100 µL PBS and trypsinize with 25 µL 10X TVP for 5 min at 37°C.
12. Add 175 µL CM+β+LIF and resuspend.
13. Transfer two 100 µL portions of cell suspension to two new 96-well plates with MEFs and 100 µL CM+β+LIF medium.
14. Culture one plate to semi-confluency (2–3 days) and process for freezing (*see* **Section 3.1.2**). The other plate is cultured as dense as possible for DNA isolation (*see* **Section 3.1.3**).

3.1.2. Freezing Procedure

1. Prepare CM (without LIF) with 20% dimethylsulfoxide (DMSO) at 4°C.
2. Aliquot 100 µL CM plus 20% DMSO in individual tubes of a 96-tube freezing box (Greiner, 975561).
3. Wash the cells with PBS and trypsinize with 25 µL 10X TVP.

4. Add 75 μL complete medium and resuspend carefully.
5. Add 100 μL of trypsinized cells to 100 μL 20% DMSO in the freezing tubes and resuspend.
6. Store in liquid nitrogen (or at –80°C for maximally 4 weeks).

3.1.3. DNA Isolation

1. Wash the cells with PBS.
2. Optional: Cells can be stored dry in 96-well at –20°C.
3. Add 50 μL of Lysis Buffer with 100 μg/mL Proteinase K.
4. Carefully seal 96-well plate and incubate for 2 h or overnight at 55°C.
5. Add to the lysates 100 μL ethanol 96% and mix in a shaker (Amersham) at RT for 30 min.
6. Centrifuge plates at 3200 rpm for 30 min at RT.
7. Remove the supernatant with a Gilson pipette, each well individually.
8. Add 150 μL 70% ethanol, shake for 10 min, and centrifuge at 3200 rpm for 15 min at RT.
9. Remove the supernatant with a Gilson pipette, allow the DNA precipitates to dry.
10. Add 100 μL TE and incubate overnight at 55°C.
11. Use 40 μL DNA for Southern analysis.

3.1.4. Thawing Procedure

1. Cut the tube with the desired clone out of the freezing box and thaw immediately at 37°C.
2. Transfer the cells to a sterile Eppendorf tube or a 15 mL Falcon tube containing 1 mL of CM.
3. Centrifuge the cell suspension (1200 rpm).
4. Resuspend the cells in CM+β+LIF and transfer the cells to a 96-well or 4-well multidish (with MEFs).
5. Culture the subclone to the appropriate surface for DNA analysis and long-term storage in liquid nitrogen.

3.2. Generation of Double-Knockout ES Cells by Selection

3.2.1. Determination of Drug Resistance Level

1. Seed single-knockout ES cells onto a gelatin-coated 6-well plate at a density of 10^5 cells per well in 3 mL of BRL+β+LIF (60% medium).
2. Refeed the cells the next day with fresh BRL+β+LIF (60% medium), adding to each well increasing concentrations of the drug to be tested:
   G418: 0 mg/mL, 0.4 mg/mL, 0.8 mg/mL, 1.2 mg/mL, 1.6 mg/mL, 2.0 mg/mL
   Hygromycin B: 0 mg/mL, 0.3 mg/mL, 0.6 mg/mL, 0.9 mg/mL, 1.2 mg/mL, 1.5 mg/mL

Puromycin: 0 μg/mL, 4 μg/mL, 8 μg/mL, 12 μg/mL, 16 μg/mL, 20 μg/mL

(if cells survive the highest concentration, higher drug levels could be tested. For example, for G418 up to 12 mg/mL in steps of 1 mg/mL).

3. Culture the cells for at least 7 days. Determine the drug concentration at which complete cell death is observed. If cells are surviving the highest achievable concentration, method 2 (*see* **Section 3.3**) should be used.

3.2.2. Selection for Spontaneous Double-Knockouts at High Drug Concentration

1. Culture a subcloned single-knockout ES cell line in CM+β+LIF on one 6-well plate with MEFs. A confluent well contains approximately 8×10^6 cells.
2. Trypsinize the cells with 10X TVP and seed onto ten 10-cm Petri dishes (6×10^5 per dish), coated with gelatin, in BRL+β+LIF (60% medium).
3. Refeed the cells the next day with fresh BRL+β+LIF (60% medium) and start selection: Add to four dishes the concentration of drug at which all cells died within 7 days; add to three dishes a somewhat lower and to the remaining three a somewhat higher concentration.
4. Refeed the cells with selective medium every 2–4 days. Surviving colonies arise after 8–14 days (*see* **Note 1**).
5. Culture selected colonies on 96-well plates (*see* **Section 3.4**).

3.3. Generation of Double-Knockouts by Consecutive Targeting

3.3.1. DNA

1. Linearize the targeting vector or purify the DNA targeting fragment (25–30 μg per experiment) from an agarose gel. Extract DNA with phenol, phenol/chloroform and chloroform, and precipitate with alcohol.
2. Dry the DNA pellet in a flow cabinet and dissolve in 100 μL of PBS (sterile).
3. Check the DNA concentration by agarose gel electrophoresis of 1–2 μL.

3.3.2. ES Cells

1. Trypsinize single-knockout ES cells cultured on one well of a 6-well plate.
2. Wash the cells with a small volume of PBS.
3. Add 0.5 mL 10X TVP, incubate 5 min at 37°C.
4. Detach the cells.
5. Add 0.5 mL of complete medium and resuspend to single cells.
6. Count the cells, approximately 6×10^6 cells are present.
7. Spin down the cells for 5 min at 1200 rpm at RT.
8. Resuspend the cell pellet in 100 μL PBS.

3.3.3. Electroporation (see Note 2)

1. Mix the cells ($\approx 6 \times 10^6$) with the DNA (≈ 25 μg) → ± 200 μL: Leave 5 min at RT.
2. Transfer the mixture to a 0.4 cm electroporation cuvette (Biorad, 165-2088).
3. Electroporate the cells with a Biorad gene pulser, model 165-2078: Voltage 0.8 kV, capacitance 1 μF. The time constant of a pulse should be 0.06 ms.
4. Leave 5 min at RT.
5. Resuspend the cells in 24 mL of BRL+β+LIF (60% medium) and spread 3×8 mL in 10-cm^2 gelatin-coated Petri dishes.
6. Incubate ON at 37°C/5% CO_2.

3.3.4. Drug Selection

1. Refeed the cells with 8 mL BRL+β+LIF (60% medium) plus the appropriate drug (*see* **Notes 3** and **4**):
   G418: 200 μg/mL

   Hygromycin B: 150 μg/mL

   L-Histidinol: 1.5–2.5 mM

   HAT: Hypoxanthine 0.1 mM, aminopterin 0.8 μM, thymidine 20 μM

   Puromycin: 1.8 μg/mL

   Blasticidin S: 5 μg/mL
2. Refeed the cells with selective medium every 2–4 days.
3. After 6–8 days individual colonies can be seen.
4. Culture selected colonies on 96-well plates (*see* **Section 3.4**).

3.4. Culturing of Selected Colonies on 96-Well Plates

1. Prepare a 96-well plate (V-bottom, Costar, 3894) with 15 μL PBS per well using the multichannel pipette.
2. Scrape off individual colonies with a Gilson P10 pipetman and transfer them to the PBS in the 96-well plate. Pick series of 24–36 colonies.
3. Add 15 μL 2X TVP to each well and incubate 5 min at 37°C.
4. Add 50 μL CM+β+LIF medium and resuspend.
5. Transfer trypsinized colonies (80 μL) to a 96-well plate with MEFs (flat bottom) containing 100 μL CM+β+LIF per well.
6. Culture the cells overnight and refresh the medium (150 μL).
7. Culture the cells to (semi)confluency (2–3 days).
8. Wash the cells with PBS and trypsinize with 25 μL 10X TVP for 5 min at 37°C.
9. Add 175 μL CM+β+LIF and resuspend.

10. Transfer two 100 µL portions of cell suspension to two new 96-well plates with MEFs containing 100 µL CM+β+LIF.
11. One plate is cultured to semi-confluency (2–3 days) and processed for freezing (*see* **Section 3.1.2**).
12. The other plate is cultured as dense as possible for DNA isolation (*see* **Section 3.1.3**, **Note 5**).

4. Notes

1. When high drug concentrations are required, selection for resistant colonies may take longer: for example, for G418 up to 20 days.
2. Electroporation can be done at a larger scale, for example, mix cells from a T75 with 50–100 µg DNA in a total volume of 600 µL PBS. Electroporate at 0.8 kV with a capacitance of 3 µF giving a time constant of ±0.1 and spread the cells onto six 10-cm dishes.
3. In some cases cross resistance can occur. For example, in the presence of a *hyg* marker, selection for *neo* requires a somewhat higher G418 concentration (250 µg/mL).
4. Double selections are not recommended as this would obscure lethality of complete gene inactivation. For example, exclusive targeting of the already targeted allele would be missed if the double-knockout is cellular lethal.
5. Double-knockout ES cell clones, as identified by Southern analysis, are karyotyped to verify the chromosome number. It is strongly recommended to subclone double-knockout ES cell lines before using them for in vitro experiments or the generation of chimeric mice.

Acknowledgments

The protocols in this chapter were designed and refined by many workers in this field. We thank our colleagues Marieke Aarts, Sandra de Vries, and Jan-Hermen Dannenberg for sharing their experiences on the various aspects of gene targeting in embryonic stem cells. We acknowledge financial support from the Dutch Cancer Society (NKI 2004-3084).

References

1. Capecchi, M.R. Altering the genome by homologous recombination. Science 1989;244:1288–92.
2. Skarnes, W.C., von Melchner, H., Wurst, W. et al. International Gene Trap Consortium. A public gene trap resource for mouse functional genomics. Nat Genet 2004;36:543–4.
3. Robanus Maandag, E.C., Van der Valk, M., Vlaar, M. et al. Developmental rescue of an embryonic-lethal mutation in the retinoblastoma gene in chimeric mice. Embo J 1994;13:4260–68.
4. Cobrinik, D., Lee, M.-H., Hannon, G., et al. Shared role of the pRB-related p130 and p107 proteins in limb development. Genes Dev 1996;10:1633–44.
5. Hilberg, F., and Wagner, E.F. Embryonic stem (ES) cells lacking functional c-jun: consequences for growth and differentiation, AP-1 activity and tumorigenicity. Oncogene 1992;7:2371–80.
6. De Wind, N., Dekker, M., Berns, A., Radman, M., and Te Riele, H. Inactivation of the mouse Msh2 gene results in mismatch repair deficiency, methylation tolerance, hyperrecombination, and predisposition to cancer. Cell 1995;82:321–30.
7. Te Riele, H., Robanus Maandag, E., and Berns, A. Highly efficient gene targeting in embryonic stem cells through homologous recombination with isogenic DNA constructs. Proc Natl Acad Sci USA 1992;89:5128–32.
8. Mortensen, R.M., Conner, D.A., Chao, S., Geisterfer-Lowrance, A.A.T., and Deidman, J.G. Production of homozygous mutant ES cells with a single targeting construct. Mol Cell Biol 1992;12:2391–5.
9. Lefebvre, L., Dionne, N., Karaskova, J., Squire, J.A., and Nagy, A. Selection for transgene homozygosity in embryonic stem cells results in extensive loss of heterozygosity. Nat Genet 2001;27:257–8.
10. Donahue, S.L., Lin, Q., Cao, S., and Ruley, H.E. Carcinogens induce genome-wide loss of heterozygosity in normal stem cells without persistent chromosomal instability. Proc Natl Acad Sci USA 2007;103:11642–46.
11. Te Riele, H., Robanus Maandag, E., Clarke, A., Hooper, M., and berns, A. Consecutive inactivation of both alleles of the pim-1 proto-oncogene by homologous recombination in embryonic stem cells. Nature 1990;348:649–51.
12. McBurney, M.W., Sutherland, L.C., Adra, C.N., Leclair, B., Rudnicki, M.A., and Jardine, K. The mouse Pgk-1 gene promoter contains an upstream activator sequence. Nucleic Acids Res 1991;19:5755–61.
13. Thomas, K.R., and Capecchi, M.R. Site-directed mutagenesis by gene targeting in mouse embryo-derived stem cells. Cell 1987;51:503–12.
14. Vara, J., Portela, A., Ortín, J., and Jiménez, A. Expression in mammalian cells of a gene from *Streptomyces* conferring puromycin resistance. Nucleic Acids Res 1986;14:4617–24.
15. Hartman, S.C., and Mulligan, R.C. Two dominant-acting selectable markers for gene transfer studies in mammalian cells. Proc Natl Acad Sci USA 1988;85:8047–51.
16. Izumi, M., Miyazawa, H., Kamakura, T., Yamagushi, I., Endo, T., and Hanaoka, F. Blasticidin S-resistance (bsr): a novel selectable marker for mammalian cells. Exp Cell Res 1991;197:229–33.
17. Van der Lugt, N., Robanus Maandag, E., Te Riele, H., Laird, P.W., and Berns, A. A pgk::hprt fusion as a selectable marker for targeting of genes in mouse embryonic stem cells: disruption of the T-cell receptor δ-chain-encoding gene. Gene 1991;105:263–5.
18. Hooper, M., Hardy, K., Handyside, A., Hunter, S. and Monk, M. HPRT-deficient (Lesh–Nyhan) mouse embryos derived from germline colonization by cultured cells. Nature 1987;326:292–5.

Chapter 12

Differentiation Analysis of Pluripotent Mouse Embryonic Stem (ES) Cells In Vitro

Insa S. Schroeder, Cornelia Wiese, Thuy T. Truong, Alexandra Rolletschek, and Anna M. Wobus

Abstract

Pluripotent embryonic stem (ES) cells are characterized by their almost unlimited potential to self-renew and to differentiate into virtually any cell type of the organism. Here we describe basic protocols for the in vitro differentiation of mouse ES cells into cells of the cardiac, neuronal, pancreatic, and hepatic lineage. The protocols include (1) the formation of embryoid bodies (EBs) followed by (2) the spontaneous differentiation of EBs into progenitor cells of the ecto-, endo-, and mesodermal germ layer and (3) the directed differentiation of early progenitors into the respective lineages. Differentiation induction via growth and extracellular matrix factors leads to titin-expressing spontaneously beating cardiac cells, tyrosine hydroxylase-expressing dopaminergic neurons, insulin and c-peptide co-expressing pancreatic islet-like clusters, and albumin-positive hepatic cells, respectively. The differentiated cells show tissue-specific proteins and electrophysiological properties (action potentials and ion channels) in cardiac and neuronal cells, glucose-dependent insulin release in pancreatic cells, or glycogen storage and albumin synthesis in hepatic cells. The protocols presented here provide basic systems to study differentiation processes in vitro and to establish strategies for the use of stem cells in regenerative therapies.

Key words: Embryonic stem cells, mouse, cardiac, neuronal, pancreatic, hepatic differentiation.

1. Introduction

Pluripotent embryonic stem (ES) cells have the potential to self-renew and to differentiate into virtually any cell type of somatic and germ cell lineages *(1)*. In the early 1980s, undifferentiated embryonic cells from the inner cell mass (ICM) of mouse blastocysts were cultured as pluripotent mouse (m) ES cell lines *(2–5)*. Pluripotent embryonic cell lines were also established from mouse primordial germ cells (EG cells, *(6)*) and life stock and laboratory

Ralf Kühn, Wolfgang Wurst (eds.), *Gene Knockout Protocols: Second Edition, vol. 530*

DOI 10.1007/978-1-59745-471-1_12 Springerprotocols.com

animals *(7, 8)*. The first derivations of human (h) ES cells from blastocysts generated by in vitro fertilization have been described in 1998 *(9)*. Since then, more than 500 hES cell lines have been established worldwide *(10–13)*.

Self-renewal capacity and multilineage differentiation ability are the unique properties of ES cells. Therefore, ES cells are suitable tools for studying pluripotency of reprogrammed cells. Although hES cells have been investigated for more than 8 years, until now, mES cells are the most efficient experimental systems with respect to ES cell derivation, long-term cultivation, genetic manipulation and multilineage differentiation. In the following, we describe methods of in vitro cultivation and differentiation of mES cells.

To maintain the undifferentiated state, murine ES cells are routinely cultured on feeder layer (FL) cells of mouse embryonic fibroblasts (MEFs) and/or in the presence of leukemia inhibitory factor (LIF) *(14)*. LIF is a member of the interleukin-6 (IL-6) family of cytokines that act via gp130 receptor-mediated pathways *(15)*. Undifferentiated mES cells are characterized by a normal (euploid) karyotype and a short generation time of 9–12 h with a short G1 cell cycle phase *(16)*. Undifferentiated mES cells express specific cell surface antigens (SSEA-1, *(17)*), and enzyme activities such as the tissue non-specific alkaline phosphatase (ALP) *(5)* and telomerase *(7, 18, 19)*.

Undifferentiated mES cells are further characterized by the expression of molecular markers that define "stemness", such as the transcription factor Oct 3/4 *(20)* and the homeodomain protein Nanog *(21, 22)* (*see also (1)*). Both Nanog and Oct 3/4 (*23*) are essential to maintain ES cell identity, but STAT3, following LIF activation, and the activation of BMP (bone morphogenetic protein)- and MEK/ERK-dependent signaling pathways play an accessory role *(15, 24)*. Besides these parameters, other pluripotency-associated markers including Sox2, Rex1, and FoxD3 have been described *(25–28)*.

To test for pluripotency or multilineage differentiation capacity, three experimental models are accepted for mES cells: (1) Contribution of a given ES cell line to all cell lineages including the germline after transfer into mouse blastocysts *(29)*, (2) the induction of benign teratomas and/or malignant teratocarcinomas containing various somatic cell types by extra-uterine transplantation of ES cells into appropriate mouse strains *(5)*, and (3) the differentiation of ES cells in vitro *(1)*.

Experimental protocols for the in vitro differentiation of ES cells have been established based on the "hanging drop" method *(30, 31)*, "mass culture" *(2)*, cultivation in methylcellulose *(32)*, or by co-culture with stromal cell line activity *(33)*, and directed differentiation induction in adherent monolayer cultures in the absence of LIF *(34)*.

Here, we present differentiation protocols based on "hanging drop" cultures, where aggregates called EBs of a defined number of mES cells are formed. After plating to adhesive substrates, cells within EBs undergo specific morphological changes. First, an outer layer of early endoderm-like cells is formed, followed by the development of an ectodermal rim and the differentiation of meso- and endodermal cells. The differentiation is promoted into a variety of specialized cell types, including cardiac (e.g., *(2, 31, 35, 36)*), smooth muscle *(37, 38)*, skeletal muscle *(39)*, hematopoietic (e.g., *(32, 40)*), adipogenic *(41)*, chondrogenic *(42)*, endothelial (e.g., *(38, 43)*), neuronal (e.g., *(33, 44–48)*), epithelial *(49)*, hepatic (e.g., *(50–54)*), and pancreatic (e.g., *(55–57)*) cells. The temporal expression of tissue-specific genes and proteins in ES-derived cells during in vitro differentiation indicates that the early processes of in vivo development into ectoderm, mesoderm and endoderm lineages are recapitulated in vitro *(1, 58, 59)*. During this process, the differentiated cell types develop functional properties like ion channels, receptors, and cell junctions (e.g. *(37, 60, 61)*).

The culture conditions and specific parameters, such as ES cell density, media components, growth factors and additives, and the quality of fetal calf serum (FCS) specifically influence the differentiation capacity of ES cells. In addition, different cell lines are characterized by different developmental properties in vitro *(30)*. Differentiation factors have to be added or, alternatively, genetic modifications have to be introduced into ES cells to increase the capacity to develop into the desired phenotype (e.g., *(30, 55, 62, 63)*).

In most cases, the media used to differentiate ES cells contain FCS. However, FCS contains variable amounts of growth factors or signaling molecules like fibroblast growth factor (FGF), Activin A or IL-6 in addition to several unknown factors, which may adversely affect or at least complicate in vitro differentiation. Furthermore, the use of FCS poses additional threats: FCS is a major source of potential contaminants like viruses, prions, and mycoplasms, all of which may affect the differentiation process. In addition, the molecular and cellular analysis of specific signaling factors for directed differentiation requires a medium that does not contain unknown components, which adversely influence the differentiation program. In 1995, Johansson and Wiles for the first time designed a chemically defined medium (CDM) in which ES cells survived in the absence of serum and differentiated in response to exogenously added signaling factors, such as BMP4 and Activin A *(64)*. Later, this CDM was used with minor modifications for mesoderm induction by Ng et al. *(65)*, and for neuronal differentiation by Bouhon et al. *(66)*.

Here, the cultivation of mES cells in CDM via hanging drop or monolayer culture is presented as an alternative to the FCS- containing medium used for spontaneous differentiation.

Moreover, this chapter describes the routine cultivation of undifferentiated mES cells and the differentiation into derivatives of the three primary germ layers, ectoderm (neuronal cells, *(67)*), mesoderm (cardiomyocytes, *(30, 35, 36)*), and endoderm (hepatic and pancreatic *(53, 55, 68)*) cells.

2. Materials

2.1. Media, Reagents, and Stock Solutions for Cell Culture

2.1.1. Solutions for ES Cell Subculture

1. Phosphate-buffered saline (PBS): Containing 10 g NaCl, 0.25 g KCl, 1.44 g Na_2HPO_4, 0.25 g KH_2PO_4 $2H_2O$/L, filter-sterilized through a 0.22-μm filter.
2. Trypsin solution: 0.2% trypsin 1:250 from bovine pancreas (Serva Electrophoresis, Heidelberg, Germany) in PBS for routine passage, 0.1% in PBS for replating of EB outgrowths, filter-sterilized through a 0.22-μm filter.
3. EDTA (ethylene diamine tetraacetic acid) solution: 0.02% EDTA in PBS for routine passage, 0.08% EDTA in PBS for replating of EB outgrowths, filter-sterilized through a 0.22-μm filter.
4. Trypsin: EDTA: Mix trypsin solution and EDTA solution at 1:1.
5. MC solution: Dissolve 2 mg mitomycin C (Serva Electrophoresis) in 10 mL PBS, filter-sterilize through a 0.22-μm filter. From this stock solution, dilute 300 μL into 6 mL of PBS (final concentration is 0.01 mg/mL). MC stock solution should be freshly prepared at weekly intervals and stored at 4°C. *Caution*: MC is carcinogenic.

2.1.2. Solutions Supporting Cell Adhesion

1. Gelatine solution: 1% gelatin (Sigma, Steinheim, Germany) in triple-distilled or Milli-Q water, autoclaved and diluted 1:10 with PBS. Incubate tissue culture dishes with 0.1% gelatine solution for 1–24 h at 4°C before use.
2. Poly-L-ornithine solution: 0.1 mg/mL poly-L-ornithine (Sigma) in 10 mM sodium borate buffer (pH 8.4) filter-sterilized through a 0.22-μm filter. Incubate tissue culture dishes at 37°C for 3 h. Wash three times with Milli-Q water and incubate at room temperature for 12 h. Wash three times with Milli-Q water and dry at 40°C.
3. Laminin solution: 0.001 mg/mL laminin (Sigma) in PBS filter-sterilized. To prepare poly-L-ornithine/laminin-coated tissue culture plates, incubate poly-L-ornithine-coated dishes at 37°C for 3 h. Wash twice with PBS before use. Store at 4°C.

4. Collagen type I solution: Dilute collagen type I (BD Biosciences, Heidelberg, Germany) in sterile 0.02 N acetic acid solution. Incubate tissue culture dishes with 10 μg/cm^2 collagen at room temperature for 1 h. Wash twice with PBS and use immediately, or store for a maximum of 1 week at 2°C–8°C.

*2.1.3. Additives and Growth Factors (see **Note 1**)*

1. LIF: 10 ng/mL (Millipore, Schwalbach, Germany).
2. Progesterone (Sigma): Prepare a 1 mM stock solution in PBS, filter-sterilize through a 0.22-μm filter. Store aliquots at –20°C.
3. Putrescine (Sigma): Prepare a 20 mM stock solution in PBS, filter-sterilize, and store at 4°C.
4. Insulin (Sigma): Prepare a stock solution from 100 mg of insulin into 10 mL Milli-Q water with 100 μL glacial acetic acid, filter-sterilize. Store aliquots at –20°C.
5. Sodium selenite (Sigma): Prepare a 1 mM stock solution in PBS, filter-sterilize and store at 4°C.
6. Fibronectin (Invitrogen, Karlsruhe, Germany): Dissolve 1 mg of fibronectin in 1 mL of sterile Milli-Q water. Do not store in dilute solution and do not freeze/thaw repeatedly.
7. Transferrin (Sigma): Prepare a stock solution with 4 mg/mL transferrin in Milli-Q water and filter-sterilize. Store aliquots at –20°C.
8. Nicotinamide (Sigma): Prepare a 5 M stock solution in Milli-Q water and filter-sterilize. Store at room temperature.
9. Monothioglycerol (3-Mercapto-1, 2-propanediol, MTG; Sigma): Prepare a stock solution from 13 μL of MTG into 1 mL of Iscove's modification of DMEM (IMDM; Invitrogen), filter-sterilize through a 0.22-μm filter. Make fresh before use.
10. β-mercaptoethanol (β-ME; Serva Electrophoresis): Prepare a stock solution from 7 μL of β-ME into 10 mL of PBS (stock concentration is 10 mM). Make fresh at weekly intervals and store at 4°C. *Caution*: β-ME is toxic.
11. Phenol red solution (0.5% in DPBS; Sigma).
12. Additives I: To 100 mL medium add 1 mL of 200 mM L-glutamine stock (100X), 1 mL of β-ME stock, 1 mL of non-essential amino acids (NEAA) stock (100X), and 1 mL of penicillin/streptomycin stock (100X) (all from Invitrogen).
13. Additives II: To 100 mL medium add 1 mL of 200 mM L-glutamine stock (100X), 300 μL MTG of stock solution (final concentration is 450 μM), 1 mL of NEAA stock (100X), and 1 mL of penicillin/streptomycin stock (100X).

14. Additives III (B1): To 200 mL medium add 100 μL of insulin stock (final concentration is 5 μg/mL), 6 μL of sodium selenite stock (final concentration is 30 nM), 2.5 mL of transferrin stock (final concentration is 50 μg/mL), 1 mg fibronectin (final concentration is 5 μg/mL), and 2 mL of penicillin/streptomycin stock (100X) (= ITSFn).
15. Additives IV (B2): To 400 mL medium add 1 mL of insulin stock (final concentration is 25 μg/mL), 12 μL of sodium selenite stock, 5 mL of transferrin stock, 8 μL of progesterone stock (final concentration is 20 nM), 2 mL of putrescine stock (final concentration is 100 μM), 400 μL laminin (final concentration is 1 μg/mL), and 4 mL penicillin/streptomycin stock (100X).
16. Additives V: To 100 mL medium add 500 mg bovine serum albumin (BSA) (final concentration is 5 mg/mL; Invitrogen), 1 ml of 100X chemically defined lipid concentrate (Invitrogen), 3.75 mL transferrin (final concentration is 150 μg/mL), 300 μL of MTG stock solution (final concentration is 450 μM), 70 μL of insulin stock solution (final concentration is 7 μg/mL), and 0.02 ng/mL LIF (final concentration; *see* **Section 2.1.3 #1**)
17. Basic FGF (bFGF; Strathmann Biotec AG, Hamburg, Germany): 10 μg/mL stock solution in sterile PBS with 0.1% BSA (Invitrogen). Store aliquots in silanized tubes at –20°C (*see* **Note 1**).
18. Epidermal growth factor (EGF; Strathmann Biotec AG): Prepare a 10 μg/mL stock solution in sterile PBS with 0.1% BSA. Store aliquots in silanized tubes at –80°C (*see* **Note 1**).
19. Survival promoting factors (SPFs): 200 pg/mL IL-1β (PeproTech, London, UK), 700 μM db-cAMP (Sigma), 2 ng/mL glial cell line-derived neurotrophic factor (GDNF; R&D Systems, Wiesbaden, Germany), transforming growth factor-β_3 (TGF-β_3; PeproTech), and 10 ng/mL neurturin (NTN; PeproTech).

2.1.4. Cultivation and Differentiation Media

1. Cultivation medium I: Dulbecco's Modified Eagle's Medium (DMEM, 4.5 g/L glucose) supplemented with 15% heat-inactivated FCS (selected batches) and additives I for cultivation of MEF FL cells.
2. Cultivation medium II: DMEM supplemented with 15% heat-inactivated FCS, 10 ng/mL LIF, and additives I for ES cell cultivation.
3. Differentiation medium I: IMDM supplemented with 20% heat-inactivated FCS and additives II for EB formation.

4. Differentiation medium II: DMEM/F12 (1:1, v/v) supplemented with additives III for EB differentiation into neuronal cells.
5. Differentiation medium III: DMEM/F12 (1:1, v/v) supplemented with additives IV, and daily addition of 10 ng/mL bFGF and 20 ng/mL EGF for differentiation of neuronal cells.
6. Differentiation medium IV: "Neurobasal" medium supplemented with 2% B27 (both Invitrogen), 10% FCS, and SPFs for differentiation into neurons.
7. Differentiation medium V: DMEM/F12 (1:1, v/v) supplemented with additives IV, 2% B27, 10 mM nicotinamide for differentiation into pancreatic cells.
8. Differentiation medium VI: Hepatocyte culture medium (HCM) composed of 500 mL hepatocyte basal medium (Lonza, Wuppertal, Germany), 0.5 mL ascorbic acid, 10 mL BSA-FAF (fatty acid free), 0.5 mL hydrocortisone, 0.5 mL transferrin, 0.5 mL insulin, 0.5 mL human epithelial growth factor (EGF), and 0.5 mL gentamycin-amphothericin (GA-1000; all from Lonza) supplemented with 10% FCS, 5 mL penicillin/streptomycin stock (100X), and 5 mL phenol red. Prepare immediately before use.
9. CDM I: IMDM/F12 (1:1, v/v, containing Glutamax; Invitrogen) supplemented with additives V, prepared on the day of use.
10. CDM II: IMDM/F12 (1:1, v/v, containing Glutamax; Invitrogen) supplemented with additives V and 1 mg/ml trypsin inhibitor (Invitrogen), prepared on the day of use.

2.2. Solutions for RT-PCR (see *Notes 2 and 3*)

1. Diethyl pyrocarbonate-treated water (DEPC-H_2O): Add 1 mL DEPC to 1 L Milli-Q water and stir overnight. DEPC is inactivated by heating to 100°C for 15 min, or autoclaving for 15 min.
2. RNA lysis buffer: Add 23.6 g of guanidinium thiocyanate to 5 mL of 250 mM Na-citrate, pH 7.0, 2.5 mL of 10% sarcosyl, and add DEPC-H_2O to a total volume of 49.5 mL and mix carefully. Make fresh at monthly intervals. Add 1% β-ME before use.
3. 2 M Na-acetate, pH 4.0: Dissolve 27.2 g of Na-acetate × 3 H_2O in 0.1% DEPC-H_2O, adjust the pH to 4.0 with glacial acetic acid, and adjust to 100 mL with DEPC-H_2O. Treat the buffer with 0.1% DEPC-H_2O at 37°C for at least 1 h and heat to 100°C or autoclave for 15 min.
4. Acidic phenol: Phenol is saturated with DEPC-H_2O instead of Tris. The saturated acidic phenol contains 0.1% hydroxyquinoline (antioxidant, partial inhibitor of RNase, and a weak

chelator of metal ions; its yellow color provides a convenient way to identify the organic phase). Store at 4°C for up to 2 months.

5. Chloroform: Isoamylalcohol (24:1).
6. Proteinase K solution: Prepare Proteinase K buffer containing 0.2 M Tris-HCl (pH 8.0), 25 mM EDTA (pH 8.0), 0.3 M NaCl, and 2% sodium dodecyl sulfate (SDS). Add Proteinase K to a final concentration of 50 μg/mL.
7. 75% Ethanol: Prepare in DEPC-H_2O.
8. 25 mM $MgCl_2$ (Fermentas, St. Leon-Rot, Germany).
9. 5X Reaction buffer for reverse transcription (Fermentas): 250 mM Tris-HCl, pH 8.3, 250 mM KCl, 20 mM $MgCl_2$, 50 mM dithiothreitol (DTT)
10. 10X PCR buffer without $MgCl_2$ (Fermentas): 750 mM Tris-HCl, pH 8.8, 200 mM $(NH_4)_2SO_4$, 0.1% Tween 20.
11. RNase inhibitor (Fermentas): 40 U/μL.
12. Oligo $d(T)_{18}$ (Fermentas): 50 μM in 10 mM Tris-HCl, pH 8.3.
13. RevertAis^{TM} M-MuLV reverse transcriptase (Fermentas): 200 U/μL.
14. Recombinant Taq DNA Polymerase (Fermentas): 5 U/μL.
15. 10 mM dNTP mix: dNTP (dGTP, dATP, dCTP, dTTP; Fermentas) dilute to 100 mM with DEPC-H_2O and freeze at –20°C. 10 mM dNTP mix is freshly made by mixing the equal volumes of 100 mM of each dNTP before use.
16. Select PCR primer pairs: Dilute synthetic oligonucleotides to 10 mM with DEPC-H_2O and freeze at –20°C.
17. Glycogen: 20 mg/mL (Fermentas).
18. 5 M NaCl: Dissolve 29.2 g of NaCl in Milli-Q water; adjust to 100 mL with water and autoclave.
19. TE buffer: 10 mM Tris-HCl, 1 mM EDTA, pH 7.5, filter-sterilized through a 0.22-μm filter.
20. 10X Loening solution: 1.8 M Tris-HCl, pH 7.7, 1.5 M $NaH_2PO_4 \times 1\ H_2O$, 50 mM EDTA and autoclave.
21. 6X Loading buffer: 10 mL 10X Loening solution, 30 mL glycerine, 10 mL 10% SDS, 20 mL 10% N-lauroyl-sarcosine, 100 mg bromocresol green (Sigma), and add 30 mL of Milli-Q water.
22. 5X TBE: Dissolve 54 g Tris-base and 27.5 g boric acid in Milli-Q water, add 20 mL of 0.5 M EDTA, pH 8.0, and adjust to 1 L with Milli-Q water.
23. Ethidium bromide aqueous solution: 1% w/v = 10 mg/mL.

24. Agarose gels: Melt electrophoresis-grade agarose in 1X TBE by gentle boiling in a microwave oven. Cool to about 60°C and pour into an agarose gel mold. Run small gels at around 80–100 V by using bromocresol green in the stop mix as an indicator of migration. *Caution*: Ethidium bromide is carcinogenic. Use nitrile gloves and dispose all contaminated tips, agarose gels and buffers separately.

2.3. Solutions for Immunohistochemical Analysis

1. 4% Paraformaldehyde (PFA): Dissolve 4 g PFA in PBS and adjust to 100 mL with PBS, heat the mixture to 60°C, stir until the solution becomes clear, and cool to room temperature. *Caution*: PFA is toxic. Work under the hood and use gloves.
2. Methanol: acetone (7:3, v/v) fixative.
3. 10% goat serum (Invitrogen) or 1% BSA in PBS for blocking unspecific binding of antibodies.
4. Hoechst 33342 (5 μg/mL in PBS)
5. Mounting medium: DakoCytomation Fluorescent Mounting Medium (Dako, Hamburg, Germany) or Vectashield (Vector Laboratories Inc., Burlingame, CA, USA).
6. 0.5% BSA in PBS for dilution of secondary antibodies.

2.4. Enzyme-Linked Immunosorbent Assay (ELISA)

1. Krebs–Ringer Bicarbonate HEPES (KRBH) buffer for sample preparation for insulin ELISA: containing 118 mM sodium chloride, 4.7 mM potassium chloride, 1.1 mM potassium dihydrogen phosphate, 25 mM sodium hydrogen carbonate, 3.4 mM calcium chloride, 2.5 mM magnesium sulphate, 10 mM HEPES, and 2 mg/mL BSA
2. For insulin ELISA, prepare solutions of 2.5 mM, 5.5 mM, and 27.7 mM glucose dissolved in KRBH buffer.
3. Acid ethanol: 1 M hydrochloric acid:absolute ethanol (1:9, v/v).
4. Insulin ELISA (Mercodia, Uppsala, Sweden).
5. Albumin ELISA (Bethyl Laboratories, Montgomery, TX, USA).
6. Bradford assay (Bio-Rad Laboratories, Munich, Germany).

2.5. Solutions for the Determination of Glycogen Storage (Hepatic Differentiation)

1. 1% periodic acid
2. Schiff's reagent (Sigma)
3. Mayer's hematoxylin (Sigma)

2.6. Equipment

1. Tissue culture plates: 35 mm, 60 mm, and 100 mm (Nunc, Wiesbaden, Germany or BD Biosciences).
2. Bacteriological Petri dishes (Greiner, Frickenhausen, Germany): 60 mm for EB mass culture, 100 mm for EB "hanging drop" culture.

3. Counting chamber (e.g., Thoma).
4. Tissue culture incubator with 37°C and 5% CO_2 atmosphere.
5. For FL culture: Sterile dissecting instruments, screen or sieve (about 0.5–1 mm diameter), Erlenmeyer flasks with stir bars, centrifuge tubes.
6. Smart Spec 3000 (Bio-Rad Laboratories).
7. PCR apparatus: Mastercycler gradient (Eppendorf, Hamburg, Germany).
8. Electrophoresis equipment (Bio-Rad Laboratories).
9. Silanized 1.5-mL microtubes (Porex Bio Products, Inc., Petaluma, CA, USA) and silanized 10 μL, 100 μL, and 1,000 μL tips (Porex Bio Products, Inc.).
10. Cover slips (18 × 18 mm and 12 mm Ø) and slides.
11. ELISA Reader (Bio-Rad Laboratories).
12. LUCIA HEART imaging system (Nikon, Düsseldorf, Germany).

3. Methods

3.1. Cultivation of Undifferentiated ES Cells on FL

3.1.1. FL Culture (see **Note 4**)

1. Remove embryos from a mouse pregnant for 15–17 days (e.g., NMRI or CD-1 outbred strains), rinse in PBS, and remove placenta and fetal membranes, head, liver, and heart. Rinse the carcasses in trypsin solution.
2. Mince the tissue in 5 mL of fresh trypsin solution and transfer to an Erlenmeyer flask containing a stir bar.
3. Stir on magnetic stirrer for 25–45 min (use longer incubation time if the embryos are older), filter the suspension through a sieve or a screen, add 10 mL of culture medium I, and spin down.
4. Resuspend the pellet in about 3 mL of culture medium I and plate on 100-mm tissue culture plates (about 2×10^6 cells per 100 mm dish) containing 10 mL cultivation medium I, incubate at 37°C and 5% CO_2 for 24 h.
5. Change the medium to remove debris, erythrocytes, and unattached cellular aggregates, cultivate for additional 1–2 days.
6. Passage the primary culture of MEFs: Split 1:2 to 1:3 on 100-mm tissue culture plates, grow in cultivation medium I for 1–3 days. The cells in passages 2–4 are most suitable as FL for undifferentiated ES cells.

7. Incubate FL cells with MC buffer for 2–3 h, aspirate the MC solution, wash 3X with PBS, trypsinize feeder cells, and replate to new gelatin (0.1%)-treated microwell plates or Petri dishes. FL cells prepared one day before ES cell subculture are optimal.

3.1.2. Culture of Undifferentiated ES Cells (see Notes 5 and 6)

It is important to passage ES cells every 24 h or 48 h. Do not cultivate longer than 48 h without passage, or the cells may differentiate and be unsuitable for differentiation studies. Selected batches of FCS have to be used for ES cell culture (*see* **Note 6**).

1. Change the medium 1–2 h before passage.
2. Aspirate the medium, wash two times with 2 mL of trypsin–EDTA mixture.
3. Remove carefully the trypsin–EDTA mixture, add 2 mL of fresh cultivation medium II.
4. Resuspend the cell population with a 2 mL glass pipette into a single-cell suspension and split 1:3 to 1:10 to freshly prepared (60 mm) FL plates.

3.2. In Vitro Differentiation Protocols

For the development of ES cells into differentiated phenotypes, cells are cultivated in three-dimensional aggregates or EBs. The differentiation of cardiac muscle, neuronal, pancreatic, or hepatic cells requires different conditions. In this chapter, differentiation protocols utilizing the “hanging drop” method are described (*see* **Note** 7).

3.2.1. Preparation of EBs (Fig. 12.1)

1. Prepare a single-cell suspension containing a defined ES cell number of 200, 400, or 600 (*see* **Section 3.2.2**) cells in 20 μL of differentiation medium I.
2. Place 20 μL drops (n = 50–60) of the ES cell suspension on the lids of 100 mm bacteriological Petri dishes containing 10 mL PBS.
3. Cultivate the ES cells in hanging drops for 2 days. The cells will aggregate and form one EB per drop.
4. Rinse the aggregates carefully from the lids with 2 mL of medium, transfer into a 60 mm bacteriological Petri dish with 5 mL of differentiation medium I and continue cultivation in suspension for 2–3 days until the time of plating.

3.2.2. Differentiation of EB Outgrowths (see Note 8)

3.2.2.1. Cardiac Muscle Cell Differentiation ((30), Fig. 12.2 and Color Plate 1)

1. Use of 400–600 ES cells for preparation of EBs is optimal for cardiac differentiation.
2. Culture with differentiation medium I.
3. Plate 20–25 EBs onto gelatine-coated 60-mm tissue culture plates at day 5. The first beating clusters in EBs could already be seen 2 days after plating (= 5+2 days).

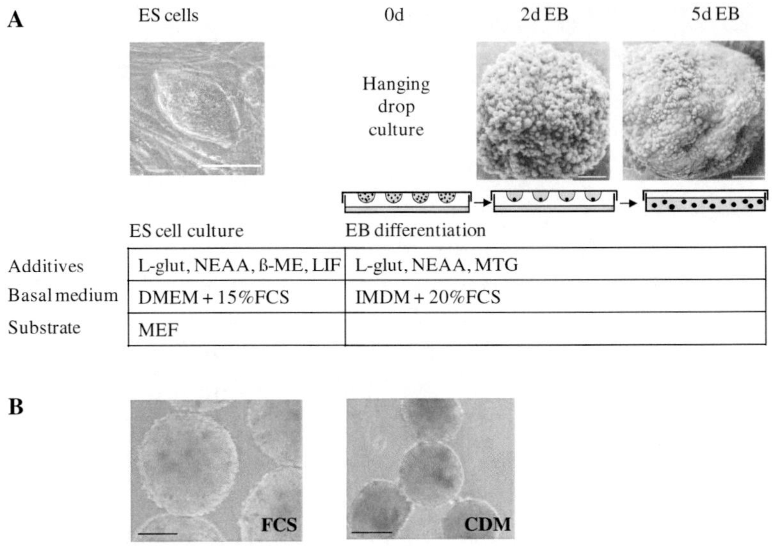

Fig. 12.1. Culture of undifferentiated mES cells and differentiation into embryoid bodies (EBs) by the hanging drop technique. (**A**) Scheme displays media, additives, and substrates used for maintenance of undifferentiated ES cells and spontaneous differentiation of EBs, which are kept in hanging drops for 2 days and in suspension culture for another 3 days. Representative microscopy pictures show mES cells at day 0 (bar = 100 μm, from (*35*)) and scanning electron microscopy of a 2 day and 5 day EB (Bars = 50 μm, from (*30*)). (**B**) Morphology of 2-day EBs cultured in FCS-containing medium (*left*) or serum-free CDM (*right*) by phase contrast microscopy, bars = 50 μm.

4. Analyze cardiac muscle cells after EB plating by RT-PCR and immunofluorescence.

3.2.2.2. Neuronal Cell Differentiation ((*67*), **Fig. 12.3** and Color Plate 2)

Because the spontaneous differentiation into neuronal cell types is rather limited (*48*), specific protocols were applied to increase the differentiation in vitro by (1) differentiation induction with retinoic acid (RA) (*69*), (2) lineage selection (*70*), (3) neuronal induction by stromal cell-derived inducing activity (*33*), and (4) lineage-restricted differentiation (*46, 47, 67*). In the following, the lineage-restricted differentiation protocol is described (*see* **Note 9**).

1. Neuronal phenotypes are obtained by using 200 ES cells for EB formation.
2. Culture with differentiation medium I.
3. Plate 20–30 EBs onto gelatine-coated 60-mm tissue culture dishes at day 4 and cultivate in differentiation medium I (= for 24 h with FCS-containing medium).
4. Exchange medium with differentiation medium II, one day after plating (4+1 days).
5. Replenish medium every two days.

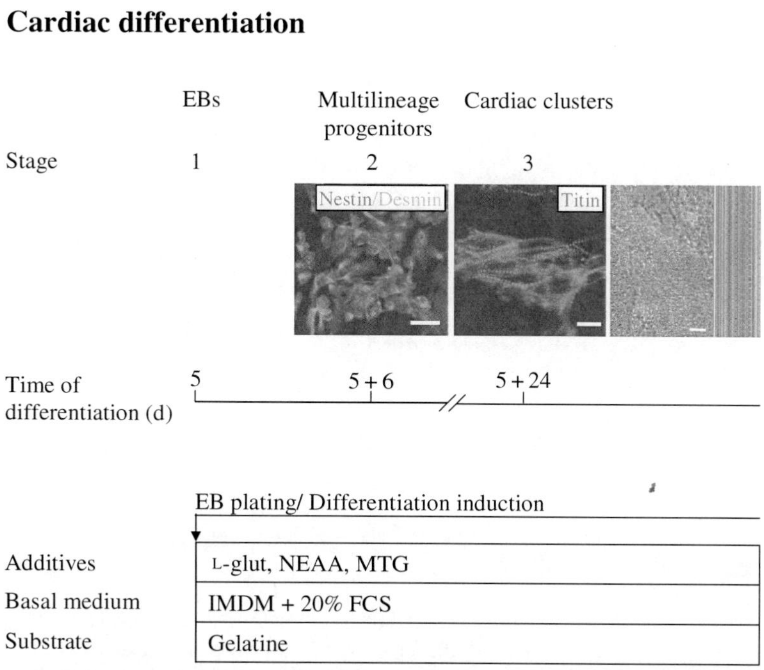

Fig. 12.2. Protocol for mES cell-derived cardiac differentiation. Five-day EBs were plated onto gelatin- or laminin-coated plates and cultured in IMDM+20%FCS supplemented with L-glutamine, NEAA, and MTG for up to 24 days. Multilineage progenitors at the intermediate stage 2 co-express nestin and desmin, while terminally differentiated cardiac clusters (stage 3) show well-organized sarcomeric staining of Z-disk epitopes of titin. Beating frequency measured from a beating cluster (phase contrast) by the LUCIA HEART imaging system is shown at the *right*, bar = 50 μm (*see* Color Plate 1).

6. At day 4+8, dissociate EBs by trypsin (0.1%): EDTA (0.08%) 1:1 solution for 1 min, collect cells by centrifugation (300*g*, 3 min), and replate onto poly-L-ornithine/laminin-coated tissue culture dishes into differentiation medium III (*see* **Note 10**).
7. Change medium every two days.
8. At day 4+14, induce neuronal cell differentiation by further cultivation with differentiation medium IV.
9. Analyze cultures for neuronal cells by RT-PCR and immunofluorescence.

3.2.2.3. Pancreatic Cell Differentiation (*see* **Note 11**, **Fig. 12.4** and Color Plate 3)

1. Pancreatic and hepatic cell types are differentiated using 600 ES cells for preparation of EBs.
2. Culture EBs in differentiation medium I.
3. Plate 20–30 EBs onto gelatin-coated 60-mm tissue culture dishes at day 5 in differentiation medium I.
4. Replenish medium every two days.

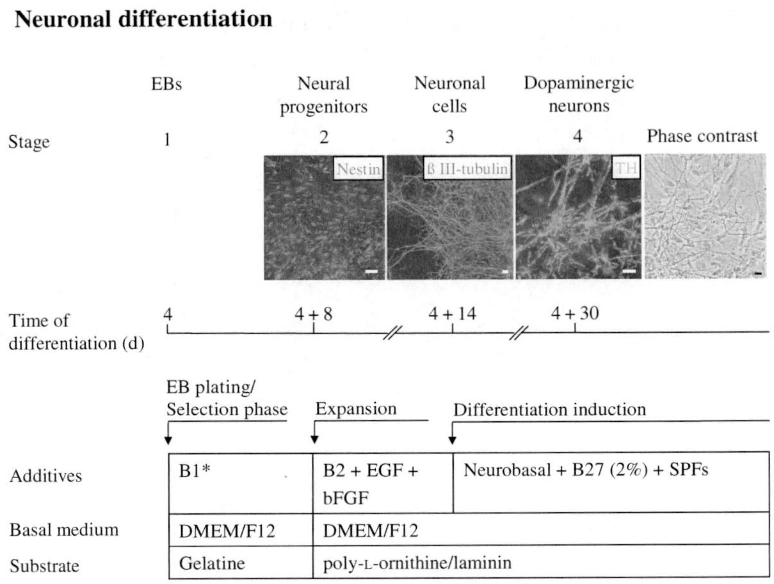

Fig. 12.3. Protocol for mES cell-derived neuronal differentiation. ES cells were cultured as EBs for 4 days. After plating onto gelatin, cells were cultured in B1 supplements and FCS-containing medium for 24 h (*). After medium change (at day 4+1), EB outgrowths were cultured until day 4+8 without FCS to select for neural progenitors. At day 4+8, EBs were dissociated and replated onto poly-L-ornithine/laminin until day 4+14, when differentiation of mature neurons was induced by "Neurobasal" medium, B27 supplement, and SPFs ("survival promoting factors"). The table shows the media, additives, and substrates used with this protocol. Differentiation led to nestin-positive neural progenitors (stage 2) followed by β-III-tubulin-expressing neuronal cells at stage 3 (4+14 d) and dopaminergic neurons expressing tyrosine hydroxylase at stage 4. A phase contrast picture shows the morphology of the ES cell-derived neurons at stage 4 (*right*) (*see* Color Plate 2).

5. At day 5+9, dissociate EBs by trypsin (0.1%): EDTA (0.08%) 1:1 solution for 1 min, collect cells by centrifugation (300*g*, 3 min), and replate onto poly-L-ornithine/laminin-coated tissue culture dishes into differentiation medium V (= replating in FCS-containing medium for 24 h to support attachment of cells).
6. Change medium every two days.
7. Analyze cultures for 7–9 d after replating by RT-PCR, immunofluorescence, and ELISA.

3.2.2.4. Hepatic Cell Differentiation (*see* **Note 11**, **Fig. 12.5** and Color Plate 4)

1. Hepatic cell types are differentiated using 600 ES cells for preparation of EBs.
2. Culture EBs in differentiation medium I.
3. Plate 20–30 EBs onto gelatin-coated 60-mm tissue culture dishes at day 5 in differentiation medium I.
4. Replenish medium every two days.

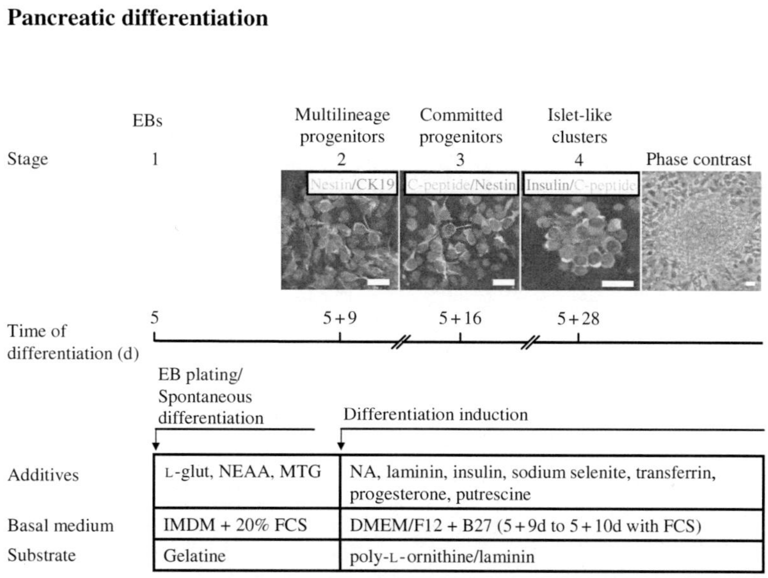

Fig. 12.4. Protocol for mES cell-derived pancreatic differentiation. Scheme displays media, additives, and substrates used during the differentiation process. Five-day EBs were plated onto gelatin for spontaneous differentiation in IMDM containing 20% FCS, L-Glut, NEAA, and MTG. At day 5+9, EBs were dissociated and replated onto poly-L-ornithine/laminin and subjected to differentiation by adding the differentiation factors niacinamide (NA), laminin, insulin, sodium selenite, transferrin, progesterone, and putrescine (and FCS for 24 h after plating). After medium change (at day 5+10), differentiation was continued (without FCS) until day 5+28. During spontaneous differentiation, nestin/CK19 co-expressing multilineage progenitors were formed (stage 2). Directed differentiation resulted in C-peptide/nestin-positive committed progenitors (stage 3) and insulin/C-peptide co-expressing islet-like clusters (stage 4; images from (*68*)). Morphology by phase contrast is shown (*right*) from (*77*). Cell nuclei were visualized by Hoechst 33342 (*blue*). Bars = 20 μm (*see* Color Plate 3).

5. At day 5+9, dissociate EBs by trypsin (0.1%): EDTA (0.08%) 1:1 solution for 1 min, collect cells by centrifugation (300*g*, 3 min), and replate onto collagen I-coated tissue culture dishes into differentiation medium VI.
6. Change medium every two days.
7. Analyze cultures for 10–30 days after replating by RT-PCR, immunofluorescence, and ELISA.

3.2.3. Differentiation of ES Cells in CDM

3.2.3.1. Differentiation via the Hanging Drop Method

1. Subculture ES cells twice on gelatin-coated plates (splitting cells 1:3) in cultivation medium II containing LIF to deplete feeder cells.
2. To remove traces of FCS and LIF, wash ES cells twice with CDM I.
3. Incubate cells for 30 min in CDM I at 37°C.

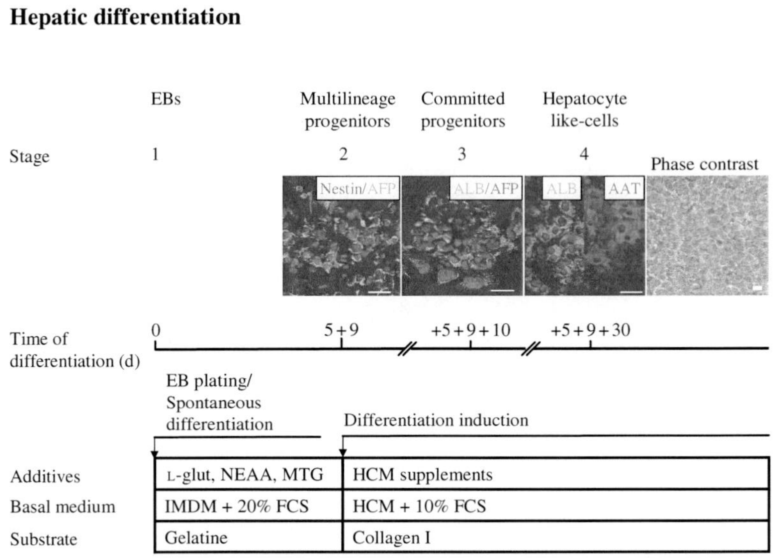

Fig. 12.5. Protocol for mES cell-derived hepatic differentiation. Scheme displays media, additives, and substrates used during the differentiation process. Five-day EBs were plated onto gelatin for spontaneous differentiation in IMDM containing 20% FCS, L-Glut, NEAA, and MTG. At day 5+9, differentiation into the hepatic lineage was induced by dissociation of the EBs and replating onto collagen I. Cells were cultured in differentiation medium (HCM) containing 10% FCS until day 5+9+30. Spontaneous differentiation led to nestin/AFP-positive multilineage progenitors (stage 2). Differentiation resulted in albumin/AFP co-expressing committed progenitors at stage 3, and albumin- and AAT-positive, partially binucleated hepatocyte-like cells (stage 4, images from (*53*)) with cuboidal morphology (phase contrast, *right*, from (*77*)) at day 5+9+30. Cell nuclei were visualized by Hoechst 33342 (*blue*). Bars = 20 μm (*see* Color Plate 4).

4. Dissociate cells by trypsin (0.2%): EDTA (0.02%) 1:1 solution for 5–7 min.
5. Add CDM II to stop trypsin activity.
6. Centrifuge cells at 1,000*g* for 5 min.
7. Resuspend pellet in CDM I (3×10^4 cells/mL).
8. Place 20 μL of the ES cell suspension ($n = 50$–60 drops) on the lids of 100 mm bacteriological Petri dishes containing 10 mL PBS.
9. To induce differentiation, cultivate the ES cells in hanging drops for 2 days. The cells will aggregate and form one EB per drop.
10. Rinse the aggregates carefully from the lids with 1.5 mL of CDM I, transfer EBs from two plates into a 60-mm bacteriological Petri dish with a total of 5 mL of CDM I and continue cultivation in suspension for up to 10 days.

3.2.3.2. Differentiation via Monolayer Culture

1. Subculture ES cells twice on gelatin-coated plates (1–10 × 10^5 cells/6 cm tissue culture dish) in cultivation medium II containing LIF to deplete feeder cells.
2. Wash ES cells once with CDM I.
3. To induce differentiation, culture cells in CDM I for up to 14 days (corresponding to day 8+6 of the hanging drop method).

3.3. Characterization of Differentiated Phenotypes

The transcripts of genes, which are specifically expressed during ES cell differentiation, are analyzed by RT-PCR with primers of tissue-specific genes (**Table 12.1**). The following steps are used to harvest differentiated cells:

3.3.1. RT-PCR Analysis (see **Note 12**)

3.3.1.1. Preparation of Cell Samples (*see* **Notes 13** and **14**)

1. Discard the medium and wash twice with PBS.
2. Add 400 μL of RNA lysis buffer per 60-mm culture dish. Allow the lysis buffer to spread across the surface of the dish and transfer the lysate into a 1.5-mL microtube.
3. Store samples at –20°C or –80°C.

3.3.1.2. Isolation of Total RNA (*see* **Note 15**)

The method described here is based on the use of a chaotropic agent (guanidine salt) for disruption of cells and inactivation of ribonucleases (*71*).

1. Thaw lysate (400 μL) and vortex for 15 s.
2. Add 40 μL (1/10 vol) of 2 M Na-acetate, pH 4.0. Mix carefully.
3. Add 400 μL of acidic phenol and vortex vigorously.
4. Add 80 μL of chloroform-isoamylalcohol (24:1) and vortex again.
5. Store for 15 min on ice.
6. Separate the organic and aqueous phases by centrifugation at 16,000*g* for 10 min at 4°C.
7. Transfer the upper aqueous phase carefully to a fresh tube, add an equal volume of isopropanol, and mix well. Store for 30 min at –80°C (*see* **Note 15**).
8. Centrifuge at 16,000*g* for 10 min at 4°C. Carefully discard the supernatant.
9. Resolve the pellet in 300 μL of Proteinase K solution. Incubate for 1 h at 56°C. Cool down, add an equal volume (300 μL) of isopropanol, and mix well. Store at –80°C for 30 min. Centrifuge at 16,000*g* for 10 min at 4°C.
10. Wash the pellet with 500 μL of 75% ice-cold ethanol (made with DEPC-H_2O), vortex briefly, centrifuge at 16,000*g* for 10 min, discard supernatant, and allow the pellet of nucleic acid to dry in the air.

Table 12.1
Primers for RT-PCR

Genes	Primer sequences (Forward/Reverse)	Annealing temperature	Product size
Definite endoderm			
Mixl1	5′-GCG GCG CCC CAC GAC TC 5′-CCT CAA CGC TCC CAG AAT CCT CAG	62°C	596 bp
Sox17	5′-CCA TAG CAG AGC TCG GGG TC 5′-GTG CGG AGA CAT CAG CGG AG	62°C	627 bp
Primitive endoderm			
GATA-4	5′-CCA TCC AGT GCT GTC TGC TCT 5′-ACT TTG CTG GCC CCC ACG TC	60°C	205 bp
HNF4α	5′-CTT CCT TCT TCA TGC CAG 5′-ACA CGT CCC CAT CTG AAG	55°C	253 bp
Ectoderm			
Sox1	5′-GGC GCC CTC GGA TCT CTG GTC 5′-CCG CGC CCT GGT AGT GCT GTG	65°C	202 bp
Pax6	5′-TCA CAG CGG AGT GAA TCA G 5′-CCC AAG CAA AGA TGG AAG	58°C	332 bp
Mesoderm			
Nodal	5′-CCG TCC CCT CTG GCG TAC ATG 5′-GAC CTG AGA AGG AAT GAC GG	55°C	321 bp
Brachyury	5′-CGC TGT GAC TGC CTA CCA GAA TG 5′-GAG AGA GAG CGA GCC TCC AAA C	59°C	230 bp
Bmp4	5′-TGA TTC CTG GTA ACC GAA TGC TGA TGG T 5′-CCT GGG ATG TTC TCC AGA TGT TCT TCG TGA	65°C	394 bp
Cardiac			
NKX 2.5	5′-CGA CGG AAG CCA CGC GTG CT 5′-CCG CTG TCG CTT GCA CTT G	60°C	180 bp
Cardiac α-MHC	5′-CTG CTG GAG AGG TTA TTC CTC G 5′-GGA AGA GTG AGC GGC GCA TCA AGG	64°C	302 bp
Cardiac β-MHC	5′-GCC AAC ACC AAC CTG TCC AAG TTC	64°C	203 bp

(continued)

Table 12.1 (continued)

Genes	Primer sequences (Forward/Reverse)	Annealing temperature	Product size
	5′-TGC AAA GGC TCC AGG TCT GAG GGC		
ANF	5′-AGG ATT GGA GCC CAG AGT GGA CTA GG 5′-TGA TAG ATG AAG GCA GGA AGC CGC	64°C	202 bp
MLC-2 V	5′-TGT GGG TCA CCT GAG GCT GTG GTT CAG 5′-GAA GGC TGA CTA TGT CCG GGA GAT GC	60°C	190 bp
Neural			
Nestin	5′-CTA CCA GGA GCG CGT GGC 5′-TCC ACA GCC AGC TGG AAC TT	60°C	219 bp
en-1	5′-TGG TCA AGA CTG ACT CAC AGC A 5′-TCT CGT CTT TGT CCT GAA CCG T	61°C	211 BP
NFM	5′-GAG GCA CTA AGG AGT CCC TG 5′-TAT TGT GAC TGA GGG CTG TCG G	60°C	295 bp
Tyrosine Hydroxylase (TH)	5′-TGT CAG AGC AGC CCG AGG TC 5′-CCA AGA GCA GCC CAT CAA AG	64°C	412 bp
Synaptophysin	5′-TAC CGA GAG AAC AAC AAA GGG C 5′-GCC TGT CTC CTT GAA CAC GAA C	60°C	288 bp
GFAP	5′-TCG GAG TTG AAA GTT ACA GG 5′-AGG ATG GTT GTG GAT TCT TC	50°C	234 bp
Pancreatic			
Isl-1	5 ′-GTT TGT ACG GGA TCA AAT GC 5′-ATG CTG CGT TTC TTG TCC TT	60°C	503 bp
Ngn3 (MATH4B)	5′-TGG CGC CTC ATC CCT TGG ATG 5′-AGT CAC CCA CTT CTG CTT CG	60°C	159 bp
Pax4	5′-ACC AGA GCT TGC ACT GGA CT 5′-CCC ATT TCA GCT TCT CTT GC	60°C	300 bp
Pdx1 (IPF-1)	5′-CTT TCC CGT GGA TGA AAT CC 5′-GTC AAG TTC AAC ATC ACT GCC	60°C	205 bp
Insulin 1/Preproinsulin 1	5′-TAG TGA CCA GCT ATA ATC AGA GAC 5′-CGC CAA GGT CTG AAG GTC	60°C	288 bp 406 bp
Insulin 2/Preproinsulin 2	5′-CCC TGC TGG CCC TGC TCT T 5′-AGG TCT GAA GGT CAC CTG CT	65°C	213 bp 701 bp

(continued)

Table 12.1 (continued)

Genes	Primer sequences (Forward/Reverse)	Annealing temperature	Product size
Glucagon	5′-CAT TCA CAG GGC ACA TTC ACC 5′-CCA GCC CAA GCA ATG AAT TCC	55°C	207 bp
Glut-2	5′-TTC GGC TAT GAC ATC GGT GTG 5′-AGC TGA GGC CAG CAA TCT GAC	60°C	556 bp
IAPP	5′-TGA TAT TGC TGC CTC GGA CC 5′-GGA GGA CTG GAC CAA GGT TG	65°C	233 bp
PP	5′-ACT AGC TCA GCA CAC AGG AT 5′-AGA CAA GAG AGG CTG CAA GT	60°C	364 bp
Somatostatin/ Preprosomato	5′-TCG CTG CTG CCT GAG GAC CT-3′ 5′-GCC AAG AAG TAC TTG GCC AGT TC	60°C	232 bp/ 897 bp
Hepatic			
Albumin	5′-GTC TTA GTG AGG TGG AGC AT 5′-ACT ACA GCA CTT GGT AAC AT	58°C	569 bp
Alpha-1-antitrypsin	5′-GAA GGG AAC CCA AGG AAA GAT AGT 5′-GGA GAG GTC AGC CCC ATT GT	55°C	472 bp
Alpha-fetoprotein	5′-CAC TGC TGC AAC TCT TCG TA 5′-CTT TGG ACC CTC TTC TGT GA	58°C	300 bp
Cyp2β9	5′-CTG GCC ACC ATG AAA GAG TT 5′-GAT GAT GTT GGC TGT GAT GC	53°C	153 bp
HNF3 β	5′-GCG GGT GCG GCC AGT AG 5′-GCT GTG GTG ATG TTG CTG CTC G	63°C	378 bp
Transthyretin	5′-CTC ACC ACA GAT GAG AAG 5′-GGC TGA GTC TCT CAA TTC	55°C	224 bp
TAT	5′-ACC TTC AAT CCC ATC CGA 5′-TCC CGA CTG GAT AGG TAG	50°C	206 bp
Housekeeping			
β5-tubulin	5′-TCA CTG TGC CTG AAC TTA CC 5′-GGA ACA TAG CCG TAA ACT GC	60°C	318 bp
GAPDH	5′-CCA TGT TTG TGA TGG GTG TGA ACC 5′-TGT GAG GGA GAT GCT CAG TGT TGG	65°C	712 bp

11. Dissolve RNA pellet in 30 μL of DEPC-H_2O and freeze at –80°C.
12. Dilute 1 μL of RNA with 99 μL of DEPC-H_2O, measure OD_{260} and the concentration of RNA using Smart Spec 3000 or a suitable spectrophotometer, adjust all samples to the same RNA concentration (i.e., 0.2 μg/μL) with DEPC-H_2O, measure again to ensure the same RNA concentration of all samples. The yield of RNA from EBs ($n = 20$) is in the range of 20–100 μg.

3.3.1.3. Reverse Transcription (RT) Reactions (*see* **Note 16**)

All RT and PCR solutions are available from commercial suppliers in ready-to-use form. RT reactions are performed in 20 μL of reaction volumes using 0.5-mL microcentrifuge tubes.

1. Label one PCR reaction tube for each sample and appropriate controls. Add the same amount of RNA (0.5–1.0 μg in 3 μL) to each tube.
2. Prepare the following RT-Mastermix for 25 reactions (or a smaller quantity as required) containing: 261.25 μL of DEPC-H_2O, 100 μL of 5X reaction buffer, 20 μL of 10 mM dNTPs mix, 12.5 μL of RNase inhibitor, 25 μL of Oligo $d(T)_{18}$, and 6.25 μL of RevertAis^{TM} M-MuLV reverse transcriptase to a total volume of 425 μL.
3. Add 17 μL of RT-Mastermix to each tube, mix carefully and centrifuge briefly.
4. Transfer tubes to thermal cycler and perform RT reactions for 1 h at 42°C and then heat to 99°C for 5 min.
5. Cool the samples to 4°C or store at –20°C until use.

3.3.1.4. Polymerase Chain Reactions (PCRs, *see* **Note 16**)

1. Prepare a PCR-Mastermix for 25 reactions (or a smaller quantity as required) containing: 798.75 μL of DEPC-H_2O, 125 μL of 10X PCR buffer, 90 μL of 25 mM $MgCl_2$, 80 μL of 10 mM dNTPs mix, 50 μL of 10 μM 5′ sense primer of target gene, 50 μL of 10 μM 3′ antisense primer of target gene, 6.25 μL of Taq DNA polymerase, to a total volume of 1200 μL.
2. Label new PCR reaction tubes and add 2.0 μL of RT reaction product to each tube as template DNA.
3. Add 48 μL of PCR-Mastermix to each tube, mix by vortexing, and centrifuge briefly.
4. Transfer tubes to thermal cycler. Amplify the cDNA through 25–40 thermal cycles. Standard conditions are denaturation at 95°C for 40 s, annealing at 50–65°C for 40 s, and extension at 72°C for 40 s. The conditions depend on the primers and thermal cycler used.

5. Run a parallel reaction containing 2.0 μL of RT reaction product and 48 μL of PCR-Mastermix with primers of the internal standard gene (i.e., β-tubulin or GAPDH) instead of the target gene.
6. Cool the samples to 4°C and store at –20°C.

3.3.1.5. Post-PCR Treatment of Samples

1. Transfer the PCR products to 1.5-mL microtubes.
2. Add 2.5 μL of a 1: 4 mixture of Glycogen: 5 M NaCl and 150 μL of ice-cold ethanol to each tube.
3. Incubate at –20°C for at least 1 h and centrifuge at 16,000*g* for 15 min.
4. Dissolve the pellet in 25 μL of TE buffer, add 5 μL of 6X loading buffer and store at 4°C.

3.3.1.6. Electrophoresis

1. Separate one-third of each PCR reaction (10 μL) by electrophoresis on a 2% agarose gel in 1X TBE containing 0.35 μg/mL of ethidium bromide at 5–10 V/cm for 70–100 min.
2. Illuminate the gel by UV light and obtain a digital image.

*3.3.2. Immunofluorescence Analysis (see **Notes 17** and **18**)*

The formation of tissue-restricted proteins in the EB outgrowths is analyzed by immunofluorescence with a normally equipped fluorescence microscope or with a confocal laser scanning microscope (CLSM) (*see* **Note 17**). For characterization of ES-derived cardiac, neuronal, pancreatic, and hepatic phenotypes, suitable primary antibodies are summarized in **Table 12.2**.

Table 12.2
Primary antibodies for the analysis of cell-specific proteins by immunofluorescence

Primary antibody	Dilution	Supplier	Fixation
Cardiac muscle markers			
Mouse anti-titin (clone T12)	1:100	Gift by D. Fuerst (*78*)	MeOH:Ac
Rabbit anti-α-actinin (clone 653)	1:20	Sigma	MeOH:Ac
Mouse anti-myomesin (MyBB78)	1:2	Gift by D. Fuerst (*79*)	MeOH:Ac
Chicken anti-sarcomeric MHC (MF-20)	Undiluted	Developmental Studies Hybridoma Bank, Iowa, USA	MeOH:Ac
Mouse anti-α-sarcomeric actin (5C5)	1:500	Sigma	4% PFA
Mouse anti-α-cardiac MHC (BA-G5)	1:500	(*80*) gift by D. Fuerst	4% PFA

(continued)

Table 12.2 (continued)

Primary antibody	Dilution	Supplier	Fixation
Neural markers			
Mouse anti-nestin IgG (clone rat 401)	1:3	Developmental Studies Hybridoma Bank, IA, USA	4% PFA
Mouse anti-β III-Tubulin (clone TU-20)	1:200	Millipore	4% PFA
Mouse anti-synaptophysin	1:150	Merck	4% PFA
Rabbit anti-tyrosine hydroxylase (TH)	1:150	Merck	4% PFA
Rabbit anti-GABA	1:100	Sigma	4% PFA
Rabbit anti-serotonin	1:100	Sigma	4% PFA
Mouse anti-glial fibrillary acidic protein (GFAP, clone G-A-5)	1:20	Millipore	MeOH:Ac
Common pancreatic and hepatic markers			
Mouse anti-cytokeratin 18 (clone KS-B17.2)	1:100	Sigma	MeOH:Ac
Mouse anti-cytokeratin 19 IgM	1:100	Cymbus, Chilworth, UK	MeOH:Ac[a] 4% PFA[b]
Pancreatic markers			
Rabbit anti-Isl-1 IgG	1:50	Abcam, Paris, France	4% PFA
Rabbit anti-carbonic anhydrase II IgG	1:200	Abcam	4% PFA
Mouse anti-insulin IgG (clone K36AC10)	1:100	Sigma	4% PFA+ 0.1% glutaraldehyde
Sheep anti-C-peptide IgG	1:100	Acris, Hiddenhausen, Germany	4% PFA
Rabbit anti-glucagon IgG	1:40	Abcam	4% PFA
Rabbit anti-somatostatin IgG	1:40	Biomeda, Foster City, CA, USA	4% PFA
Rabbit anti-PP IgG	1:40	Dako	4% PFA
Hepatic markers			
Sheep anti-albumin	1:100	Serotec, Düsseldorf, USA	MeOH:Ac
Rabbit anti-α-1-antitrypsin	1:100	Sigma	MeOH:Ac

(continued)

Table 12.2 (continued)

Primary antibody	Dilution	Supplier	Fixation
Goat anti-α-fetoprotein	1:100	Santa Cruz, Heidelberg, Germany	MeOH:Ac
Goat anti-amylase	1:100	Santa Cruz	MeOH:Ac
Mouse anti-cytokeratin 14 (clone CKB1)	1:100	Sigma	MeOH:Ac
Goat anti-dipeptidyl peptidase IV	1:100	Santa Cruz	MeOH:Ac

MeOH:Ac: methanol:acetone (7:3, vol:vol) fixation at –20°C for 10 min.
PFA: 4% paraformaldehyde fixation at room temperature for 20 min.
[a]Filament structures.
[b]Dot-like structures.

1. Rinse cover slips containing EB outgrowths twice with PBS.
2. Fix cells onto cover slips with acetone at –10°C for 6 min, methanol: acetone (7:3) at –20°C for 10 min, or alternatively, with 4% PFA in PBS at room temperature for 20 min (depending on the antibody used).
3. Rinse cover slips twice with PBS at room temperature for 5 min.
4. Incubate the cells on cover slips with 10% goat serum or 1% BSA in PBS in a humidified chamber at room temperature for 30 min to prevent unspecific immunostaining (*see* **Note 18**).
5. Incubate cover slips with the primary antibody at 37°C for 60 min, or at 4°C overnight (final concentration according to manufacturer's instructions).
6. Rinse cover slips with PBS three times at room temperature for 5 min.
7. Incubate cover slips with fluorescence-labeled specific secondary antibody (depending on the primary antibody, *see* **Table 12.3**) diluted in PBS with 0.5% BSA in a humidified chamber at 37°C for 45–60 min.
8. Incubate cover slips with 5 μg/mL Hoechst 33342 in PBS for 10 min at room temperature to label the cell nuclei.
9. Rinse cover slips twice with PBS at room temperature for 5 min.
10. Rinse cover slips quickly with distilled water at room temperature.

Table 12.3
Fluorescence-labeled secondary antibodies

Secondary antibody	Dilution	Supplier
Cy3™-conjugated donkey anti-goat IgG	1:600	Jackson ImmunoResearch Laboratories, Bar Harbor, ME, USA
Cy3™-conjugated goat anti-mouse IgG	1:600	
Cy3™-conjugated goat anti-rabbit IgG	1:600	
Cy3™-conjugated goat anti-mouse IgM	1:600	
ALEXA 488-conjugated donkey anti-sheep IgG	1:100	Invitrogen
ALEXA 488-conjugated goat anti-mouse IgG	1:100	

11. Embed cover slips in mounting medium and analyze immunolabeled cells with a conventional fluorescence or CLSM.

3.3.3. ELISA

3.3.3.1. Insulin ELISA (*see* **Note 19**)

1. Prior to the ELISA assay incubate cells in differentiation medium V without insulin for 24–48 h.
2. Wash cells with PBS (5X) and pre-incubate in freshly prepared KRBH buffer with 2.5 mM glucose for 90 min at 37°C.
3. For control, replace KRBH buffer with 2.5 mM glucose by KRBH buffer with 5.5 mM glucose. For induction of insulin release, replace KRBH buffer with 2.5 mM glucose by KRBH buffer containing 27.7 mM glucose. Incubate for 15 min at 37°C.
4. Collect supernatant for immediate determination of insulin release or store supernatant at –20°C.
5. Dissociate cells by treatment with 0.2% trypsin : 0.02% EDTA 1:1 solution for 3 min and centrifuge.
6. Extract proteins from the cells with acid ethanol overnight at 4°C, followed by cell sonification. Add 500 μL Aqua dest., spin down cell pellet, and store supernatant at –20°C for determination of total cellular insulin and protein content.
7. Perform ELISA according to manufacturer's recommendations.
8. Determine total protein content with protein Bradford assay according to manufacturer's recommendations.

3.3.3.2. Albumin ELISA

1. Wash cells with PBS (5X) and incubate in differentiation medium VI in the absence of BSA and FCS at 37°C for 24 h.
2. Collect supernatant for immediate measurement of albumin release or store supernatant at –20°C.

3. Dissociate cells by treatment with 0.2% trypsin:0.02% EDTA 1:1 solution for 3 min, centrifuge, and add PBS for the determination of total cell number.
4. Perform Albumin ELISA according to manufacturer's recommendations.

3.3.4. Determination of Glycogen Storage (Hepatic Differentiation)

1. Oxidize cells from stages 5+9 days and 5+9+30 days grown on slides for 5 min with 1% periodic acid.
2. Wash cells three times with distilled water.
3. Incubate slides for 15 min in Schiff's reagent.
4. Wash slides in distilled water for 5–10 min.
5. Stain with Mayer's hematoxylin solution for 1 min and wash slides in distilled water.

4. Notes

1. Thawing and repeated freezing of growth factors must be avoided; otherwise, the biological activity will be decreased.
2. DEPC is a suspected carcinogen and should be handled with care.
3. If possible, the solutions should be treated with 0.1% DEPC at 37°C for 1 h and then heated to 100°C for 15 min or autoclaved for 15 min. DEPC reacts rapidly with amines and cannot be used to treat solutions containing buffers, such as Tris.
4. FL cells used for ES cell cultivation should be plated at an appropriate density and cell cycle-inactivated through mitomycin C treatment or gamma irradiation. They provide LIF and other still unknown components essential for the prevention of mES cell differentiation.
5. Murine ES cells need FL cells for growth in the undifferentiated state, or FL cells can be replaced by conditioned medium containing LIF or by recombinant LIF. LIF is commercially available (Millipore), or may be prepared from LIF expression vectors (*72, 73*).
6. Good-quality FCS is critical for long-term culture of ES cells, and failure to acquire good-quality serum may be one reason why ES cells fail to differentiate appropriately. Extensive serum testing is therefore necessary to achieve good results. The most sensitive tests for sera include (1) comparative plating efficiencies at 10%, 15%, and 30% serum concentrations using ES and embryonic carcinoma cells, (2) ALP activity in

undifferentiated ES cells (e.g., Vector Blue Substrate Kit, Vector Laboratories, Burlingame, CA, USA), and (3) test of in vitro differentiation capacity after 3–5 passages in selected serum batches (*30, 74*).

7. For preparation of EBs, three different protocols may be used: (1) the "hanging drop" method (*1, 5, 30*), (2) the "mass culture" (*2*), or (3) the "methylcellulose" technique (*32*). The "hanging drop" method, as described here, generates EBs of a defined cell number (and size). This technique is especially used for developmental studies, because the differentiation pattern is dependent on the number of ES cells that differentiate within the EBs. A precondition for the usage of this method is the dissociation of ES cells. Plate 5×10^5 to 2×10^6 mES cells into 60-mm bacteriological Petri dishes containing 5 mL differentiation medium. After 2 days, let the aggregates settle in a centrifuge tube, remove medium, and carefully transfer the aggregates with 5 mL fresh differentiation medium into a new bacteriological dish. Change the medium every second day. Mass cultures of EBs may also be used for differentiation of a large number of cells. For hematopoietic differentiation, the "methylcellulose" method is preferred (*32*). Methylcellulose (e.g., MethoCult H4100; Stem Cell Technologies, Vancouver, BC, Canada) is added to the differentiation medium at a final concentration of 0.9%.

8. To achieve efficient differentiation, cells should be plated on appropriate extracellular matrix proteins. Generally, poly-L-ornithine/laminin is used for neuronal and pancreatic differentiation, whereas gelatine and collagen I are more suitable for cardiac and hepatic differentiation, respectively.

9. An efficient protocol for growth factor-mediated differentiation of neuronal cells is described here. The protocol includes (1) formation of cells of all three primary germ layers in the EBs, (2) selective differentiation of neuroectodermal cells by growth factor removal (serum depletion), (3) proliferation and maintenance of nestin-positive neural precursor cells in the presence of bFGF and EGF, and (4) the differentiation induction of functional neurons by withdrawal of bFGF/EGF, and the addition of neuronal differentiation factors and SPFs (*47, 67*).

10. Replating of differentiating cells is a critical step. The cell density after replating should be optimal to prevent either overgrowth resulting in metabolic starvation, necrosis, and cell death or poor differentiation efficiency because of reduced cell-to-cell contacts.

11. During pancreatic and hepatic differentiation, the cells show significant morphological changes. After induction of pancreatic differentiation, committed progenitor cells form

islet-like cell clusters; after hepatic differentiation, binucleated large cuboidal, epithelial-like cells similar to primary hepatocytes develop.

12. Use gloves and filtertips throughout the whole procedure.
13. Do not leave RNA lysis buffer in culture dishes longer than 5 min, because polystyrene is not resistant to lysis buffer.
14. mRNA isolation from small samples of cells can be performed using the Dynabeads® mRNA DIRECT™ Micro kit (Dynal Biotech, Oslo, Norway).
15. Never mix and disturb the organic and the aqueous phases.
16. For RT-PCR, rTth DNA polymerase can also be used as both reverse transcriptase and DNA polymerase (*75*). In this case, the components of both RT- and PCR-Mastermix are different from using MuLV reverse transcriptase and Taq DNA polymerase.
17. CLSM analysis can be applied to study EBs or EB outgrowths. For immunofluorescence analysis of EBs, the CLSM is used, because the three-dimensional EBs require an extended depth of focus. Some EBs are up to 400 μm in diameter. EBs are scanned in thin sections (0.5–10 μm) using the appropriate filter combinations depending on the fluorescent dyes used.
18. Unspecific binding of primary and especially secondary antibodies to the cells should be blocked by incubation with serum proteins of animal species, which were not used for the generation of the primary antibody. Specificity of immunostaining could be demonstrated by the absence of signals after incubation with PBS or with control antibodies instead of the specific primary antibody.
19. The analysis of differentiated pancreatic endocrine cells should include the determination of insulin production as a functional assay. The intracellular insulin content can be measured by the commercially available insulin ELISA. Due to reports showing the uptake of insulin from culture medium (*76*), additionally glucose responsiveness should be tested. For this purpose, insulin release in the presence of low (5.5 mM, control) and high (27.7 mM) glucose concentration is determined. Tolbutamide (10 μM), a sulfonylurea known to stimulate insulin secretion together with 5.5 mM glucose is used alternatively for the measurement of insulin release after induction. However, failure of glucose response may be dependent on insufficient maturation during differentiation. Such effects were described during pancreatic differentiation of mES cells, where insulin was secreted in response to glucose at an advanced stage of 32 days of differentiation, but not at day 28 (*55*).

Acknowledgments

We are grateful to Sabine Sommerfeld, Oda Weiss, and Karla Meier for technical assistance and Drs. Przemyslav Blyszczuk and Gabriela Kania for their expert help in the establishment of pancreatic and hepatic differentiation conditions. We are grateful to Dr. Dieter Fuerst, University of Bonn, Germany, for kindly providing antibodies. We thank the German Research Foundation (DFG) and the Ministry of Education and Research (BMBF) for funding our stem cell projects.

References

1. Wobus, A.M. and Boheler, K.R. Embryonic stem cells: prospects for developmental biology and cell therapy. Physiol Rev 2005;85:635–78.
2. Doetschman, T.C., Eistetter, H., Katz, M., Schmidt, W., and Kemler, R. The in vitro development of blastocyst-derived embryonic stem cell lines: formation of visceral yolk sac, blood islands and myocardium. J Embryol Exp Morphol 1985;87:27–45.
3. Evans, M.J. and Kaufman, M.H. Establishment in culture of pluripotential cells from mouse embryos. Nature 1981;292:154–6.
4. Martin, G.R. Isolation of a pluripotent cell line from early mouse embryos cultured in medium conditioned by teratocarcinoma stem cells. Proc Natl Acad Sci USA 1981;78:7634–8.
5. Wobus, A.M., Holzhausen, H., Jakel, P., and Schoneich, J. Characterization of a pluripotent stem cell line derived from a mouse embryo. Exp Cell Res 1984;152:212–9.
6. Stewart, C.L., Gadi, I., and Bhatt, H. Stem cells from primordial germ cells can reenter the germ line. Dev Biol 1994;161:626–8.
7. Prelle, K., Vassiliev, I.M., Vassilieva, S.G., Wolf, E., and Wobus, A.M. Establishment of pluripotent cell lines from vertebrate species – present status and future prospects. Cells Tissues Organs 1999;165:220–36.
8. Familari, M. and Selwood, L. The potential for derivation of embryonic stem cells in vertebrates. Mol Reprod Dev 2006;73:123–31.
9. Thomson, J.A., Itskovitz-Eldor, J., Shapiro, S.S. et al. Embryonic stem cell lines derived from human blastocysts. Science 1998;282:1145–7.
10. Guhr, A., Kurtz, A., Friedgen, K., and Loser, P. Current state of human embryonic stem cell research: an overview of cell lines and their use in experimental work. Stem Cells 2006;24:2187–91.
11. Hovatta, O. Derivation of human embryonic stem cell lines, towards clinical quality. Reprod Fertil Dev 2006;18:823–8.
12. Loeser, P. and Wobus, A.M. Aktuelle Entwicklungen in der Forschung mit humanen embryonalen Stammzellen. Naturwiss. Rundsch. 2007;60:229–37.
13. Ware, C.B., Nelson, A.M., and Blau, C.A. A comparison of NIH-approved human ESC lines. Stem Cells 2006;24:2677–84.
14. Williams, R.L., Hilton, D.J., Pease, S. et al. Myeloid leukaemia inhibitory factor maintains the developmental potential of embryonic stem cells. Nature 1988;336:684–7.
15. Burdon, T., Chambers, I., Stracey, C., Niwa, H., and Smith, A. Signaling mechanisms regulating self-renewal and differentiation of pluripotent embryonic stem cells. Cells Tissues Organs 1999;165:131–43.
16. Rohwedel, J., Sehlmeyer, U., Shan, J., Meister, A., and Wobus, A.M. Primordial germ cell-derived mouse embryonic germ (EG) cells in vitro resemble undifferentiated stem cells with respect to differentiation capacity and cell cycle distribution. Cell Biol Int 1996;20:579–87.
17. Solter, D. and Knowles, B.B. Monoclonal antibody defining a stage-specific mouse embryonic antigen (SSEA-1). Proc Natl Acad Sci USA 1978;75:5565–9.
18. Armstrong, L., Lako, M., Lincoln, J., Cairns, P.M., and Hole, N. mTert expression correlates with telomerase activity during the differentiation of murine embryonic stem cells. Mech Dev 2000;97:109–16.
19. Prelle, K., Zink, N., and Wolf, E. Pluripotent stem cells – model of embryonic

development, tool for gene targeting, and basis of cell therapy. Anat Histol Embryol 2002;31:169–86.

20. Scholer, H.R., Hatzopoulos, A.K., Balling, R., Suzuki, N., and Gruss, P. A family of octamer-specific proteins present during mouse embryogenesis: evidence for germline-specific expression of an Oct factor. EMBO J 1989;8:2543–50.
21. Chambers, I., Colby, D., Robertson, M. et al. Functional expression cloning of Nanog, a pluripotency sustaining factor in embryonic stem cells. Cell 2003;113:643–55.
22. Mitsui, K., Tokuzawa, Y., Itoh, H. et al. The homeoprotein Nanog is required for maintenance of pluripotency in mouse epiblast and ES cells. Cell 2003;113: 631–42.
23. Niwa, H., Miyazaki, J., and Smith, A.G. Quantitative expression of Oct-3/4 defines differentiation, dedifferentiation or self-renewal of ES cells. Nat Genet 2000;24:372–6.
24. Ying, Q.L., Nichols, J., Chambers, I., and Smith, A. BMP induction of Id proteins suppresses differentiation and sustains embryonic stem cell self-renewal in collaboration with STAT3. Cell 2003;115:281–92.
25. Hoffman, J.A. and Merrill, B.J. New and renewed perspectives on embryonic stem cell pluripotency. Front Biosci 2007;12:3321–32.
26. Rao, S. and Orkin, S.H. Unraveling the transcriptional network controlling ES cell pluripotency. Genome Biol 2006;7:230.
27. Wang, J., Rao, S., Chu, J., Shen, X., Levasseur, D.N., Theunissen, T.W., and Orkin, S.H. A protein interaction network for pluripotency of embryonic stem cells. Nature 2006;444:364–8.
28. Boiani, M. and Scholer, H.R. Regulatory networks in embryo-derived pluripotent stem cells. Nat Rev Mol Cell Biol 2005;6:872–84.
29. Bradley, A., Evans, M., Kaufman, M.H., and Robertson, E. Formation of germ-line chimaeras from embryo-derived teratocarcinoma cell lines. Nature 1984;309:255–6.
30. Wobus, A.M., Guan, K., Yang, H.T., and Boheler, K.R. Embryonic stem cells as a model to study cardiac, skeletal muscle, and vascular smooth muscle cell differentiation. Methods Mol Biol 2002;185:127–56.
31. Wobus, A.M., Wallukat, G., and Hescheler, J. Pluripotent mouse embryonic stem cells are able to differentiate into cardiomyocytes expressing chronotropic responses to adrenergic and cholinergic agents and Ca2+ channel blockers. Differentiation 1991;48:173–82.
32. Wiles, M.V. and Keller, G. Multiple hematopoietic lineages develop from embryonic stem (ES) cells in culture. Development 1991;111:259–67.
33. Kawasaki, H., Mizuseki, K., Nishikawa, S. et al. Induction of midbrain dopaminergic neurons from ES cells by stromal cell-derived inducing activity. Neuron 2000;28:31–40.
34. Ying, Q.L., Stavridis, M., Griffiths, D., Li, M., and Smith, A. Conversion of embryonic stem cells into neuroectodermal precursors in adherent monoculture. Nat Biotechnol 2003;21:183–6.
35. Boheler, K.R., Czyz, J., Tweedie, D., Yang, H.T., Anisimov, S.V., and Wobus, A.M. Differentiation of pluripotent embryonic stem cells into cardiomyocytes. Circ Res 2002;91:189–201.
36. Maltsev, V.A., Rohwedel, J., Hescheler, J., and Wobus, A.M. Embryonic stem cells differentiate in vitro into cardiomyocytes representing sinusnodal, atrial and ventricular cell types. Mech Dev 1993;44:41–50.
37. Drab, M., Haller, H., Bychkov, R. et al. From totipotent embryonic stem cells to spontaneously contracting smooth muscle cells: a retinoic acid and db-cAMP in vitro differentiation model. FASEB J 1997;11:905–15.
38. Yamashita, J., Itoh, H., Hirashima, M. et al. Flk1-positive cells derived from embryonic stem cells serve as vascular progenitors. Nature 2000;408:92–6.
39. Rohwedel, J., Maltsev, V., Bober, E., Arnold, H.H., Hescheler, J., and Wobus, A.M. Muscle cell differentiation of embryonic stem cells reflects myogenesis in vivo: developmentally regulated expression of myogenic determination genes and functional expression of ionic currents. Dev Biol 1994;164:87–101.
40. Nishikawa, S.I., Nishikawa, S., Hirashima, M., Matsuyoshi, N., and Kodama, H. Progressive lineage analysis by cell sorting and culture identifies FLK1+VE-cadherin+ cells at a diverging point of endothelial and hemopoietic lineages. Development 1998;125:1747–57.

41. Dani, C., Smith, A.G., Dessolin, S. et al. Differentiation of embryonic stem cells into adipocytes in vitro. J Cell Sci 1997;110:1279–85.
42. Kramer, J., Hegert, C., Guan, K., Wobus, A.M., Muller, P.K., and Rohwedel, J. Embryonic stem cell-derived chondrogenic differentiation in vitro: activation by BMP-2 and BMP-4. Mech Dev 2000;92:193–205.
43. Risau, W., Sariola, H., Zerwes, H.G. et al. Vasculogenesis and angiogenesis in embryonic-stem-cell-derived embryoid bodies. Development 1988;102:471–8.
44. Bain, G., Kitchens, D., Yao, M., Huettner, J.E., and Gottlieb, D.I. Embryonic stem cells express neuronal properties in vitro. Dev Biol 1995;168:342–57.
45. Fraichard, A., Chassande, O., Bilbaut, G., Dehay, C., Savatier, P., and Samarut, J. In vitro differentiation of embryonic stem cells into glial cells and functional neurons. J Cell Sci 1995;108:3181–8.
46. Lee, S.H., Lumelsky, N., Studer, L., Auerbach, J.M., and McKay, R.D. Efficient generation of midbrain and hindbrain neurons from mouse embryonic stem cells. Nat Biotechnol 2000;18:675–9.
47. Okabe, S., Forsberg-Nilsson, K., Spiro, A.C., Segal, M., and McKay, R.D. Development of neuronal precursor cells and functional postmitotic neurons from embryonic stem cells in vitro. Mech Dev 1996;59:89–102.
48. Strubing, C., Ahnert-Hilger, G., Shan, J., Wiedenmann, B., Hescheler, J., and Wobus, A.M. Differentiation of pluripotent embryonic stem cells into the neuronal lineage in vitro gives rise to mature inhibitory and excitatory neurons. Mech Dev 1995;53:275–87.
49. Bagutti, C., Wobus, A.M., Fassler, R., and Watt, F.M. Differentiation of embryonal stem cells into keratinocytes: comparison of wild-type and beta 1 integrin-deficient cells. Dev Biol 1996;179:184–96.
50. Chinzei, R., Tanaka, Y., Shimizu-Saito, K. et al. Embryoid-body cells derived from a mouse embryonic stem cell line show differentiation into functional hepatocytes. Hepatology 2002;36:22–9.
51. Hamazaki, T., Iiboshi, Y., Oka, M. et al. Hepatic maturation in differentiating embryonic stem cells in vitro. FEBS Lett 2001;497:15–9.
52. Jones, E.A., Tosh, D., Wilson, D.I., Lindsay, S., and Forrester, L.M. Hepatic differentiation of murine embryonic stem cells. Exp Cell Res 2002;272:15–22.
53. Kania, G., Blyszczuk, P., Jochheim, A., Ott, M., and Wobus, A.M. Generation of glycogen- and albumin-producing hepatocyte-like cells from embryonic stem cells. Biol Chem 2004;385:943–53.
54. Yamada, T., Yoshikawa, M., Kanda, S. et al. In vitro differentiation of embryonic stem cells into hepatocyte-like cells identified by cellular uptake of indocyanine green. Stem Cells 2002;20:146–54.
55. Blyszczuk, P., Czyz, J., Kania, G., Wagner, M., Roll, U., St Onge, L., and Wobus, A.M. Expression of Pax4 in embryonic stem cells promotes differentiation of nestin-positive progenitor and insulin-producing cells. Proc Natl Acad Sci USA 2003;100:998–1003.
56. Hori, Y., Rulifson, I.C., Tsai, B.C., Heit, J.J., Cahoy, J.D., and Kim, S.K. Growth inhibitors promote differentiation of insulin-producing tissue from embryonic stem cells. Proc Natl Acad Sci USA 2002;99:16105–10.
57. Vaca, P., Martin, F., Vegara-Meseguer, J.M., Rovira, J.M., Berna, G., and Soria, B. Induction of differentiation of embryonic stem cells into insulin-secreting cells by fetal soluble factors. Stem Cells 2006;24:258–65.
58. Leahy, A., Xiong, J.W., Kuhnert, F., and Stuhlmann, H. Use of developmental marker genes to define temporal and spatial patterns of differentiation during embryoid body formation. J Exp Zool 1999;284:67–81.
59. Rohwedel, J., Guan, K., Hegert, C., and Wobus, A.M. Embryonic stem cells as an in vitro model for mutagenicity, cytotoxicity and embryotoxicity studies: present state and future prospects. Toxicol In Vitro 2001;15:741–53.
60. Maltsev, V.A., Wobus, A.M., Rohwedel, J., Bader, M., and Hescheler, J. Cardiomyocytes differentiated in vitro from embryonic stem cells developmentally express cardiac-specific genes and ionic currents. Circ Res 1994;75:233–44.
61. Rohwedel, J., Kleppisch, T., Pich, U. et al. Formation of postsynaptic-like membranes during differentiation of embryonic stem cells in vitro. Exp Cell Res 1998;239:214–25.
62. Rohwedel, J., Horak, V., Hebrok, M., Fuchtbauer, E.M., and Wobus, A.M. M-twist expression inhibits mouse embryonic

stem cell-derived myogenic differentiation in vitro. Exp Cell Res 1995;220:92–100.

63. Prelle, K., Wobus, A.M., Krebs, O., Blum, W.F., and Wolf, E. Overexpression of insulin-like growth factor-II in mouse embryonic stem cells promotes myogenic differentiation. Biochem Biophys Res Commun 2000;277:631–8.

64. Johansson, B.M. and Wiles, M.V. Evidence for involvement of activin A and bone morphogenetic protein 4 in mammalian mesoderm and hematopoietic development. Mol Cell Biol 1995;15:141–51.

65. Ng, E.S., Azzola, L., Sourris, K., Robb, L., Stanley, E.G., and Elefanty, A.G. The primitive streak gene Mixl1 is required for efficient haematopoiesis and BMP4-induced ventral mesoderm patterning in differentiating ES cells. Development 2005;132:873–84.

66. Bouhon, I.A., Kato, H., Chandran, S., and Allen, N.D. Neural differentiation of mouse embryonic stem cells in chemically defined medium. Brain Res Bull 2005;68:62–75.

67. Rolletschek, A., Chang, H., Guan, K., Czyz, J., Meyer, M., and Wobus, A.M. Differentiation of embryonic stem cell-derived dopaminergic neurons is enhanced by survival-promoting factors. Mech Dev 2001;105:93–104.

68. Blyszczuk, P., Asbrand, C., Rozzo, A. et al. Embryonic stem cells differentiate into insulin-producing cells without selection of nestin-expressing cells. Int J Dev Biol 2004;48:1095–104.

69. Rohwedel, J., Guan, K., and Wobus, A.M. Induction of cellular differentiation by retinoic acid in vitro. Cells Tissues Organs 1999;165:190–202.

70. Li, M., Pevny, L., Lovell-Badge, R., and Smith, A. Generation of purified neural precursors from embryonic stem cells by lineage selection. Curr Biol 1998;8:971–4.

71. Chomczynski, P. and Sacchi, N. Single-step method of RNA isolation by acid guanidinium thiocyanate-phenol-chloroform extraction. Anal Biochem 1987;162:156–9.

72. Smith, D.B. and Johnson, K.S. Single-step purification of polypeptides expressed in *Escherichia coli* as fusions with glutathione S-transferase. Gene 1988;6:31–40.

73. Gearing, D.P., Nicola, N.A., Metcalf, D., Foote, S., Willson, T.A., Gough, N.M., and Williams, R.L. Production of leukemia inhibitory factor in *Escherichia coli* by a novel procedure and its use in maintaining embryonic stem cells in culture. BioTechnology 1989;7:1157–61.

74. Robertson E.J. (1987) Embryo-derived stem cell lines, in Robertson, E.J. (ed.) Teratocarcinoma and Embryonic Stem Cells – A Practical Approach. IRL Press, Oxford, pp. 71–112.

75. Myers, T.W. and Gelfand, D.H. Reverse transcription and DNA amplification by a *Thermus thermophilus* DNA polymerase. Biochemistry 1991;30:7661–6.

76. Rajagopal, J., Anderson, W.J., Kume, S., Martinez, O.I., and Melton, D.A. Insulin staining of ES cell progeny from insulin uptake. Science 2003;299:363.

77. Blyszczuk, P., Kania, G., and Wobus, A.M. (2006) In vitro differentiation of mouse ES cells into pancreatic and hepatic cells, in Notarianni, E. and Evans, M.J. (eds.) Embryonic Stem Cells. Oxford University Press, Oxford, pp. 218–37.

78. Furst, D.O., Osborn, M., Nave, R., and Weber, K. The organization of titin filaments in the half-sarcomere revealed by monoclonal antibodies in immunoelectron microscopy: a map of ten nonrepetitive epitopes starting at the Z line extends close to the M line. J Cell Biol 1988;106: 1563–72.

79. Obermann, W.M., Gautel, M., Steiner, F., van der Ven, P.F., Weber, K., and Furst, D.O. The structure of the sarcomeric M band: localization of defined domains of myomesin, M-protein, and the 250-kD carboxy-terminal region of titin by immunoelectron microscopy. J Cell Biol 1996; 134:1441–53.

80. Rudnicki, M.A., Jackowski, G., Saggin, L., and McBurney, M.W. Actin and myosin expression during development of cardiac muscle from cultured embryonal carcinoma cells. Dev Biol 1990;138:348–58.

Chapter 13

Cloning of ES Cells and Mice by Nuclear Transfer

Sayaka Wakayama, Satoshi Kishigami, and Teruhiko Wakayama

Abstract

We have been able to develop a stable nuclear transfer (NT) method in the mouse, in which donor nuclei are directly injected into the oocyte using a piezo-actuated micromanipulator. Although the piezo unit is a complex tool, once mastered it is of great help not only in NT experiments, but also in almost all other forms of micromanipulation. Using this technique, embryonic stem (ntES) cell lines established from somatic cell nuclei can be generated relatively easily from a variety of mouse genotypes and cell types. Such ntES cells can be used not only for experimental models of human therapeutic cloning but also as a means of preserving mouse genomes instead of preserving germ cells. Here, we describe our most recent protocols for mouse cloning.

Key words: Nuclear transfer, reprogramming, ntES cell, cloning, embryo, stem cell, oocyte.

1. Introduction

We have developed a nuclear injection method for mouse cloning using a piezo impact drive unit (hereafter piezo unit) *(1)*, which skips several steps of the original nuclear transfer (NT) procedure, such as zona pellucida cutting, donor cell insertion into the perivitelline space, moving cells and enucleated oocytes to an electrofusion machine, applying electrofusion and confirmation of cell fusion 30 min later. On the other hand, piezo injection method is technically very difficult and irritating; it will probably take several months to attain sufficient technical skill to get useful data. Without hard practice, production of cloned mice or establishment of NT embryonic stem cell (ntES) lines is impossible. However, once the piezo unit is properly set up on the micromanipulator, it will greatly help not only in NT, but in other forms of micromanipulation such

Ralf Kühn, Wolfgang Wurst (eds.), *Gene Knockout Protocols: Second Edition, vol. 530*

DOI 10.1007/978-1-59745-471-1_13 Springerprotocols.com

as sperm injection into oocytes *(2, 3)* or ES cell injection into blastocysts *(4)*. Moreover, the piezo unit simplifies pipette preparation, as it allows one to use blunt-tip pipettes without any additional treatment.

This approach has made it possible to create new types of ntES cell lines from adult somatic cells *(5, 6)*. We have shown that such ntES cell lines have the same differentiation potential as ES cells from fertilized blastocysts *(7)*. Moreover, cloned mice can be obtained from these ntES cell lines using a second NT procedure *(6, 8)*, which can be used as a backup of the donor cell genome and help increase the overall success rate of mouse cloning *(8)*. Importantly, those techniques are also potentially applicable to the preservation of genetic resources of any mouse strain instead of preserving them as embryos or gametes *(5, 8)*. At present, this technique is the only one available for the preservation of valuable genetic resources from infertile mutant mice or even frozen dead bodies without the use of germ cells *(9, 10)*.

Recently, we have found that the efficiency of mouse cloning and ntES cell establishment can be enhanced up to sixfold by adding the histone deacetylation inhibitor, trichostatin A (TSA), into the oocyte activation medium *(11–13)*. The oocyte activation method was also improved as convenient *(13)*. Here we describe our improved approaches for the production of cloned mice and establishment of ntES cell lines from somatic cells.

2. Materials

2.1. Preparation of Mice and Oocytes

1. Mouse strains: B6D2F1 (C57BL/6 × DBA/2) or B6C3F1 (C57BL/6 × C3H/He) mice are used for donor cell and oocyte collection (about 2–3 months old; *see* **Note 1**). The ICR (CD-1) strain is used for pseudopregnant surrogate and for lactating foster mothers and for vasectomized males.
2. For superovulation, equine chorionic gonadotrophin (eCG or PMSG, Sigma-Aldrich catalog #G4527) and human chorionic gonadotrophin (hCG, Sigma-Aldrich C8554) are dissolved in normal saline, stored in aliquots at –30°C, and used at 5 IU per mouse.

2.2. Media for Oocyte or Embryo Culture

1. Purchase distilled water (DW, Sigma-Aldrich W1503), mineral oil (Sigma-Aldrich M5310), KSOM medium (Specialty Media MR-106-D, http://www.specialtymedia.com/), M2 medium (Specialty Media MR-015), M2 + hyaluronidase (Specialty Media MR-051), and polyvinylpyrrolidone (PVP; 360 kD) (IrvineScientific No. 99311, www.irvinesci.com).

2.3. Media for Nuclear Transfer

1. Cytochalasin B (1 mg, Sigma-Aldrich C6762) and TSA (1 mg, Sigma-Aldrich T8552) are used to make a stock solution. Add 6.621 mL DMSO (Sigma-Aldrich D2650) into a vial with 1 mg TSA (500 μM: TSA 10^5 stock). Add 2 mL DMSO into a vial with 1 mg CB, and then add 2 μL of TSA 10^5 stock solution and store in aliquots at –80°C. We call this CBTSA 100X stock solution (TSA 500 nM and CB 500 μg).
2. $SrCl_2$ $6H_2O$ (Sigma-Aldrich S0390) is dissolved in DW at 100 mM and stored in aliquots at room temperature (20 × stock solution).
3. EGTA (Sigma-Aldrich E8145) is dissolved in DW at 200 mM and stored in aliquots at 4°C (100X stock solution).
4. For enucleation medium, add 2 μL of CBTSA 100X stock solution to 198 μL of M2 medium.
5. For oocyte activation medium, add 10 μL of $SrCl_2$ stock solution, 2 μL of EGTA stock solution, and 2 μL of CBTSA stock solution to 186 μL of KSOM medium *(14)*.

2.4. Preparation of Nuclear Transfer Tools

1. Inverted microscope with Hoffman optics from Olympus (IX71).
2. Micromanipulator set from Narishige (MMO-202ND, http://www.narishige.co.jp/).
3. Microforge (Narishige MF-900).
4. Piezo impact drive system from Prime Tech (P MM-150FU, http://www.primetech-jp.com/english/newproduct.htm).
5. Pipette puller (P-97) and pipette (B100-75-10) from Sutter instrument company, http://www.sutter.com.
6. Micropipettes can be ordered from several companies (e.g., Humagen Fertility Diagnostics, http://www.humagenivf.-com/Holding_Micropipets.htm), but we prefer to make them ourselves. For the holding pipette, the outside diameter (OD) should be smaller than that of the oocyte (e.g., OD 80 μm; inner diameter (ID) 10 μm). The ID of the enucleation pipette is 7–8 μm. The ID of the injection pipette depends on cell type: 5–6 μm for cumulus cells and 6–7 μm for fibroblasts or ES cells. Ask the supplying company to bend all pipettes close to the tip (about 300 μm back) at 15–20° using a microforge.
7. Backload a small amount (about 3 mM long column) of mercury (Fisher Scientific M-140, https://www1.fishersci.-com/index.jsp) into the enucleation/injection pipette using a 1-mL syringe and store in 10-cm dish at room temperature. *Caution*: Mercury is toxic if absorbed through breathing or the skin. Wear appropriate gloves, and always use mercury in a working fume hood.

2.5. ntES Cell Establishment and Culture

1. CultiCell Medium (Dainippon Sumitomo Pharma S2211101, http://www.ds-pharma.co.jp/english/index.ht mL) is used for ntES cell establishment. ES cell medium is used for maintaining ntES cells (Specialty media, R-ES-101). PBS Ca/Mg free (Sigma-Aldrich BSS-1006-B). Purchase acidic Tyrode's solution (Specialty Media MR-004-D) and trypsin solution (Specialty Media, SM-2003).

3. Methods

Currently, cumulus cells *(1)*, tail-tip cells (probably fibroblasts) *(15)*, Sertoli cells *(16)*, fetal cells *(17, 18)*, and ES cells *(19)* have been generally used to produce cloned mice. Freeze-dried and preserved cumulus cells or ES cells at 4°C also can be use as donor *(20)*. NKT cells *(21)*, primordial germ cells *(22)*, hematopoietic stem cell *(23)* and fetal neuronal cells *(24)* or newborn neuronal stem cells *(25, 26)* have also been used. Interestingly, although fresh adult brain cannot use to produced cloned mice, once brain was frozen and thawed, it can be possible to use *(10)*. The genetic background of the mouse strain is also very important *(18, 27)*. So far, only hybrid mice and the 129 strain have been used successfully as sources of donor nuclei (either somatic or ES cells) or recipient oocytes. Therefore, each researcher must choose the donor cell type carefully according to need. For example, ES cells are the most popular for NT experiments because they produce the best success rate of producing full-term offspring. However, these are pluripotent cells, not differentiated somatic cells, so they are not appropriate for genomic reprogramming experiments. Recently, we increased the efficiency of mouse cloning and ntES cell establishment up to sixfold by adding the histone deacetylation inhibitor, TSA, into the oocyte activation medium (**Fig. 13.1**) *(12)*. This new protocol has allowed us to generate cloned mice from inbred strains for the first time *(11)*. Because information on the effectiveness of TSA treatment is very new, it is possible that the protocols may change, such as the concentration of TSA (5–100 nM) or timing (pre or post treatment) *(28, 29)*. A flowchart is provided to allow planning of timing (**Fig. 13.2**).

3.1. Set Up the Micromanipulator

1. Place 3–5 approximately 15-μL droplets of three different media (M2, M2+CBTSA, PVP) on the top of a 10-cm dish and then cover this with mineral oil. This chamber can be used for both enucleation and microinjection. Draw a line on dish to distinguish these media.

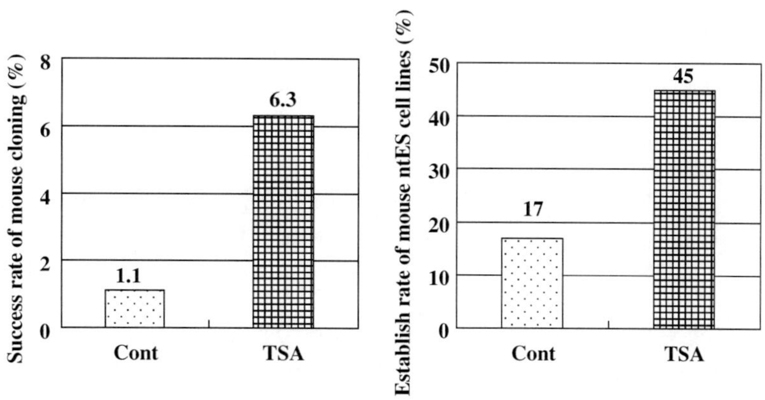

Fig. 13.1. Effect of trichostatin A (TSA) on mouse nuclear transfer. The success rate of mouse cloning and rate of establishment of ntES cell lines increased up to sixfold for cloning (**A**) and threefold for ntES cell derivation (**B**) by adding this histone deacetylation inhibitor into the oocyte activation medium.

2. Attach the enucleation pipette to the pipette holder of the piezo unit. The top of the pipette holder must be screwed in tightly (*see* **Note 2**). Hang the piezo unit on the micromanipulator.
3. Expel any air and oil and a few drops of mercury from the enucleation pipette in the PVP medium (*see* **Note 3**). Wash both the inside and outside of the pipette using PVP medium.
4. Adapt and fit the pipette to the piezo unit. While expelling the air and mercury from the pipette in the PVP droplet, the piezo unit must be applied with high power (more than 10 units) and high speed (more than 10 units) for at least 1 min continuously.
5. Attach the holding pipette on the other side of the micromanipulator.

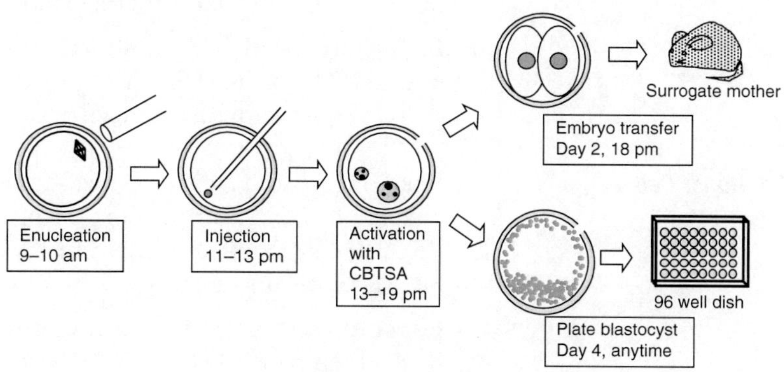

Fig. 13.2. The mouse cloning procedure. Without diligent practice it is very difficult to complete this procedure within the allotted time. This method is referred to as the "Honolulu method" because it was developed in Honolulu in 1998. After activation with CBTSA, additional treatment with only TSA for 2–4 h reinforces the reprogramming and increases the success rate of either production of cloned mice or establishment of ntES cell lines.

3.2. Collection of Oocytes

1. Inject eCG (or PMSG; 5 IU) and hCG (5 IU) 48 h apart (i.p.). Usually we inject mice at 5–6 p.m.
2. Collect oocyte–cumulus cell complexes from the oviduct ampullae at 14–15 h after hCG injection (usually we collect oocytes at 8–9 a.m.) and move them into M2+Hyaluronidase medium.
3. After 5 min, pick up the good oocytes, wash them in M2 medium three times and place in KSOM medium; cover with mineral oil at 37°C in a 5% CO_2 incubator.

3.3. Enucleation of Oocytes

1. Place about 10–20 oocytes into a M2+CBTSA droplet in the micromanipulation chamber and wait at least 7 min before starting enucleation.
2. Find the metaphase II spindle inside the oocyte. It can be recognized without any staining using Nomarski or Hoffman optics (*see* **Note 4**). Rotate the oocyte to place the spindle at between 2 o'clock and 4 o'clock or 8 o'clock and 10 o'clock and then attach it firmly to the holding pipette (**Fig. 13.3A**).
3. Cut through the zona pellucida using a few piezo pulses (*see* **Note 2**). To avoid damaging the oocyte, ensure that there is a large space between the zona pellucida and the oolemma: approximately as large as the thickness of the zona pellucida.
4. Insert the enucleation pipette into the oocyte without breaking the oolemma and remove the metaphase II spindle by aspiration with a minimal volume of cytoplasm. The oocyte membrane and spindle must be pinched off slowly: do not apply piezo pulses to cut the membrane (**Fig. 13.3B**). The MII spindle is harder than the cytoplasm, so you can feel it through the micromanipulator.
5. Wash the enucleated oocytes three times in KSOM to remove the CB completely and keep them in KSOM medium for at least 30 min in the incubator before starting donor cell injection.
6. If you are tired, take a short break until you recover the enucleated oocytes. From the next step, there is no respite, and it requires intense concentration.

3.4. Donor Cell Preparation

1. For noncultured fresh cells, wash them in PBS to remove any enzyme by centrifugation (100 rpm or 3*g* for 10 min) at least three times (*see* **Note 5**), except for cumulus cells, which can be used without washing.
2. For cultured cells, remove any culture medium from dish, and wash them in PBS (Ca^{++}, Mg^{++} free). Remove the PBS, add trypsin medium, and then incubate 5–20 min. Add culture medium (including serum) and take pipette in and out several times to select single cells. Wash cells with PBS by centrifugation at least three times as above.

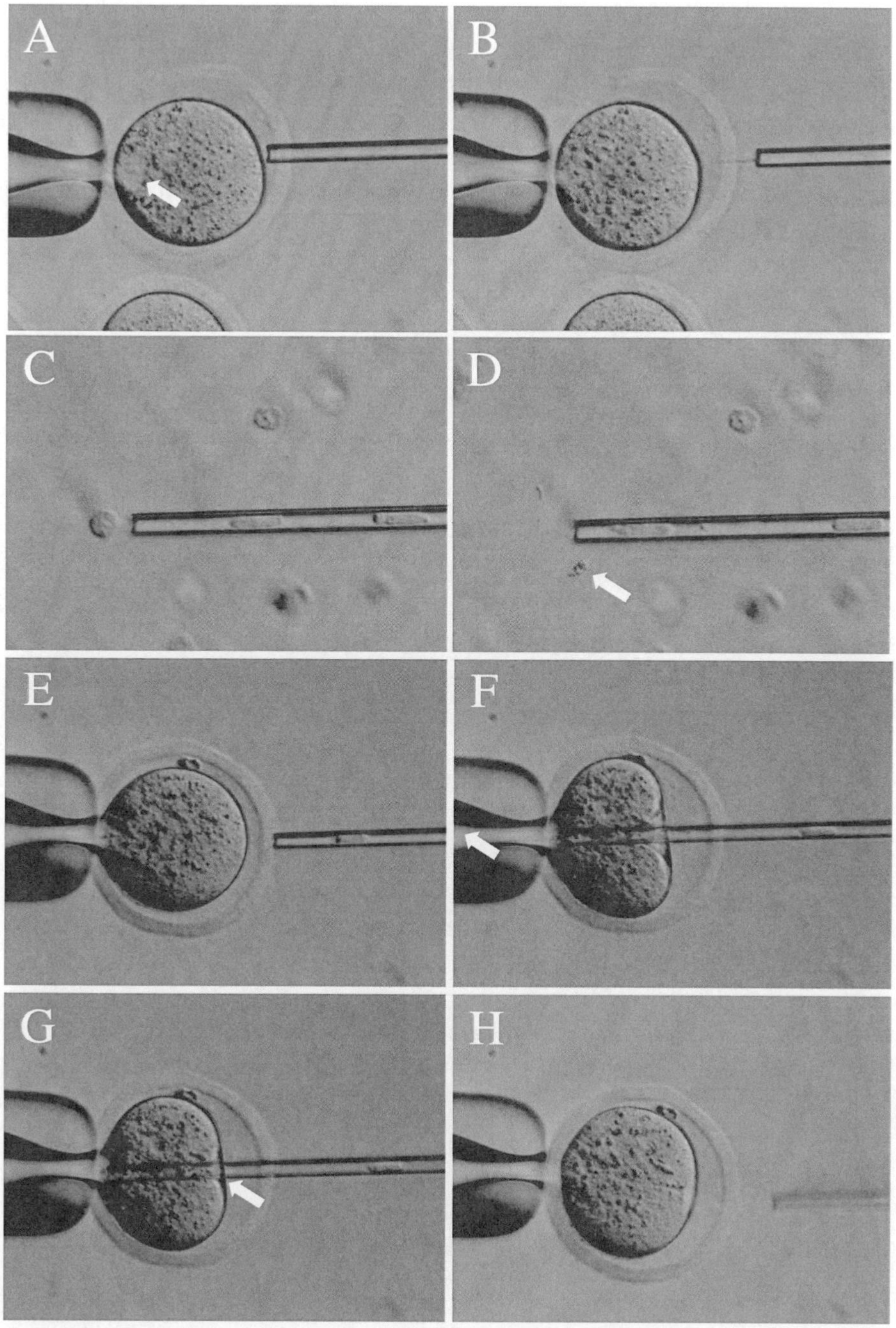

Fig. 13.3. Nuclear transfer using the piezo unit. (**A**) Rotate the oocyte, locate the metaphase II spindle, and place it between the 8 o'clock and 10 o'clock or 2 o'clock and 4 o'clock position (*see arrow*). Then stabilize this oocyte on the holding pipette. (**B**) Remove the spindle by suction without breaking the plasma membrane and gently pull the pipette away from the oocyte. (**C** and **D**) Donor nuclei are gently aspirated in and out of the injection pipette until their nuclei are largely devoid of visible cytoplasmic material (*see arrow*). (**E**) Hold the enucleated oocyte and cut the zona pellucida using piezo pulses. (**F**) Insert the injection pipette into the enucleated oocyte. (**G**) Apply a single piezo pulse to break the membrane, then inject the donor nucleus immediately. (**H**) Gently withdraw the injection pipette from oocyte.

3. Make a very concentrated cell suspension in the medium (*see* **Note 6**). The final volume should be 10–20 μL.
4. Pick up 1–3 μL of condensed donor cell suspension and introduce into a PVP droplet in the micromanipulation chamber. Mix the donor cells with PVP medium using sharp tweezers, gently but completely (*see* **Note 7**). Do not scratch the bottom of the chamber.

3.5. Donor Nuclei Injection

1. Place about 10–20 enucleated oocytes into M2 medium. The numbers of oocytes depend on each individual's skill level. Each group should be finished within 15 min.
2. Remove the donor nuclei from the cells by gently aspirating them in and out of the injection pipette until each nucleus is clearly separate from any visible cytoplasmic material (**Fig. 13.3C, D**) Take up a few nuclei into the injection pipette.
3. Stabilize an enucleated oocyte using a holding pipette. Cut the zona pellucida using a few piezo pulses (*see* **Note 2**; **Fig. 13.3E**).
4. Reduce the power level of the piezo unit (power level of piezo unit should be 1–2 and the speed is 1). The oolemma is weaker than the zona pellucida and the survival rate of oocytes after injection will be better with this reduced power.
5. Push one nucleus forward until it is near the tip of the pipette and advance the pipette until it almost reaches the opposite side of the oocyte's cortex (*see* **Note 8**; **Fig. 13.3F**).
6. Apply one weak piezo pulse to puncture the oolemma at the pipette tip. This is indicated by a rapid relaxation of the oocyte membrane (**Fig. 13.3G**). Expel the donor nucleus into the enucleated oocyte cytoplasm immediately with a minimal amount of PVP medium (*see* **Note 9**). Gently withdraw the injection pipette from the oocyte (*see* **Note 10**; **Fig. 13.3H**).
7. Wash the injection pipette with PVP medium by expelling some mercury and applying power from the piezo unit. This washing step is essential (*see* **Note 3**).
8. Keep the injected oocytes in the M2 drop for at least 10 min then transfer them into KSOM medium and culture for at least 30 min in the incubator before activation (*see* **Note 11**).

3.6. Activation and Embryo Culture

1. Make up the oocyte activation medium using the stock solution. Prepare the culture drops on the dish at least 30 min before use and equilibrate in a CO_2 incubator (*see* **Note 12**).

2. Transfer and culture each group of oocytes into drops of activation medium for 6 h in a 5% CO_2 incubator at 37°C (*see* **Note 13**). If embryos are cultured an additional 2–4 h in KSOM with TSA without CB, the success rate will increase, but you must change the medium at midnight.
3. All embryos must be washed several times in KSOM medium to remove the chemicals completely.
4. Examine the rate of oocyte activation. If NT and activation are done properly, those oocytes should each possess two or three pseudopronuclei (*see* **Note 14**) *(1)*.
5. Culture these clone embryos until they develop to the two-cell stage (next day) or to the blastocyst stage (3 days later; *see* **Note 15**).

3.7. Embryo Transfer

1. Mate estrous ICR female mice with normal males on the same day or 1–2 days before the experiment: these will be used as foster mothers.
2. Mate estrous ICR female mice with vasectomized males on the same day as the experiment: these will be used as pseudopregnant (surrogate) mothers.
3. Transfer the two-cell clone embryos into oviducts or morulae/ blastocysts into the uteri of 0.5 days post copulation (dpc) or 2.5 dpc pseudopregnant female mice, respectively *(30)*.
4. Caesarean section is required to recover the cloned mice securely (*see* **Note 16**). Euthanize the recipient female at 18.5 dpc or 19.5 dpc. Remove the uterus from the abdomen and dissect out the cloned pups with their placentas. Wipe away the amniotic fluid from skin, mouth, and nostril and stimulate the pups to breathe by rubbing the back or pinching the body of pup with blunt forceps.
5. Transfer the cloned pups to the cage of a naturally delivered foster mother. Mix the cloned pups with bedding material from the foster mother's cage. Remove some pups from the foster mother's litter and then mix the cloned pups with the original pups.

3.8. Establishing ntES Cell Lines from Cloned Embryos

1. To make embryonic feeder cells, collect day 12.5–13.5 dpc fetuses from the pregnant mother and then cut off the head and internal organs in a 10-cm petri dish containing PBS. Place the embryos into a new 10-cm dish and mince those embryos into very small pieces with sterile scissors. Add 25 mL DMEM medium and plate into large (175 cm^2) tissue culture flasks. One or two days later, split the cells 1:5 by trypsinization and allow them to grow to confluence.

2. Mitomycin C treatment. When the cells become confluent, treat with 10 μg/mL mitomycin C for 2 h in an incubator. Wash the flask several times using PBS to remove mitomycin C and then collect the cells by trypsinization. Pellet the cells by centrifugation (1000 rpm or 3*g* for 10 min). Aspirate the supernatant and gently resuspend the cell pellet in freezing medium (final concentration about 1×10^6 cells/mL) and divide into cryovials. Place the vials into a –80°C freezer overnight; next day the vials can be transferred to liquid nitrogen for long-term storage.
3. Thaw the vial quickly by warming and transfer the cells into a 15-mL tube filled with ES medium. Pellet the cells by centrifuging in the same way as before and then aspirate the supernatant and resuspend in fresh ES medium. Plate into a 96-well multidish and culture in the incubator until needed. These feeder cells should be prepared at least one day in advance.
4. Remove the zona pellucida from cloned blastocysts (day 4) using acidic Tyrode's solution (*see* **Note 17**). The zona pellucida will dissolve within 30 s. Wash the cloned blastocysts several times in M2 medium and then plate them into 96-well multidishes containing a preformed cell feeder layer (*see* above sentence) and filled with at least 200 μL CultiCell medium *(31)* in each well (*see* **Note 18**).
5. Culture the multidish for 10–14 days in incubator without changing the medium. During this period, the cloned blastocyst will attach to the surface of the feeder layer and the inner cell mass (ICM) can be seen to grow (**Fig. 13.4A, B**).
6. Some of the wells should have clumps of large ICMs. When those clumps appear, treat them with trypsin and disaggregate the cells using a 200-μL pipette. Then replate the suspension into another well (preformed cell feeder) of the same multidish.
7. When ES-like cell colonies dominate the well (**Fig. 13.4C**), the cells should be expanded gradually to 48-well, 24-well, 12-well and then 12.5 cm^2 flask and 25 cm^2 flask by repeat passages several times. We do not use feeder cells from 48-well dishes and bigger. After the cell numbers have increased, the cells should be frozen-stored as usual for ES cells *(32)*.

3.9. Production of Cloned Mice from ntES Cell Nuclei

1. NT should be done using these ntES cells as donors, repeating Steps 3.1–3.7. Unlike somatic cells, ntES cells (like ES cells in general) *(7)* will divide indefinitely, so you can use them without limitation. Moreover, the overall success rates of cloning from individuals are increased when ntES cell lines are used as intermediate nuclear donors *(8)*.

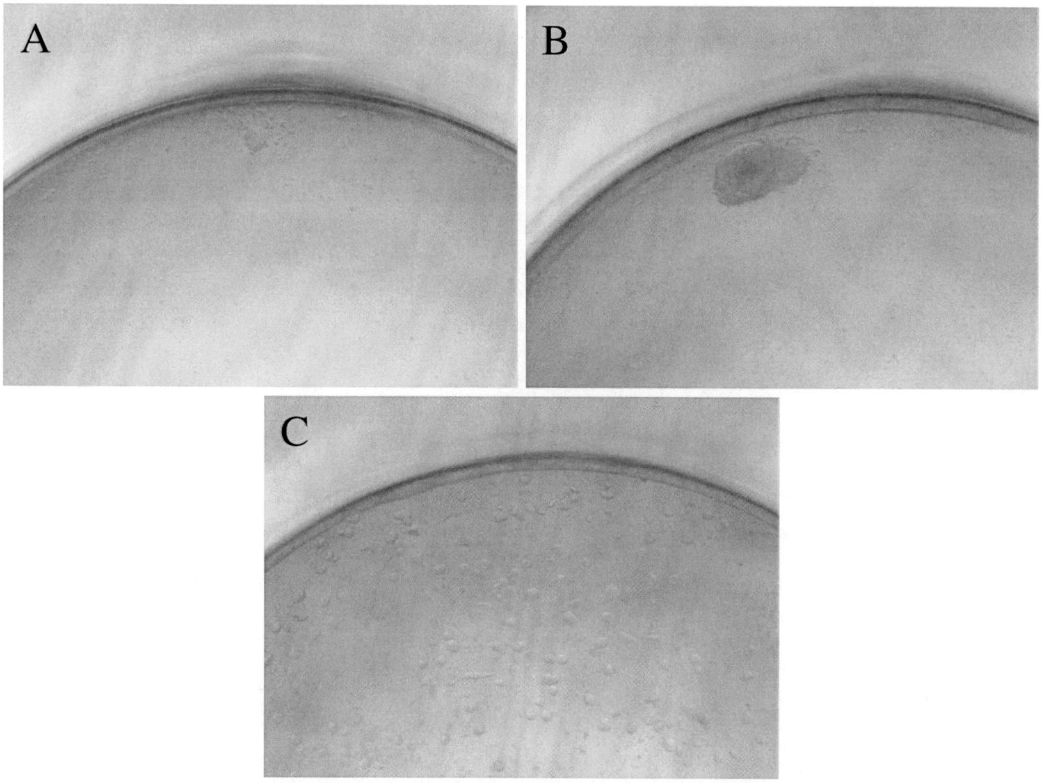

Fig. 13.4. Establishment of ntES cell lines. (**A**) Cloned blastocysts are attached onto the feeder cells in the 96-well dish, then the trophoblast cells spread and the inner cell mass (ICM) appears 7–9 days after plating. (**B**) The ICMs grow almost 5–10 times as large at 11–14 days after plating. (**C**) Two days after trypsinization, some wells show newly established ntES cell lines.

4. Notes

1. The mouse strain is important for several reasons. B6D2F1 oocytes are very clear and the metaphase spindle is easy to find. Oocytes from B6D2F1 mice are stronger for in vitro manipulation and culture. Somatic cells from B6D2F1 mice are much better donor cells for the production of cloned mice.

2. If you cannot cut the zona pellucida, check the connection between the pipette and pipette holder. The top of the pipette holder must be screwed in tightly. Expel all oil inside the pipette, as oil may have reduced the piezo power. There should be a slight negative pressure inside the pipette to enhance the piezo power.

3. PVP will cover both the inside and outside of the pipette to keep the surface nonsticky. Washing the pipette in PVP medium is very important and will affect not only oocyte survival rate, but also embryo development after NT. Without this step, the pipette soils rapidly and needs to be changed.
4. The room temperature is very important as the spindle microtubules will disperse and become unclear at room temperature. However, the spindle will become visible again if the oocytes are cultured at 37°C for 30 min before enucleation. Oocyte transparency also depends on the mouse strain: BDF1 is better than others.
5. As trypsin is very toxic at the time of nuclear injection, the donor cells must be washed by centrifugation at least three times.
6. Cell concentration is important. If the final concentration is too low, it is difficult to find an appropriate donor cell and the delay can be detrimental to the recipient oocyte awaiting NT.
7. If the cells are not mixed sufficiently they will aggregate and it will be difficult to isolate single cells. ES cells are especially sensitive and fragile in PVP medium. It is better to make new ES cell suspension drops every 30 min.
8. Do not apply the piezo unit's power until the pipette reaches the opposite side. If the piezo power is applied with the tip of the pipette in the middle of the oocyte, the oocyte will die after injection.
9. If it is difficult to release nuclei from the pipette, probably the pipette is too dirty. It must be washed frequently using PVP medium by expelling some mercury and applying power from the piezo unit.
10. This step is one of the most difficult and affects the oocyte survival rate after NT. If you are beginner, all oocytes will lyse immediately after injection. One month after starting practice, about half of your oocytes may survive. One year later, about 80% of oocytes will survive, if you continue to practice diligently. Other factors include too large a pipette, excessive room temperature or making the pipette insertion too shallow. Warm temperatures increase the rate of oocyte lysis, and the injection pipette must be inserted very deep into the oocyte before applying the piezo pulse.
11. If oocytes are transferred to KSOM medium immediately after injection, 10%–20% of them undergo lysis from the damage of injection. The oocyte membrane must be allowed to recover: this takes about 10 min.
12. Information on the effectiveness of TSA treatment is very new, and we may need to change our protocols, such as the concentration of TSA (5–100 nM) or timing (pre or post

treatment) *(28, 29)*. If ES cells are used as sources of donor nuclei, TSA does not work well. If G2/M phase ES cells are used as donor nuclei, CB must be omitted from the medium.

13. The activation medium must be checked before use, using intact oocytes. During strontium treatment, up to 10% of the oocytes will die and the medium will become dirty. This is normal and the survived oocytes are usually undamaged.
14. There are several reasons why oocytes do not form pronuclei. Usually it is because of failure to break the donor cell membrane or failure of oocyte activation. The injection pipette must be smaller than the donor cell. If the donor cell has a tough cell membrane (e.g., tail-tip fibroblasts), apply piezo power to break the donor cell membrane at the time of cell pickup.
15. Some of the chemicals used for activation can diffuse to other drops through the mineral oil and are embryotoxic. Therefore, all embryos should be moved to different culture dishes for long-term culture. Some batches of mineral oil are toxic and need to be tested before starting experiments.
16. All cloned mice to date have been born with abnormal and hypotrophic placenta, but this may not be related to the success of getting the pups to breathe. If you have no success in getting full-term development, change the donor cell type from somatic cell to ES cell or try using somatic cells from other hybrid mouse strains as well as making up new PVP and embryo culture media. However, the most important solution is to keep practicing. Technical skill is essential. Do not give up.
17. Before dissolving the zona pellucida completely, you must pick up the embryos and wash them several times in M2 medium. The remaining thinned zonae are easy broken by repeated pipetting. Prolonged exposure to acid Tyrode's solution will decrease the quality and survival of embryos.
18. CultiCell media do not contain fetal calf serum, which contains potential differentiation factors. Therefore, it is important to use this medium for establishment of new ntES or ES cell lines.

Acknowledgments

This work was supported by grants for Scientific Research in Priority Areas and the Project for the Realization of Regenerative Medicine (research field: technical development of stem cell manipulation) to T.W. from the Ministry of Education, Culture, Sports, Science and Technology of Japan.

References

1. Wakayama T, Perry AC, Zuccotti M, Johnson KR, Yanagimachi R. Full-term development of mice from enucleated oocytes injected with cumulus cell nuclei. Nature 1998;394:369–74.
2. Kimura Y, Yanagimachi R. Intracytoplasmic sperm injection in the mouse. Biol Reprod 1995;52:709–20.
3. Wakayama T, Yanagimachi R. Development of normal mice from oocytes injected with freeze-dried spermatozoa. Nat Biotechnol 1998;16:639–41.
4. Kawase Y, Iwata T, Watanabe M, Kamada N, Ueda O, Suzuki H. Application of the piezo-micromanipulator for injection of embryonic stem cells into mouse blastocysts. Contemp Top Lab Anim Sci 2001;40:31–4.
5. Wakayama S, Ohta H, Kishigami S, et al. Establishment of male and female nuclear transfer embryonic stem cell lines from different mouse strains and tissues. Biol Reprod 2005;72:932–6.
6. Wakayama T, Tabar V, Rodriguez I, Perry AC, Studer L, Mombaerts P. Differentiation of embryonic stem cell lines generated from adult somatic cells by nuclear transfer. Science 2001;292:740–3.
7. Wakayama S, Jakt ML, Suzuki M, et al. Equivalency of nuclear transfer-derived embryonic stem cells to those derived from fertilized mouse blastocysts. Stem Cells 2006;24:2023–33.
8. Wakayama S, Mizutani E, Kishigami S, et al. Mice cloned by nuclear transfer from somatic and ntES cells derived from the same individuals. J Reprod Dev 2005;51:765–72.
9. Wakayama S, Kishigami S, Van Thuan N, et al. Propagation of an infertile hermaphrodite mouse lacking germ cells by using nuclear transfer and embryonic stem cell technology. Proc Natl Acad Sci USA 2005;102:29–33.
10. Wakayama S, Ohta H, Hikichi T, et al. Production of healthy cloned mice from bodies frozen at −20°C for 16 years. Proc Natl Acad Sci USA 2008;105:17318–22.
11. Kishigami S, Bui HT, Wakayama S, et al. Successful mouse cloning of an outbred strain by trichostatin A treatment after somatic nuclear transfer. J Reprod Dev 2007;53:165–70.
12. Kishigami S, Mizutani E, Ohta H, et al. Significant improvement of mouse cloning technique by treatment with trichostatin A after somatic nuclear transfer. Biochem Biophys Res Commun 2006;340:183–9.
13. Kishigami S, Wakayama S, Thuan NV, et al. Production of cloned mice by somatic cell-nuclear transfer. Nat Protoc 2006;1:125–38.
14. Kishigami S, Wakayama T. Efficient Strontium-Induced Activation of Mouse Oocytes in Standard Culture Media by Chelating Calcium. J Reprod Dev 2007;53:1207–15.
15. Wakayama T, Yanagimachi R. Cloning of male mice from adult tail-tip cells. Nat Genet 1999;22:127–8.
16. Ogura A, Inoue K, Ogonuki N, et al. Production of male cloned mice from fresh, cultured, and cryopreserved immature Sertoli cells. Biol Reprod 2000;62: 1579–84.
17. Ono Y, Shimozawa N, Ito M, Kono T. Cloned mice from fetal fibroblast cells arrested at metaphase by a serial nuclear transfer. Biol Reprod 2001;64:44–50.
18. Wakayama T, Yanagimachi R. Mouse cloning with nucleus donor cells of different age and type. Mol Reprod Dev 2001;58:376–83.
19. Wakayama T, Rodriguez I, Perry AC, Yanagimachi R, Mombaerts P. Mice cloned from embryonic stem cells. Proc Natl Acad Sci U S A 1999;96:14984–9.
20. Ono T, Mizutani E, Li C, Wakayama T. Nuclear transfer preserves the nuclear genome of freeze-dried mouse cells. J Reprod Dev 2008 In press.
21. Inoue K, Wakao H, Ogonuki N, et al. Generation of cloned mice by direct nuclear transfer from natural killer T cells. Curr Biol 2005;15:1114–8.
22. Miki H, Inoue K, Kohda T, et al. Birth of mice produced by germ cell nuclear transfer. Genesis 2005;41:81–6.
23. Inoue K, Ogonuki N, Miki H, et al. Inefficient reprogramming of the hematopoietic stem cell genome following nuclear transfer. J Cell Sci 2006.
24. Yamazaki Y, Makino H, Hamaguchi-Hamada K, et al. Assessment of the developmental totipotency of neural cells in the cerebral cortex of mouse embryo by nuclear transfer. Proc Natl Acad Sci USA 2001;98:14022–6.
25. Inoue K, Noda S, Ogonuki N, et al. Differential developmental ability of embryos

cloned from tissue-specific stem cells. Stem Cells 2007;25:1279–85.

26. Mizutani E, Ohta H, Kishigami S, et al. Developmental ability of cloned embryos from neural stem cells. Reproduction 2006;132:849–57.
27. Inoue K, Ogonuki N, Mochida K, et al. Effects of donor cell type and genotype on the efficiency of mouse somatic cell cloning. Biol Reprod 2003;69:1394–400.
28. Kishigami S, Van Thuan N, Hikichi T, et al. Epigenetic abnormalities of the mouse paternal zygotic genome associated with microinsemination of round spermatids. Dev Biol 2006;289:195–205.
29. Rybouchkin A, Kato Y, Tsunoda Y. Role of histone acetylation in reprogramming of somatic nuclei following nuclear transfer. Biol Reprod 2006;74:1083–9.
30. Nagy A, Gertsenstein M, Vintersten K, Behringer R. Manipulating the mouse embryo. 3rd ed. New York: Cold Spring Harbor Laboratory Press; 2003.
31. Ogawa K, Matsui H, Ohtsuka S, Niwa H. A novel mechanism for regulating clonal propagation of mouse ES cells. Genes Cells 2004;9:471–7.
32. Schatten G, Smith J, Navara C, Park JH, Pedersen R. Culture of human embryonic stem cells. Nat Methods 2005;2:455–63.

Part III

Genetically Engineered Mice

Chapter 14

Isolation, Microinjection and Transfer of Mouse Blastocysts

Susan W. Reid and Lino Tessarollo

Abstract

Genetically modified mice by means of homologous recombination in embryonic stem (ES) cells are generated by injection of manipulated ES cells into recipient blastocysts. The injected blastocysts, following reintroduction into recipient foster mice, will produce chimeric mice in which the manipulated ES cells populate the germline and transmit the induced mutation to the offspring. Crossing of the chimeras' offspring bearing the targeted mutation in heterozygosis will ultimately produce mice homozygous for the specific genetic mutation. Here we describe the steps and procedures required to generate the chimeric mice leading to the transfer of a genetic mutation to the mouse germline.

Key words: Embryonic stem cells, chimeras, microinjections, mice, blastocysts.

1. Introduction

In the 1980s, the efforts of several pioneering developmental biologists established the conditions to culture, manipulate, and reintroduce mouse embryos that will successfully develop to term *(1–4)*. The isolation of teratocarcinoma cell lines, first, and later of blastocyst-derived embryonic stem (ES) cell lines that can contribute to the germline of recipient mouse embryos has provided the basis for a system to introduce targeted mutations into the mouse genome *(5–7)*. Compared to those initial efforts in which improvised equipment was used, today the technology to produce genetically manipulated mice has been made more accessible to investigators by the emergence of an industry that provides high-quality instrumentation, materials and reagents. However, specific technical skills still need to be acquired by the investigator in order to successfully produce genetically engineered mouse models.

Ralf Kühn, Wolfgang Wurst (eds.), *Gene Knockout Protocols: Second Edition, vol. 530,*

DOI 10.1007/978-1-59745-471-1_14 Springerprotocols.com

The generation of mice with targeted mutations includes the introduction of altered ES cells into blastocyst-stage host embryos. For this technique to be successful, ES cells injected into the blastocoel of a blastocyst must integrate within the inner cell mass, which gives rise to the embryo proper. These embryos will result in F0-generation chimeric mice that are partially derived from the modified ES cells. If part of the germline is derived from the modified ES cells, the chimeras will produce mice uniformly heterozygous for the mutation of interest (F1 generation). Subsequent interbreeding of these heterozygous mice results in F2-generation mice that are homozygous for the intended mutation. Although the injection procedure results in mice that are chimeric with varying degrees of ES cell contribution, it is still the approach most widely used in the generation of mutant mice with a targeted mutation. Aggregation of ES cells with morulae-stage embryos further enhanced by laser-assisted microinjection, has provided an alternative to blastocyst microinjection *(8–10)*. However, it is still unclear whether this procedure will become widely accepted and of general use.

Here, we describe a relatively straightforward method to isolate mouse blastocysts, manipulate them by injecting totipotent ES cells and to generate chimeric mice with an ES cell-derived contribution to the germline. Upon appropriate manipulation of the ES cells in vitro, this system should enable the transmission of new genetic traits to the chimera's offspring.

2. Materials

2.1. Mice

2.1.1. Mice for Blastocyst Injection Production

The choice of the appropriate mouse strain as a source of embryos used for injection, and as recipient females to reintroduce the manipulated blastocysts, is critical to the success of the outlined procedure.

Two criteria must be followed in choosing the donor strain of mouse blastocysts. The recipient embryo should be permissive to the ES cell lines used for injection; that is, ES cells must be able to colonize and contribute to the germline of the recipient embryo. Secondly, the recipient embryo should have coat-color markers that indicate the contribution of the ES cells in the resulting chimera. The majority of ES cell lines used in genetic manipulation experiments are derived from 129/Sv mouse strains because, for reasons that are still unclear, they are the easiest to derive and keep their totipotency even after extensive in vitro manipulations. These strains have agouti coat color. The most successful, and hence the most utilized, embryo donor mice are of the BALB/c and the C57BL/6 strains. These two mouse strains are permissive to the 129/Sv-derived ES cells and have coat-color-determining genes (white and black, respectively) that are recessive to the 129/Sv agouti gene. Furthermore, these strains breed well and produce a relatively large number of embryos. Since the use of ES cell lines of the C57BL/6 strain have become more common, BALB/c and albino C57BL/6 strains have been successfully used as embryo donors (*see* **Note 1** and *(11)*). To allow coat-color screening, we use co-isogenic C57BL/6 mice that are albino because of a point mutation in the tyrosinase gene (i.e., C57BL/6 J-*Tyr* ^{cBrd} *(12)* available from the Charles River NCI animal production facility-Frederick, MD; or C57BL/6 J-Tyrc-2 J, Jackson Laboratory, Bar Harbor, MA, stock #000058) *(13–15)* (*see* **Note 2**).

2.1.2. Mice Recipient of Manipulated Blastocysts

Different strains of mice can be used as recipients of manipulated blastocysts as long as they tolerate foreign blastocysts. Usually, hybrid strains are favored since they are physically stronger and mate at a fairly high frequency (*see* **Note 3**). Recipient females must be able to support the implanted embryos. The hormonal changes necessary for pregnancy and initial development of manipulated embryos are induced by mating with vasectomized mice from hybrid strains that can now conveniently be purchased from commercial suppliers (e.g., Charles Rivers and Jackson Laboratory).

2.2. Media and Supplies

1. Isolation and injection medium:

   For isolating and injecting of blastocysts we use KO DMEM (Dulbecco Modified Eagle Medium; Invitrogen) supplemented with
   (a) 15% Fetal Calf Serum (FCS; e.g., Hyclone Laboratories and Invitrogen)
   (b) 2 mM L-glutamine (e.g., 100X sol. Invitrogen) or Glutamax-1 (100X solution, Invitrogen) (*see* **Note 4**)
   (c) Penicillin/Streptomycin (100X solution, Invitrogen)
   (d) 25 mM HEPES buffer solution (e.g., Invitrogen).
2. Embryo-tested mineral oil: Light Mineral Oil ES cell qualified (Specialty Media).
3. Tribromoethanol (Avertin) anesthesia: Dissolve 25 g of 2,2,2-Tribromoethanol (Aldrich cat #T4840-2) in 15.5 mL of Ter-amyl Alcohol (Aldrich cat #24048-6) at 37°C. This solution is stable for up to a year if stored at 4°C in dark bottles wrapped in aluminum foil. Keep bottles well sealed and handle quickly to avoid evaporation of ter-amyl alcohol and consequent change in tribromoethanol concentration.

Dilute 0.5 mL of the stock solution with 40 mL of 0.9 % NaCl solution (this solution is stable for up to 4–5 months when stored at 4°C in dark bottles wrapped in aluminum foil). Inject intraperitoneally 0.25 mL/10 g body weight.

4. 1 mm capillaries: For both needle and holding pipettes use 30-μL or 60-μL microcaps (depending on the length required) (Drummond cat #7695D43); for holding pipettes different types of glass can also be used such as Ziptrol tubes (Drummond cat #7690H12).
5. Pasteur pipettes: These are used to make mouth-controlled transfer pipettes. Pull Pasteur pipette on a Bunsen burner to make a capillary. Snap the tip such to obtain a blunt edge. Connect the pipette to plastic tubing with a filter to prevent contamination during mouth pipetting.
6. Miscellanea: G30 and G25 needles with 1-mL and 5-mL syringes used, respectively, to administer the anesthetic to the mice and to flush the uterine horns for blastocyst recovery; also needed are scalpels and rubber sheets to make injection needles and 6-cm tissue culture dishes.

2.3. Dissecting Tools

1. Tools for mouse surgery: Scissors for microdissections (e.g., Roboz cat #RS-5910 and RS-5940); Straight and curved forceps (e.g., Roboz cat #RS-5130 and RS-5137); Tweezers for microdissections (e.g., Roboz cat #RS-5055); Micro clamps (e.g., Roboz cat #RS-7439); 9-mm wound clip applier (e.g., Roboz cat #RS-9260); 9-mm wound clips (e.g., Roboz cat #RS-9262); clip-removing forceps (e.g., Roboz cat #RS-9268).
2. Tools for uterus dissection: Operating scissors (e.g., Roboz cat # RS-6814); Micro dissecting scissors (e.g., Roboz cat # RS-5960); Micro dissecting spring scissors (e.g., Roboz cat # RS-5650); Straight forceps (e.g., Roboz cat # RS-5130); Tweezers for microdissections (e.g., Roboz cat # RS-5055).
3. Warming tray: Needed to keep the mice warm after surgery for blastocysts transfer.

2.4. Microscope for Injection

Microinjection is most efficiently performed using an inverted microscope with x/y mechanical stage and movable objectives (*see* **Fig. 14.1**). Good-quality microscopes suitable for microinjection are available from different vendors (e.g., Zeiss, Olympus, Leica, Nikon; *see* **Note 5**). Magnifications of phase contrast objectives should be 5-, 10-, and 20-fold. An additional tenfold magnification is provided by the eye piece. The low-power magnification (50 ×) aids in manipulation of the embryos whereas the injection can be performed at 100 × or 200 × . With phase contrast objectives it is possible to discriminate between dead and live ES cells. However, Nomarski optics allow for good-quality images of the cells (*see* **Note 6**).

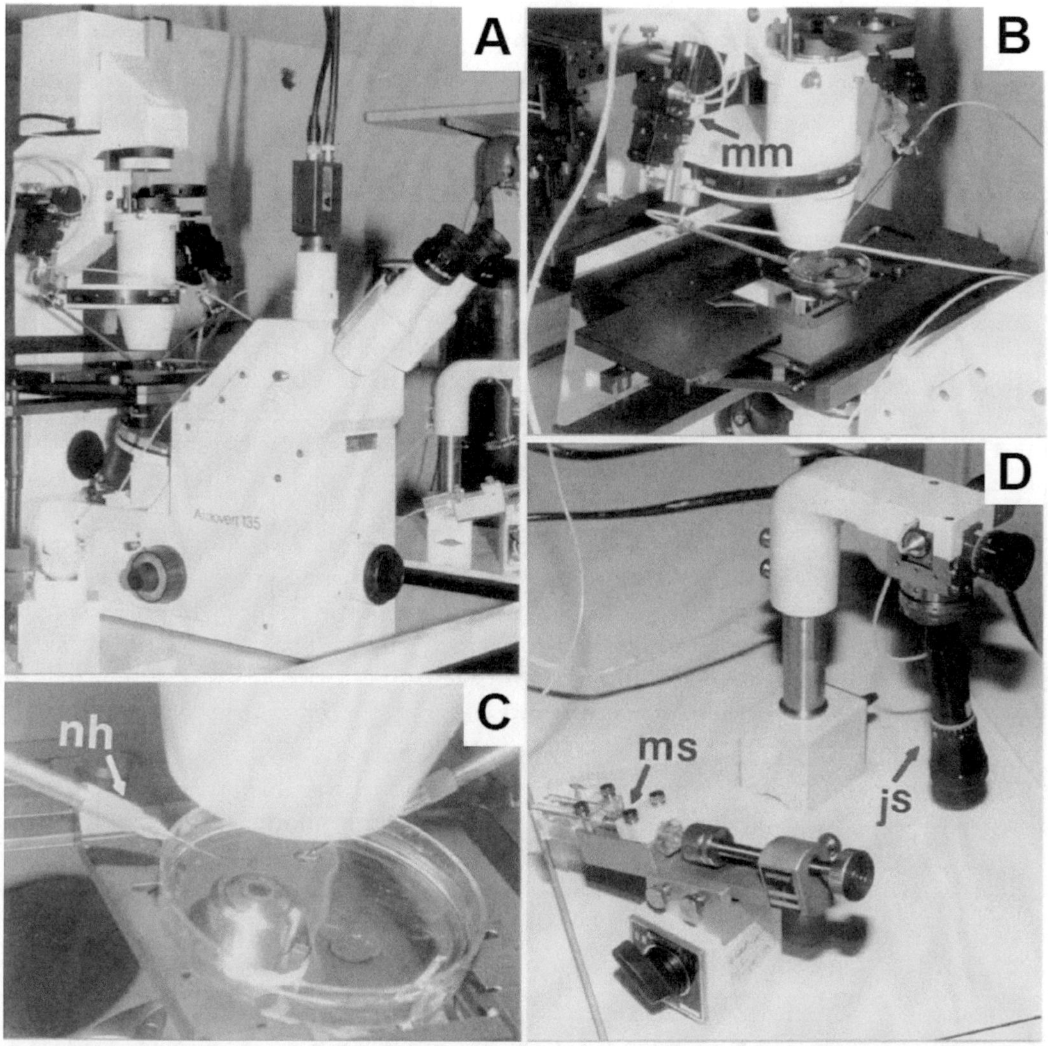

Fig. 14.1. Set up for blastocyst injection. (**A**) Inverted microscope; (**B**) micromanipulators attached to the light pillar, and x/y movable stage; (**C**) injection chamber and needle holders with needles; (**D**) hanging joystick to control micromanipulators; micrometer syringe to control suction and injection pressure of holding and injection pipettes. mm, micromanipulator; nh, needle holder; ms, micro syringe; js, hanging joystick.

2.5. Micromanipulators

Two micromanipulators (e.g., Narishige, Leitz) are required to move the holding and injection pipettes in all three dimensions. Mechanical joysticks are preferred over electric devices as they offer a greater flexibility in manipulation of the injection pipette (*see* **Fig. 14.1B,D**). The micromanipulators should not be attached to the microscope stage. Several vendors provide attachments to the illumination pillar of the microscope (**Fig. 14.1B**). The microinjection and the holding pipettes are attached to a micrometer syringe through flexible tubing and an injection or holding pipette holder (**Fig. 14.1B,C**). The tubing and the pipettes should be filled with light paraffin oil to avoid air bubbles. The very end of both pipettes should be filled with injection media.

2.6. Air Table

For stability, the microinjection set-up should be mounted on an air table that buffers environmental vibrations. Air tables can be purchased from different vendors; however, they can also be supplied by the microscope vendor. The latter will ensure perfect fitting of the microscope and microinjection apparatus.

2.7. Surgical and Dissecting Microscope

A dissecting microscope that allows 25 × and 50 × magnifications (*see* **Fig. 14.2A**) is used to handle blastocysts, for instance to collect them after flushing the uterine horns (*see* **Section 3.2** and **Fig. 14.5**) or for grouping injected blastocysts to reduce the amount of media used during transfer to recipient females (*see* **Section 3.6** and **Fig. 14.7**).

A surgical microscope (**Fig. 14.2B**) is used for transfer of manipulated blastocysts into recipient females. It should allow about 7 × and 15 × magnifications in order to be able to see the hole in the uterine horn created by the needle and to introduce the tip of the mouth-controlled pipette for blastocyst transfer (for details *see* **Section 3.6** and **Fig. 14.7E,F**). Also, it should allow sufficient clearance to perform the operation.

2.8. Pipette Puller and Microforge

Different pullers are commercially available. We prefer a vertical pipette puller (e.g., Kopf instruments, model 720) because needles are pulled by gravity and once an optimal temperature setting is established, it is easy to reproduce needles with specific characteristics (*see* **Fig. 14.3A,B**).

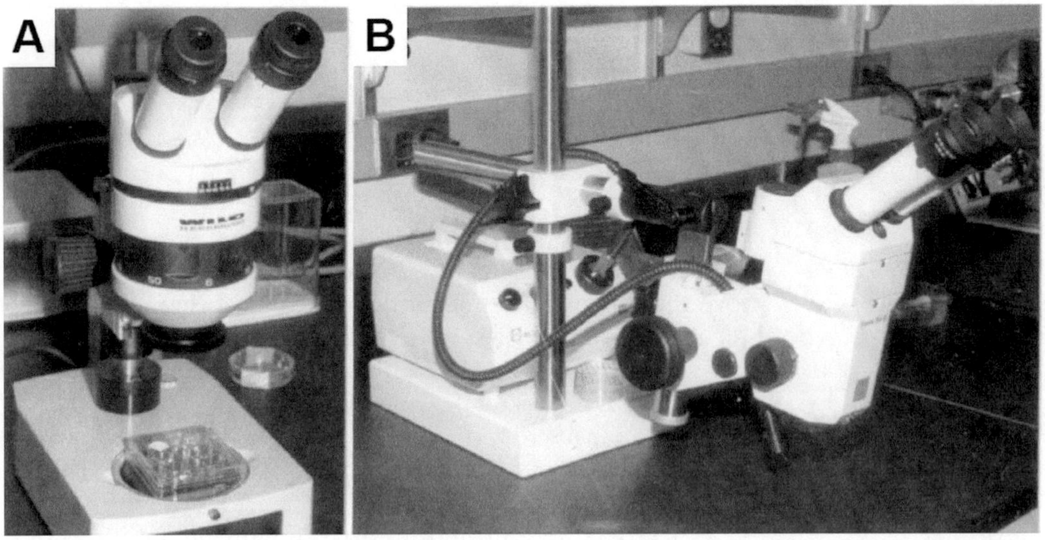

Fig. 14.2. Typical dissecting (**A**) and surgical (**B**) microscopes.

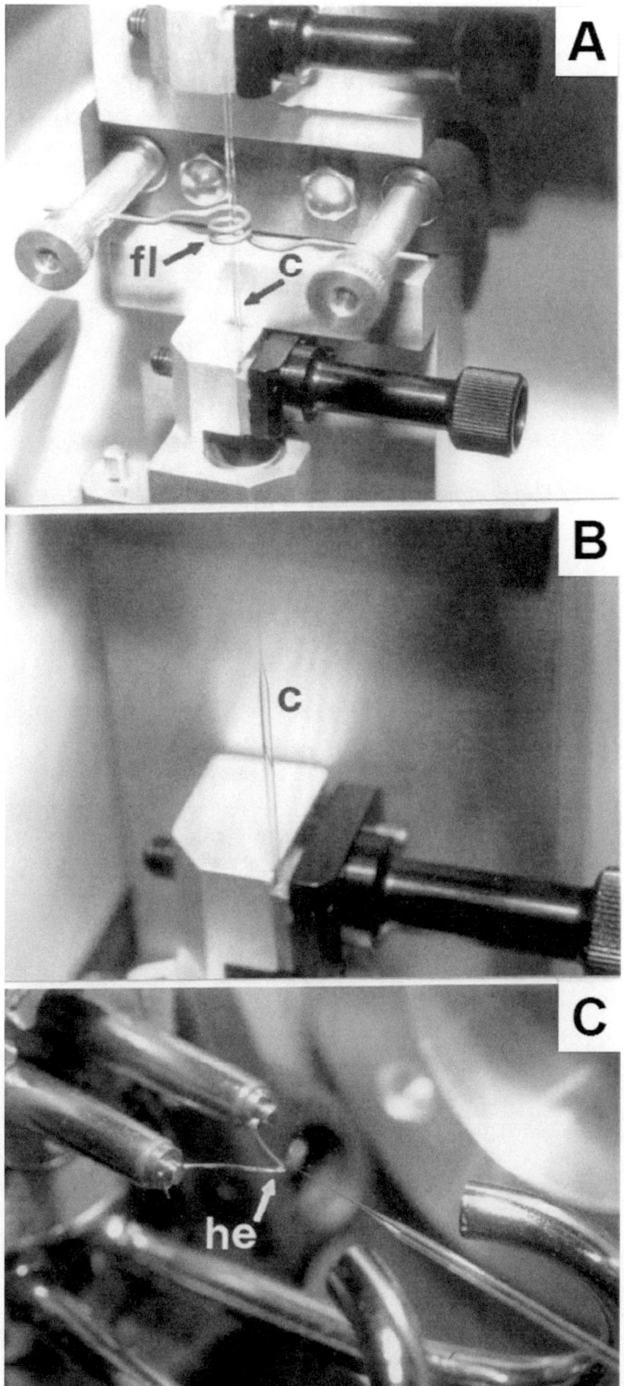

Fig. 14.3. Preparation of needles. (**A**) A capillary (c) is placed through the pipette puller filament (fl). (**B**) By regulating the temperature of the filament, pulled capillaries of different lengths will be generated. (**C**) View of a holding pipette during the bending by the microforge (also, see Fig. 14.4E). he, heating element.

Most commonly available microforges are supplied with two heating elements of different sizes. We use a de Fonbrune-type microforge. Settings vary according to manufacturer's protocols and individual preferences (**Fig. 14.3C**).

3. Methods

3.1. Preparation of Needles

The microinjection procedure requires a holding and an injection pipette that can be prepared from 10–15 cm long 1-mm glass capillary. Both the holding and injection pipettes are made from similarly pulled capillaries. Holding pipettes are obtained from the thick part of the capillary (diameter of about 100 μm; the blastocysts have a diameter of about 180–200 μm) whereas the needle of the injection pipette is obtained from the tip of the capillary measuring about 20–25 μm in diameter (ES cells have a diameter ranging between 15 μm and 20 μm). The capillaries can be pulled consistently after appropriate setting of the pipette puller. This makes it fairly easy to identify the appropriate region to make a holding or an injection pipette (*see* **Note 7**).

3.1.1. Holding Pipettes

Holding pipettes are required to keep the blastocysts immobilized in place during ES cell injection. The blastocyst is held by gentle suction through the opening (20–40 μm) of the holding pipette. Holding pipettes are made from pulled capillaries by using a microforge: (**Fig. 14.3C**).

1. The heating filament of the microforge carries a glass bead at the tip. The pipette should be positioned horizontally above the glass bead at the appropriate level of the capillary (about 80–100 μm external diameter).
2. Turning on the heat element will cause the glass bead to expand and move along side of the pipette (*see* **Fig. 14.4A**). The heating has to be controlled carefully to ensure that the capillary will not bend or melt when contacting the heated glass bead. Exact parameters must be adjusted depending on the microforge used.
3. As soon as the pipette fuses slightly with the glass bead and extends in length, the heat should be switched off. Contraction of the glass causes the pipette to snap vertically (**Fig. 14.4B**).
4. The opening of the holding pipette is then placed in front of the filament with the glass bead at an appropriate distance to avoid contact during the polishing of the tip.

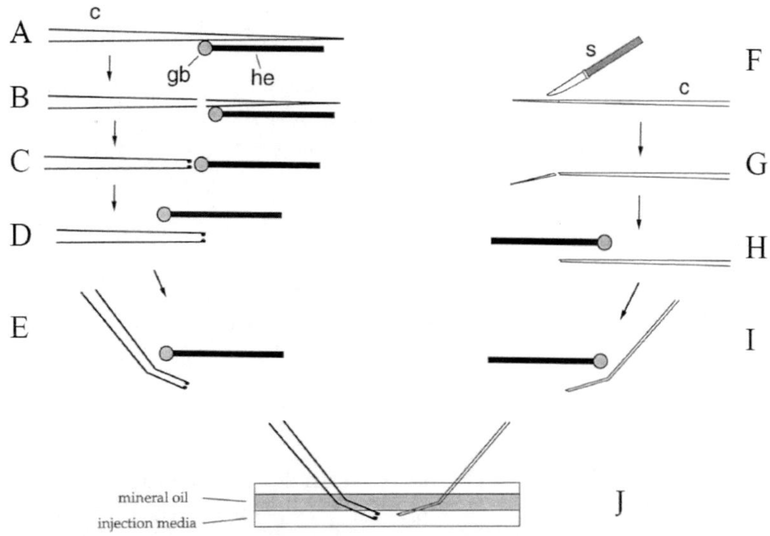

Fig. 14.4. Schematic illustrating the procedure to make holding (**A–E**) and injecting (**F–I**) pipettes (see text for details). Representation of the injecting chamber (**J**). he, heating element; gb, glass bead; c, pulled capillary; s, scalpel.

5. The pipette is moved carefully towards the glowing glass bead and the edges will start to melt inward creating a smoother, narrower tip (**Fig. 14.4C**).
6. The holding pipette is then bent (between 30 degrees and 45 degrees depending on the setting of the pipette holders) to allow an almost parallel position of the tip relative to the bottom of the injection chamber. The bend is introduced near the tip of the holding pipette by moving the capillary toward the filament.
7. Carefully controlling the temperature of the filament and its distance from the pipette, the tip is allowed to bend until the appropriate angle is achieved (**Fig. 14.4D,E**). Holding pipettes can be used for several injections until breakage or blockage occurs.

3.1.2. Injection Pipettes

The injection pipette has the shape of a very fine hypodermic needle (**Fig. 14.3A,B**). The inner diameter of the injection pipette must be slightly larger than the ES cells to be injected so that they can be collected without clogging the pipette. A reproducible method to make an injection pipette involves crafting it with a microforge. However, this requires experience and skill. A faster method requires only pulled capillaries, a scalpel, and a transparent rubber sheet.

1. Place the capillary on the rubber sheet under the dissecting microscope.

2. Cut the capillary with the scalpel at the level where the diameter is sufficient to allow the passage of ES cells (**Fig. 14.4F,G**). While the choice of glass used for making holding pipettes is not critical, it is a crucial factor in obtaining injection needles that form the right type of tip (shape of a hypodermic needle) when cut (Drummond microcaps).
3. Introduce a bend as previously done with the holding pipette but such that the opening faces the side of the injection chamber (**Fig. 14.4H,J**). This will allow the collection of ES cells without damaging the tip.
4. Store the injection and holding pipettes such that the tips do not contact any surface.

3.2. Isolation of Mouse Blastocysts

1. Set up female donor mice (e.g., of the C57BL/6 strain; *see* **Section 2.1.1**) for natural matings.
2. The following morning they are checked for vaginal plugs. At this point, the plugged female is considered 0.5 days post coitus (dpc).
3. Three days later, sacrifice donor females (3.5 dpc) by CO_2 asphyxiation.
4. Place the animal on its back on a gauze pad and expose the reproductive tract by cutting the abdominal wall with a pair of scissors (**Fig. 14.5A**).
5. Isolate the uterus by holding the cervix with tweezers and cutting the connective vaginal side (**Fig. 14.5B**).
6. After isolation, place the uterus on a gauze pad (all uteri isolated should be laid on the pad with the same orientation) (**Fig. 14.5C**).
7. Drop some media on the uterus to keep it moist and carefully remove all the fat tissue (**Fig. 14.5D**). Trimming of fat tissue is very helpful since residual fat cells will obstruct vision while collecting the blastocysts.
8. Clip off the uterine horns very close to the cervix (leave the cervix next to the uteri for orientation) and also cut off a small piece at the level of the ovaries (**Fig. 14.5E**).
9. Insert a G25 needle (the tip of which has been rounded with sand paper) of a syringe full of ES media at the ovary side of the uterine horn and flush the embryos into a 6-cm Petri dish with about 0.5 mL of media (**Fig. 14.5F,G**). The flushed embryos should then be checked under a binocular lens for a visible blastocoel cavity and intact zona pellucida and collected with a mouth pipette for injection (*see* **Note 8**).

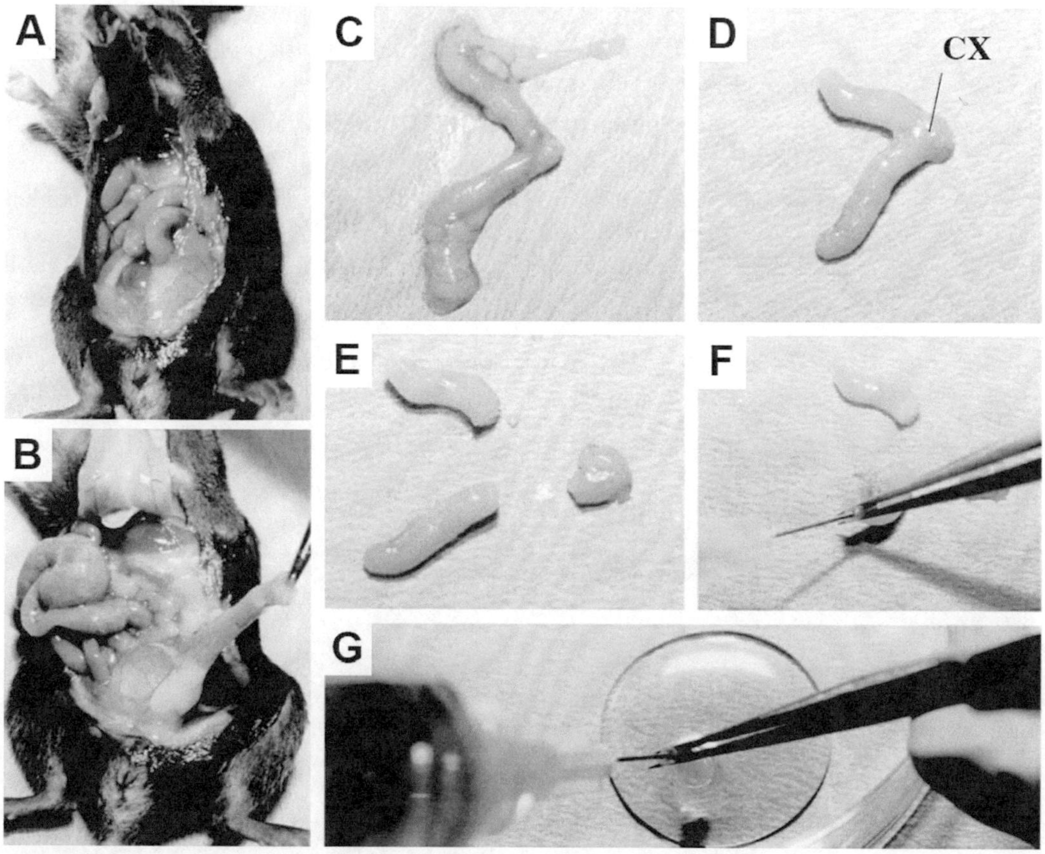

Fig. 14.5. Method employed to isolate and dissect the uterus from a pregnant female to recover mouse blastocysts (see text for detailed description of the procedure). cx, cervix.

3.3. Preparation of ES Cells for Injection

ES cells from a 60%–70% confluent 6-cm plate are trypsinized as described in **Chapter 9** and resuspended in 3–4 mL of ES media. The ES cell suspension should be fairly concentrated such that during the injection procedure a few drops can provide a sufficient number of cells for a specific injection field. Cells can be kept at room temperature until they are ready to be injected.

3.4. Microinjection of Blastocysts

During the initial phase of the injection procedure the quality of the needle will be apparent. Lack of sharpness and blockage of the needle by accumulation of cell debris can severely impact the success of injection. Cells should not escape from the opening produced by the tip of the injection pipette.

1. Microinjection of blastocysts is performed on a 6-cm Petri dish lid. Put about 3 mL of blastocyst injection media on the plate and cover the surface with embryo-tested mineral oil (**Fig. 14.4 J**).

2. Add a few drops of ES cells to the media followed by the addition of the blastocysts by a mouth pipette being careful to keep them grouped on the plate.
3. Lastly, set and position the holding and injection pipettes on the plate and fill their terminal endings with media.
4. Load individual ES cells with good refraction (for viability) into the injection pipette. Any cells that look dark are dying and should not be used. To keep the volume of injected media as small as possible, cells in the injection pipette should be loaded at the very tip such as to form trains of cells (**Fig. 14.6A**). A total of 15–25 healthy looking cells are picked up at a time.

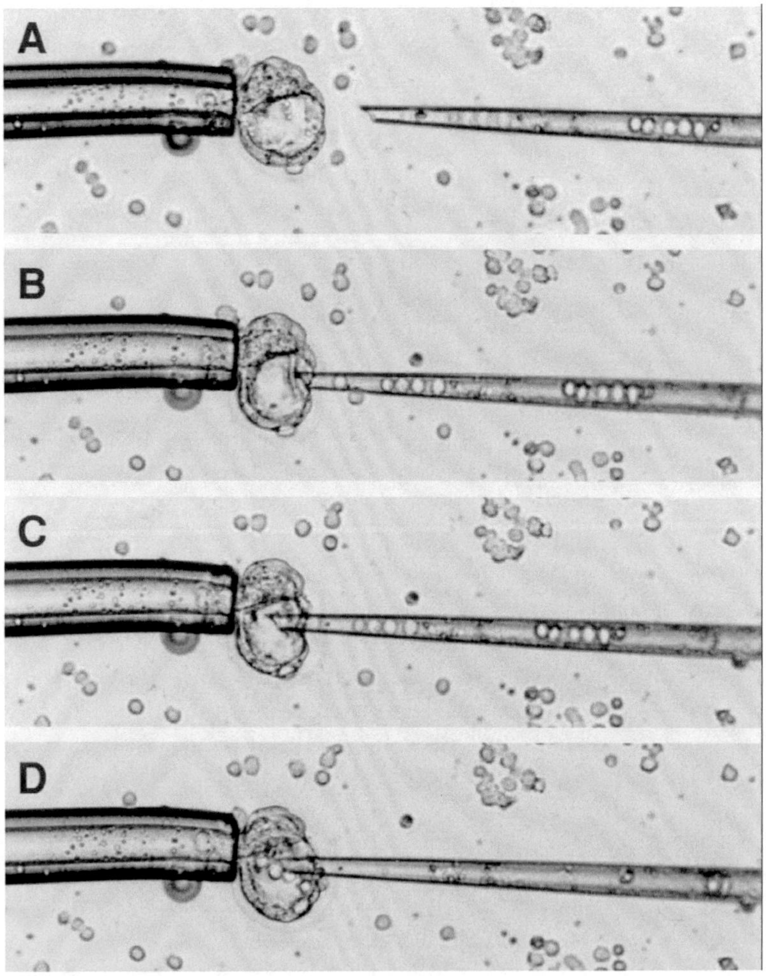

Fig. 14.6. Microinjection of blastocyst. (**A**) Cells are loaded into the injection pipette (*right*) and a blastocyst is immobilized near the inner cell mass by the holding pipette (*left*); note the trains of ES cells in the injection pipette. (**B**) The needle is moved close to the blastocyst and is now puncturing its wall between trophectoderm cells. (**C**) The tip is now positioned at the center of the blastocoel. (**D**) Cells are injected.

5. Select a blastocyst for injection.
6. Using the holding pipette maneuver the blastocyst into the correct position and then gently immobilize by suction.
7. Select the correct position to inject the blastocyst. The inner cell mass should be close to the opening of the holding pipette but a slightly tilted position is also desirable (**Fig. 14.6**). The best spot for penetration into the blastocoel is between trophectoderm cells.
8. With one smooth push, bring the tip of the injection pipette into the embryo without collapsing or puncturing of the opposite wall (**Fig. 14.6B,C**). Once in the blastocyst, the injection needle should not touch the inner cell mass.
9. When the tip of the injection pipette is clearly visible inside the blastocoel, expel the ES cells (15–25 cells per blastocyst) by positive pressure (**Fig. 14.6D**).
10. Following injection, withdraw the needle from the embryo.
11. Slightly lift the holding pipette and move the injected blastocyst to a different field, such that injected and non-injected embryos are kept separated.
12 Repeat the process with a new blastocyst.
13. Collect injected blastocysts for their transfer to recipient females.

3.5. Preparation of Pseudopregnant Recipients

The injected embryos must be implanted into recipient females. Mating of recipients with male mice is required for the hormonal changes necessary to establish pregnancy and for proper development of manipulated embryos. Since recipients should not carry embryos of their own, females must be mated with vasectomized males (see also **Section 2.1.2**). To assure availability of greater than five pseudopregnant females for a specific day of injection we set up matings with about 30 females that are 6–8-weeks old. Pseudopregnant females in excess can be used again after a 2-week resting period. Females at 2.5 dpc are used for transfer of the injected 3.5-dpc blastocysts. This 1-day difference is adopted to compensate for the delay in embryonic development caused by in vitro manipulations. Empirically, this results in a higher percentage of implantations.

3.6. Transfer of Blastocysts to Pseudopregnant Females

We usually carry out the transfer of manipulated blastocysts soon after the injection.

1. Anesthetize recipient females that are at 2.5 dpc by intraperitoneal injection of about 0.5 mL of Avertin.
2. Place the animal on its belly and shave its back.

3. Under a surgical microscope, using a pair of scissors, make an incision through the skin and the peritoneum on the right side of the back just along the bottom of the rib cage, avoiding cutting any blood vessels (**Fig. 14.7A,B**). The ovary should be visible right underneath the incision (**Fig. 14.7C**).
4. Grasp the ovary, oviduct, and uterus with tweezers by the fat pad that is attached to them avoiding any pinching of the reproductive organs.
5. Carefully pull the uterus through the incision in the peritoneum and place it on a drape.
6. Attach a small clamp to the fat pad to hold the ovarian end of the uterus in place and prevent the uterus from sliding back into the peritoneum (**Fig. 14.7D**).
7. Pick up the blastocysts with a mouth-controlled transfer pipette. Up to 15 embryos can be transferred into one uterine horn.
8. Grasp the uterus very close to the uterus–oviduct junction with a pair of microdissection tweezers and make a small hole using a 30-gauge hypodermic needle (**Fig. 14.7E**).

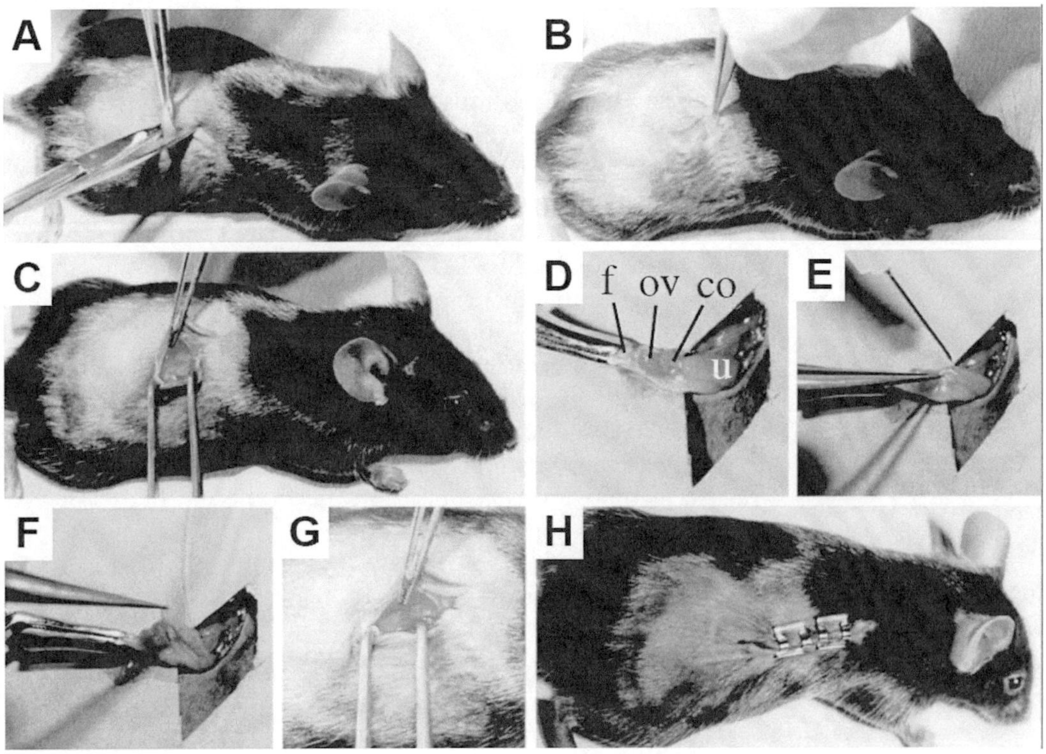

Fig. 14.7. Surgical procedure for transferring blastocysts into pseudopregnant females. See text for detailed description. f, fat pad; ov, ovary; co, coiled oviduct; u, uterus.

9. Insert the tip of the transfer pipette and gently mouth-pipette the embryos while monitoring the appearance of few air bubbles inside the uterus (**Fig. 14.7F**). Any resistance indicates that the pipette is touching the other wall of the uterine horn and the pipette should be pulled back gently.
10. When air bubbles are seen, all the embryos are assumed to be transferred into the lumen of the uterus.
11. Remove the clamp and gently push the uterus back into the peritoneum (**Fig. 14.7G**).
12. Staple the skin with two 9-mm surgical wound clips (**Fig. 14.7H**).
13. Lastly, return the recipients to their cages onto a warm tray to ensure that they are kept warm until they regain consciousness. Pups from manipulated embryos should be born 16–18 days after the day of transfer.

4. Notes

1. Since 129 strains do not breed well, and are reported to have abnormal anatomy and behavior, ES cells of the C57BL/6 background are preferred. Moreover, C57BL/6 is the most widely used inbred strain because it is permissive for expression of most mutations (http://jaxmice.jax.org/strain/000664.html).
2. It is important to note that when injecting C57BL/6 ES cells into C57BL/6 albino blastocysts, the contribution of these ES cells to the chimeras is lower (10%–40%) than that of 129/SV ES cells (50–100%). However, even low-percentage chimeras (30%–50%) can transmit the ES cell-derived genome to the progeny.
3. Although different facilities may use different hybrid F1 recipient females, we routinely introduce manipulated C57BL/6 blastocysts into C57BL/6xDBA/2 mice of the F1 generation.
4. We have now switched completely to Glutamax because it provides a very stable source of glutamine.
5. Because these instruments are expensive and are subjected to significant changes over time we recommend checking with different vendors before committing to a specific brand. However, it should also be noted that most vendors provide microscopes and microinjection systems of very good quality, which greatly reduces the chances of making the "wrong" purchase.

6. Some investigators recommend cooling of the embryo during injection and therefore recommend to purchase a cooling device. We found that no significant difference is observed in the injection and viability of the embryos when performing the procedure at room temperature.
7. Although, some protocols recommend washing the capillaries in HNO_3 before use, we found that to be unnecessary.
8. Some investigators recommend superovulation protocols to increase the embryo recovery. However, embryos produced by superovulation have the disadvantage of being less synchronous with regard to their developmental stage.

Acknowledgments

We thank Eileen Southon for critical reading of the manuscript. This research was supported by the Intramural Research Program of the NIH, National Cancer Institute. The contents of this publication do not necessarily reflect the views or policies of the Department of Health and Human Services, nor does mention of trade names, commercial products, or organizations imply endorsement by the U.S. Government. The Center for Cancer Research, NCI-Frederick has filed an Animal Welfare Assurance with the Office for Protection from Research Risks (OPRR). The protocols herein described have been approved by the NCI-Frederick Institutional Animal Care and Use Committee.

References

1. Robertson E, Bradley A, Kuehn M, Evans M. Germ-line transmission of genes introduced into cultured pluripotential cells by retroviral vector. Nature 1986; 323: 445–8.
2. Bradley A, Evans M, Kaufman MH, Robertson E. Formation of germ-line chimaeras from embryo-derived teratocarcinoma cell lines. Nature 1984; 309:255–6.
3. Martin GR. Isolation of a pluripotent cell line from early mouse embryos cultured in medium conditioned by teratocarcinoma stem cells. Proc Natl Acad Sci USA 1981; 78:7634–8.
4. Evans MJ, Kaufman MH. Establishment in culture of pluripotential cells from mouse embryos. Nature 1981; 292:154–6.
5. Smithies O. Forty years with homologous recombination. Nat Med 2001; 7:1083–6.
6. Evans MJ. The cultural mouse. Nat Med 2001; 7:1081–3.
7. Capecchi MR. Generating mice with targeted mutations. Nat Med 2001; 7: 1086–90.
8. Poueymirou WT, Auerbach W, Frendewey D, et al. F0 generation mice fully derived from gene-targeted embryonic stem cells allowing immediate phenotypic analyses. Nat Biotechnol 2007; 25:91–9.
9. Eakin GS, Hadjantonakis AK. Production of chimeras by aggregation of embryonic stem cells with diploid or tetraploid mouse embryos. Nat Protoc 2006; 1:1145–53.
10. Wood SA, Allen ND, Rossant J, Auerbach A, Nagy A. Non-injection methods for the production of embryonic stem cell–embryo chimaeras. Nature 1993; 365:87–9.

11. Rivera J, Tessarollo L. Genetic background and the dilemma of translating mouse studies to humans. Immunity 2008; 28:1–4.
12. Liu P, Zhang H, McLellan A, Vogel H, Bradley A. Embryonic lethality and tumorigenesis caused by segmental aneuploidy on mouse chromosome 11. Genetics 1998; 150:1155–68.
13. Seong E, Saunders TL, Stewart CL, Burmeister M. To knockout in 129 or in C57BL/6: that is the question. Trends Genet 2004; 20:59–62.
14. Schuster-Gossler K, Lee AW, Lerner CP, et al. Use of coisogenic host blastocysts for efficient establishment of germline chimeras with C57BL/6 J ES cell lines. Biotechniques 2001; 31:1022–4, 6.
15. Auerbach W, Dunmore JH, Fairchild-Huntress V, et al. Establishment and chimera analysis of 129/SvEv- and C57BL/6-derived mouse embryonic stem cell lines. Biotechniques 2000; 29:1024–8, 30, 32.

Chapter 15

Aggregation Chimeras: Combining ES Cells, Diploid, and Tetraploid Embryos

Mika Tanaka, Anna-Katerina Hadjantonakis, Kristina Vintersten, and Andras Nagy

Abstract

During the past 40 years, mouse chimeras have served as invaluable tools for studying not only genetics but also embryonic development, and the path from undifferentiated cell populations to fully committed functional cell types. This chapter gives a description of the early events of cell commitment and differentiation in the pre-and postimplantation-stage embryo. Next, a discussion follows highlighting the most commonly used as well as more recently developed applications of various cell types and origins used in the production of chimeras. Finally, detailed protocols and trouble-shooting suggestions will be presented for each of the steps involved.

Key words: ES cells, mouse, embryo, chimera, tetraploid, aggregation.

1. Introduction

The word "chimera" stems from ancient times, describing creatures composed of cells with more then one embryonic origin. In modern science, mythical creatures like the Egyptian sphinx have been replaced by sophisticated combinations of cells and embryos providing us with powerful tools to study mouse development *(1)*. Before entering into a deep discussion on the various chimera combinations, we must take a detailed look at the early steps of differentiation in the developing embryo. As depicted in **Fig. 15.1**, the blastomeres of morula-stage mouse embryos are all totipotent, that is, each of them is capable of giving rise to all embryonic and extraembryonic tissues. However, at the blastocyst stage, the very first distinct cell types arise: the inner cell mass now consisting of primitive endoderm and primitive ectoderm, and an

Ralf Kühn, Wolfgang Wurst (eds.), *Gene Knockout Protocols: Second Edition, vol. 530*

DOI 10.1007/978-1-59745-471-1_15 Springerprotocols.com

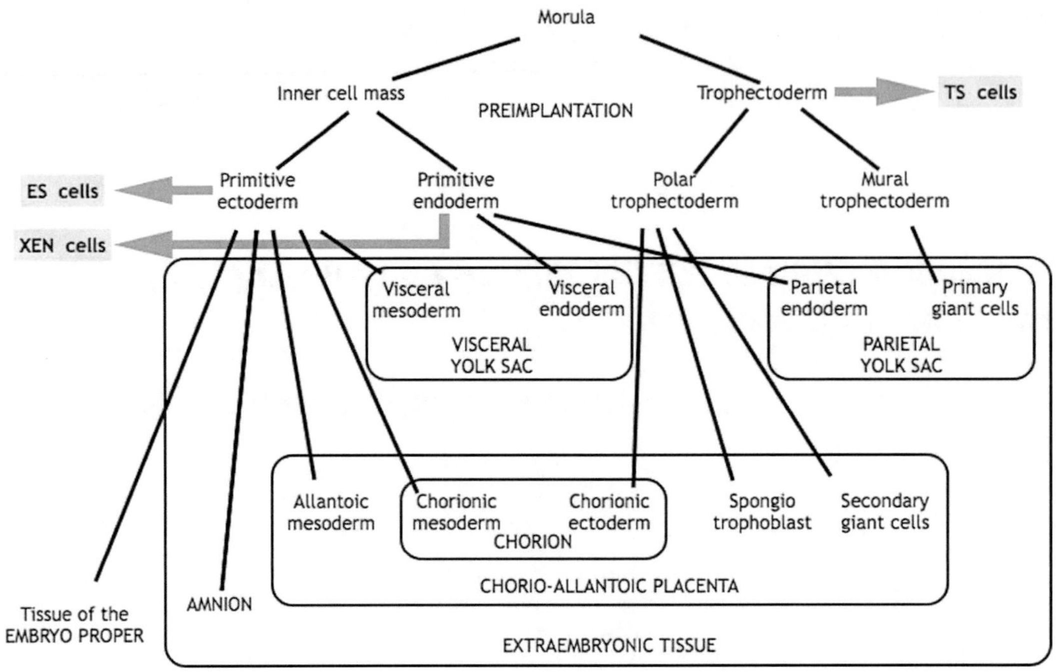

Fig. 15.1. Schematic presentation of the various embryo proper and extraembryonic lineages, as well as their relation to each other.

outer layer of trophectoderm. Each of these three compartments has its unique potential as well as limitation. Trophectoderm cells are committed to the development of the trophoblast cells in the placenta. Primitive endoderm cells are capable of forming the outer layers of the yolk sac, while primitive ectoderm cells will contribute to the embryo proper. It is important to note that this potential is strictly accompanied by a limitation. Each of the three cell types of the blastocyst is restricted to the contribution listed above *(2)*.

Embryonic stem (ES) cells *(3–5)* are isolated from the primitive ectoderm. Although the concept of referring to "stem cells" in the context of ES cells is very well established, it is somewhat misleading. The basic definition of the stem cell property includes very quiescent and slowly self-renewing cells that reside for a long time in the organism. Primitive ectoderm cells however are a very transient population, existing no longer then a few hours in the blastocyst-stage embryo. It is more useful to think about ES cells as an in vitro artifact; cells derived from the primitive ectoderm and artificially prohibited to follow their natural differentiating program by specific culture conditions. Similar to ES cells, permanent cell lines have recently been established also from the other two compartments of the blastocyst-stage embryo: TS cells from the trophectoderm *(6)* and XEN cells from the primitive endoderm *(7)*. Interestingly, all of these faithfully recapitulate both the potential as well as the restriction of the cell types of their origin.

TS and XEN cells differentiate exclusively into the trophoblast and extraembryonic endoderm lineages, respectively, when introduced into a blastocyst-stage embryo. ES cells combined with preimplantation-stage embryos, and allowed to develop in vivo, are excluded from the trophoblast of the placenta and primitive endoderm derivatives (*8*). However, the embryo proper and the resulting chimeric animal will possess a mixture of cells originating from both the original components. A complete separation of ES cells to the embryo proper and host embryo to the placenta can be achieved by the use of tetraploid embryos as hosts. A tetraploid embryo *(8, 9)*, which can be made by electrofusing the cells of a two-cell-stage diploid embryo *(8, 10, 11)*, can efficiently form the trophoblast compartment of the placenta as well as the endoderm layer of the yolk sac. However, it is virtually incapable of contributing to the embryo proper *(12)*. Therefore, in ES cell–tetraploid embryo chimeras, the embryo proper, the amnion, the yolk sac mesoderm, the allantois, and the embryonic mesoderm component of the placenta are completely ES cell-derived, whereas the yolk sac endoderm and the trophoblast cell lineages originate from the tetraploid embryo *(8, 11, 13)*. ES cell–tetraploid embryo aggregates have an attractive feature in that they are a reliable and simple way of producing completely ES cell-derived embryos and animals from developmentally competent cell lines *(4, 5, 8–11, 13–16)*. This feature is promoting their application in an increasing number of studies. In addition, chimeras between diploid cells (both diploid embryo–diploid embryo, and ES cell–diploid embryo chimeras) are going through a renaissance in addressing specific biological questions *(1, 17)*. **Figure 15.2** illustrates the various possible combinations of ES cells and diploid versus tetraploid embryos.

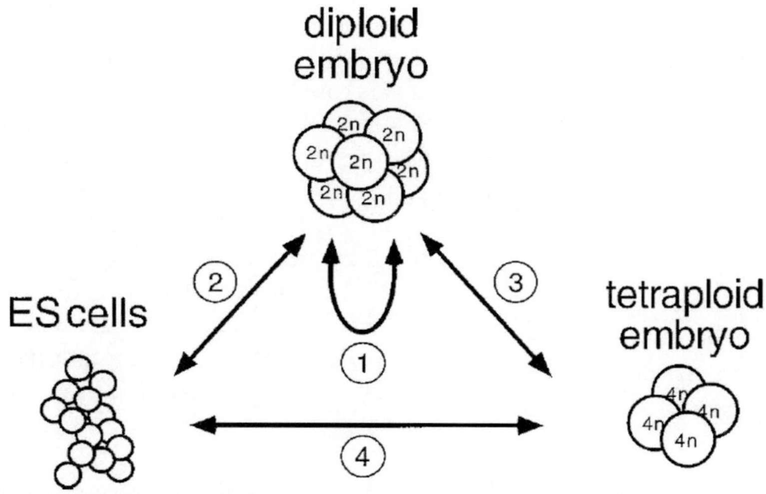

Fig. 15.2. The three components that can be combined to produce chimeras: diploid embryos, tetraploid embryos, and ES cells. (1) Diploid embryodiploid embryo, (2) diploid embryo–ES cells, (3) diploid embryo–tetraploid embryos, and (4) tetraploid embryos–ES cells.

Embryo–embryo chimeras are made by aggregating two blastomere-stage embryos. Chimeras with an ES cell-derived component have traditionally been produced by injecting the cells into blastocyst-stage embryos using micromanipulators. However, we have learned that a more simple way of producing such chimeras can be performed by aggregating ES cells with blastomere-stage embryos *(18)*. This alternative method also gives a high efficiency in chimera production as long as optimal culture conditions can be provided. Some combinations, such as embryo–embryo chimeras are only possible to make through the aggregation technique. Hence, in this chapter, we discuss the production and use of aggregation chimeras.

1.1. Uses of Chimeras

As outlined in **Fig. 15.2**, there are several possible combinations of embryos and cells to produce chimeras. A suitable aggregation combination should be chosen depending on the aim of the experiment.

1.1.1. Generation of Mutants; Germline Transmission and ES Cell-Derived Embryos/Animals

To obtain germline transmission of an ES cell genome, the cells can be aggregated with diploid embryos, resulting in viable and fertile chimeras *(18)*. Due to frequent X chromosomal instability in female ES cells, male ES cells are used for this purpose almost exclusively. In most cases, male chimeras that have a high contribution of ES cell derivatives, scored by using coat color markers, are ideal for obtaining germline transmission of an ES cell genome *(19)*. It should be noted that there are cases in which male chimeras with strong ES cell contribution may not be ideal. With some ES cell lines, high ES cell contribution positively correlates with sterility and decreased viability. The success of germline transmission mostly depends on the quality of ES cells. It is essential to perform genome alteration on ES cells that have a high developmental potential and germline compatibility. However, even in this case, a minor ratio of subclones will lose their original capabilities, and give rise to sterile or non-transmitting chimeras. The most permissive genetic background for germline-competent ES cell lines is inbred 129. Therefore the 129-derived ES cell lines were dominating the ES cell-based mouse mutagenesis approaches for almost 20 years. As the culture conditions have been more defined, ES cells derived from C57Bl/6 embryos can now be used with high efficiency for germline transmission after genetic modifications. Therefore, the international high-throughput gene knockout projects *(20)* (EUCOMM, NorCOMM, and KOMP) decided to derive mutant ES cells for all the known genes in the C57BL/6 background, which is considered the gold standard among the inbred stains of mice. The developmental potential of inbred lines are generally not high enough to efficiently yield viable completely ES-derived animals following aggregation with tetraploid embryos. However, some F1 hybrid ES cell lines have proven to possess superior developmental potential when applied to this technique *(21)*. Hence, this technology has become highly feasible to analyze in vivo phenotypes directly from mutant ES cells *(16)*.

Certain types of mutations, such as dominant mutations or mutations in haploinsufficient or X-chromosome-linked genes, may directly affect the developmental potential of ES cells and are only able to contribute to viable chimeras at a low level. However, even in this situation, it is not impossible to obtain germline transmission through chimeras exhibiting weak contribution of the ES cells *(22)*. Haploinsufficiency could also create an apparently surprising phenomenon, when in the case of germline transmission of the ES cell genome no heterozygous F1 animals are observed. In this case, the heterozygotes might start developing but then die in utero. As a consequence, no mutant mouse line can be established. The phenotype of the haploinsufficiency can be analyzed through the chimera-fathered embryos. The only possible way to access the homozygous null phenotype in a severe haploinsufficiency case is the production of homozygous null ES cells followed by the production of completely ES cell-derived embryos *(22)*. Genomic imprinting could create a similar situation. Imprinting is the phenomenon in which the activity of a gene shows a difference depending on the parental origin. Maternally imprinted genes require transmission through the paternal germline for activation. Therefore, one may never find viable progeny carrying the knockout allele of such a gene from a male chimera. In this case, again, female ES cells and chimeras would be a choice to obtain germline transmission. The opposite, the knockout allele of a paternally imprinted gene, does not have germline transmission problems through male chimeras *(14, 23, 24)*.

Other potential difficulties could arise if the mutation introduced into the ES cell line itself is the cause of the sterility, resulting from, for example, a defect in spermatogenesis. A possible way to circumvent this problem can be provided by the occasional germline transmission through female chimeras *(25)*. However, one must keep in mind that the Y chromosome usually is lost or becomes nonfunctional in fertile transmitting female chimeras.

1.1.2. Determination of Cell Autonomy of Particular Mutations

Chimeric analysis has proven to be a powerful method for studying the cell-autonomous requirements of genes of interest *(26–29)*. Mutant cells can be ES cells as well as diploid and/or tetraploid embryos depending on the question that one would like to address. The suitable aggregation combination for chimeric analysis to address cell autonomy will be discussed in further detail in **Section 1.2**.

1.1.3. Separation of Embryonic and Extraembryonic Phenotypes

Chimeras also provide an excellent way to separate embryonic from any extraembryonic phenotypes of a genetic alteration *(30)*. The complementary restricted developmental potential of ES cells and tetraploid embryos as mentioned in the introduction makes this feasible. An example highlighting this scenario is the relatively common case where mutant embryos die from placental failure or other extraembryonic defects. By aggregating mutant diploid with wild-type tetraploid embryos, one can rescue the defect in the extraembryonic compartment, and gain access to mutant embryos for further phenotypic analysis. In this method – often referred to as the tetraploid complementation assay – the wild-type tetraploid embryos provide functionally normal placentae, while being excluded from the embryo proper *(14, 25, 31, 32)*. The reverse scenario is also possible: an embryonic phenotype can be rescued in order to allow the study of extraembryonic defects by aggregating mutant diploid embryos with wild-type ES cells *(29)*. In this case, the ES cells will provide primitive ectoderm derivatives, but will never contribute to the primitive endoderm or trophoblast lineages. The aggregation combination for this use will be discussed further in **Section 1.2.4**.

1.1.4. Accessing Phenotypes Without Germline Transmission

ES cells carrying dominant genome alterations that may cause a phenotype in primitive ectoderm cell lineages can be aggregated with wild-type tetraploid embryos to study the phenotype directly, without going through the potentially problematic and time-consuming germline transmission breeding scheme *(22)*. In the case of recessive mutations, the production of ES cell lines homozygous for the mutation is required *(33, 34)*. However, dominant-negative, gain-of-function mutations or stable transgenesis-based RNAi knockdowns can be analyzed directly from ES cell-derived embryos/animals after a single genetic alteration *(16, 35)*.

1.2. Aggregation Combinations

As discussed above, different aggregation combinations are required depending on the aim of each experiment. In this section, all possible aggregation combinations using ES cells, diploid and tetraploid embryos are listed and the expected contribution of mutant cells in resulting chimeras from each aggregation combination is discussed with examples of their practical use.

1.2.1. Wild-Type Diploid Embryo–Mutant Diploid Embryo

The contribution of mutant diploid cells is expected to be present in all cell lineages in the resulting chimeras made by this aggregation combination (*see* **Fig. 15.3**) unless cells from the mutant embryo have developmental restrictions. This makes it possible to assess the question of cell autonomy of mutations (*13*).

1.2.2. Wild-Type Diploid Embryo–Mutant Tetraploid Embryos

In this scenario, the contribution of mutant tetraploid cells is limited to the trophoblast and primitive endoderm derivatives (*see* **Fig. 15.3**). The resulting chimeras are expected to have chimeric extraembryonic tissues with no contribution of mutant cells in the primitive ectoderm derivatives, such as the embryo proper. This combination could be used in order to address cell autonomy

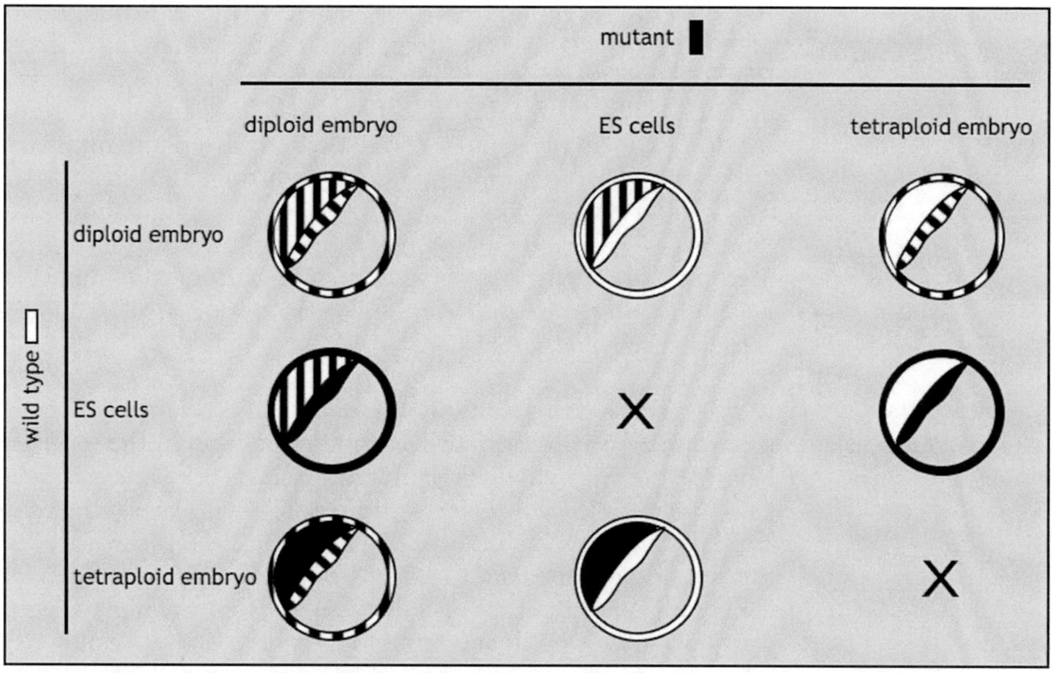

Fig. 15.3. Tissue contributions and lineage restrictions associated with the three components of chimeras.

of the mutation specifically in the extraembryonic lineages, when there are multiple cell-autonomous defects in both the extraembryonic as well as the embryonic lineages.

1.2.3. Mutant Diploid Embryo–Wild-Type Tetraploid Embryo

There is no restriction for the contribution of mutant diploid cells, whereas the contribution of wild-type tetraploid cells is limited to the extraembryonic tissues (*see* **Fig. 15.3**). This will result in chimeras that have chimeric extraembryonic tissues and exclusively mutant-embryo-derived primitive ectoderm derivatives. Such an approach will be the choice to study embryonic phenotypes while rescuing the extraembryonic defects *(14, 29)*.

1.2.4. Wild-Type ES Cells–Mutant Diploid Embryo

In this case, the extraembryonic tissue, that is, the trophoblast and primitive endoderm lineages, will be derived solely from the mutant embryo, while chimerism will be found in all other lineages (*see* **Fig. 15.3**).

If mutant embryos of a gene of interest show phenotypes in both the embryonic and extraembryonic lineages, this aggregation would be the choice to address whether the placental defects are cell autonomous or secondary to the embryonic defects. This will be possible as the aggregation with wild-type ES cells can rescue the embryonic phenotype depending on the degree of chimerism without having any ES cell contribution in the extraembryonic tissues.

1.2.5. Mutant ES Cell–Wild-Type Diploid Embryos

In chimeras resulting from this scenario, the contribution of mutant ES cells is restricted to primitive ectoderm derivatives (*see* **Fig. 15.3**). This aggregation combination is suitable for all uses described in **Section 1.1.1**. For the study of cell autonomy during the development of primitive ectoderm derivatives, mutant ES cell lines carrying haploinsufficient or X-chromosome-linked or dominant mutations, or homozygous for recessive mutations, are required. Chimeric embryos from this aggregation can also be used to investigate phenotypes resulting from the mutation depending on the degree of ES cell contribution in the primitive ectoderm cell lineages.

1.2.6. Wild-Type ES Cells–Mutant Tetraploid Embryos

In this case, the contribution of mutant cells is solely restricted to the trophoblast and primitive endoderm lineages in the resulting chimeras (*see* **Fig. 15.3**) If the mutation affects both lineages, such chimeras can allow for the extraembryonic phenotypes to be separated from embryonic ones *(29)*. This is the clearest way to assess this question compared to other combinations, as there is no concern about the degree of chimerism due to the complementary distribution of ES cells and tetraploid embryo derivatives.

1.2.7. Mutant ES Cells–Wild-Type Tetraploid Embryos

If mutant ES cells are available, the aggregation with wild-type tetraploid embryos provides a powerful and quick way to analyze embryonic phenotypes without germline transmission. Normally, wild-type ES cells are capable of developing to form the primitive ectoderm derivatives (*see* **Fig. 15.3**) with help of wild-type tetraploid embryos, which provide functional placenta and yolk sac. In the case of mutant ES cells, the phenotype is manifested in the completely ES cell-derived embryo proper. This aggregation also makes it possible to assess embryonic phenotypes with no influence from the extraembryonic lineages.

2. Materials

1. ES cell medium Dulbecco's Modified Eagle's Medium (DMEM) High Glucose (Life Technologies 61965-026) supplemented with the following:
   a. 0.1 mM Nonessential Amino Acids (NEAA) (100X stock, Life Technologies 11140-035)
   b. 1 mM sodium pyruvate (100X stock, Life Technologies 11360-039)
   c. 100 μM β-mercaptoethanol (Sigma M 7522)
   d. 2 mM L-Glutamine (100X stock, Life Technologies 25030-024)

e. 15% Fetal Bovine Serum (FBS). ES-cell tested.

f. Penicillin and streptomycin, final concentration 50 μg/mL each (100X stock, Life Technologies 15140-122)

g. 1000–2000 U/mL Leukemia inhibitory factor (LIF) (Chemicon/Millipore ESG 1107)

2. 0.1% gelatin (Sigma G-2500).
3. PBS without calcium and magnesium. The solution is autoclaved and stored at 4°C.
4. Trypsin/EDTA (1 × solution, Life Technologies 25200).
5. M2 embryo culture medium (Specialty Media MR-015P-5F).
6. KSOM embryo culture medium (Specialty Media MR-020P-5F).

All embryo culture media should be stored at 4°C until use. The solution is equilibrated at 37°C/5% CO_2overnight, just prior to use. M2 and KSOM can also be prepared from individual reagents according to **Table 15.1**. If this is done,

Table 15.1
Composition of embryo culture media

	KSOM	M2
Component	Concentration (g/L)	Concentration (g/L)
NaCl	5.55	5.53
KCl	0.186	0.356
KH2PO4	0.0476	0.162
MgSO4	0.0493	0.293
Nalactate	1.12 or 1.87 g of 60% syrup	4.349 g of 60% syrup
D(+)glucose	0.036	1.00
Na.pyruvate	0.022	0.036
NaHCO3	2.10	0.349
$CaCl_2$ (dihydrous)	0.251	0.252
L-glutamine (Gibco, 25030)	0.146	–
EDTA (tetrasodium salt)	0.0038	–
BSA (Sigma, A3311)	1.0	4.00
Penicillin-G	0.060	0.060
Streptomycinsulfate	0.050	0.050
Phenol Red	–	0.01
HEPES Buffer	–	4.969

care should be taken that the water is of the highest quality grade and embryo-tested. All chemicals should also be embryo-tested individually. The use of plastic pipettes and containers is recommended, since glassware often contains residues that can be deleterious for embryos. Concentrated stock solutions can be stored at –80°C for a few months. Ready-to-use medium should be filter-sterilized and stored at +4°C for no longer than 2 weeks.

7. 0.3 M mannitol

   Mannitol (Sigma M4125) prepared in water with 0.3% BSA (Sigma A3311). Filter-sterilized and stored in aliquots at –20°C.

8. Light mineral oil, embryo-tested (Sigma, M8410)
9. Acid Tyrode's solution (Sigma T-1788)
10. Tissue-culture-treated plasticware (for cells and embryos): We routinely use Nunc, Corning, and Falcon plasticware.
11. Humidified incubators: Separate incubators for ES cell and embryo in vitro culture. Maintained at 37°C and 5% CO_2.
12. Upright stereo dissecting microscopes with illumination from below. These are required for preimplantation embryo work, such as flushing embryos form oviducts/uteri, setting up the aggregations, and for the transfer into recipient females.
13. Fine surgical instruments: Required for preimplantation-stage embryo recovery and embryo transfer into recipient females.
14. Darning needles: Suitable for making depressions in plates for aggregations. These should have the correct beveling such that a smooth depression is produced. Specially made needles can be purchased from BLS Ltd., H-1165 Budapest, Zsélyi Aladár u. 31, Hungary, http://www.bls-ltd.com/.
15. Pipette for handling embryos: Mouth pipette fitted with a drawn-out Pasteur pipette.
16. Electrofusion apparatus for tetraploid embryo production (BLS #CF-150).

3. Methods

3.1. Preparing the Aggregation Plate

1. Place four rows of drops of KSOM (approx. 3 mm in diameter) into a 35-mm tissue culture dish using a 1-mL syringe fitted with a 26G needle, with the first and fourth row comprising three drops and the second and third having five (*see* **Fig. 15.4; Note 1**).
2. Overlay the drops with mineral oil, so that they are totally submerged.

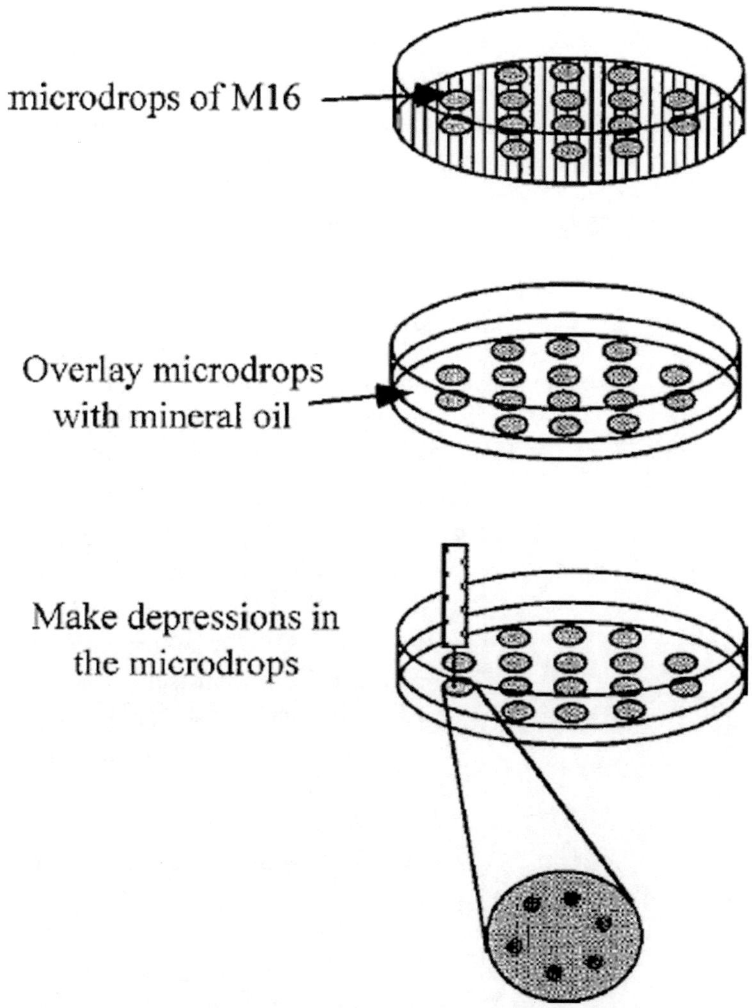

Fig. 15.4. Preparation of an aggregation plate involves making microdrops of media, overlaying them with mineral oil, followed by forging depressions for the placement of the aggregates. The plate once set up is placed in a temperature-controlled humidified incubator with 5% CO_2 content.

3. Sterilize the aggregation needle with ethanol, and immediately use it to make approximately six depressions per microdrop (*see* **Fig. 15.4**, **Note 2**).
4. Put the plate into the incubator overnight, or at least 3 h (*see* **Note 3**).

3.2. Obtaining the Embryos

1. Remove both oviducts from 1.5 days post coitum (dpc) females (for tetraploids) and 2.5 dpc (for diploids), and transfer to a drop of M2 medium in a Petri dish (*see* **Note 4**).
2. Flush the oviducts by inserting a flushing needle attached to a 1-mL syringe filled with M2 into the infundibulum (*see* **Notes 5, 6, 7**).

3. Collect embryos and wash them free of any debris in several drops of M2 using a mouth pipette.
4. Wash embryos in several drops of KSOM. Transfer the embryos to an organ culture dish with KSOM medium, and place in the incubator until all embryos and cells are prepared.

3.3. Electrofusion to Generate Tetraploid Embryos

1. Place the electrode in a plastic 10-cm Petri dish (*see* **Fig. 15.5**). Connect the cables from the electrode to the pulse generator, adjust all the parameters, and place on a dissecting stereomicroscope. We routinely use two pulses ("repeat" set to 2) of 100 V and 40 μs duration with 250 μm electrodes. These parameters, however, may vary between machines and genetic background of the embryos. Therefore, the optimum settings should be experimentally determined.
2. Put a large drop of 0.3 M mannitol over the electrodes.
3. Place two drops of M2 medium and one drop of 0.3 M mannitol on the surface of the Petri dish, outside the electrodes (*see* **Note 8**).
4. Introduce 50–100 embryos to one drop of M2. From there, take 20–25 embryos at a time into a drop of the mannitol. After they have settled, place them between the electrodes.
5. Carefully apply the orienting electric field. If some embryos do not properly align, correct their orientation manually (*see* **Note 9**).
6. When all embryos lie in the correct orientation, apply the pulse.

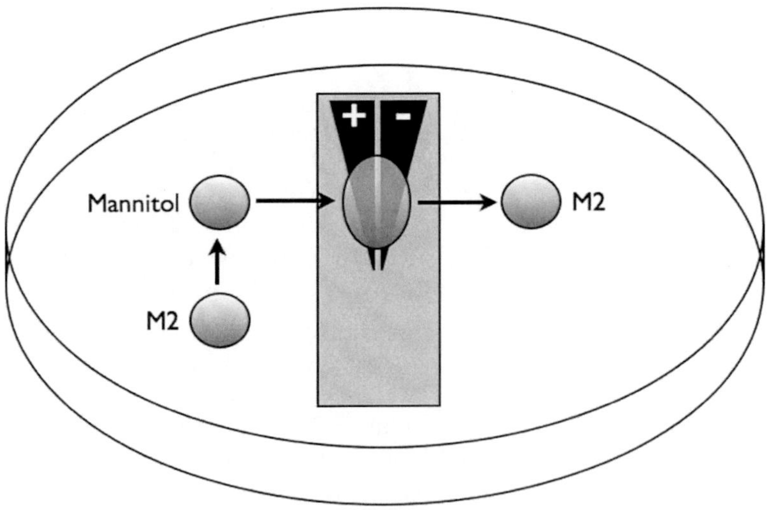

Fig. 15.5. Dish setup for electrofusion. The electrodes are placed in the middle. Drops of M2 media are placed on the dish for washing embryos. Mannitol solution is placed on the electrodes for the fusion process.

7. Immediately transfer the embryos back into a drop of M2. It is very important to keep the time the embryos spend in mannitol to a minimum. Until experience has been gained, it may be advisable to keep the number of embryos in each group lower.
8. Repeat Steps 5–8 until all the embryos have been subjected to the fusing pulses.
9. When all embryos have been treated, rinse them briefly in M2 and then in KSOM.
10. Transfer the embryos to an organ culture dish containing KSOM or microdrops under mineral oil, and place in the incubator (*see* **Note 10**).
11. After approximately 1 h, separate the fused embryos from those that are damaged or have remained diploid. This separation is a crucial point in the protocol. Only at this time-point is it possible to distinguish successfully fused (now tetraploid) embryos. Failing to carry out this step correctly could lead to the wrong interpretation of the chimera experiment. Embryos with a single round blastomere and with no evidence of cell lysing (debris under the zona pellucida) are successfully fused and should be separated from all others (*see* **Fig. 15.6a**). Embryos with a single but small blastomere and debris present

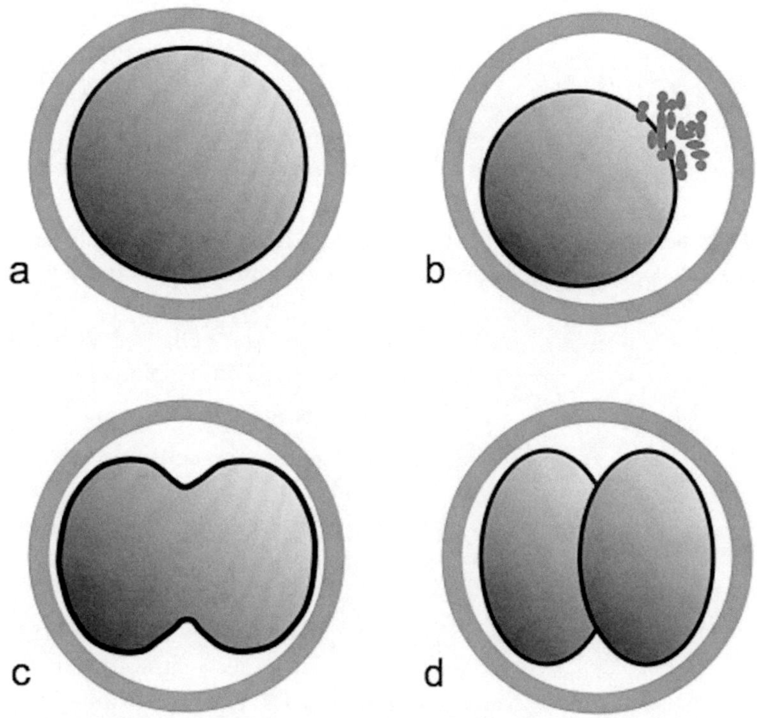

Fig. 15.6. Embryos following the fusion procedure. (**A**) Correctly fused, tetraploid embryo, (**B**) one blastomere has lysed, the embryo remains diploid, (**C**) embryo that has not yet completed the fusion, and (**D**) non-successful fusion, the embryo remains diploid.

under the zona are not tetraploid; the second blastomere has lysed during the fusion process (*see* **Fig. 15.6b**). Some embryos will show an elongated appearance (*see* **Fig. 15.6c**). They are under the process of fusion and should be allowed to complete the process in the incubator for an additional 10 min. Embryos that remain having two blastomeres are also still diploid and can be subjected to a second round of fusion (*see* **Fig. 15.6d**). If doing so, the same care should be taken to sort them after an additional hour in culture.

12. Incubate the embryos overnight.
13. The tetraploid embryos should now have developed to form four blastomeres, and are ready for aggregation (*see* **Note 11**). Embryos that only have two blastomeres can be left for a few more hours in the incubator as they may still divide.

3.4. Removing the Zona Pellucida from the Embryos

1. Place a few drops of M2, KSOM, and acid Tyrode's in a non-tissue culture-treated plastic Petri dish (if tissue culture dishes are used, it is important to use the lid so that the "naked" embryos will not stick to the surface).
2. Transfer a group of embryos into the first drop of acid, rinse briefly, and then transfer them to the second drop.
3. Continually observe the embryos, and note when their zona has dissolved (*see* **Note 12**). At that point, immediately transfer them to a drop of M2, and subsequently wash them in several drops of M2 (*see* **Note 13**).
4. Wash the embryos in KSOM. Embryos are now ready for transfer to the aggregation plate.

3.5. Preparing the ES Cells

Extensive protocols for the maintenance and culture of ES cells are beyond the scope of this chapter and are described elsewhere (*36*, *37*).

1. Day 1: Thaw the cells 3 days before the aggregation onto a feeder cell containing plate (*see* **Note 14**).
2. Day 2: Change the medium.
3. Day 3: Passage the cells onto gelatinized plates but instead of the usual 1:5 ratio, make 3–5 dishes with an increasingly higher dilution of 1:50–1:500 (*see* **Note 15**).
5. Day 4: Choose one of the dishes for preparation. It should be the one with a density showing small colonies with an average size of approximately 20 cells. Trypsinize the cells briefly, just until the colonies begin to detach from the plate. Stop the action of the trypsin by adding ES cell medium to the plate. Select clumps of 10–15 loosely attached cells for the aggregation, and transfer them into KSOM microdrops contained on the aggregation plate (*see* **Note 16**).

3.6. Setting Up the Aggregation

3.6.1. Diploid Embryo–Diploid Embryo Aggregation

1. Transfer the zona-free embryos to the aggregation plate and into a microdrop without depressions.
2. From there, place individual embryos with the first genotype into the individual depressions of the central two rows of microdrops (*see* **Notes 17, 18**).
3. Repeat Steps 1 and 2 with the embryos of the second genotype.
4. After all the embryos have been assembled into aggregates (*see* **Fig. 15.7**), return the plate into the incubator and incubate overnight (*see* **Note 19**).
5. The next day, most of the aggregates should have formed a single embryo that has progressed to the blastocyst stage and therefore ready for transfer into recipient females.

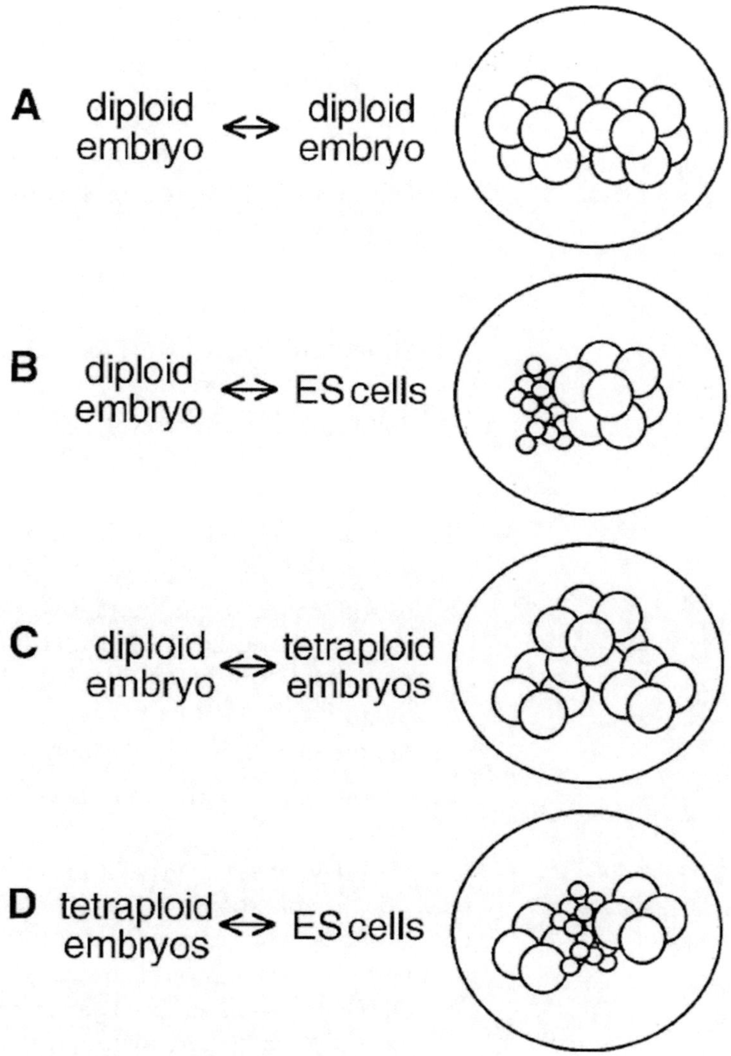

Fig. 15.7. Different aggregate combinations.

3.6.2. Diploid Embryo–ES Cells Aggregation

1. Select several clumps of approx. 10–15 ES cells (*see* **Fig. 15.7B**) and transfer them to the microdrops without depressions.
2. From there, place individual clumps into the microdrops harboring the depressions.
3. Set up the aggregation either by first placing the embryo in the well and then overlaying the ES cell clump or by putting in the cells first and then the embryo.
4. Follow Steps 4 and 5, as in **Section 3.6.1**.

3.6.3. Tetraploid Embryo–Diploid Embryo Aggregation

1. Place embryos into the microdrops of the aggregation plate that do not contain depressions. From here, move two tetraploid embryos into each depression.
2. Place the diploid embryos into the microdrops of the aggregation plate that do not contain depressions (nor any tetraploid embryos).
3. Carefully place a single diploid embryo next to the tetraploid embryos that are already positioned in the depression.
4. Follow Steps 4 and 5 as in **Section 3.6.1**.

3.6.4. Tetraploid Embryo–ES Cell Aggregation

1. Place the tetraploid embryos in the drop without depressions in the aggregation plate.
2. From here, move individual tetraploid embryos and place each one in a depression.
3. Following this, select several clumps of approx. 10–15 ES cells and transfer them to a microdrop without depressions (nor any embryos).
4. Take a loosely attached clump of cells, and place it carefully next to the embryo already positioned within a depression.
5. Introduce the second tetraploid embryo into the depression so that it lies at the side of the cells opposing the first embryo. This is best done by gently rolling the second embryo over the rim of the depression.
6. Repeat Steps 2–5 until all the "sandwiches" have been set up.
7. Follow Steps 4 and 5, as in **Section 3.6.1**.

3.7. Transfer of Blastocyst-Stage Aggregates

On the day after aggregation, the embryos should have reached the blastocyst stage (corresponding to 3.5 dpc of development) and are ready for transfer to pseudopregnant recipient females.

1. Optimally, 8–10 embryos are transferred into each uterine horn of a 2.5 dpc pregnant female. We routinely use CD1 or ICR outbred mice as recipients.

2. In the event of a shortage of recipients, the number can be increased to 14 per uterine horn (a total of 28). Another alternative is to transfer the embryos into the oviduct of 0.5 dpc pregnant females.
3. The embryo transfer procedure is detailed elsewhere (*38*).

4. Notes

1. A few microdrops are usually left without depressions (upper and lower rows) so that they can be used to introduce and briefly rinse ES cells and/or embryos just before assembly of the aggregation.
2. The depression should have a clear smooth wall and be deep enough to hold the aggregates without a risk of disassembly of the aggregate or spilling over as the plate is placed in the incubator. They should however also not be made too deep, as this will make the recovery of the chimeric blastocysts difficult without damaging them.
3. The plates should be prepared at least a few hours in advance (better the day before) and kept in the incubator until use. This allows enough time for the media to equilibrate to the correct temperature and pH level.
4. Oviducts are removed by making incisions in the upper part of the uterus and right below the ovary.
5. Flushing is performed from the infundibulum, resulting in embryos being expelled from the short length of uterus.
6. Superovulation of female mice results in a much higher yield of embryos than can be expected from naturally ovulating animals. However, the hormone treatment has to be carefully adjusted to the genetic background of the mice as well as the light cycle in the vivarium. Detailed protocols for superovulation can be found elsewhere (*38*).
7. Flushing needles are made by cutting the tip off a 30 G1/2 needle, and then beveling the end with a sharpening stone.
8. Electrofusion can be performed in an electrolyte or nonelectrolyte solution. We favor – and have provided the protocol for – the nonelectrolyte method as it allows for multiple embryos to be fused at the same time, as well as the possibility to automatically orient the embryos with the high-frequency AC field.
9. The adjustable AC field is applied in order to allow for the correct orientation of the embryos. Only the minimal necessary voltage should be used. If the field is too high it can cause lysis of the cells.

10. It usually takes approximately half an hour to an hour for the blastomere fusion to occur. It is important to select only the fused (and therefore tetraploid embryos) approximately an hour after the pulsing. We recommend that embryos that have fused be transferred to a new organ culture dish or microdrop, and then cultured overnight.
11. After overnight culture in KSOM medium, tetraploid embryos will have developed to the four-cell stage, which is equivalent to the eight-cell stage of diploid embryos. Tetraploids should be aggregated at the four-cell stage, as it is at this time that they will initiate compaction.
12. The zona pellucida is a glycoprotein coat that encapsulates the embryo. Late blastocyst-stage embryos will usually hatch out of their zona prior to implantation in the uterine wall. The zona is refractory to aggregation, as an intimate contact is needed between the cells and/or embryos.
13. Even though it does not matter whether embryos are in M2 or KSOM prior to their Tyrode's treatment, M2 is used right after as it has a superior buffering capacity.
14. Some ES cell clones may exhibit a lower than usual grow rate. These can be thawed one or two additional days earlier and stated in the protocol.
15. A highly diluted plating of single cells is required in order to produce optimally sized clumps (10–15 cells). Clumps in which cells are loosely connected are favored for setting up the aggregate. If clumps of the right size cannot be obtained, larger clumps can be reduced in size by careful pipetting. If the clumps are too small, two or three clumps can be used instead of one.
16. Care should be taken so as not to disaggregate the cell clumps by pipetting too vigorously or by using extensive trypsin treatment.
17. Aggregates should be set up in such a way that there is maximal contact between the cells and embryos.
18. Aggregations involving tetraploid embryos are set up as a "sandwich," where two tetraploid embryos are used to flank either the ES cells or the diploid embryo. However, if there is a shortage of tetraploid embryos, a one-to-one ratio can also be used.
19. When setting up the aggregates, especially if they are tetraploid "sandwich" types, take care not to jolt the plate and dissociate the intimately contacted embryos and/or cells.

4.1. Genotyping Segregating Embryo Components in Chimeras

Genotyping of chimeras can sometimes be a problem because of the mix of mutant and wild-type cell populations. One can avoid dealing with this issue by isolating the tissue that is expected to be solely mutant in origin, for example, the yolk sac endoderm in the case of "mutant diploid embryo–ES cells" or "mutant tetraploid embryos–ES cells," and embryo proper in the case of "mutant diploid embryo–tetraploid embryos." In other cases, that is, "mutant diploid embryo–wild-type diploid embryo" and "mutant tetraploid embryos–wild-type diploid embryo," the contribution of mutant cells are mixed with wild-type cells and that makes genotyping very difficult. Here, practical approaches to solving this problem will be discussed.

4.1.1. Genotyping of Potentially Chimeric Tissues

In the case of recessive mutations, homozygotes for the mutation have to be obtained from a cross between heterozygous females and males. Genotyping chimeras can be performed by preparing genomic DNA followed by genomic Southern or PCR. Problem occurs if mutant cells are mixed with wild-type cells. As a result, phenotyping will be indistinguishable between chimeras containing heterozygous cells and mutant cells (*see* **Fig. 15.8**).

One solution for this problem is taking advantage of using two alleles for either wild-type alleles (*see* **Fig. 15.8**) or mutant alleles (*see* **Fig. 15.8**). In this way, one can distinguish chimeras made by mutant embryos from heterozygous litter mates.

4.1.2. Isolation of Potentially Mutant Cells by the Use of Markers

It is possible to isolate potentially mutant cells (populations derived from het × het crosses) from wild-type cells if one can design the cross to introduce a gene as an ubiquitous reporter. For this purpose, markers such as lacZ or a fluorescent protein (GFP, YFP, CFP or RFP) are suitable. In the case of lacZ, genotyping is performed with lacZ-stained tissues and/or by PCR. The use of fluorescent proteins (FPs) is more straightforward because one can detect them without the use of a chromogenic substrate. FP-positive cells can be collected manually using a drawn-out glass pipette (A. N., unpublished observation) or by the use of fluorescence-activated cell sorting (FACS) (*39*, *40*). In general, the use of the ubiquitously expressed reporters makes genotyping chimeras more feasible and reliable.

5. Conclusion

Chimeric studies have been a feature of modern mouse embryology since its inception more than 50 years ago. During the first half of this period, embryo–embryo chimeras answered many questions about basic events during embryogenesis, such as cell movement and clonality. The 1980s brought a new component to chimera analyses: mouse ES cells. A vast knowledge about gene function has since been generated by the introduction of mutations in ES cells and subsequent passage of the mutant phenotype

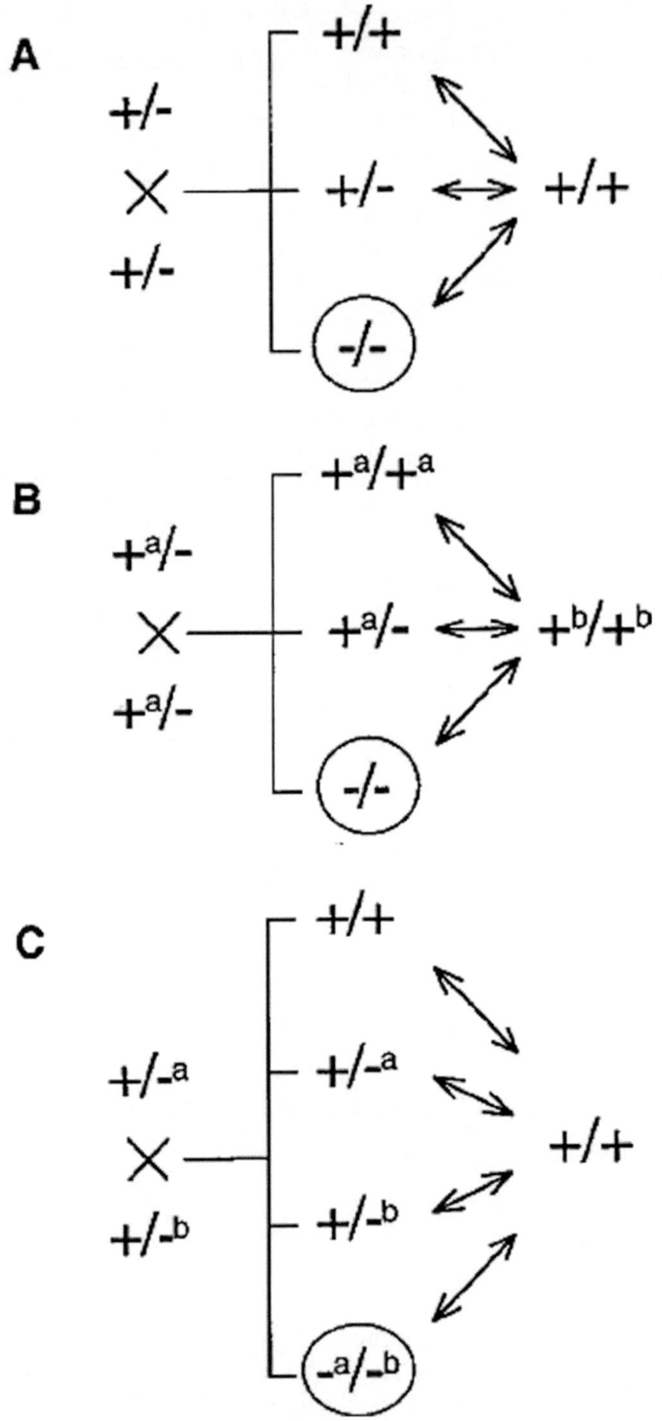

Fig. 15.8. If mutant cells are intermingled with wild-type cells in the tissues of a chimera (**A**) the result obtained from genotyping will be indistinguishable between chimeras containing heterozygous cells and mutant cells. To distinguish chimeras made by mutant embryos from heterozygous littermates, one can employ two different tags for either wild-type alleles (**B**) or mutant alleles (**C**).

through the germline of ES cell chimeras. In the 1990s, the classic use of chimeras as tools employed for answering biological questions re-entered the scientific scene. As the means for generating homozygous mutant ES cells became available, recessive phenotypes could be represented in cell culture and the resulting chimeras.

During the past decade, we have learned that ES cells are developmentally restricted, so that they are not able to differentiate into trophoblast or primitive endoderm lineages, but have full potential in the primitive ectoderm lineage, for example, in the embryo proper. A further component came into play after it was demonstrated that tetraploid embryos could provide a normal extraembryonic environment to an ES cell-derived embryo, such that tetraploid cells are selected against in the embryo proper if diploid cells (ES or embryo) are present.

Thus, today the three chimera components, diploid embryos, tetraploid embryos, and ES cells, whether mutant or wild type, open up a variety of possible combinations for creating specific chimeras, each tailor-made to address any relevant biological question. As a consequence, over the past few years, there has been an upsurge in the number of such studies reported in the literature. We have now entered the twenty-first century with a fully updated version of a classical tool that can be applied in many laboratories using genetic technologies in order to understand normal development, adult life and disease. This approach is going to have a particular importance in the massive effort of functionally annotating the mammalian genome, particularly when all the mouse genes will be mutated by the ongoing international efforts.

References

1. Nagy A, Rossant J. Chimaeras and mosaics for dissecting complex mutant phenotypes. Int J Dev Biol 2001; 45:577–82.
2. Rossant J. Investigation of the determinative state of the mouse inner cell mass. II. The fate of isolated inner cell masses transferred to the oviduct. J Embryol Exp Morphol 1975; 33:991–1001.
3. Evans MJ, Kaufman MH. Establishment in culture of pluripotential cells from mouse embryos. Nature 1981; 292:154–6.
4. Martin GR. Isolation of a pluripotent cell line from early mouse embryos cultured in medium conditioned by teratocarcinoma stem cells. Proc Natl Acad Sci USA 1981; 78:7634–8.
5. Nagy A, Rossant J, Nagy R, Abramow-Newerly W, Roder JC. Derivation of completely cell culture-derived mice from early-passage embryonic stem cells. Proc Natl Acad Sci USA 1993; 90:8424–8.
6. Tanaka S, Kunath T, Hadjantonakis AK, Nagy A, Rossant J. Promotion of trophoblast stem cell proliferation by FGF4. Science 1998; 282:2072–5.
7. Kunath T, Arnaud D, Uy GD, et al. Imprinted X-inactivation in extraembryonic endoderm cell lines from mouse blastocysts. Development 2005; 132:1649–61.
8. Nagy A, Gocza E, Diaz EM, et al. Embryonic stem cells alone are able to support fetal development in the mouse. Development 1990; 110:815–21.
9. Kaufman MH, Webb S. Postimplantation development of tetraploid mouse embryos produced by electrofusion. Development 1990; 110:1121–32.

10. Kubiak JZ, Tarkowski AK. Electrofusion of mouse blastomeres. Exp Cell Res 1985; 157:561–6.
11. Nagy A, Rossant J. Production of ES-cell aggregation chimeras. In: Joyner A, ed. Gene Targeting: A Practical Approach. Oxford, UK: IRL Press at Oxford University 1999; 177–205.
12. Lu TY, Markert CL. Manufacture of diploid/tetraploid chimeric mice. Proc Natl Acad Sci USA 1980; 77:6012–6.
13. Tanaka M, Gertsenstein M, Rossant J, Nagy A. Mash2 acts cell autonomously in mouse spongiotrophoblast development. Dev Biol 1997; 190:55–65.
14. Guillemot F, Nagy A, Auerbach A, Rossant J, Joyner AL. Essential role of Mash-2 in extraembryonic development. Nature 1994; 371:333–6.
15. Ueda O, Jishage K, Kamada N, Uchida S, Suzuki H. Production of mice entirely derived from embryonic stem (ES) cell with many passages by coculture of ES cells with cytochalasin B induced tetraploid embryos. Exp Anim 1995; 44:205–10.
16. George SH, Gertsenstein M, Vintersten K, et al. Developmental and adult phenotyping directly from mutant embryonic stem cells. Proc Natl Acad Sci USA 2007; 104:4455–60.
17. Draper JS, Nagy A. Improved embryonic stem cell technologies. Handb Exp Pharmacol 2007:107–28.
18. Wood SA, Allen ND, Rossant J, Auerbach A, Nagy A. Non-injection methods for the production of embryonic stem cell–embryo chimaeras. Nature 1993; 365:87–9.
19. Hogan B, Beddington R, Constantini F, Lacy E. Manipulating the Mouse Embryo: Cold Spring Harbor Laboratory Press; 1994.
20. Collins FS, Finnell RH, Rossant J, Wurst W. A new partner for the International Knockout Mouse Consortium. Cell 2007; 129:235.
21. Eggan K, Akutsu H, Loring J, et al. Hybrid vigor, fetal overgrowth, and viability of mice derived by nuclear cloning and tetraploid embryo complementation. Proc Natl Acad Sci USA 2001; 98:6209–14.
22. Carmeliet P, Ferreira V, Breier G, et al. Abnormal blood vessel development and lethality in embryos lacking a single VEGF allele. Nature 1996; 380:435–9.
23. Zhang P, Liegeois NJ, Wong C, et al. Altered cell differentiation and proliferation in mice lacking p57KIP2 indicates a role in Beckwith–Wiedemann syndrome. Nature 1997; 387:151–8.
24. Yan Y, Frisen J, Lee MH, Massague J, Barbacid M. Ablation of the CDK inhibitor p57Kip2 results in increased apoptosis and delayed differentiation during mouse development. Genes Dev 1997; 11:973–83.
25. Riley P, Anson-Cartwright L, Cross JC. The Hand1 bHLH transcription factor is essential for placentation and cardiac morphogenesis. Nat Genet 1998; 18:271–5.
26. Partanen J, Puri MC, Schwartz L, Fischer KD, Bernstein A, Rossant J. Cell autonomous functions of the receptor tyrosine kinase TIE in a late phase of angiogenic capillary growth and endothelial cell survival during murine development. Development 1996; 122:3013–21.
27. Shalaby F, Ho J, Stanford WL, et al. A requirement for Flk1 in primitive and definitive hematopoiesis and vasculogenesis. Cell 1997; 89:981–90.
28. Varlet I, Collignon J, Robertson EJ. Nodal expression in the primitive endoderm is required for specification of the anterior axis during mouse gastrulation. Development 1997; 124:1033–44.
29. Damert A, Miquerol L, Gertsenstein M, Risau W, Nagy A. Insufficient VEGFA activity in yolk sac endoderm compromises haematopoietic and endothelial differentiation. Development 2002; 129:1881–92.
30. Rossant J. Mouse mutants and cardiac development: new molecular insights into cardiogenesis. Circ Res 1996; 78:349–53.
31. Duncan SA, Nagy A, Chan W. Murine gastrulation requires HNF-4 regulated gene expression in the visceral endoderm: tetraploid rescue of Hnf-4(−/−) embryos. Development 1997; 124:279–87.
32. Rossant J, Guillemot F, Tanaka M, Latham K, Gertenstein M, Nagy A. Mash2 is expressed in oogenesis and preimplantation development but is not required for blastocyst formation. Mech Dev 1998; 73:183–91.
33. Braun T, Arnold HH. ES-cells carrying two inactivated myf-5 alleles form skeletal muscle cells: activation of an alternative myf-5-independent differentiation pathway. Dev Biol 1994; 164:24–36.
34. Mortensen RM, Conner DA, Chao S, Geisterfer-Lowrance AA, Seidman JG. Production of homozygous mutant ES cells with a single targeting construct. Mol Cell Biol 1992; 12:2391–5.
35. Kunath T, Gish G, Lickert H, Jones N, Pawson T, Rossant J. Transgenic RNA

interference in ES cell-derived embryos recapitulates a genetic null phenotype. Nat Biotechnol 2003; 21:559–61.

36. Wurst W, Joyner A. Embryonic stem cell, creating transgenic animals. In: Joyner A, ed. Gene Targeting: A Practical Approach. Oxford, UK: IRL Press at Oxford University; 1993:33–62.

37. Pirity M, Hadjantonakis A-K, Nagy A. Cell Culture for Cell and Molecular Biologist. In: Mather JP, Barnes D, eds. San Diego: Academic Press 1998; 279–93.

38. Nagy A. Manipulating the mouse embryo: A Laboratory Manual. 3rd ed: Cold Spring Harbor Laboratory Press 2003.

39. Gagneten S, Le Y, Miller J, Sauer B. Brief expression of a GFP cre fusion gene in embryonic stem cells allows rapid retrieval of site-specific genomic deletions. Nucleic Acids Res 1997; 25:3326–31.

40. Hadjantonakis AK, Nagy A. FACS for the isolation of individual cells from transgenic mice harboring a fluorescent protein reporter. Genesis 2000; 27:95–8.

Chapter 16

VelociMouse: Fully ES Cell-Derived F0-Generation Mice Obtained from the Injection of ES Cells into Eight-Cell-Stage Embryos

Thomas M. DeChiara, William T. Poueymirou, Wojtek Auerbach, David Frendewey, George D. Yancopoulos, and David M. Valenzuela

Abstract

With the completion of the human and mouse genome sequences and the development of high-throughput knockout mouse technologies, there is now a need for equally high-throughput methods for the production of mice for phenotypic studies. In response to this challenge, we recently developed a new method termed VelociMouse for the production of F0-generation mice that are fully derived from gene-targeted ES cells. In the version of the VelociMouse method described here, laser ablation of a portion of the zona pellucid (zp) of a normal eight-cell-stage embryo facilitates ES cell injection. Upon gestation in a surrogate mother, the injected embryos produce F0 mice that carry no detectable host embryo contribution (<0.1%). The fully ES cell-derived mice are normal, healthy, and fertile and exhibit 100% germline transmission for optimal breeding efficiency. The VelociMouse method accommodates both inbred or hybrid ES cells and either inbred or outbred eight-cell host embryos. Because the F0 mice produced are suitable for direct phenotyping studies, the VelociMouse method, coupled with high-throughput ES cell targeting technologies, such as VelociGene, offers an accelerated path to new drug target discovery and validation and a revolutionary approach to realize the full value of large-scale functional genomic efforts, such as the NIH Knockout Mouse Project *(1)* and the European Conditional Mouse Mutagenesis Project *(9)*.

Key words: Eight-cell-stage embryos, VelociGene, VelociMouse, laser-assisted eight-cell embryo injections.

1. Introduction

The established method for the elucidation of mammalian gene function is targeted mutagenesis by homologous recombination in embryonic stem (ES) cells *(2–4)*, which is commonly referred to as knockout mouse technology because most of the mutations are

Ralf Kühn, Wolfgang Wurst (eds.), *Gene Knockout Protocols: Second Edition, vol. 530*

DOI 10.1007/978-1-59745-471-1_16 Springerprotocols.com

designed to ablate gene function. The value of knockout mouse technology was recently recognized with the awarding of the 2007 Nobel Prize in Physiology or Medicine to the pioneers in the field *(2–4)*. Knockout mouse technology has been an important tool for the discovery and validation of new drug targets and for the creation of mouse models of human diseases. But the traditional methods were too slow and labor-intensive to take full advantage of the wealth of new information revealed by the completion of the human and mouse genome sequences. To be able to screen many more genes in a shorter period of time we developed VelociGene *(5)*, an assembly line method for the creation of precise, gene-targeted mutations in ES cells at unprecedented speed and throughput. VelociGene shifted the bottleneck of mouse functional genomics from the generation of gene-targeted ES cells to the production of genetically altered mice ready for phenotypic studies. To address this new challenge, we developed VelociMouse *(6)*, a novel method for the generation of F0-generation mice that are fully derived from gene-targeted ES cells.

In standard knockout mouse technology genetically altered ES cells are injected into blastocyst-stage embryos, which develop in a surrogate mother to produce chimeric founder mice (F0 generation). Chimeras are useful only as breeders to transmit the mutant allele through the germline to heterozygous progeny in the F1 generation. The efficiency of germline transmission varies according to the extent of the ES cell contribution to the germ cells. Another round of breeding between F1 heterozygous mice produces homozygous F2-generation mice that are used for phenotypic studies. Multiple rounds of breeding are time consuming: assuming the chimeras are able to transmit the mutant allele through the germline, it takes a minimum of 9 months from the birth of F0 chimeras to obtain F2 mice. In addition to a lengthy timeline, the breeding required for the production of F2 study cohorts requires extensive husbandry efforts and mouse housing that are expensive in labor, materials, and space. To eliminate the time and costs of the inefficient breeding required to produce F2-generation mice by traditional knockout methods, our goal in the development of the VelociMouse method was to develop a simple and universal procedure to directly produce mice in the F0 generation that were genetically identical to and fully derived from genetically modified ES cells.

The VelociMouse method described employs a laser to make a hole in the zp of a diploid eight-cell-stage host embryo to facilitate the injection of ES cells without damaging the embryo (*see* **Fig. 16.1**). To promote ES cell contribution to the inner cell mass (ICM) (epiblast) of the developing embryo, we culture the injected embryos resulting in a preponderance of fully ES-cell-derived F0-generation mice. We have used inbred and hybrid ES cells to efficiently generate healthy and fertile F0

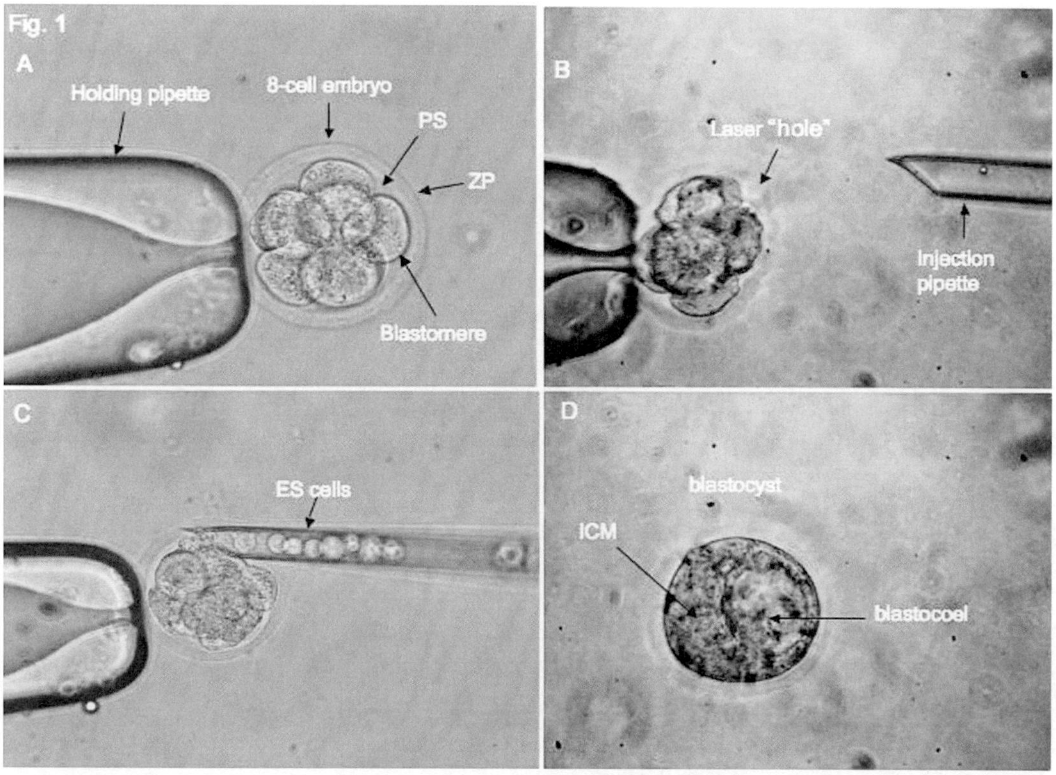

Fig. 16.1. The eight-cell embryo injection process. (**A**) An uncompacted eight-cell embryo held on the holding pipette prior to being injected with ES cells. The zona pellucida (ZP) should be clearly in focus. Perivitelline space (PS). (**B**) Following one pulse of the laser, a hole in the zp is easily seen with the 50 $\times$ objective. (**C**) The injection pipette containing about ten ES cells is gently pushed through the hole in the ZP and into the PS to deposit 7–9 ES cells at the "10 o'clock" position. (**D**) Following overnight culture, the injected eight-cell embryos will advance to a blastocyst-stage embryo with a well-defined inner cell mass (ICM) and blastocoel.

mice that contain no detectable host embryo contribution (<0.1%). When injecting inbred ES cells into host embryos from the outbred Swiss Webster (SW) strain, we observe some chimeras among the fully ES cell-derived mice in the F0 litters. Thus, if the research goal is to obtain F0 cohorts for phenotyping from ES cells targeted to have heterozygous or homozygous mutations, we have demonstrated that it is more efficient to use eight-cell-stage host embryos from inbred mouse strains *(5)*. The advantages of the VelociMouse method are (1) inbred or hybrid ES cells can be injected into either inbred or less expensive outbred host embryos to obtain F0 mice suitable for phenotyping; (2) XY and isogenic XY-derived XO ES cells carrying heterozygous mutations can be used to produce both male and female F0 mice that are 100% germline transmitters, which can then be intercrossed to obtain homozygous mice in the F1 generation; (3) ES cells engineered to carry multiple genetic alterations (e.g., a knockout allele, paired with a conditional allele and accompanied by an inducible Cre gene) can produce F0 mice identical in genotype without the lengthy and inefficient breeding required to produce the same mice by standard methods; (4) the production of F0 mice from gene targeted or transgenic inbred ES cells obviates the need for extensive backcrossing to obtain a pure inbred background; and (5) in addition to fully ES cell-derived F0 mice, the overall quality of the F0 chimeras that are sometimes produced from the injection of eight-cell-stage embryos have better ES cell contribution than those obtained from the standard blastocyst injection method *(5)*. Because the fully ES cell-derived F0 mice provide a faster route to the homozygous allele and can be phenotyped directly, we refer to them as VelociMice.

2. Materials

2.1. ES Cells

1. VGF1 is a hybrid F1 cell line derived from an individual blastocyst-stage embryo obtained from a mating between a 129S6Sv/Ev female to a C57BL/6 N male (*see* **Note 1**).
2. The ES cell lines were established by previously published methods *(7)*.

2.2. Mouse Strains

1. SW mice were purchased from Taconic Farms, Germantown, NY.
2. C57BL/6Tyr$^{c\text{-}2j}$ (B6 albino) mice were purchased from The Jackson Labs, Bar Harbor, ME.
3. All mice were housed in isolator units and maintained on a 10-h dark/14-h light cycle.

2.3. ES Cell Culture

1. ES cell growth medium contained high-glucose Dulbecco's Minimal Essential Medium (DMEM) (#11971, Gibco, Bethesda, MD) supplemented with 0.1 mM non-essential amino acids, 1 mM sodium pyruvate, 0.1 mM 2- mercaptoethanol, 2 mM L-glutamine, penicillin and streptomycin (50 μg/mL each) (Gibco), 15% fetal bovine serum (FBS, Hyclone, Odgen UT) and 2000 U/mL LIF (Millipore, Billerica, MA).
2. 0.25% Trypsin-EDTA (#25200, Gibco).
3. Dulbecco's phosphate-buffered saline 1 × (#14190, Gibco).
4. Primary embryonic fibroblasts prepared from E13.5 mouse embryos *(8)*.
5. 24-well tissue culture plates (#142475 Nunc A/S, Roskilde, Denmark).

2.4. Superovulation of Embryo Donor Females

1. Pregnant mare serum gonadotrophin (PMSG) (5 IU/mouse, Calbiochem, San Deigo, CA).
2. Human chorionic gonadotrophin (hCG) (7.5 IU/mouse, Calbiochem).
3. One-mL syringe fitted with a 27.5-gage needle.

2.5. Harvesting Eight-Cell-Stage Embryos

1. ES cell medium (without LIF, *see* **Section 2.3.1** above).
2. CO_2 chamber.
3. Dissection scissors and blunt forceps.
4. Mineral oil (#M8410, Sigma-Aldrich, St. Louis, MO).
5. 10-mm tissue culture dish (Falcon® 3001 Becton Dickinson, Franklin Lakes, NJ).
6. 37°C incubator maintained at 7.5% CO_2.
7. MZ6 dissection microscope (Leica, Wetzlar, Germany).

2.6. Embryo Injection Setup

1. Inverted microscope (model IX50, Olympus America, Center Valley, PA) with 4 × and 20 × objectives and phase contrast optics to visualize ES cells and embryos.
2. To manipulate ES cells and embryos, glass microcapillary cell transfer pipettes (20° capillary angle and 15 μm, inner diameter, Eppendorf, North America) fitted on to micromanipulators (Leica) with manually controlled joysticks.
3. Mineral oil was used to hydraulically control the movement of ES cells in and out of the transfer pipette, and for the catch and release of the embryo on the holding pipette.

2.7. Post-injected Embryo Culture

1. Injected eight-cell-stage embryos.
2. KSOM (Millipore).
3. 10-mm tissue culture dish.
4. Mineral oil (#M8410, Sigma-Aldrich, St. Louis, MO).
5. 37°C incubator maintained at 7.5% CO_2.
6. MZ6 dissection microscope (Leica, Wetzlar, Germany).

2.8. XYClone Laser System

1. Computer-controlled device utilizing an infrared laser (1480 nm) fired through a 50 × objective using a software interface to control the alignment, temperature and delivery of a laser pulse to ablate a small portion of the zp without causing embryonic lethality (Hamilton Thorne Biosciences, Beverly, MA).

2. For injection of ES cells into eight-cell-stage embryos, the laser intensity was fixed at 100% power with an 800-μs pulse duration to minimize heat conduction to the adjacent embryonic blastomeres.

2.9. Embryo Transfer to Recipient Females

1. Dissection scissors
2. Blunt forceps
3. Surgical mask and gown
4. Gloves
5. Draping and gauze
6. Avertin (2-2-2 Tribromoethanol, Fluka 90710, Aldrich Chemical Co., Milwaukee, WI)
7. Tert-amyl alcohol
8. Glass pipettes (#14672-380, 9 inch, VWR Scientifc, www.vwr.com)
9. Mouth piece and rubber tubing (Aspirator tube assembly, #A5177-SEA, Sigma-Aldrich, St. Louis, MO)
10. Bunsen burner
11. Suture clips
12. Heat pad

2.10. Genomic DNA Prepared from Mouse Tissues

1. Tissue lysis solution: 0.1 M Tris-HCl (pH 7.6), 0.005 M ethylenediamine tetraacetic acid (EDTA), 0.2% (w/v) sodium dodecyl sulfate (SDS), 0.2 M NaCl, approximately 0.3 mg/mL proteinase K (Roche, cat. no. 03115844001).
2. 96-well deep well (1 ml) block plate (VWR, cat. no. 40002-009) with cap mat (VWR, cat. no. 40002-009), or conventional microcentrifuge tubes.
3. DNA resuspension buffer contains 10 mM Tris-HCl, 1 mM EDTA (pH 7.5).

2.11. Polymerase Chain Reaction Genotype Analysis of F0 Mouse Tissue DNA

1. Taq DNA polymerase with 10X reaction buffer (Sigma-Aldrich, cat. no. D1806-250UN).
2. Four deoxynucleotide triphosphate solutions (10 mM each of dATP, dGTP, dCTP, and TTP; Sigma, cat. no. DNTP10-1KT).
3. Polymerase chain reaction (PCR) primers
   a. Assay of the *Tyr* gene mutation in C57BL/6Tyr$^{c\text{-}2j}$ albino mice:
      Albino BL/6 WTF 5′-TCAAAGGGGTGGATGACCG-3′
      Albino BL/6 MUTF 5′-TCAAAGGGGTGGATGACCT-3′
      TYR Rev2 5′-ACAAAGAGGTCGTAGATGTTG-3′

b. Assay of the Tyr gene mutation in SW albino mice:
SW WTF 5′-GGTTTCAACTGCGGAAACTG-3′
SW MUTF 5′-GGTTTCAACTGCGGAAACTC-3′
TYR Rev2 5′-ACAAAGAGGTCGTAGATGTTG-3′

3. Methods

3.1. Harvesting Eight-Cell-Stage Embryos

1. For a consistent yield of eight-cell-stage embryos from inbred mouse strain C57BL/6Tyr$^{c\text{-}2j}$ (B6 albino), 3–4-week-old female mice are superovulated by hormone injection (*see* **Note 2**). We normally inject and mate 30 females to obtain approximately 25 females with copulation plugs, and a yield of about 150 eight-cell-stage embryos at 2.5 days post coitum (dpc).
2. B6 albino stud males are singly housed and are used twice weekly as breeders. B6 albino females are generated from our in-house colony maintained in semi-rigid isolator units (Plastic Design) in pathogen-free conditions. The females are weaned at 21 days of age and either hormone-injected immediately or housed for up to 1 week for use.
3. PMSG is administered intraperitoneally (i.p.) (5 IU/mouse) using a 1-mL syringe fitted with a 27.5-gage needle at 17:00–18:00 hours approximately 48 h prior to mating. HCG is similarly administered (7.5 IU/mouse) two hours prior to mating.
4. Thirty mating pairs are set up at approximately 16:00–17:00 hours and the females are visually inspected for a copulation plug the following morning (0.5 dpc).
5. At 2.5 dpc (~10:00) the females are sacrificed by CO_2 asphyxiation. The mice are placed on a clean draping with their ventral sides up. The abdominal fur is wiped with 70% ethanol. Using a clean pair of dissecting scissors, an incision is made along the ventral midline from the pubis to the rib cage. The skin and mesentery are resected and the viscera are moved aside to expose the ovaries, oviducts and uterine horns.
6. Being careful not to damage the oviduct, it is cut away from the ovary and the uterine horn. Using a 1-mL syringe fitted with a bent 30.5-gage needle inserted into the infundibulum, the embryos are flushed from the oviduct with ES media (without LIF) into a clean 35-mm culture dish (*see* **Note 3**).
7. After visual inspection under a dissecting microscope only the uncompacted eight-cell-stage embryos are selected for ES cell injection. These embryos are placed into a new 35-mm dish

containing a droplet of ES cell medium (without LIF) that has been overlaid with filtered mineral oil. The embryos are incubated at 37.0°C in a 7.5% CO_2 atmosphere prior to use.

3.2. Preparation of ES Cells for Injection into Eight-Cell-Stage Embryos

1. ES cells are grown to 50% confluence in 24-well tissue culture plates (Falcon) containing a monolayer of γ-irradiated primary embryonic fibroblasts *(8)*.
2. In preparation for harvesting the ES cells, the medium is removed and the cells are washed with 1 mL phosphate-buffered saline (Gibco). The cells are incubated for 10 min at 37°C with 0.5 mL 0.25% Trypsin-EDTA (Gibco, #25200).
3. Using a pre-sterilized RT-L1000F tip (Molecular BioProducts) the ES cells are pipetted up and down to a single-cell suspension.
4. The suspension is mixed with 1 mL of ES cell growth medium (*see* **Note 2**) to inhibit the trypsin and the cells are collected by centrifugation at 2,000*g* for 5 min.
5. The cells harvested from one well of a 24-well plate are suspended in 1 mL of medium (*see* **Note 2**) and placed on ice with occasional gentle mixing for the length of the embryo injection period.

3.3. Injecting Eight-Cell Embryos (*see* Fig. 16.1)

1. All injections are performed in the lid of a 35-mm dish in ES cell medium (without LIF) overlaid with filtered mineral oil.
2. Injections are normally begun at about 12:30 hrs.
3. Approximately ten healthy ES cells are collected (*see* **Note 4**) with an Eppendorf ES transfer pipette for each eight-cell embryo to be injected.
4. Using gentle suction through the Eppendorf holding pipette, an eight-cell embryo is maneuvered (*see* **Note 5**) to identify a region of the embryo suitable for the laser perforation of the zp. The appropriate drilling site should be chosen such that the distance between the nearest blastomere and target site in the zp site is maximized (*see* **Note 6**).
5. The drilling site is oriented at the 1 o'clock position of the embryo (*see* **Note 7**). The innermost isotherm ring (red) should be centered over the zp site and a single laser pulse is produced using either a foot pedal or a single left mouse click on the "Fire" button on the user interface.
6. The injection needle is introduced into the opening of zp and inserted under and along the zp to minimize damage to the blastomeres (*see* **Note 8**). A total of 7–9 ES cells are deposited into each embryo (*see* **Note 9**). The ES cells should be deposited the maximum possible distance from perforation site to

help prevent "backflow" of the ES cells out of the opening in the zp. Depositing the ES cells at a "10 o'clock" location is preferred (*see* **Note 10**).

3.4. Culturing Embryos Following ES Cell Injection

1. The injected eight-cell embryos are cultured in KSOM.
2. The medium drops containing the embryos are overlaid with mineral oil, and the embryos are cultured for 20–24 h in a 37°C incubator with a 7.5% CO_2 atmosphere.
3. Following the post-injection culture of the embryos, those that progressed to the morula or blastocyst stage and were deemed to be viable based on visual inspection in a dissection microscope were transferred into recipient females.

3.5. Transferring Injected Embryos to Recipient Females

1. SW females deemed to be in estrus as judged by visual examination of the external genitalia are mated at 15:00 hours with vasectomized SW males. The following morning, females with an observed copulation plug (designated as 0.5 dpc) are used as recipients on 2.5 dpc for the uterine horn transfer of injected embryos.
2. Following embryo culture, only healthy embryos are selected for transfer (*see* **Note 11**). The procedure is performed at approximately 11:00 hours on the day following injection.
3. Surgical equipment is autoclaved prior to use and treated with 70% ethanol between surgeries. The surgery takes place on a clean/sterile draping in a laminar flow hood and the surgeon wears clean gloves, a mask, and a cap and gown.
4. Avertin 2.5% made up in tert-amyl alcohol is administered (i.p.) (*see* **Note 12**) at a dose of 0.1 ml/10 g body weight to anesthetize the recipient female mouse.
5. Following the administration of the anesthetic, the lack of a flinching response by the mouse to a firm paw pinch with a blunt forceps is a useful method to determine if the mouse is sufficiently anesthetized (*see* **Note 13**). The approximate duration of the surgery is 10 min per mouse.
6. The surgical site is liberally treated with 70% alcohol and after the incision is made, the excess fur is wiped clear of the site with an alcohol pad. A 1-cm incision is made in the skin over the lateral lumbar region with a sharp scissors to create a "window" to visualize the underlying ovarian fat pad and attached ovary. An alcohol pad is used to remove loose hair. A 0.5-cm incision is made in the body wall overlying the ovarian fat pad and then a blunt forceps is used to gently remove the fat pad along with the attached ovary and uterine horn.

7. While holding the uterine horn nearly vertical, a 27.5-gage needle is used to make a puncture into the lumen beginning at a point 2–3 mm distal to the oviduct.
8. Prior to the embryo transfer procedure, prepare transfer pipettes by quickly passing a glass pipette tip over a Bunsen burner flame, and drawing it to a diameter just small enough to fit into the uterine puncture.
9. Using an aspirator tube assembly fitted onto the prepared glass pipette, 4–7 embryos are transferred into the uterine lumen. Prior to collecting the embryos, a small air bubble is taken into the pipette to locate the position of the embryos as they are deposited into the uterine lumen. The uterine horn, ovary and fat pad are then gently returned into the body cavity and the skin incision is sealed with suture clips. The procedure is repeated on the other uterine horn.
10. Following the surgery, recipient females are placed on a heating pad. The mice are observed for several minutes to ensure that their breathing appears normal for their anesthetized condition. They will usually stay down for 15–20 min following the surgical procedure. Following recovery from the uterine transfer surgery, the recipient females are housed two per cage.

3.6. Preparation of Mouse Tissue Genomic DNA

1. Harvest tissue samples from post-natal F0-generation mice.
2. Place immediately in 0.5 mL (or sufficient to cover the sample) of tissue lysis solution.
3. Incubate at 55°C for at least 16 h and then hand-shake the lysates for 1 min.
4. Centrifuge the samples at ≥15,000*g* for 10 min to pellet undigested debris.
5. Place 0.5 mL of isopropyl alcohol in microcentrifuge tubes.
6. Carefully decant or pipette 0.5 mL of the lysate supernatant from each sample into a tube containing the isopropyl alcohol.
7. Invert tubes until a visible DNA precipitate appears.
8. Centrifuge the samples at ≥15,000*g* for 5 min.
9. Remove the supernatants and gently wash the precipitates with 0.1 mL of 70% ethyl alcohol. (A brief centrifugation step is optional.)
10. Remove the alcohol, wash and allow the precipitate to air-dry of 30 min or until all the alcohol has evaporated.
11. Add 0.05–0.1 mL of DNA resuspension buffer to the dry precipitates and incubate the tubes in 65°C water bath for 30 min.

12. Gently shake the samples to resuspend the DNA and centrifuge briefly to collect the condensate to the bottom of the tube.

3.7. Assessment of ES Cell/Host Embryo Contribution to F0 Mice

1. To determine if F0-generation mice derived from ES cells injected into eight-cell embryos retain any genetic contribution from host embryos contributed by C57BL/6Tyr$^{c\text{-}2j}$ albino strain mice, perform a PCR that detects either the wild-type *Tyr* gene sequence or the albino allele (*see* **Note 14**).
2. To assay for contribution from C57BL/6Tyr$^{c\text{-}2j}$ host embryos, perform one PCR with the Albino BL/6 WTF and TYR Rev 2 primers and a second PCR with the Albino BL/6 MUTF and TYR Rev 2 primers. Both PCRs generate products of 319 bp on templates that are recognized by the WTF or MUTF primers.
3. PCR conditions. Prepare a 2.5 mM mixed dNTP solution by combining equal volumes of the four individual 10 mM dATP, dGTP, dCTP, and TTP stock solutions. Prepare primers as 100 μM stock solutions. The PCRs, which discriminate single-bp differences between the wild-type *Tyr* and albino alleles, work best with non-proofreading Taq DNA polymerases. Assemble 25 μL PCRs with 2.5 μL 10X PCR buffer, 2 μL 2.5-mM dNTPs, 0.125 μL Taq DNA polymerase (5 units/μL), 0.5 μL WTF or MUTF primer, 0.5 μL TYR Rev2 primer, 5 μL of tissue sample DNA (2.5–5 ng), and 14.325 μL of de-ionized water. Use the following temperature cycling protocol: 1 cycle at 94°C for 4 min; 35 cycles at 94°C for 30 sec, 60°C for 30 s, 72°C for 30 s; and ending with one extension cycle at 72°C for 2 min.

4. Notes

1. We have demonstrated the other inbred (C57BL/6) or outbred (SW) mouse strains can be used as donors of eight-cell-stage embryos *(6)*.
2. We have demonstrated that the VelociMouse method also works with inbred ES cells lines derived from 129, C57BL/6 and Balb/C mouse strains *(6)*.
3. A person can normally harvest the embryos from 10 to 12 female mice in about 1 h.

4. Only healthy looking ES cells are selected for injection. These are defined as being rounded, refractive and smooth. We try to avoid irregularly shaped cells and those containing abundant dark pigmentation, which is indicative of ES cells that are beginning to differentiate.
5. To hold the embryo in the proper position with the holding pipette, the embryo can be maneuvered by rolling it along the bottom of the dish, while using suction to catch and release the embryo.
6. We have found that maximizing this distance will minimize heat damage to blastomeres resulting in greater embryo survival, in addition to facilitating the insertion of the injection needle into the perivitelline space without damaging the blastomeres.
7. Selecting the "1 o'clock" position will limit embryo roll during injection and will allow for the proper placement of ES cells within the embryo.
8. To help minimize damage to the blastomeres during injection process, the tip of a new injection pipette can be dulled by gently striking against the holding pipette.
9. We have found that the injection of fewer than six ES cells more often results in chimeric F0 mice. The injection of more than nine cells has no added benefit because the excess cells tend to leak out through the opening in the zp.
10. If necessary, gentle downward pressure can be applied with the injection pipette to the zp directly over the deposited ES cells in order to prevent them from escaping though the laser "hole".
11. The embryos that have advanced beyond the uncompacted eight-cell stage during the culture period and judged to be viable by observation in a dissecting scope (4 × power) are transferred to the uterine horns of pseudo-pregnant SW females. In general, we find that >90% of injected embryos are suitable for transfer.
12. The only adverse effects on the survival of surrogate females that we have encountered are thought to have occurred from the use of an Avertin preparation stored longer than 2 months. In these cases, the recipient females died within days following the surgery. To eliminate this adverse effect, a fresh preparation of Avertin is stored at 4°C in a foil-wrapped 50-mL Falcon tube and used for a maximum of 2 months. If used properly recipient females rarely suffer any ill effects (1 in about 150 females).
13. Should the mouse respond to the paw pinch, an additional 0.1 mL of the Avertin preparation is administered i.p. and after 1 min the paw pinch test is repeated.

A

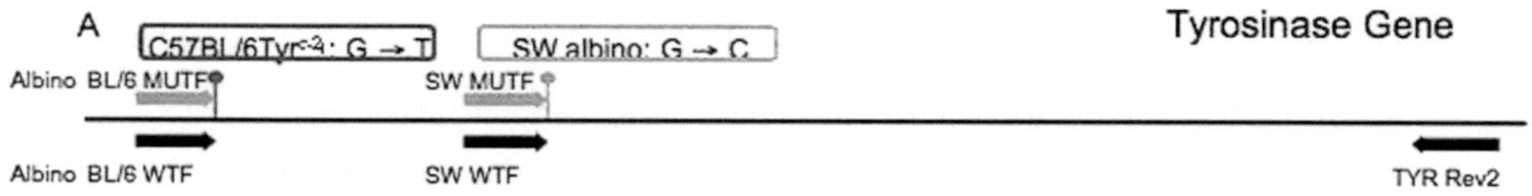

B

PCR	Forward Primer	Reverse Primer	PCR product (bp)
Tyr WT allele	Albino BL/6 WTF TCAAAGGGGTGGATGACCG	TYR Rev2 ACAAAGAGGTCGTAGATGTTG	319
C57BL/$6^{Tyrc\text{-}2j}$ allele	Albino BL/6 MUTF TCAAAGGGGTGGATGACCT	TYR Rev2 ACAAAGAGGTCGTAGATGTTG	319
Tyr WT allele	SW WTF GGTTTCAACTGCGGAAACTG	TYR Rev2 ACAAAGAGGTCGTAGATGTTG	242
SW albino allele	SW MUTF GGTTTCAACTGCGGAAACTC	TYR Rev2 ACAAAGAGGTCGTAGATGTTG	242

C

PCR Using Albino BL/6 Primers

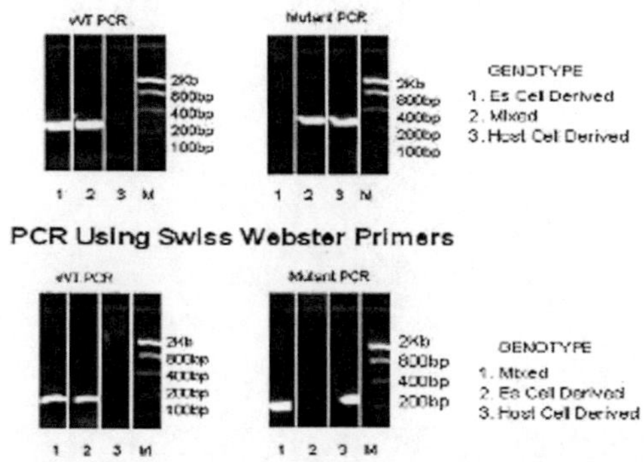

Primer	PCR Product Detected		
WTF	No	Yes	Yes
MUTF	Yes	Yes	No
Result	host embryo derived	mixed contribution	ES cell derived

Fig. 16.2. Genetic analysis to determine the presence of host embryo contribution to F0 mice. (**A**) Schematic representation of the mouse tyrosinase gene showing the relative locations of the inactivating mutations in the C57BL/$6^{c\text{-}2j}$ and Swiss Webster (SW) strains and the respective wild-type (WT) and mutant (MUT) forward PCR primers and the WT Tyr reverse primer. (**B**) Forward and reverse primer pair sequences and the PCR product sizes in base pairs (bp) to distinguish the ES cell-derived WT Tyr gene from the C57BL/$6^{c\text{-}2j}$ or SW host embryo-derived mutant Tyr genes. (**C**) PCR results from both WT and MUT primer pairs. Top panels show an example of F0 mouse tail DNA analysis from the injection of a C57BL/$6^{c\text{-}2j}$ host embryo. Bottom panels show an example of F0 mouse tail DNA analysis from the injection of a SW host embryo. The accompanying table shows the possible outcomes of the PCR reactions and the derivation of the F0 mice.

14. Suggested genotyping assays for the detection of C57BL/$6^{c\text{-}2j}$ and SW host embryo contribution are provided in **Fig.** 16.2.

Acknowledgments

We thank Yongli Chang for preparing ES cells, Joseph Hickey and Jennifer Escaravage for microinjection, and Lakeish Esau for performing genotyping assays. WTP wants to thank JJP for his support.

References

1. Austin CP, Battey JF, Bradley A, et al. The knockout mouse project. Nat Genet 2004; 36:921–4.
2. Capecchi MR. Generating mice with targeted mutations. Nat Med 2001; 7:1086–90.
3. Evans MJ. The cultural mouse. Nat Med 2001; 7:1081–3.
4. Smithies O. Forty years with homologous recombination. Nat Med 2001; 7:1083–6.
5. Valenzuela DM, Murphy AJ, Frendewey D, et al. High-throughput engineering of the mouse genome coupled with high-resolution expression analysis. Nat Biotechnol 2003; 21:652–9.
6. Poueymirou WT, Auerbach W, Frendewey D, et al. F0 generation mice fully derived from gene-targeted embryonic stem cells allowing immediate phenotypic analyses. Nat Biotechnol 2007; 25:91–9.
7. Auerbach W, Dunmore JH, Fairchild-Huntress V, et al. Establishment and chimera analysis of 129/SvEv- and C57BL/6-derived mouse embryonic stem cell lines. Biotechniques 2000; 29:1024–8, 30, 32.
8. Hogan, B. et al. in "Manipulating the Mouse Embryo: A Laboratory Manual". Cold Spring Harbor Press 1994; pp. 260–262.
9. Auwerx J, Avner P, Baldock R, et al. The European dimension for the mouse genome mutagenesis program. Nat Genet 2004; 36:925–7.

Chapter 17

Generation of Cre Recombinase-Expressing Transgenic Mice Using Bacterial Artificial Chromosomes

Jan Rodriguez Parkitna, David Engblom, and Günther Schütz

Abstract

Generation of genetically modified mice is one of the primary methods for understanding gene function. In particular, approaches that allow for restricting the effects of a mutation to defined cell-types are fundamental for understanding the roles of genes in specific cells or tissues. The Cre/loxP recombination system is the most robust approach to produce cell-type-specific gene inactivation. When the Cre recombinase is expressed from a transgene containing a tissue-type-specific promoter it will delete genomic segments flanked by loxP sequences in this tissue only. In this regard, the selectivity and reproducibility of Cre expression is absolutely critical for the result. To meet these requirements large constructs based on bacterial artificial chromosomes (BACs) have been successfully used. Here we present a protocol for the generation of constructs in which the Cre recombinase or a tamoxifen-inducible Cre fusion protein, are inserted at the translation start sequence of a BAC-derived gene. We describe all the critical steps, including construct-design, recombineering, and preparation of the transgene-containing genomic fragment for pronuclear injection and identification of "founder" animals among the resulting offspring. In our experience, the use of this protocol typically results in specific and transgene copy number-dependent expression of the Cre recombinase.

Key words: Cre recombinase, Cre/loxP, conditional gene deletion, recombineering, bacterial artificial chromosome, tamoxifen.

1. Introduction

The Cre/loxP recombination system is the most widely used system for cell-type-specific gene inactivation *(1–3)*. The core of this system is the Cre recombinase, a protein from bacteriophage P1, which mediates recombination of specific sequences called loxP sites. When two loxP sites are introduced into the genome in the same orientation and the Cre recombinase is expressed, the Cre will

Ralf Kühn, Wolfgang Wurst (eds.), *Gene Knockout Protocols: Second Edition, vol. 530*

DOI 10.1007/978-1-59745-471-1_17 Springerprotocols.com

recombine the loxP sites and delete the sequence between the sites. In this way the Cre/loxP system can be used to delete genes or critical parts of genes. The system is also widely used for conditional gene activation. In this setting a cassette that is blocking transcription, a so-called "stop-cassette" that is flanked by loxP sites, is introduced between a promoter and the translational start site of a gene. Thus, the gene is silent under basal conditions but activated upon Cre-mediated removal of the stop-cassette. For a review of advanced applications of the Cre/loxP recombination system see *(4, 5)*. Cre fusion proteins have been developed which are inactive in the basal state since they are fused to a modified ligand-binding domain of the estrogen receptor and thus kept in the cytoplasm *(6–9)*. Upon binding of tamoxifen, a synthetic steroid, the fusion protein is translocated to the nucleus and the Cre is able to perform the recombination of the loxP sites. Furthermore, in order to restrict gene inactivation in a certain cell-type or tissue, the Cre fusion protein can be expressed under a tissue-type-specific gene promoter.

The usefulness of this approach is critically dependent on the quality of the mouse line carrying the Cre transgene. Thus, the recombinase must be expressed in the vast majority of the cells that should be targeted (high efficiency) and no recombination should be observed in other cells, or in the absence of the ligand (low leakiness). Classic approaches, based on plasmid-derived constructs containing short (10–15 kb) genomic fragments, often do not meet these requirements. In addition they commonly suffer from inconsistent expression patterns between different lines harboring the same construct and the expression patterns often change between generations as well as between different mouse backgrounds. A major step forward for the generation of high-quality transgenic mice was the development of methods for using large pieces of genomic DNA propagated by yeast artificial chromosomes (YACs) *(10, 11)* and bacterial artificial chromosomes (BACs) *(12, 13)*. A second breakthrough along this line was the development of systems for manipulating BAC-based genomic DNA fragments by homologous recombination in bacteria *(14)*. These systems allow the manipulation of large genomic fragments, which due to their considerable size (typically 100–250 kb) in most cases contain all of the regulatory elements necessary for the expected expression pattern of the transgene. Collectively, this typically makes the expression pattern independent of the site of integration and the strength of expression is mainly dependent on the number of transgenes integrated into the genome ("copy-number"). These advantages have made BAC-based transgenesis the state-of-the-art method for the generation of Cre-expressing lines. Here we describe a detailed protocol for the generation of constructs for Cre lines using BAC recombineering. In our experience, the use of this protocol typically results in specific, reproducible and copy number-dependent expression of the Cre.

2. Materials

2.1. Software and Web Resources

1. Vector NTI software package is available from Invitrogen and is free for academic use (registration required). pDraw is distributed by Acaclone Software (www.acaclone.com) and is provided as "freeware".
2. The interface for searching BAC clones is provided at www.ncbi.nlm.nih.gov/projects/genome/clone/. A similar functionality is also provided by the Ensembl genome browser: www.ensembl.org
3. The website of CHORI, which provides a comprehensive collection of mouse BAC clones for academic use worldwide is found at bacpac.chori.org

2.2. Plasmids and Bacterial Strains

1. Bacterial strains EL250 and SW105 harboring inducible recombinases are available from the NCI (*see* http://recombineering.ncifcrf.gov/).
2. The pIndu and pConst plasmids were cloned by Erich Greiner and are available to academic institutions from our laboratory. Maps of the plasmids are shown in **Fig. 17.1**.

2.3. Equipment

1. Shaking incubators for bacterial cultures set to 37°C and 32°C.
2. Incubators set to 37°C and 32°C for LB agar plates.
3. A water bath set to 42°C for steps requiring the induction of the temperature-sensitive expression of the λ prophage.

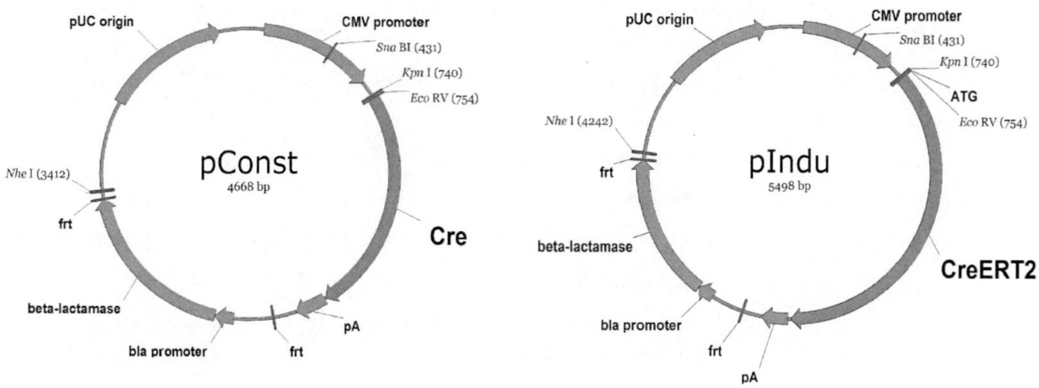

Fig. 17.1. Maps of plasmids pConst and pIndu. The Cre in the pConst plasmid is the codon-optimized ("improved") Cre variant by Shimshek and colleagues (2002) *(22)*. The CreER^{T2} on the pIndu plasmid is the fusion of the improved Cre and a modified ligand-binding domain of the human estrogen receptor *(23)*. Abbreviations: CMV, cytomegalovirus; pA, sequence containing the polyadenylation signal; bla, beta-lactamase; frt, binding sites for the Flp-recombinase.

4. Refrigerated centrifuge(s) and rotors for 1.5-mL tubes and 50-mL tubes.
5. A spectrophotometer for measuring DNA concentration and monitoring growth of bacteria (wavelengths 260 nm and 600 nm, respectively).
6. A complete setup for running DNA agarose gels with ethidium bromide.
7. An electroporation device and 0.1-cm electroporation cuvettes (i.e., BIORAD Gene Pulser and Gene Pulser Cuvettes 0.1 cm #1652089).
8. Equipment for pulse-field electrophoresis (i.e., BIORAD CHEF-DR III System).

2.4. Reagents

1. LB medium.
2. Chloramphenicol 25 mg/mL in ethanol.
3. Ampicillin 100 mg/mL in sterile deionized water.
4. LB agar plates containing 25 μg/mL chloramphenicol.
5. LB agar plates containing 100 μg/mL ampicillin (*Note*: Do not use ampicillin plates that were stored longer than 2 months).
6. LB agar plates containing both 25 μg/mL chloramphenicol and 100 μg/mL ampicillin.
7. 10% (v/v) glycerol in deionized water sterilized by autoclaving. Pre-chill the solution to 0°C for several hours before use.
8. Liquid nitrogen, a plastic device for removing tubes after freezing and appropriate protective gear (goggles and gloves).
9. Buffers P1, P2, and P3 from Qiagen (available with DNA purification kits or separately).
10. Qiagen Qiafilter Plasmid Midi Kit for plasmid DNA isolation (or equivalent).
11. BD Falcon serological 5-mL pipette #357529 (for the gel filtration column).
12. Sepharose CL-4B from Amersham Biosciences.
13. 10% (w/v) L-arabinose in water, filter-sterilized.
14. Injection buffer: 10 mM Tris-HCl pH 7.5, 0.1 mM EDTA pH 8.0, 100 mM NaCl, autoclaved and filtered.
15. TE buffer pH 8.0 (10 mM Tris-HCl pH 8.0, 1 mM EDTA, filter-sterilized or autoclaved.
16. Filter paper for removing precipitated proteins during BAC purification.
17. Restriction enzymes for analysis of the BACs and preparation of the construct for injection.

3. Methods

The protocol below is based on the methods described by *(14–19)* and the lab protocols of Tim Wintermantel, Erich Greiner and Stefan Berger.

3.1. Guidelines for Constructing the Plasmid with the Cre

The Cre (or CreER^{T2}) encoding sequence is introduced into the genomic DNA fragment by recombineering. Therefore, it has to be flanked by "homology arms" to allow for homologous recombination mediated by the λ prophage recombinase system (extensive description of the bacterial strains harboring the temperature-sensitive λ prophage is found in *(15, 20)*). Correct recombination in the target locus is essential. Introducing any additional nucleotides in the promoter region may affect the specificity or efficiency of transcription. We use the following guidelines for cloning the homology arms.

1. Each homology arm should be around 200–500 bp. The "left" homology arm (upstream, 5′) should include the genomic fragment 5′ from the start codon. On the 3′ end it should include an ATG sequence followed by an *Eco*RV "half-site" (*see* **Fig. 17.2**). The ATG sequence will correspond to the start codon of the Cre recombinase. The 5′ end point of the genomic sequence included should be a "half-site" for a blunt cutting enzyme (i.e., *Sna*BI). Additionally, a restriction site (i.e., *KpnI*) should be introduced on the 5′ end for subsequent cloning into the plasmid (*see* **Note 1**).

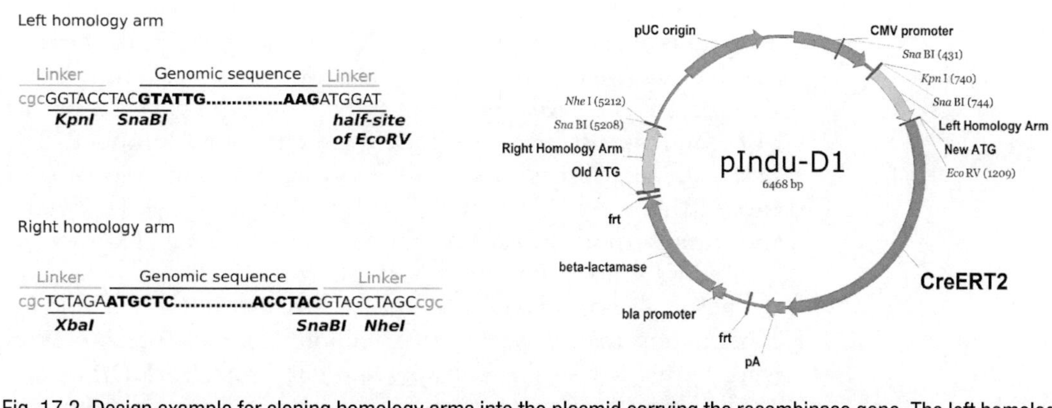

Fig. 17.2. Design example for cloning homology arms into the plasmid carrying the recombinase gene. The left homology arm was amplified by PCR, phosphorylated with polynucleotide kinase and then cloned into the BAC using the *KpnI* and *Eco*RV sites. A particular advantage of the *Eco*RV site is the presence of a guanosine residue at the +4 position, which should enhance translation rate ("strong" Kozak sequence). The right homology arm was amplified by PCR and digested with *XbaI* and *NheI*. The resulting sticky ends are compatible with each other, and were cloned into the *NheI* site in the vector. The advantage of using the *XbaI* and *NheI* versus *NheI* sites alone is the greatly facilitated screening for orientation of the insert after ligation (the *NheI* site will be reconstituted only on one side of the insert). The pIndu-D1 plasmid shown on the right is the final product of the cloning procedure.

2. The "right" homology arm (downstream, 3′) should contain the genomic sequence starting with the ATG and ending with a "half-site" for a blunt cutting enzyme (i.e., *Sna*BI) (*see* **Fig. 17.2**). The arm needs to be flanked with restriction sites that will be used for cloning (i.e., *Nhe*I/*XbaI*).
3. An example of a plasmid construct with homology arms is shown in **Fig. 17.2**.

3.2. Selecting the BAC

While screening BAC libraries of the mouse genome has been the classic approach to selecting genomic fragments of interest for the last decade, recently the indexing (sequencing of the BAC "arms") has been almost completed. The collection of indexed BAC clones is available among others from CHORI (bacpac.chori.org) and several commercial suppliers (*see* **Note 2**). Two tools allow for easy finding of a BAC clone containing the promoter region of the gene of interest: the Clone Registry at www.ncbi.nlm.nih.gov/projects/genome/clone/ and the Ensembl genome browser at www.ensembl.org. The BAC display may be enabled while in Contig View by selecting an appropriate option from the DAS resources pull-down menu. An advantage of using both databases for querying BAC clones is that they complement each other. On independent genome assembly projects and thus do not necessarily provide identical results.

When selecting the optimal genomic fragment on a BAC clone, consider having large stretches (>30 kb) upstream and downstream from the gene used to drive the Cre expression. At the same time, avoid including other genes on the BAC clone, as this may result in a phenotype associated with overexpression of the genes included. An example of a BAC clone used for constructing a transgene is shown in **Fig. 17.3**. In the example, the gene of interest is *Drd1a*, the mouse D1 dopamine receptor gene. Upstream is the *Sfxn1* gene, but only part of the gene fits on the clone and we assume it will yield no functional product. The *Drd1a* gene consists of two exons, with the translational start encoded in exon 2. In some cases a BAC meeting the criteria above might not be available and then ET-cloning *(14, 16)* may be used to remove unwanted portions of the BAC.

Select one or preferably two BACs that would be suitable for constructing the transgene. It is recommended to prepare exact maps of the BACs using software like Vector NTI or pDraw, as it will be essential to analyze restriction enzyme cleavage patterns at several of the subsequent steps.

3.3. Purification of the BAC for Restriction Analysis and Electroporation

1. Use a sterile pipette tip to gently scratch the stab, and then transfer the tip to a sterile 1.5-mL centrifuge tube containing 1 mL of LB medium.

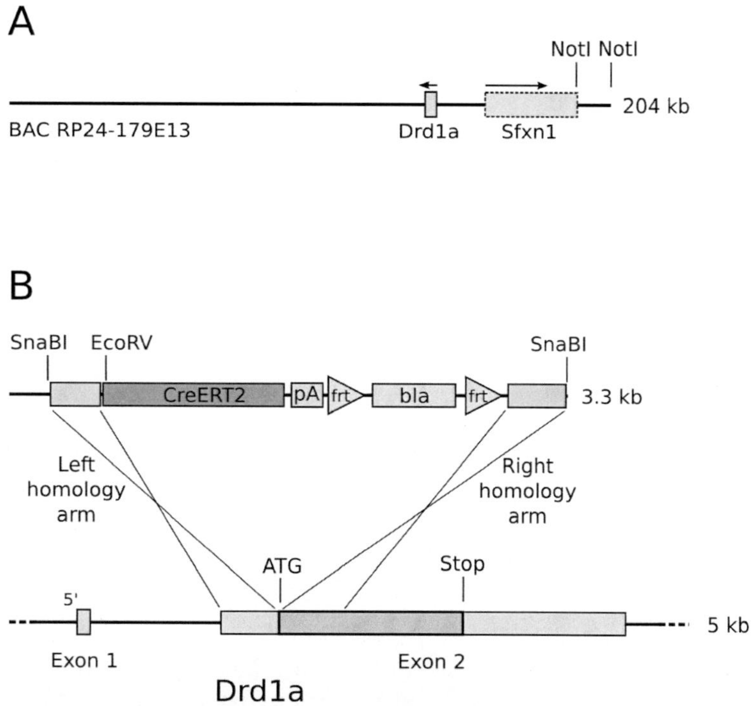

Fig. 17.3. Recombineering of the Cre-encoding sequence into the target genomic sequence. (**A**) Example of a BAC clone (RP24-179E13) used for construction of the transgene. The boxes represent the *Drd1a* and *Sfxn1* genes. The dashed line around the *Sfxn1* gene indicates that the sequence is incomplete: the last three exons are missing. The arrows above the boxes indicate the direction of transcription. The BAC "backbone"" containing genes and sequences necessary for replication and selection in bacteria lies between the *NotI* sites indicated. (**B**) Schematic of recombineering of the CreER^{T2} sequence into the coding region of the *Drd1a* gene. The plasmid fragment containing the recombinase is inserted by homologous recombination catalyzed by the λ prophage recombinases into the site corresponding to the start of translation in exon 2 of the *Drd1a* gene.

2. Perform serial dilutions so you have 10^{-3}, 10^{-5} and 10^{-7} dilutions of the BAC clone-harboring bacteria.
3. Plate 100 μl of the dilutions on LB agar plates containing 25 μg/mL chloramphenicol. Let the plates dry and then incubate them overnight at 37°C.
4. On the next day pick a few single colonies and use them to inoculate 5 mL overnight cultures (37°C with shaking). Use up to 4 mL of the culture for the next step, but keep 1 mL in case a clone will be selected for making a glycerol stock.
5. Pour ~1.4 mL of the BAC clone overnight culture into a centrifuge tube (1.5 mL) and spin it down 5 min at ~10,000*g*.

6. Discard the supernatant, add more culture and spin down. Repeat these steps until 4 mL of culture had been spun down.
7. Carefully remove traces of LB and then add 300 μl of Qiagen buffer P1. Resuspend the bacteria by gentle pipetting or mixing.
8. Add 300 μl of Qiagen buffer P2, mix 6–8 times by inverting the tube and incubate for 5 min at room temperature (do not extend this step).
9. Add 300 μl of Qiagen buffer P3, mix by inverting and place the tube for 5′ on ice. Spin 10′ at maximum speed (>10,000*g*) at 4°C.
10. Carefully remove 800 μl of the supernatant and transfer it to a new tube. To precipitate the BAC DNA, add 540 μl of isopropanol and mix gently by inverting.
11. Spin the tube 10′ at maximum speed (>10,000*g*) at 4°C. Carefully remove supernatant without disturbing the pellet.
12. Add 500 μl 70% (v/v) ethanol to wash the pellet, and spin 5′ at maximum speed (>10,000*g*) at 4°C. Remove supernatant, spin shortly again at room temperature, and remove remaining traces of supernatant.
13. Leave the tubes open for 2–3′ to allow the pellet to dry, do not use a speedvac.
14. Dissolve the isolated BAC in 40 μl TE pH 8.0. The BAC is much more difficult to dissolve than plasmids, leave the tube at 4°C for 1–2 h and mix it by tipping occasionally. Do not try to suspend the pellet by pipetting since this will result in shearing of the BAC. Store the BACs at 4°C.
15. The purified BACs may be used for a diagnostic restriction cut, in order to confirm that a predicted pattern from the map of the BAC matches the actual result. Select a restriction enzyme (or a combination of enzymes) that will produce several easy-to-separate products in the range of 300–3000 bp apart from larger bands. Larger bands will resolve poorly, but when necessary they may be resolved by pulse-field electrophoresis (*see* **Fig. 17.4**).
16. Cut the BAC with selected restriction enzymes and run the products of the reaction on a 1% (v/w) agarose gel.
17. Allow the bromophenol blue band (corresponding to ~200 bp on 1% agarose gel) to migrate two-thirds of gel length. When the electrophoresis is finished, post-stain the gel with EtBr or any other dye. The post-staining step is often necessary to visualize relatively faint bands.

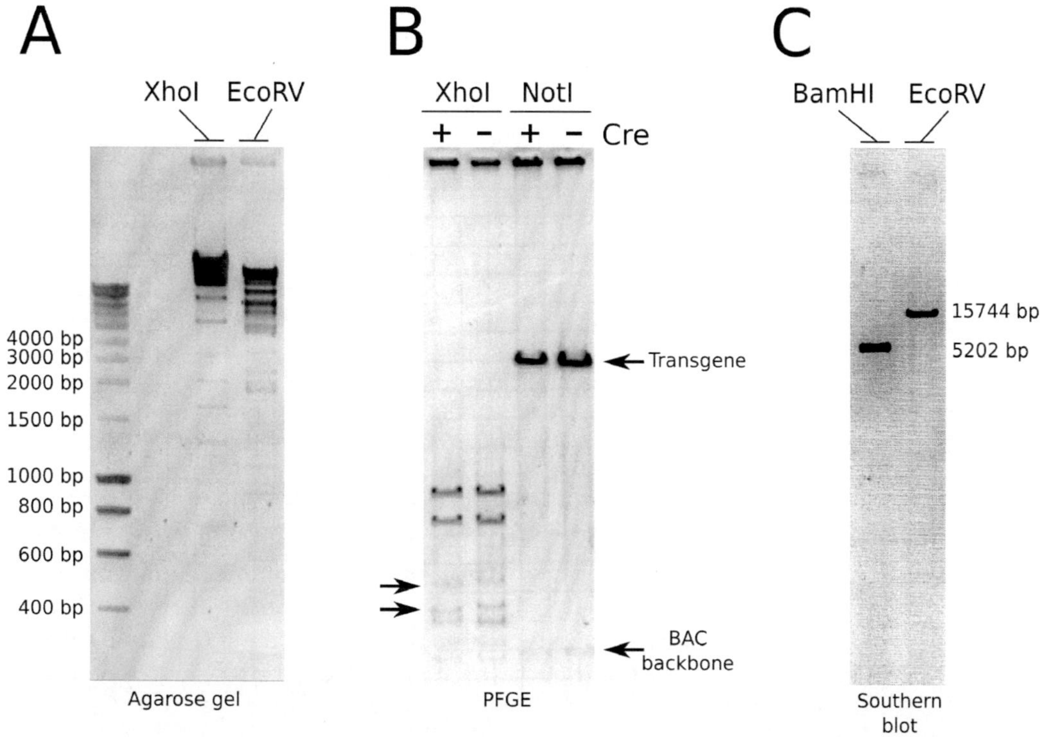

Fig. 17.4 Examples of BAC clone analysis. (**A**) Restriction fragment pattern analysis on agarose gel. Enzymes used for cleavage of the BAC clone are indicated above the gel image.(**B**) Pulse-field gel electrophoresis (PFGE) analysis of restriction fragment patterns. Compared to agarose gels this method allows for resolving the larger fragments. The image shown as the example represents the restriction patterns of the initial BAC clone (Cre–) and the final result of cloning with inserted Cre and removed beta-lactamase (Cre+). The presence of only two bands in the lanes corresponding to the *NotI* cleavage products (the lower one is very faint) confirms removal of the beta-lactamase. The difference in the two bands after *XhoI* treatments confirms the CreER^{T2} integration. (**C**) Detection of Cre insertion site by Southern blotting. The presence of a single band of predicted size in the Southern blotting result confirms a correct and unique integration of the CreER^{T2}.

18. Acquire an image of the band patterns on the gel and compare the results against the theoretically predicted pattern. In case there is a perfect match then the BAC clone is suitable for further work.
19. Make a glycerol stock (preferably more than one) from the culture with validated BAC. The purified and verified BAC will be used for electroporation into cells allowing for inducible recombination of large DNA fragments.

3.4. Preparation of Competent Cells for Electroporation

1. Prepare serial dilutions from SW105 or EL250 bacterial cells (*see* **Note 3**) and plate them on LB agar plates (no antibiotic). Incubate for ~20 h at 32°C.
2. Inoculate a 5 mL overnight culture with a single colony picked from one of the plates.

3. In the morning, cool 500 mL sterile 10% (v/v) glycerol in H_2O on ice (for at least 3 h before use) (*see* **Note 4**). Transfer 2 mL of the overnight cultures to Erlenmeyer flasks containing 50 mL of LB. Grow the cultures 3–5 h, until they reach absorbance at 600 nm of 0.45–0.5. Then transfer them to 50-mL Falcon tubes and spin down at 4000*g* for 10′ at 0°C in a pre-cooled rotor.
4. Carefully discard the supernatant, and place the inverted tube on a stack of paper tissues in a 4°C refrigerator, to dry remaining supernatant. Do not leave the tubes unattended, after 2–3 min the supernatant will be drenched and the pellets will start "sliding" down.
5. Take the tubes back on ice, add 50 mL of ice-cold 10% (v/v) glycerol and resuspend the cells (*see* **Note 5**). Spin down again at 4000*g* for 10′ at 0°C and perform the washing with 10% (v/v) glycerol two more times for a total of three washes.
6. After the last wash the remaining volume should be around 0.6–0.7 mL. Mix gently and aliquot 50 μl into pre-cooled 1.5-mL centrifuge tube tubes. Flash freeze in liquid nitrogen for future use, or proceed immediately to electroporation.

3.5. Electroporation of the BAC into Competent Bacterial Cells

1. Place clean 100 μl electroporation cuvettes on ice 5 min before use. Put the tube with purified BAC and aliquots of competent bacteria on ice as well.
2. Prepare the electroporations by pipetting 50 μl of the competent bacteria and 1 μl of the purified BAC into the cuvette. Put the cuvette in the electroporation device and deliver a pulse at 2.3 kV with 25 μF and pulse controller set to 200 Ω (BioRAD Gene Pulser). After the pulse, add 1 mL LB to the cuvette and then transfer the whole volume to a 1.5-mL centrifuge tube. Perform additional two electroporations, using 2 μl and 5 μl of the purified BAC.
3. The three tubes containing electroporated bacteria in LB should be incubated for 1 h at 32°C with shaking.
4. Plate from each tube 100 μl of the culture from each tube and then spin the tube shortly to pellet the bacteria, remove 800 μl medium, resuspend the bacteria in the remaining medium, and plate them on LB agar with 25 μg/mL chloramphenicol. Incubate the plates for at least 20 h at 32°C.
5. Pick colonies, use them to inoculate 5-mL cultures and purify the BACs as described above (**Section 3.3**). Verify the integrity of the BAC by restriction analysis and prepare glycerol stocks of the BAC clone in the SW105 (or EL250) cells.

3.6. Preparation of Competent Cells for Homologous Recombination

Once the BAC is transformed into the SW105/EL250 cells it is necessary to induce the λ prophage recombinase system that will mediate the homologous recombination between the targeting construct containing the Cre and the site in the genomic sequence containing the start of translation of the gene of interest. The procedure follows the same steps as the standard protocol for preparing electrocompetent cells (**Section 3.4**) with the following modifications:

1. When the cells reach absorbance at 600 nm of ~0.5, transfer the culture to a shaking water bath set to 42°C for 15′ (we simply place the flask in a water bath and shake it by hand) to induce the bacterial recombinase activity.
2. Move the flask to an ice-slurry and shake it gently to cool the medium to 0°C. Leave the culture on ice for ~20′. Further steps are performed according to the standard protocol (**Section 3.5**).

3.7. Preparation of a Construct for Recombination with the BAC

1. Purify ~0.5 pmol (typically ~1.5 μg) of the plasmid harboring the construct prepared for recombination (a standard miniprep is usually enough).
2. Set up a reaction to excise the fragment with the blunt-cutting enzyme of choice (i.e., *Sna*BI). Set the conditions in a way that there is a considerable "over-cut"; for example, extend the time of the reaction to overnight (*see* **Note 6**). Isolate the excised fragment by preparative gel electrophoresis.
3. Electroporate 0.1 pmol of the fragment into competent cells with induced λ prophage recombinase activity and harboring the target BAC clone. Use identical conditions as described in **Section 3.5** (2.3 kV). Add 1 mL of LB medium to the electroporation cuvette, then transfer the culture to a 1.5-mL tube and incubate the bacteria for 1 h at 32°C.
4. Plate the bacteria on LB agar containing 25 μg/mL chloramphenicol and 50 μg/μl ampicillin. Incubate the plates for ~20 h at 32°C.
5. Pick several colonies from the plates and prepare small overnight cultures for BAC isolation.
6. Perform restriction analysis on isolated BACs from each of the cultures to verify which of the clones appear to have correctly integrated the Cre cassette (*see* **Fig. 17.4** and **Note 7**). Select a validated BAC clone for further steps.

3.8. Removal of Antibiotic Resistance Gene

The beta-lactamase gene conferring ampicillin resistance is flanked by frt sites and may be removed inducing the Flp-recombinase expression (*see* **Fig. 17.3**), which is placed under an arabinose operon-controlled promoter in the genome of SW105 or EL250 cells.

1. Prepare an overnight culture (LB + 25 μg/mL chloramphenicol) from a recombined and validated BAC clone containing the Cre cassette.
2. In the morning, inoculate 50 mL of LB supplemented with 25 μg/mL chloramphenicol with 1 mL of the overnight culture. Grow the cells at 32°C with shaking until they reach absorbance at 600 nm of ~0.5. Add 0.5 mL 10% (w/v) L-arabinose.
3. Incubate the culture for one more hour and then dilute it by transferring 5 mL of the culture to a new flask containing 50 mL of LB supplemented with 25 μg/mL chloramphenicol. Grow the diluted culture for one more hour at 32°C.
4. Prepare serial dilutions from the culture and plate 100 μl from 10^{-3}, 10^{-5} and 10^{-7}dilutions on LB agar plates containing 25 μg/mL chloramphenicol. Allow the colonies to grow for ~20 h at 32°C.
5. Pick several colonies to prepare small-scale BAC preparations and perform restriction fragment analysis (*see* **Note 8**). Select a couple of clones *(2* to *4)* that display the desired band pattern and perform a medium-scale BAC preparation from each, using for example Qiagen large-construct kits (*see* **Note 9**).

Before proceeding to the next step it is necessary to perform sequencing of the candidate BAC, in particular regions containing the ends of the recombination arms introduced from the plasmid as well as the Cre coding sequence. This can be accomplished by using PCR to amplify fragments of the BAC or direct sequencing of the BAC with primers against the regions of interest.

3.9. Preparation of the BAC Fragment for Injection

1. Use 20–50 μg of the final BAC for restriction enzyme (i.e., *Not*I) treatment; the total volume should not exceed 100 μl (*see* **Note 10**). After the reaction is finished, denature the restriction enzyme by heating at 70°C for 10 min. Spin down the samples and store them at 4°C for further use.
2. Purification of the BAC is performed using a Sepharose CL-4B column. Pour ~20 mL of Sepharose CL-4B into a 250-mL beaker. Add 150 mL of injection buffer (*see* **Note 11**), mix by gentle swirling, and allow the gel to settle completely. Decant the solution, add another 150 mL of injection buffer and perform the procedure three more times for a total of four washes. These steps are necessary for removal of ethanol and sodium azide that are used as gel preservatives.
3. Resuspend the Sepharose in ~100 mL of injection buffer and transfer to a vessel that allows to degas the gel. Connect the Erlenmeyer flask to the pump and evacuate the gel for 15′, swirling occasionally. Use protective eyewear; the pressure may cause the flask to shatter explosively. Transfer the degassed gel to a beaker without introducing air bubbles.

4. Prepare the column using a 5-mL Falcon serological pipette. Remove most of the cotton bud and push the remaining bit to the tip of the pipette using compressed air; it will prevent the gel from flowing out. If the piece of cotton is too big it will slow the flow of buffer through the column.
5. Attach the column firmly to a holder and check that it is perfectly vertical. Close the bottom of the column (i.e., a pipette tip with a piece of parafilm at the end). Fill the column with injection buffer up to two-thirds of height.
6. Take the beaker with the gel and swirl it gently to resuspend the Sepharose. Apply the gel slurry with a Pasteur pipette to the column and fill it up to the top. The gel will start to set on the bottom of the column.
7. Once ~1 cm of gel matrix settles down, open the outflow and place a beaker (50–100 mL is usually convenient) below the column to collect the eluate.
8. Keep applying the gel on top of the column (swirl the beaker with the gel occasionally). Pack the column until the gel is 2–3 cm from the top of the column (*see* **Note 12**).
9. Once the packing is finished connect with a piece of parafilm a 50-mL syringe (without the piston) to the top of the column and fill it with injection buffer without allowing the gel to dry or disturbing its surface.
10. Wash the column with at least 30 mL of injection buffer before starting the BAC purification. Once the column is washed, wait until the upper reservoir is empty and then remove it.
11. When there is almost no buffer left on top of the gel (be careful, the gel will start shrinking before drying) apply the BAC sample (~100 μl) without disturbing the gel surface. Let the sample enter the column, and just before the column dries apply 100 μl of the injection buffer.
12. Again, allow the solution to enter the column, and repeat this step one more time. This should prevent the sample from diluting during loading and is important for good separation.
13. Load the column with injection buffer till the top of the pipette, re-attach the syringe-reservoir and fill it with injection buffer. Start collecting 300-μl fractions.
14. Collect ~35–40 fractions and stop the column. Measure the absorbance at 260 nm of all fractions (if the spectrophotometer allows for measuring in ~60-μl volume, you may measure the samples directly).
15. In most cases a first smaller peak of absorbance will separate from a second larger one, corresponding to the backbone and transgene, respectively. Load the fractions corresponding

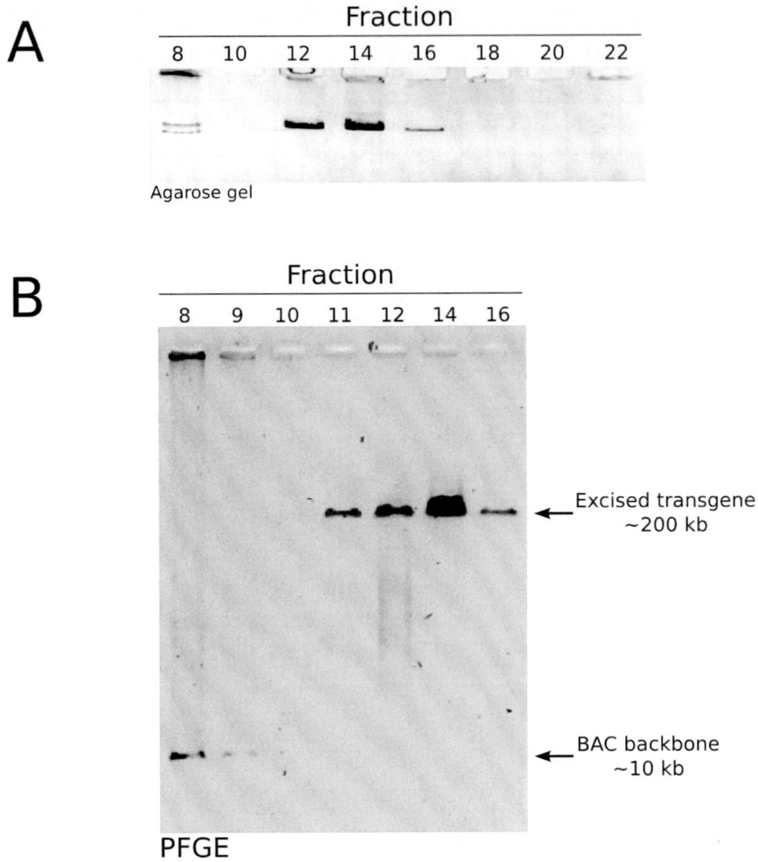

Fig. 17.5. Analysis of fractions collected from the gel column. Aliquots (20 μl) from the fractions (as indicated above the images) were run on a normal agarose gel (**A**) and pulse-field gel electrophoresis (**B**, PFGE). On panel **B** the *upper band* corresponds to the genomic fragment with the inserted Cre, and the *lower band* corresponds to the excised BAC backbone.

to the second peak (or every other fraction if absorbance measurements give no clear conclusion) on a 1% (w/v) agarose gel.

16. Load 20-μl aliquots of selected samples onto a 1% (w/v) pulse-field agarose gel and separate the fragments. After the electrophoresis stain the gel with ethidium bromide. If based on this analysis the separation was successful, and the transgene does not appear degraded, the selected fraction(s) are suitable for injection. An example of pulse-field result is shown in **Fig. 17.5** (*see* **Note 13**).

3.10. Guidelines for identification and characterization of founder animals

The genotyping of animals for identification of founders can be performed by diagnostic PCR. For the detection of the Cre or CreER^{T2} recombinases the following primers may be used *(21)*:

Cre-forward:	ctg cca ggg aca tgg cca gg
Cre-reverse:	gca cag tcg agg ctg atc agc
Product size:	316 bp

and for amplification of the CreER^{T2}exclusively:

CreER^{T2}-forward:	ggc tgg tgt gtc cat ccc tga a
CreER^{T2}-reverse:	ggt caa atc cac aaa gcc tgg ca
Product size:	406 bp

Reaction conditions are initial denaturation for 3′ at 95°C followed by 30 cycles of 30″ at 95°C, 30″ at 60°C and 60″ at 72°C with additional 10′ at 72°C at the end of the PCR.

4. Notes

1. The genomic sequence in the BAC clone is a convenient template for amplifying the "homology-arms". Use a high-fidelity polymerase (i.e., Pfu) for amplification of the arms.
2. BACs ordered from CHORI arrive as agar-stabs and are stable for days at 4°C, nevertheless take steps to preserve the BAC immediately upon arrival. Validate the BAC by analyzing restriction enzyme cleavage patterns and/or sequencing of the "arms".
3. Incubations of SW105 or EL250 cells are done at ≤ 32°C, to prevent the induction of the bacterial recombinase. As a result these bacterial strains tend to grow slowly and it is often necessary to wait for more than 20 h to obtain suitable colonies on agar plates. For the purpose of the procedure described here the two strains are equivalent. A more complete description of the strains is found in *(15, 19)*.
4. While preparing electroporation-competent bacteria all steps have to be performed as close to 0°C as possible. When using glass pipettes pre-cool them by pipetting a 0°C cold solution before pipetting bacteria. Do not remove the bacteria from ice. If possible, work in a cold room (and still keep everything on ice). The glycerol used in the procedure allows for storing the bacteria for future use. If the competent cells will be only used freshly, it is possible to omit it.

5. After subsequent glycerol (or water) washes the bacteria will be forming an increasingly less solid pellet. While it makes resuspending them easier, discarding the supernatant without disturbing the pellet becomes difficult.
6. When recombineering the Cre-containing fragment into the genomic sequence, removing traces of the plasmid is absolutely essential. The intact plasmid is greatly preferred by cells for acquiring resistance to ampicillin. It is not unreasonable to "re-cut" the agarose-purified fragment with the appropriate restriction enzyme and perform another preparative agarose gel (though make sure that the final yield will be sufficient).
7. Typical conditions for pulse-field electrophoresis would be 6 V/cm, included angle 120, initial switching time 0.5″, final switching time 20″, run time 14 h and cooling set to 14°C.
8. It is often convenient to use *Not*I for verifying the presence of the beta-lactamase gene in the BAC construct, as a *Not*I site is present in its sequence. It might be necessary to use pulse-field electrophoresis to separate the large fragments generated by *Not*I treatment. Additionally, Southern blotting may be performed using a probe targeting the Cre-encoding sequence to confirm that the recombinase was integrated in the correct location, though in this case a restriction enzyme(s) producing several fragments from the BAC is recommended.
9. The purification of BACs may be performed using commercially available kits (i.e., Qiagen large-construct kits), with the following modifications: (a) Be very careful while pipetting the BAC-containing solutions to prevent shearing. Cut away the ends of pipette tips to widen them a little. Pipette slowly. (b) Do not use syringe-mounted filters, remove the precipitate by passing the solution through a piece of filter-paper mounted on top of the purification column. (c) Do not overdry the DNA pellet after precipitation. Allow the BAC to resuspend for 1 – 2 hours or overnight at 4°C.
10. In most cases *Not*I will remove the backbone of the BAC without fragmenting the genomic sequence. Usually an overnight restriction reaction at 37°C using 1 U of *Not* I per 1 μg of DNA will result in efficient cleavage of the BAC, which should be verified on an analytical pulse-field gel electrophoresis (PFGE).
11. The injection buffer must be of highest purity; any kind of contamination might result in toxicity to the oocytes. We cannot stress enough how important meticulous care is in its preparation.
12. The packing of the Sepharose column has to be continuous: the gel should not be allowed to settle completely until the packing is finished. Do not introduce bubbles into the

column, as they would impair the separation. Typical flow rate through the column varies around 5–10 mL/h. You may leave the column overnight for washing, but make sure that the amount of buffer is sufficient and the column will not dry out. If the flow through the column is stopped, the washing should be performed again. Do not store the column at 4°C, as it will cause bubbles to appear.

13. The purification of the BAC involves a gel filtration matrix but the BAC fragments probably exceed the size of exclusion of the pores of the matrix and the order of elution of the fragments is opposite to what could be expected. Despite the fact that the separation is obviously not due to size exclusion, the method allows to separate the BAC fragments without the need of preparative pulse-field electrophoresis followed by buffer exchange.

Acknowledgments

We would like to thank Stefan Berger for critically reading and correcting the manuscript, Joachim Elzer for compiling the "PAC" protocol used previously in the lab and Antonio Caputti for sharing the gel filtration method.

References

1. Abremski K, Hoess R, Sternberg N. Studies on the properties of P1 site-specific recombination: evidence for topologically unlinked products following recombination. Cell 1983; 32:1301–11.
2. Gu H, Marth JD, Orban PC, Mossmann H, Rajewsky K. Deletion of a DNA polymerase beta gene segment in T cells using cell type-specific gene targeting. Science 1994; 265:103–6.
3. Thomas KR, Capecchi MR. Site-directed mutagenesis by gene targeting in mouse embryo-derived stem cells. Cell 1987; 51:503–12.
4. Branda CS, Dymecki SM. Talking about a revolution: the impact of site-specific recombinases on genetic analyses in mice. Dev Cell 2004; 6:7–28.
5. Dymecki SM, Kim JC. Molecular neuroanatomy's "Three Gs": a primer. Neuron 2007; 54:17–34.
6. Feil R, Brocard J, Mascrez B, LeMeur M, Metzger D, Chambon P. Ligand-activated site-specific recombination in mice. Proc Natl Acad Sci USA 1996; 93: 10887–90.
7. Zhang Y, Riesterer C, Ayrall AM, Sablitzky F, Littlewood TD, Reth M. Inducible site-directed recombination in mouse embryonic stem cells. Nucleic Acids Res 1996; 24:543–8.
8. Brocard J, Warot X, Wendling O, et al. Spatio-temporally controlled site-specific somatic mutagenesis in the mouse. Proc Natl Acad Sci USA 1997; 94:14559–63.
9. Logie C, Stewart AF. Ligand-regulated site-specific recombination. Proc Natl Acad Sci USA 1995; 92:5940–4.
10. Burke DT, Carle GF, Olson MV. Cloning of large segments of exogenous DNA into yeast by means of artificial chromosome vectors. Science 1987; 236:806–12.
11. Schedl A, Beermann F, Thies E, Montoliu L, Kelsey G, Schutz G. Transgenic mice generated by pronuclear injection of a yeast artificial chromosome. Nucleic Acids Res 1992; 20:3073–7.

12. Shizuya H, Birren B, Kim UJ, et al. Cloning and stable maintenance of 300-kilobase-pair fragments of human DNA in *Escherichia coli* using an F-factor-based vector. Proc Natl Acad Sci USA 1992; 89:8794–7.
13. Yang XW, Model P, Heintz N. Homologous recombination based modification in *Escherichia coli* and germline transmission in transgenic mice of a bacterial artificial chromosome. Nat Biotechnol 1997; 15:859–65.
14. Zhang Y, Buchholz F, Muyrers JP, Stewart AF. A new logic for DNA engineering using recombination in *Escherichia coli*. Nat Genet 1998; 20:123–8.
15. Yu D, Ellis HM, Lee EC, Jenkins NA, Copeland NG, Court DL. An efficient recombination system for chromosome engineering in *Escherichia coli*. Proc Natl Acad Sci USA 2000; 97:5978–83.
16. Angrand PO, Daigle N, van der Hoeven F, Scholer HR, Stewart AF. Simplified generation of targeting constructs using ET recombination. Nucleic Acids Res 1999; 27:e16.
17. Liu P, Jenkins NA, Copeland NG. A highly efficient recombineering-based method for generating conditional knockout mutations. Genome Res 2003; 13:476–84.
18. Muyrers JP, Zhang Y, Testa G, Stewart AF. Rapid modification of bacterial artificial chromosomes by ET-recombination. Nucleic Acids Res 1999; 27:1555–7.
19. Warming S, Costantino N, Court DL, Jenkins NA, Copeland NG. Simple and highly efficient BAC recombineering using galK selection. Nucleic Acids Res 2005; 33:e36.
20. Copeland NG, Jenkins NA, Court DL. Recombineering: a powerful new tool for mouse functional genomics. Nat Rev Genet 2001; 2:769–79.
21. Parlato R, Otto C, Begus Y, Stotz S, Schutz G. Specific ablation of the transcription factor CREB in sympathetic neurons surprisingly protects against developmentally regulated apoptosis. Development 2007; 134:1663–70.
22. Shimshek DR, Kim J, Hubner MR, et al. Codon-improved Cre recombinase (iCre) expression in the mouse. Genesis 2002; 32:19–26.
23. Feil R, Wagner J, Metzger D, Chambon P. Regulation of Cre recombinase activity by mutated estrogen receptor ligand-binding domains. Biochem Biophys Res Commun 1997; 237:752–7.

Chapter 18

Inducible Cre Mice

Susanne Feil, Nadejda Valtcheva, and Robert Feil

Abstract

The Cre/lox site-specific recombination system has emerged as an important tool for the generation of conditional somatic mouse mutants. This method allows one to control gene activity in space and time in almost any tissue of the mouse, thus opening new avenues for studying gene function and for establishing sophisticated animal models of human diseases. A major technical advance in terms of in vivo inducibility was the development of ligand-dependent Cre recombinases that can be activated by administration of tamoxifen to the animal. Here we describe how tamoxifen-dependent Cre recombinases, so-called CreER recombinases, work and how they can be used to generate time- and tissue-specific mouse mutants. The focus will be on the $CreER^{T2}$ recombinase, which is currently the most successful CreER version. We will give an overview of available $CreER^{T2}$ transgenic mouse lines and present protocols that detail the generation of experimental mice for inducible gene knockout studies, the induction of recombination by tamoxifen treatment, and the analysis of the quality and quantity of recombination by reporter gene and target gene studies. Most of the protocols can also be used as general guidelines for the generation and characterization of Cre/lox-mediated genome modifications in mice.

Key words: Transgenic mice, inducible gene knockout, somatic mutagenesis, CreER recombinase, tamoxifen, in utero, ROSA26, R26R, X-Gal staining, mouse models of human disease.

1. Introduction

In order to understand the role of a given gene product in a given cell type at a given developmental stage, genetic techniques have been developed that allow for the introduction of defined mutations into the mouse genome at will, in a specific cell type and at a chosen time *(1, 2)*. Most current conditional gene-targeting systems are based on the use of the site-specific recombinase Cre (*cy*clization *re*combination) which catalyzes recombination between two 34-bp DNA recognition sites named loxP (*lo*cus of

Ralf Kühn, Wolfgang Wurst (eds.), *Gene Knockout Protocols: Second Edition, vol. 530*

DOI 10.1007/978-1-59745-471-1_18 Springerprotocols.com

crossing [*x*-ing]-over of bacteriophage *P*1). The basic strategy for Cre/lox-directed gene knockout experiments is to flank, or "flox", an essential exon of the gene of interest with two loxP sites (by homologous recombination in embryonic stem cells), and then to "deliver" Cre that excises the intervening DNA including the exon from the chromosome, thus generating a null allele in all cells where Cre is active. Delivery of Cre can be achieved by crossing mice carrying the floxed target gene with transgenic Cre-expressing mice.

Clearly, key to successful conditional gene targeting is the availability of Cre transgenic mouse strains in which Cre activity is tightly controlled in space and time. To add inducibility to the Cre/lox system, ligand-dependent chimeric Cre recombinases, so-called CreER recombinases, have been developed *(3–6)*. They consist of Cre fused to mutated hormone-binding domains of the estrogen receptor. The CreER recombinases are inactive, but can be activated by the synthetic estrogen receptor ligand 4-hydroxytamoxifen (OHT), therefore allowing for external temporal control of Cre activity. Indeed, by combining tissue-specific expression of a CreER recombinase with its tamoxifen-dependent activity, the excision of floxed chromosomal DNA can be controlled both spatially and temporally by treating the mouse with tamoxifen, which is metabolized to OHT. The operating mode of inducible CreER mice is outlined in **Fig. 18.1**. It is important to note that inducible Cre/lox systems also allow one to limit unwanted Cre activity and associated side effects – for instance, ectopic recombination due to transient Cre expression during development or potential toxic effects due to prolonged high levels of Cre activity *(7, 8)*.

The properties of CreER recombinases were continuously improved to decrease the background activity in the absence of inducer and to increase the sensitivity and efficiency of tamoxifen-induced recombination in mice. The $CreER^{T2}$ recombinase, which contains the human estrogen receptor ligand-binding domain with a G400V/M543A/L544A triple mutation, is currently the sharpest tool in the CreER box and its use is highly recommended for inducible mutagenesis in the mouse *(4, 9, 10)*. **Table 18.1** lists transgenic mouse lines that express $CreER^{T2}$ in specific cell types. Many of them have already proven useful in addressing biological questions. A convenient way to characterize the recombination properties of a given Cre transgenic mouse line is the use of Cre reporter mice, the most popular one being the so-called R26R line that produces β-galactosidase after Cre-mediated excision of a STOP cassette from the broadly expressed ROSA26 locus *(11)*. The following sections describe the generation and analysis of inducible mouse mutants using $CreER^{T2}$ transgenic mice and R26R reporter mice. However, the protocols can readily be adapted to perform Cre/lox-assisted

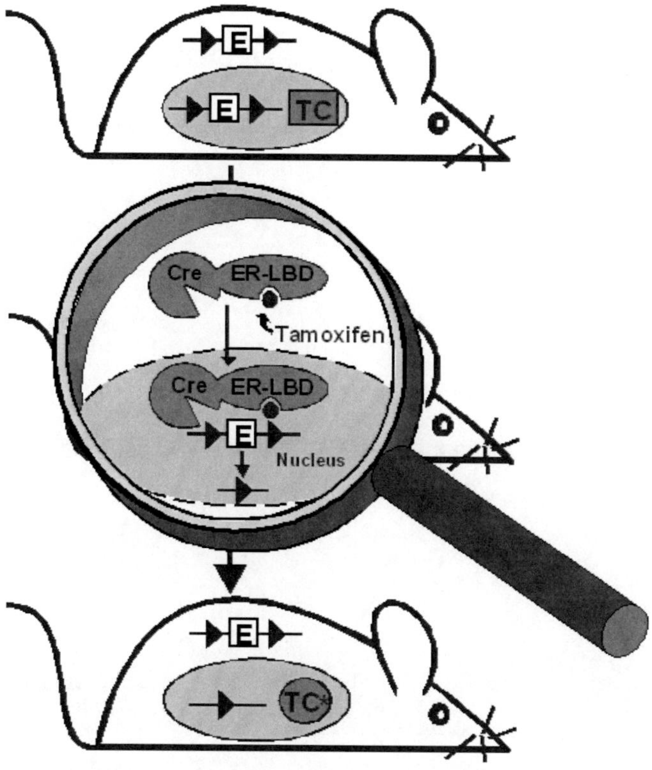

Fig. 18.1. How do CreER recombinases work? Inducible gene inactivation is based on tamoxifen-inducible excision of a loxP (triangle)-flanked exon (E) in cells (shaded oval) expressing a tamoxifen-dependent CreER recombinase (TC). The current model of CreER function is shown under the magnifying glass in the middle part of the diagram. A CreER recombinase consists of Cre fused to a mutated ligand-binding domain (LBD) of the estrogen receptor (ER). In the absence of tamoxifen, CreER is retained in the cytoplasm (boxed TC in the upper mouse). Binding of tamoxifen or OHT to the LBD results in the translocation of the recombinase into the nucleus (circled TC* in the lower mouse), where it can recombine its loxP-flanked DNA substrate; in other words, ligand binding appears to regulate primarily the localization of the recombinase rather than its enzymatic activity per se. Spatiotemporally controlled somatic mutagenesis can be achieved by tissue-specific expression (shaded oval) of a CreER recombinase.

genome modifications with other inducible or tissue-specific Cre mice. Today, hundreds of Cre transgenic mouse lines and about 15 Cre-responsive reporter lines *(1)* are available. Many of them are listed in the Cre mouse line database (http://www.mshri.on.ca/nagy/) and can be purchased from The Jackson Laboratory (http://jaxmice.jax.org/). Further information about ligand-responsive site-specific recombinases as well as about alternative tools for inducible gene manipulation, such as tetracycline-regulated expression systems, can be found elsewhere *(1, 2, 12, 13)*.

Table 18.1.
Examples of mouse lines expressing the CreER^{T2} recombinase

Tissue specificity	Mouse line	Promoter	Reference(s)
Bone			
Osteoblasts and odontoblasts	Col1a1-CreER^{T2}	Collagen 1 α1 chain	*(22)*
Chondrocytes	Col2a1-CreER^{T2}	Collagen 2 α1 chain	*(23)*
Endothelium	Tie2-CreER^{T2}	Tie2 receptor tyrosine kinase	*(24)*
	VECad-CreER^{T2}	Vascular endothelial cadherin	*(21)*
Epithelium			
Intestinal epithelium	Vil-CreER^{T2}	Villin	*(25)*
Internal epithelial organs	K18-CreER^{T2}	Keratin 18	*(26)*
Renal epithelium	KspCad-CreER^{T2}	Kidney-specific cadherin	*(27)*
Fat (adipocytes)	aP2-CreER^{T2}	Adipocyte fatty acid binding protein	*(28)*
Liver (hepatocytes)	SA-CreER^{T2}	Serum albumin	*(29)*
Nervous system			
Astrocytes	GFAP-CreER^{T2}	Glial fibrillary acidic protein	*(30)*
	GFAP-CreER^{T2}	Glial fibrillary acidic protein	*(31)*
	GLAST-CreER^{T2}	Astrocyte-specific glutamate transporter	*(32)*
Neural stem cells	Nes-CreER^{T2}	Nestin	*(33)*
Schwann cells and oligodendrocytes	PLP-CreER^{T2}	Proteolipid protein	*(34)*
Schwann cells	P0Cx-CreER^{T2}	P0 fused to connexin 32	*(34)*
Skeletal muscle	HAS-CreER^{T2}	Skeletal muscle α-actin	*(35)*
Skin			
Keratinocytes	K5-CreER^{T2}	Keratin 5	*(9)*
	K14-CreER^{T2}	Keratin 14	*(36)*
Melanocytes	Tyr-CreER^{T2}	Tyrosinase	*(37)*
	Tyr-CreER^{T2}	Tyrosinase	*(38)*
Smooth Muscle	SM-CreER^{T2}	SM22α	*(10)*
Widespread	Rosa26-CreER^{T2}	Rosa26	*(39)*
	Rosa26-CreER^{T2}	Rosa26	*(40)*
	CMVβactin-CreER^{T2}	CMV enhancer/chicken β-actin	*(41)*
	Ubc-CreER^{T2}	Ubiquitin C	*(42)*

2. Materials

2.1. Generation of Mice

1. At least three mouse lines are required: the CreER^{T2} mouse (*see* **Table 18.1** for a selection), the R26R Cre reporter mouse *(11)*, and the target mouse carrying a loxP-flanked version of

the target gene. In the following these mouse lines are referred to as "Cre mouse", "reporter mouse", and "floxed mouse", respectively. At least 5 male and 5–10 female mice of each mouse line should be available.

2. Mouse facility with standard equipment and room capacity for approx. 30 type III and 50 type II cages, in which approx. 10 and 4 mice can be housed, respectively.

2.2. Induction of Recombination

1. Tamoxifen-free base (Sigma T5648)
2. Ethanol
3. Sunflower oil (supermarket)
4. Syringes (graded in 100-μL intervals) and 22-gauge needles

2.3. Analysis of Recombination

1. CO_2 for euthanasia
2. Perfusion pump (e.g., Perfusor, Braun Medical AG, Germany)
3. 27-gauge needles
4. Dissecting tools (forceps, scissors, etc.)
5. Collecting tray for fixative solution
6. Thermocycler
7. Standard equipment for agarose gel electrophoresis

2.4. Detection of Recombination in Cre Reporter Mice by X-Gal Staining

1. Phosphate-buffered saline (PBS): 135 mM NaCl, 3 mM KCl, 8 mM Na_2HPO_4 , 2 mM KH_2PO_4, pH 7.4. Prepare at least 2 L and sterilize by autoclaving; store at RT.
2. Fixative solution: 2% formaldehyde, 0.2% glutaraldehyde in PBS ($\approx$50 mL per mouse); store at 4°C.
3. X-Gal stock solution: 40 mg/mL 5-bromo-4-chloro-3-indolyl β-D-galactoside (X-Gal) in dimethylsulfoxide. Prepare 10 mL and store in aliquots of 1 mL at –20°C.
4. X-Gal staining solution: 2 mM $MgCl_2$, 2.5 mM $K_3Fe(CN)_6$, 2.5 mM $K_4Fe(CN)_6$ in PBS; prepare 500 mL and store in the dark at RT. Before use add X-Gal stock solution (40 mg/mL, see above) to a final concentration of 1 mg/mL of X-Gal. Make fresh as required ($\approx$10 mL per mouse).

2.5. Detection of Target Gene Recombination by PCR

1. Lysis buffer: 50 mM Tris-HCl, pH 7.4, 5 mM EDTA, 1% SDS, 200 mM NaCl, 0.5 mg/mL proteinase K (add freshly before use; stock solution: 50 mg/mL in 10 mM Tris-HCl, pH 8.0, 1 mM EDTA, pH 8.0)
2. Phenol/chloroform (50%/50%)
3. Ethanol (100%, 70 %)
4. Oligonucleotide primers: each 25 pmol/μL (dissolved in H_2O)

5. 10X PCR buffer (including dNTPs): 500 mM KCl, 100 mM Tris-HCl, pH 8.0, 15 mM $MgCl_2$, 2 mM of each dNTP (dGTP, dATP, dTTP, dCTP)
6. Taq polymerase

3. Methods

3.1. Generation of Mice

For the generation of an inducible tissue-specific knockout mouse at least two basic mouse lines are required: one that expresses the CreER^{T2} recombinase under the control of a tissue-specific promoter (in the following referred to as "Cre mouse"; for a selection of available lines, *see* **Table 18.1**); and the other line carrying loxP sites in the gene of interest (in the following referred to as "floxed mouse"). Besides the generation of the inducible knockout mice of interest (**Section 3.1.2**), it is highly recommended to validate the recombination properties of the Cre mouse, that is, the sites of recombinase expression and tamoxifen-dependent Cre activity, by crossing it to a Cre reporter line (**Section 3.1.1**). Some general guidelines for efficient mouse breeding are presented in **Note 1**.

3.1.1. Generation of Reporter Mice for Detection of Cre Activity

The most popular reporter mouse line is the so-called R26R line (in the following referred to as "reporter mouse") that produces β-galactosidase after Cre-mediated excision of a STOP cassette from the broadly expressed ROSA26 locus *(11)*. The activity of β-galactosidase can easily be detected with single-cell resolution by staining tissues with X-Gal (**Section 3.3.1**). Thereby, Cre activity eventually results in a blue dye precipitate. The Cre mouse is crossed to the reporter mouse to generate mice with the genotype cre/+;+/R, where cre stands for the Cre transgene, R for the R26R reporter allele, and + for the respective wild-type allele. Experimental mice with a single copy of both the Cre and the reporter transgene can be derived in a single breeding step (**Fig. 18.2**). In order to obtain about six experimental animals at least three breeding cages should be set up (*see* **Note 1**). The mice are injected with vehicle or tamoxifen (**Section 3.2**) and analyzed for recombination by X-Gal staining (**Section 3.3.1**).

More breeding steps are required to generate mice with two copies of the reporter gene, which can lead to a stronger X-Gal signal, and/or with two copies of the Cre transgene, which can result in a higher recombination rate. In addition to the characterization of Cre mice, Cre reporter lines are increasingly used to genetically label wild-type or knockout cells in order to perform so-called cell lineage tracing or fate mapping experiments (*see* **Note 2**).

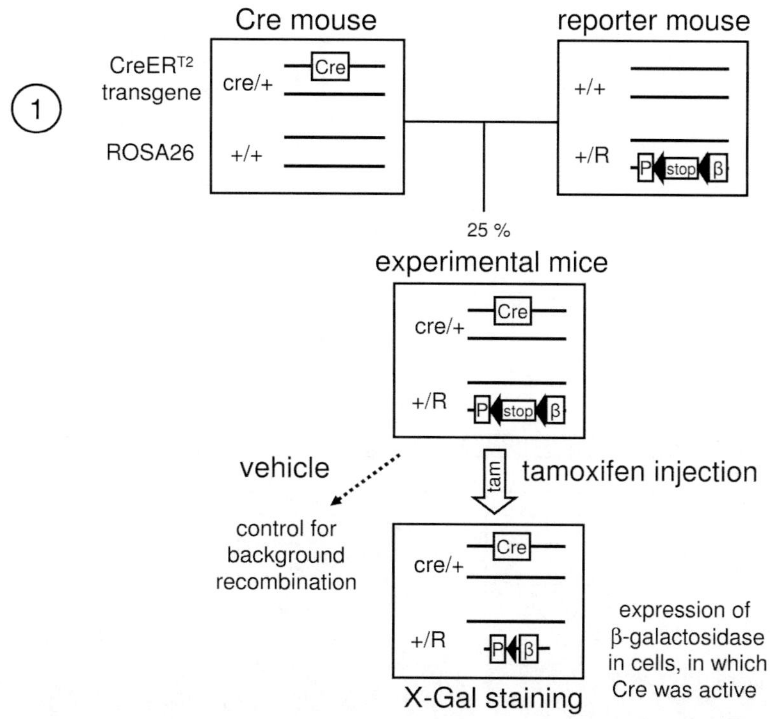

Fig. 18.2. Breeding scheme for the generation of reporter mice for the detection of Cre activity. The genotype of each mouse with respect to the CreER^{T2} transgene and the ROSA26 reporter gene is indicated in the boxes. The number of breeding steps required to derive experimental animals and the expected yield (in percent of offspring) according to Mendelian inheritance is also given. Note that injection of tamoxifen usually results in mosaic activation of the reporter gene in only a subset of the target cells, but not in 100% of the cells as pretended by the scheme. Abbreviations: Cre, CreER^{T2} transgene driven by a tissue-specific promoter; R, R26R Cre-responsive reporter consisting of a β-galactosidase transgene (β) that is expressed from the ROSA26 promoter (P) after excision of a loxP-flanked (triangles) stop cassette (stop); wild-type alleles are denoted by +. For further explanations see text.

3.1.2. Generation of Inducible Knockout Mice

For target genes whose heterozygous (+/−) knockout does not cause a phenotype, which is normally the case, it is recommended to start breeding with the so-called "L– mouse". The L– mouse line can be generated by Cre-mediated excision of the floxed target exon (L2 allele with two loxP sites) to produce an excised knock-out gene (L– allele with one loxP site) in germ cells (*see* **Note 3**). Alternative breeding strategies are possible, but might compromise the efficiency of the inducible gene knockout and complicate the interpretation of the experimental results (*see* **Notes 4** and **5**). The standard breeding scheme is depicted in **Fig. 18.3**. In the first step, Cre mice (cre/+) are crossed to L– mice (L–/+). In the second step, their Cre/L– offspring (cre/+; L–/+) is mated with floxed target mice (L2/L2) to generate the experimental mice. In order to obtain sufficient experimental animals for initial analyses

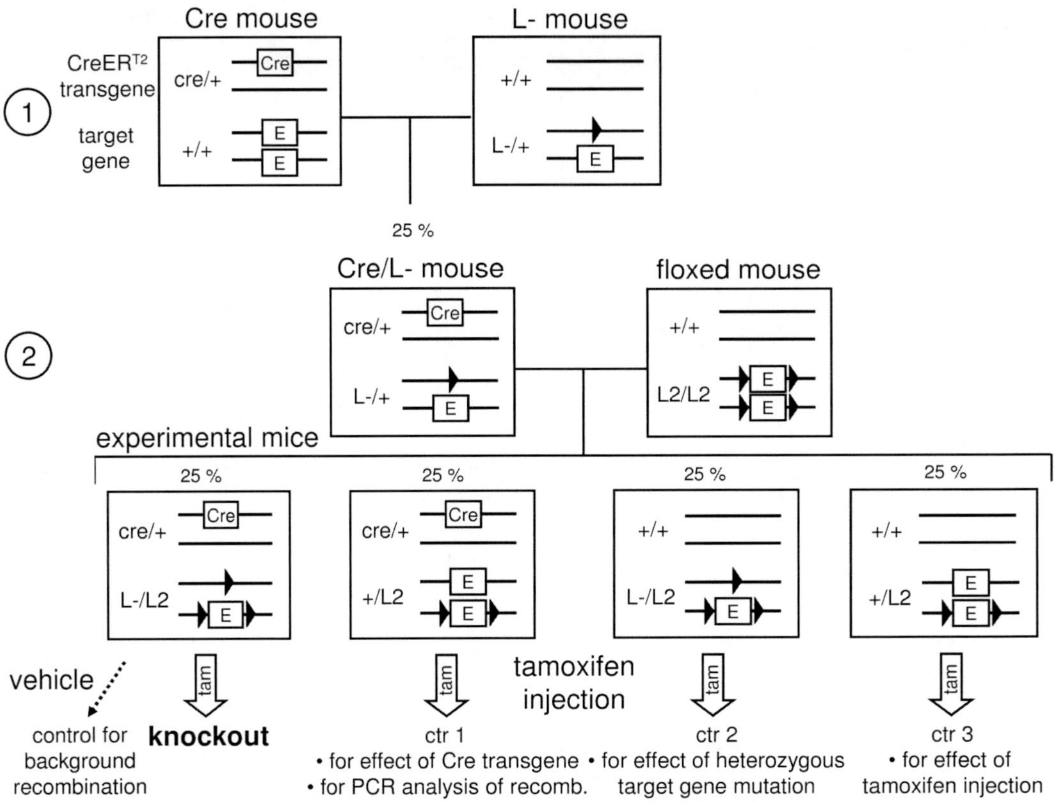

Fig. 18.3. Breeding scheme for the generation of inducible knockout mice and control animals. The genotype of each mouse with respect to the CreER^{T2} transgene and the target gene is indicated in the boxes. The number of breeding steps required to derive experimental animals and the expected yield (in percent of offspring) according to Mendelian inheritance is also given. Injection of pre-mutant mice (lower left box, genotype cre/+;L–/L2) with tamoxifen results in the excision of the floxed exon and, thus, in a gene knockout. Note that the genetic configuration of the experimental mice after tamoxifen treatment is not shown in the diagram. Note also that recombination is usually mosaic, so that only a subset of the target cells, but not 100%, are recombined. Abbreviations: Cre, CreER^{T2} transgene driven by a tissue-specific promoter; E, exon of target gene; triangles, loxP sites; +, L2, and L– denote wild-type, floxed and excised allele, respectively. For further explanations see text.

(approx. six inducible knockout mice plus controls) at least three breeding cages should be set up (*see* **Note 1**). Various useful genotypes are obtained in the same litter (**Fig. 18.3**) and all groups are injected with tamoxifen (**Section 3.2**): "pre-mutant" mice (cre/+;L–/L2) that are supposed to become knockout mice after Cre induction as well as three types of control mice to test for potential phenotypes due to the presence of the Cre transgene (ctr 1), due to the heterozygous loss of the target gene (ctr 2), or due to side effects of tamoxifen treatment which are not related to the gene knockout (ctr 3). Furthermore, some of the pre-mutant animals (cre/+;L–/L2) are injected with vehicle (**Section 3.2**) to estimate the background recombination in the absence of inductor. The initial analyses should include all genotypes and controls; however, the main studies will usually be performed with the tamoxifen-injected pre-mutant and ctr 1 mice. Some extra vehicle- and tamoxifen-treated ctr 1 mice (cre/+;+/L2) are required for the PCR analysis of recombination (**Section 3.3.2**).

Note that the two breeding strategies depicted in **Figs. 18.2** and **18.3** can be linked in order to combine the floxed target gene with the reporter gene (*see* **Note 6**). By this means mice are derived, in which Cre activity results in target gene inactivation and reporter gene activation simultaneously in the same cells, so that knockout cells can be easily visualized by X-Gal staining (**Section 3.3.1**). Such a configuration can also be used to follow the fate of knockout cells in a wild-type environment via cell fate mapping (*see* **Note 2**).

3.2. Induction of Recombination

Many reports describe the systemic or local administration of tamoxifen or OHT to induce recombination at defined stages during pre- and postnatal development. Administration schemes that have been used with CreER^{T2} mouse lines are summarized in **Table 18.2**. Below the standard protocol for the systemic administration of tamoxifen to adult mice is described.

1. Suspend 100 mg of tamoxifen-free base (Sigma T5648) in 0.5 mL of ethanol in a 15-mL tube and add 9.5 mL of sunflower oil (do not autoclave). Mix the suspension thoroughly using a vortexer and sonicate for 5 min at 37°C (*see* **Note 7**). In addition, prepare 1.0 mL of sunflower oil containing 5% ethanol for vehicle injection.
2. The tamoxifen stock solution (10 mg/mL) and dilutions thereof in sunflower oil can be stored at –20°C for up to 4 weeks.
3. Before use, thaw and sonicate the tamoxifen solution at 37°C for 5 min.
4. The experimental mice (reporter mice, *see* **Section 3.1.1** and **Fig. 18.2**, or pre-mutant and control animals, *see* **Section 3.1.2** and **Fig. 18.3**) are injected with tamoxifen at an age of 4–6 weeks. Inject each mouse intraperitoneally (i.p.) with 100 μL of tamoxifen stock solution (corresponds to 1 mg of tamoxifen) or with vehicle for five consecutive days (*see* **Note 8**). Do not house the tamoxifen-treated mice together with vehicle-treated or untreated animals in the same cage (*see* **Note 9**).
5. Analyze the mice for recombination (**Section 3.3**) and/or phenotypes at the earliest 3 days after the last injection.

3.3. Analysis of Recombination

Reporter mouse lines are useful tools to analyze the tissue specificity and the recombination efficiency of the Cre mouse line of interest. However, the recombination sensitivity of the Cre reporter must not necessarily correlate with that of the floxed target gene. Therefore, it is strongly recommended to monitor the excision of the target gene at the DNA level and, most importantly, the expression

Table 18.2
Examples of induction schemes for CreER^{T2} transgenic mice

Recombination	Mode of ligand administration	Time of analysis	Target tissue	Reference(s)
In utero in embryos[1]	To pregnant females i.p. 1 mg Tam (at E12.5)	E14.5	CNS	*(34)*
	To pregnant females oral 3-4 mg Tam (at E8.5–E11.5)	E9.5–E16.5	Limb development	*(43)*
	To pregnant females i.p. 1 mg OHT for 3 consecutive days (at E12.5–14.5)	E16.5	Bone	*(22)*
	To pregnant females oral 5-10 mg Tam depending on embryonic stage (max. 3 times 10 mg Tam at E12.5–E13.5)	Neonatal	Skeletal development	*(41)*
Postnatal in pups	To lactating mothers i.p. 1 mg Tam for 5 consecutive days (at P1–P5)	8-week-old offspring	CNS	*(34)*
	To pups (P12–P17) i.p. 0.25-1 mg OHT for 5 consecutive days	P19	Bone	*(22)*
	To pups (P1–P3) oral or intragastric injection of 0.05–0.1 mg Tam topical skin exposure to OHT for 3 consecutive days	1 week after last treatment	Endothelium	*(21)*
Postnatal in adults	To 8–10-week-old mice i.p. 0.01–1 mg OHT for 5 consecutive days	10 days and 25 days after last treatment	Skin	*(9)*
	To 5-week-old mice i.p. 0.01–1 mg Tam or OHT for 5 consecutive days oral 1 mg Tam for 5 consecutive days	3 days-16 weeks after last treatment	Smooth muscle	*(10)*
	To 4–8-week-old mice oral 1–5 mg Tam for 5 consecutive days	3 days after last treatment	Ubiquitous	*(39)*

Local	Perivascular delivery device, 0.1%–10% (w/w) OHT for 7 days or 14 days	1 week post application	Femoral and carotid artery	*(44)*
	Topical exposure to the back skin, 1 mg OHT in 0.2 mL EtOH for 5 consecutive days	Up to 5 weeks after treatment	Skin	*(45)*
	Matrigel plug containing 50 nM OHT	5 days after matrigel implantation	Endothelium of tumor vasculature	*(21)*
In cultured cells	1 μM OHT for 48 h	48 h post application	Vascular smooth muscle cells	*(10)*
	500 nM OHT for 4.5 days	4.5 days post application	Embryonic stem cells	*(39)*
	100 nM OHT for 48 h	48 h post application	Rat cortical collecting duct cells	*(46)*

E, embryonic day; i.p., intraperitoneal; OHT, 4-hydroxytamoxifen; P, postnatal day; Tam, tamoxifen.
[1] Note that treatment of mice during pregnancy with tamoxifen or OHT can impair normal delivery of the pups. To obtain live offspring, caesarean section might be necessary.

of the target protein itself (*see* **Note 10**). The following sections describe the analysis of recombination by reporter gene studies at the cellular level (**Section 3.3.1**) as well as by target gene studies at the DNA level (**Section 3.3.2**). Note that recombination rates are affected by many parameters, including the particular genomic location and distance between the loxP sites and the ability of tamoxifen to reach a given organ (*see* **Note 11**). A thorough analysis of recombination is very important, because the tamoxifen-induced gene knockouts in a given cell type or tissue are not complete in most of the cases; in other words, the induced mice are usually mosaics with varying fractions of co-existing wild-type and knock-out cells.

3.3.1. Detection of Recombination in Cre Reporter Mice by X-Gal Staining

The R26R reporter transgene *(11)* is highly sensitive to Cre-mediated activation and is thereafter widely expressed; that is, relatively low levels of Cre activity can lead to β-galactosidase expression and X-Gal-stained cells in a wide range of different tissues (*see* **Note 9**). However, note that the ROSA26 promoter that drives β-galactosidase expression in this reporter line might not be active in every tissue and cell type. Thus, it cannot be excluded that the β-galactosidase is not expressed, although the reporter transgene has been recombined at the DNA level, leading to false-negative results of the X-Gal staining. To test whether β-galactosidase can principally be expressed in the tissue of interest, the R26R reporter mouse can be crossed to a "deleter" Cre mouse in order to generate mice that carry the recombined, that is, activated, reporter transgene in every cell. This procedure is analogous to that described for the generation of L– mice (*see* **Note 3**).

1. Sacrifice the mice with CO_2 (*see* **Note 12**).
2. Perfusion fixation (not mandatory):
   - The goal of the perfusion with the fixative solution is to clear the vascular system of blood and to achieve uniform fixation, especially in bigger organs, and to get tissue preparations that are appropriate for sectioning.
   - Prepare the perfusion pump.
   - Place the mouse in a collecting tray and arrange the whole set up under a fume hood (*see* **Note 13**).
   - Open the thoracic cavity and insert a 27-gauge needle about 2 mm into the left ventricle of the heart oriented towards the ascending aorta. Be careful not to pull out the needle throughout the perfusion procedure. Perfuse with the fixative solution at 60 mL/h for 10 min. Effective perfusion is indicated by movements of the tail and legs.
3. Collect the organs and tissues of interest (*see* **Notes 14** and **15**), transfer them into a 15-mL tube with 10 mL of fixative solution and post-fix them for 30 min at RT with gentle shaking. Tissues of the same mouse can be pooled in the same tube.

4. Wash the fixed tissues three times with 10 mL of PBS for 15 min at RT with gentle shaking.
5. Incubate the samples overnight in 10 mL of X-Gal staining solution in the dark at RT with gentle shaking (*see* **Note 16**).
6. Wash the X-Gal-stained tissues three times with 10 mL of PBS for 15 min at RT with gentle shaking.
7. Analyze the samples in a Petri dish using a dissecting microscope (*see* **Note 17**). For photodocumentation different lighting conditions should be explored. Reflections can be avoided by taking pictures from samples totally covered with PBS.
8. Store the X-Gal-stained samples in PBS in the dark at 4°C (*see* **Note 18**).

3.3.2. Detection of Target Gene Recombination by PCR

An essential part of the generation of an inducible knockout mouse is the verification of target gene excision at the DNA level. This method has two advantages over reporter gene studies. First, it tests the recombination of the target gene directly and, second, it is not confounded by the uncertainties associated with the expression profile of the β-galactosidase driven by the ROSA26 promoter (*see* **Section 3.3.1**). PCR analysis is a convenient way to screen the experimental animals for recombination at the DNA level. Three primers (A, B and C) are combined, so that the wild-type (+), floxed (L2), and knockout (L–) version of the target gene can be co-detected in a single reaction (*see* **Fig. 18.4A** and **Note 19**). To analyze the efficiency of tamoxifen-induced recombination, it is recommended to use vehicle- and tamoxifen-treated mice with the genotype cre/+;+/L2 mice (*see* **Fig. 18.4B** and **Note 20**). The same "three-primer PCR" can be employed to genotype the mice during breeding (*see* **Fig. 18.3** and **Note 21**).

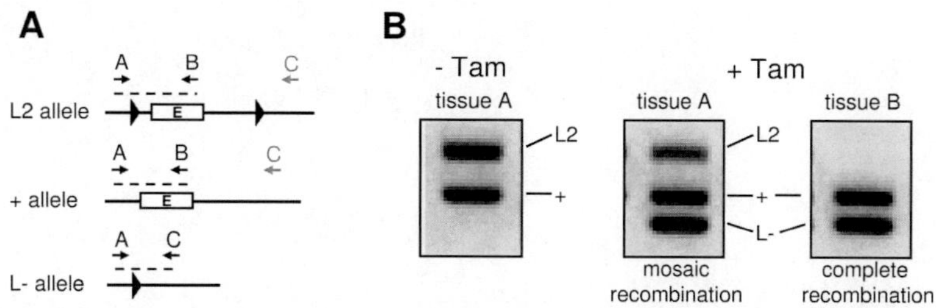

Fig. 18.4. PCR analysis of target gene recombination based on a "three-primer" strategy. (**A**) The positions of the three primers (A, B, and C) used to co-detect the floxed (L2), wild-type (+), and excised (L–) allele of the target gene are indicated. The amplified products are represented by dashed lines; E and triangles denote the target exon and loxP sites, respectively. (**B**) Examples of PCR products detected by agarose gel electrophoresis. Template DNA was obtained from tissues of cre/+;+/L2 mice that have been injected with vehicle (-Tam, *left panel*) or tamoxifen (+Tam, *right panels*). Depending on the tissue analyzed, the tamoxifen-treated mouse shows mosaic or complete recombination as indicated by the level of conversion of the L2 into the L– allele. For further explanations see text.

1. Sacrifice the mice with CO_2 (*see* **Note 12**).
2. Collect the organs and tissues of interest in a Petri dish with PBS (*see* **Note 14** and **20**). Transfer each sample (max. 50 mg) into a 10–15-mL tube (round bottom) with 500 μL of lysis buffer.
3. Homogenize the tissue with an ultraturrax for 1 min and transfer it into a 1.5-mL tube.
4. Incubate overnight at 55°C in a water bath.
5. Add 500 μL of phenol/chloroform (50%/50%) and extract samples by mixing for 30 s with a vortexer.
6. Centrifuge (17,000*g*, 5 min, RT) and transfer the upper aqueous phase that contains the DNA (≈0.5 mL) into a new 1.5-mL tube.
7. Precipitate the DNA by adding 1 mL of 100% ethanol followed by gentle mixing. Centrifuge (17,000*g*, 5 min, RT), wash pellet twice with 1 mL of 70% ethanol, and dry the pellet at RT.
8. Resuspend the pellet in 300 μL of H_2O.
9. Dilute DNA solution 1:100 in H_2O and measure OD_{260} (an OD_{260} of 1.0 corresponds to a DNA concentration of 50 μg/mL). Depending on the tissue, yields of approx. 0.5–10 μg of DNA per mg tissue are expected. Calculate DNA concentration and dilute an aliquot to 0.02 μg/μL. Use 0.5 μg of DNA as template for the PCR.
10. As a starting point use the following standard PCR conditions (*see* **Note 22**):

DNA (0.02 μg/μL)	25 μL
Primer A (25 pmol/μL)	0.5 μL
Primer B (25 pmol/μL)	0.5 μL
Primer C (25 pmol/μL)	0.5 μL
10x PCR buffer	5 μL
H_2O	18 μL
Taq polymerase (5 U/μL)	0.5 μL

- Standard PCR program: Initial denaturation step for 5 min at 95°C; then 35 cycles of denaturation (10 s at 95°C), annealing (30 s at 55°C) and polymerization (30 s at 72°C); and a final polymerization step for 5 min at 72°C; store at 4°C.

11. Analyze PCR products by agarose gel electrophoresis (*see* also **Fig. 18.4B**).

4. Notes

4.1. Generation of Mice

1. In general, breeding cages are set up by placing two female mice (>6-weeks old) to one male mouse (>6-weeks old; if possible use an experienced breeder). The average litter size is 5–10 pups. Note that each breeding step requires 2–3 months, resulting from ≈3 weeks of pregnancy plus 6–9 weeks until the offspring reaches sexual maturity, so that the next breeding step can be started. The basic mouse lines (i.e., the Cre line, the reporter line, and the floxed target line) that are crossed to obtain the experimental animals should be maintained by breeding with wild-type mice of the same genetic background. If the breeding scheme for the generation of inducible knockout mice does not yield pre-mutant animals for tamoxifen injection (*see* **Fig. 18.3**), then this might be related to the fact that both the Cre transgene as well as the target gene are located on the same chromosome. If so, another Cre mouse line which carries the transgene on a different chromosome should be used.
2. Cre reporter mice can also be used to genetically label Cre-expressing cells and their derivatives for cell lineage tracing or fate mapping experiments. Depending on the promoter driving Cre expression, the reporter will be activated only in the respective cells, so that specific cell types can be labeled and their fate can be followed by X-Gal staining, for instance, during embryogenesis, angiogenesis or pathological processes like tumor growth and atherosclerosis *(1, 2, 14–16)*.
3. The L– mouse line can be derived from floxed mice by crossing the latter to a "deleter" Cre mouse that expresses Cre in germ cells, such as the EIIa-Cre mouse *(17)* or the CMV-Cre mouse *(18)*. The Cre-mediated excision of the loxP-flanked exon results in the conversion of the floxed target allele (L2 allele) into an excised null allele (L– allele) in the germ cells of the F1 progeny. After crossing the F1 progeny to wild-type mice, a L– mouse line can be established that does not carry the Cre transgene and transmits the L– allele through the germline. An alternative strategy to obtain a L– allele is the transfection of embryonic stem cell clones that carry the L2 allele with a Cre-expressing plasmid. Last but not least, a mouse line that carries a conventional null allele of the target gene can be used instead of the L– mouse.
4. An alternative strategy to obtain inducible knockout mice is to mate in the first breeding step the floxed L2 mouse (instead of the L– mouse) with the Cre mouse to generate Cre/L2 mice (instead of Cre/L– mice). This strategy should be used

when the heterozygous knockout of the target gene already produces a phenotype or is likely to do so. Compared to the standard breeding scheme as detailed in **Section 3.1.2** and **Fig. 18.3**, the alternative strategy leads to experimental pre-mutant mice that carry two floxed L2 alleles (cre/+;L2/L2 instead of cre/+;L–/L2). Thus, not only one but two L2 alleles must be excised by Cre in order to produce a homozygous gene knockout. The requirement for two recombination events should not pose a problem for experiments with conventional Cre mice, which express a permanently active Cre enzyme. However, it could reduce the knockout efficiency in inducible Cre mice, in which the recombinase is only transiently active during the induction period. Another drawback of the alternative breeding scheme that starts with floxed L2 mice is linked to the potential background activity of the inducible Cre recombinase in the absence of ligand. Recombination background could cause the excision of the floxed exon already in the first generation. If the germline is affected, germ cells carrying the L– allele will be produced. Accordingly, some of the experimental mice obtained after the second breeding step might carry the L–allele in all cells instead of the L2 allele before induction of Cre activity, which might complicate the interpretation of the results. However, it should be noted that the background activity of CreER^{T2} in transgenic mice appears to be very low or virtually absent *(9, 10, 16)*.

5. The breeding schemes shown in **Figs. 18.2** and **18.3** can also be used with conventional tissue-specific Cre mice instead of inducible Cre mice. In this case it is even more important to use L– mice in the first breeding step to avoid uncontrolled germline recombination in the experimental animals.

6. To generate animals, which carry the R26R reporter transgene in addition to the floxed target gene (e.g., for fate mapping experiments), the floxed (L2/L2) mice should first be crossed to reporter (+/R) mice to obtain L2/L2;+/R animals. The latter can then be used instead of the "conventional" floxed (L2/L2) mice in the second breeding step of the scheme outlined in **Fig. 18.3**.

4.2. Induction of Recombination

7. Tamoxifen is not soluble in water. Oil should be added after suspending tamoxifen in a small volume of ethanol. Note that after sonication the tamoxifen stock solution might still be slightly turbid. Variations of the standard protocol include the use of peanut oil or Miglyol instead of sunflower oil or the use of OHT (minimum 70% of Z isomer, Sigma H6278) instead of tamoxifen. Note that OHT is presumably the actual ligand that binds with high affinity to the estrogen receptor

ligand-binding domain of CreER recombinases. However, because tamoxifen is metabolized to OHT *(19)* and is much cheaper than OHT, it is the drug of choice for most induction protocols that are based on systemic drug administration. OHT should be used directly, whenever limited conversion of tamoxifen to OHT is anticipated – for instance, when recombination is to be induced by locally-restricted drug application or in cultured cells.

8. The injection of mice with tamoxifen is an animal experiment and should be performed in accordance with the local guidelines for animal welfare. The recombination background in the absence of tamoxifen should always be controlled in vehicle (oil/ethanol)-treated mice.

9. Cross-contamination with tamoxifen can take place, if treated and untreated animals are housed in the same cage. Licking of oily tamoxifen suspension, grooming or coprophagous behavior can already cause recombination *(20)*.

4.3. Analysis of Recombination

10. There is no doubt that the disappearance of the gene product, which is in most cases a protein, is crucial for the success of an inducible knockout experiment. Thus, it is mandatory to monitor the expression of the target protein itself, for example, by Western blot analysis of tissue extracts and, if possible, at the cellular level by immunohistochemistry. Note that the half-life of individual mRNAs and proteins can vary greatly, so that it can take days to weeks after tamoxifen injection until the target protein is lost.

11. The conditions for inducible gene inactivation should be optimized for every application. Among the variables that can be changed in order to improve the efficiency of recombination are the mode of drug administration (*see* also **Table 18.2**) as well as the age, gender, and genetic background of the experimental mice. In general, tamoxifen-induced recombination appears to be more efficient in younger than in older mice (SF and RF, unpublished data 2007) *(21)*. Therefore, it is recommended to inject the experimental animals with tamoxifen at an age of 4–6 weeks. Moreover, tamoxifen and OHT, respectively, might not easily cross the blood–brain barrier, thus limiting recombination in the brain.

4.3.1. Detection of Recombination in Cre Reporter Mice by X-Gal Staining

12. Alternatively, other methods can be used to sacrifice the mice. However, the method of choice should be compatible with downstream processing of the tissue samples. For example, cervical dislocation is not recommended before perfusion fixation, because vessels can be destroyed and the fixative solution will not reach every organ.

13. Formaldehyde is carcinogenic! Therefore, the experiment should be performed under a fume hood and the mouse placed in a dish to collect the fixative solution that drips from the animal.
14. Collect not only tissues that are expected to show recombination, but also some not expected to be recombined. Tissues should be obtained from tamoxifen-treated mice as well as from vehicle-treated mice to control for background recombination. An overview of the experimental animals used for X-Gal staining (**Section 3.3.1**) and PCR analysis (**Section 3.3.2**) is shown in **Figs. 18.2** and **18.3**, respectively.
15. In the case of X-Gal staining, it is also important to test organs of wild-type mice for endogenous β-galactosidase activity, which often appears to be greenish-blue. Note that some tissues show a relatively high β-galactosidase background – for instance, some regions of the gut, the intestinal mucosa, and the testes.
16. X-Gal stock solution should always be freshly added to the staining solution. The exact staining time depends on the size of the samples and the level of β-galactosidase activity. A similar protocol can be used for X-Gal staining of mouse embryos and cultured cells, with a fixation time of 15 min and 5 min, respectively. Cultured cells are stained with X-Gal at 37°C without shaking.
17. The X-Gal staining solution penetrates only the outer layer of bigger organs, for example, 2–3 mm of the heart. For uniform staining of deeper regions in organs or embryos the samples can be cut into halves or several slices before X-Gal staining is performed. X-Gal-stained tissues can also be embedded in paraffin and sectioned. It is not recommended to leave the tissues in organic solvents longer than necessary, because the blue precipitate may leach out. It is also feasible to prepare frozen sections at first and to stain them with X-Gal.
18. Unstained tissues may become greenish-blue after 1–2 weeks of storage in PBS.

4.3.2. Detection of Target Gene Recombination by PCR

19. For a successful "three-primer PCR" the following criteria should be met (*see* also **Fig. 18.4A**). The length of the amplified products corresponding to the +, L2, and L– alleles should be in the range of approx. 150–300 bp. Primer C should be located such that the distance between primer A and C on the L– allele is much smaller (<200 bp) than on the + and L2 allele (>500 bp). Thus, the L– allele, if present, is preferentially amplified versus + or L2. These conditions can be fulfilled as soon as the loxP sites flanking the target exon are located more than ≈300 bp apart, which is normally the case.

20. For PCR analysis of recombination it is recommended to use tamoxifen- and vehicle-treated mice with the genotype cre/+;+/L2 (**Fig. 18.3**, ctr 1). The advantage of using these mice rather than the actual pre-mutant mice (cre/+;L–/L2) is that successful recombination is unequivocally indicated by the appearance of an additional PCR product representing the L– allele. Note that the L– allele is already present in untreated pre-mutant mice, which would complicate the recombination analysis.
21. By using the "three-primer PCR" for mouse genotyping, all possible alleles (+, L2, L–) can be detected in the same reaction. Here, every possible genotype, even +/+, results in at least one PCR product. Thus, this PCR strategy excludes the possibility of false-negative results, which can happen when only two primers are used to detect a single allele.
22. The PCR conditions depend on the length of the products and the composition of the primers. The duration and temperature of the PCR steps must be optimized for each particular set of primers and template DNA. Altering the primer concentrations is a particular effective means to adjust the relative abundance of the PCR products.

Acknowledgments

We thank the members of the Feil laboratory for critical discussion. Work in the authors' laboratory was supported by grants from the VolkswagenStiftung and the Deutsche Forschungsgemeinschaft.

References

1. Branda CS, Dymecki SM. Talking about a revolution: the impact of site-specific recombinases on genetic analyses in mice. Dev Cell 2004; 6:7–28.
2. Feil R. Conditional somatic mutagenesis in the mouse using site-specific recombinases. Handb Exp Pharmacol. 2007:3–28.
3. Feil R, Brocard J, Mascrez B, LeMeur M, Metzger D, Chambon P. Ligand-activated site-specific recombination in mice. Proc Natl Acad Sci USA 1996; 93: 10887–90.
4. Feil R, Wagner J, Metzger D, Chambon P. Regulation of Cre recombinase activity by mutated estrogen receptor ligand-binding domains. Biochem Biophys Res Commun 1997; 237:752–7.
5. Metzger D, Clifford J, Chiba H, Chambon P. Conditional site-specific recombination in mammalian cells using a ligand-dependent chimeric Cre recombinase. Proc Natl Acad Sci USA 1995; 92:6991–5.
6. Zhang Y, Riesterer C, Ayrall AM, Sablitzky F, Littlewood TD, Reth M. Inducible site-directed recombination in mouse embryonic stem cells. Nucleic Acids Res 1996; 24:543–8.
7. Loonstra A, Vooijs M, Beverloo HB, et al. Growth inhibition and DNA damage induced by Cre recombinase in mammalian cells. Proc Natl Acad Sci USA 2001; 98:9209–14.
8. Schmidt EE, Taylor DS, Prigge JR, Barnett S, Capecchi MR. Illegitimate Cre-dependent

chromosome rearrangements in transgenic mouse spermatids. Proc Natl Acad Sci USA 2000; 97:13702–7.

9. Indra AK, Warot X, Brocard J, et al. Temporally-controlled site-specific mutagenesis in the basal layer of the epidermis: comparison of the recombinase activity of the tamoxifen-inducible Cre-ER(T) and Cre-ER(T2) recombinases. Nucleic Acids Res 1999; 27:4324–7.
10. Kuhbandner S, Brummer S, Metzger D, Chambon P, Hofmann F, Feil R. Temporally controlled somatic mutagenesis in smooth muscle. Genesis 2000; 28:15–22.
11. Soriano P. eneralized lacZ expression with the ROSA26 Cre reporter strain. Nat Genet 1999; 21:70–1.
12. Gossen M, Bujard H. Studying gene function in eukaryotes by conditional gene inactivation. Annu Rev Genet 2002; 36:153–73.
13. Sprengel R, Hasan MT. Tetracycline-controlled genetic switches. Handb Exp Pharmacol 2007:49–72.
14. Feil S, Hofmann F, Feil R. SM22alpha modulates vascular smooth muscle cell phenotype during atherogenesis. Circ Res 2004; 94:863–5.
15. Lewandoski M. Analysis of mouse development with conditional mutagenesis. Handb Exp Pharmacol 2007:235–62.
16. Wolfsgruber W, Feil S, Brummer S, Kuppinger O, Hofmann F, Feil R. A proatherogenic role for cGMP-dependent protein kinase in vascular smooth muscle cells. Proc Natl Acad Sci USA 2003; 100:13519–24.
17. Lakso M, Pichel JG, Gorman JR, et al. Efficient in vivo manipulation of mouse genomic sequences at the zygote stage. Proc Natl Acad Sci USA 1996; 93:5860–5.
18. Dupe V, Davenne M, Brocard J, et al. In vivo functional analysis of the Hoxa-1 3' retinoic acid response element (3'RARE). Development 1997; 124:399–410.
19. Furr BJ, Jordan VC. The pharmacology and clinical uses of tamoxifen. Pharmacol Ther 1984; 25:127–205.
20. Brake RL, Simmons PJ, Begley CG. Cross-contamination with tamoxifen induces transgene expression in non-exposed inducible transgenic mice. Genet Mol Res 2004; 3:456–62.
21. Monvoisin A, Alva JA, Hofmann JJ, Zovein AC, Lane TF, Iruela-Arispe ML. VE-cadherin-CreERT2 transgenic mouse: a model for inducible recombination in the endothelium. Dev Dyn 2006; 235:3413–22.
22. Kim JE, Nakashima K, de Crombrugghe B. Transgenic mice expressing a ligand-inducible cre recombinase in osteoblasts and odontoblasts: a new tool to examine physiology and disease of postnatal bone and tooth. Am J Pathol 2004; 165:1875–82.
23. Chen M, Lichtler AC, Sheu TJ, et al. Generation of a transgenic mouse model with chondrocyte-specific and tamoxifen-inducible expression of Cre recombinase. Genesis 2007; 45:44–50.
24. Forde A, Constien R, Grone HJ, Hammerling G, Arnold B. Temporal Cre-mediated recombination exclusively in endothelial cells using Tie2 regulatory elements. Genesis 2002; 33:191–7.
25. el Marjou F, Janssen KP, Chang BH, et al. Tissue-specific and inducible Cre-mediated recombination in the gut epithelium. Genesis 2004; 39:186–93.
26. Wen F, Cecena G, Munoz-Ritchie V, Fuchs E, Chambon P, Oshima RG. Expression of conditional cre recombinase in epithelial tissues of transgenic mice. Genesis 2003; 35:100–6.
27. Lantinga-van Leeuwen IS, Leonhard WN, van de Wal A, et al. Transgenic mice expressing tamoxifen-inducible Cre for somatic gene modification in renal epithelial cells. Genesis 2006; 44:225–32.
28. Imai T, Jiang M, Chambon P, Metzger D. Impaired adipogenesis and lipolysis in the mouse upon selective ablation of the retinoid X receptor alpha mediated by a tamoxifen-inducible chimeric Cre recombinase (Cre-ERT2) in adipocytes. Proc Natl Acad Sci USA 2001; 98:224–8.
29. Schuler M, Dierich A, Chambon P, Metzger D. Efficient temporally controlled targeted somatic mutagenesis in hepatocytes of the mouse. Genesis 2004; 39:167–72.
30. Hirrlinger PG, Scheller A, Braun C, Hirrlinger J, Kirchhoff F. Temporal control of gene recombination in astrocytes by transgenic expression of the tamoxifen-inducible DNA recombinase variant CreERT2. Glia 2006; 54:11–20.
31. Casper KB, Jones K, McCarthy KD. Characterization of astrocyte-specific conditional knockouts. Genesis 2007; 45:292–9.
32. Mori T, Tanaka K, Buffo A, Wurst W, Kuhn R, Gotz M. Inducible gene deletion in astroglia and radial glia – a valuable tool for functional and lineage analysis. Glia 2006; 54:21–34.
33. Imayoshi I, Ohtsuka T, Metzger D, Chambon P, Kageyama R. Temporal

regulation of Cre recombinase activity in neural stem cells. Genesis 2006; 44:233–8.

34. Leone DP, Genoud S, Atanasoski S, et al. Tamoxifen-inducible glia-specific Cre mice for somatic mutagenesis in oligodendrocytes and Schwann cells.Mol Cell Neurosci 2003; 22:430–40.

35. Schuler M, Ali F, Metzger E, Chambon P, Metzger D. Temporally controlled targeted somatic mutagenesis in skeletal muscles of the mouse. Genesis 2005; 41:165–70.

36. Li M, Indra AK, Warot X, et al. Skin abnormalities generated by temporally controlled RXRalpha mutations in mouse epidermis. Nature 2000; 407:633–6.

37. Yajima I, Belloir E, Bourgeois Y, Kumasaka M, Delmas V, Larue L. Spatiotemporal gene control by the Cre-ERT2 system in melanocytes. Genesis 2006; 44:34–43.

38. Bosenberg M, Muthusamy V, Curley DP, et al. Characterization of melanocyte-specific inducible Cre recombinase transgenic mice. Genesis 2006; 44:262–7.

39. Seibler J, Zevnik B, Kuter-Luks B, et al. Rapid generation of inducible mouse mutants. Nucl Acids Res 2003; 31:e12.

40. Hameyer D, Loonstra A, Eshkind L, et al. Toxicity of ligand-dependent Cre-recombinases and generation of a conditional Cre-deleter mouse allowing mosaic recombination in peripheral tissues. Physiol Genomics 2007.

41. Santagati F, Minoux M, Ren SY, Rijli FM. Temporal requirement of Hoxa2 in cranial neural crest skeletal morphogenesis. Development 2005; 132:4927–36.

42. Gruber M, Hu CJ, Johnson RS, Brown EJ, Keith B, Simon MC. Acute postnatal ablation of Hif-2alpha results in anemia. Proc Natl Acad Sci USA 2007; 104:2301–6.

43. Ahn S, Joyner AL. Dynamic changes in the response of cells to positive hedgehog signaling during mouse limb patterning. Cell 2004; 118:505–16.

44. Zadelaar SM, Boesten LS, Pires NM, et al. Local cre-mediated gene recombination in vascular smooth muscle cells in mice. Transgenic Res 2006; 15:31–6.

45. Stratis A, Pasparakis M, Markur D, et al. Localized inflammatory skin disease following inducible ablation of I kappa B kinase 2 in murine epidermis. J Invest Dermatol 2006; 126:614–20.

46. Ouvrard-Pascaud A, Puttini S, Sainte-Marie Y, et al. Conditional gene expression in renal collecting duct epithelial cells: use of the inducible Cre-lox system. Am J Physiol Renal Physiol 2004; 286:F180–7.

Chapter 19

Creation and Use of a Cre Recombinase Transgenic Database

Andras Nagy, Lynn Mar, and Graham Watts

Abstract

In many cases, a gene "knockout" results in early embryonic lethality, which obscures the study of potential later functions. In other cases, the "knockout" does not show any phenotype due to the compensation of the gene deficiency by other family members. These limitations have called for further development of the powerful gene-targeting technology. One of the critical tools now being efficiently combined with gene-targeting is site-specific recombination. As the site-specific recombinase technology developed further in the mouse system, it became evident that this tool was going to have a significant impact on the power of mammalian genetics. The number of transgenic mouse lines expressing Cre recombinase with different specificities has steadily increased in the past 15 years and has now surpassed 500. Efficient utilization of this community-generated resource calls for a user-friendly database with all necessary information available about the properties of the Cre transgenic lines. The "CreXmice" database was created to meet these needs and has evolved over the past 4 years from flat file listings of transgenic lines into its current form, a professionally hosted SQL-driven web application. With hundreds of transgenic mouse lines, CreXmice is enriched by its presence on the World Wide Web allowing visitors the opportunity to search or contribute to the global effort by submitting the specific lines being developed by their laboratories.

Key words: Cre recombinase, transgenic mice, database.

1. Introduction

Embryonic stem (ES) cell-mediated transgenic approaches have revolutionized mammalian genetics during the last two decades. To date, close to 5000 genes have been targeted. Analysis of these genetic alterations has provided an unprecedented understanding of critical gene functions that underlie normal developmental and disease mechanisms in mammals. We have also learned, however, that mammalian genetic determination is complex: genes have multiple functions during the life of an individual. In addition,

Ralf Kühn, Wolfgang Wurst (eds.), *Gene Knockout Protocols: Second Edition, vol. 530*

DOI 10.1007/978-1-59745-471-1_19 Springerprotocols.com

mammalian genetic determination often utilizes gene families, in which the members have similar structure and overlapping expression and functions. These two phenomena contribute to some limitations of the gene-targeting approach. In many cases, a gene "knockout" results in early embryonic lethality, which obscures the study of potential later functions. In other cases, the "knockout" does not show any phenotype due to the compensation of the gene deficiency by other family members. These limitations have called for further development of the powerful gene-targeting technology *(1–3)*.

One of the critical tools now being efficiently combined with gene-targeting is site-specific recombination *(4)*. In fact, to date, three site-specific recombinases have been shown to work efficiently in mammalian systems. Two of them, the yeast-derived Flip (FLP) recombinase *(5)* and the P1 phage-derived Cre recombinase *(6)* recombine DNA between two 34-base pair (bp) homotypic sites, FRT and *loxP*, respectively *(7)*. Since the 34-bp consensus sequences are asymmetrical, their orientation is an important determining factor for the outcome of the recombination, so that recombination between similarly oriented recognition sites in the same DNA strand results in an excision of the intervening sequence (*see* **Fig. 19.1**). The recombination maintains the loxP or FRT sequence, which then could be further substrates for the recombinases. The third recombinase, PhiC31, is an integrase of the *Streptomyces* phage, which recombines between two

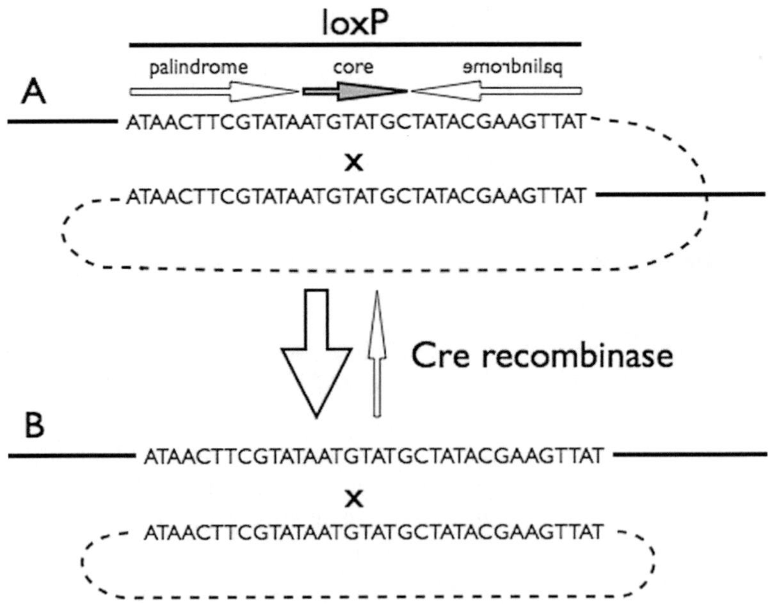

Fig. 19.1. **Principle mechanism of the Cre/loxP recombination system.** (**A**) Cre recombinase catalyzes recombination in the middle of the core sequence between two loxP sites. If the loxP sites are similarly oriented, the recombination results in an excision of the intervening sequence (**B**). The reaction is bidirectional, favoring the excision.

heterotypic sites, attB and attP. The resulting sites (attL and attB) are different sequences from attB and attP, and so they are no longer substrates for the enzyme. Although this system seems to be the least efficient of the three *(8)*, it is ideal for mediating site-specific integrations into attP or attB sites placed into the genome *(9)* or existing pseudo att sites that can be found in the human genome *(10)*.

The Cre/*loxP* system has proven to be the most efficient in mammalian systems, including ES cells. Therefore, it has been much more extensively applied than the FLP/FRT and PhiC31/attP/attB systems.

2. Cre Recombinase Action

Cre recombinase is an essential enzyme required for the reproductive cycle of the phage P1 *(7, 11)*. During the lysogenic phase, multiple copies of the phage genome are produced in the form of circular concatemers. Cre resolves these concatemers into multiple copies of single-phage genomes by restricting and covalently ligating two adjacent *loxP* sites, which flank a single genome equivalent (*see* **Fig. 19.1**).

loxP sites are not present in the mouse genome; it therefore has to be artificially introduced in order to provide a substrate to the enzyme. An important and intensive characterization of the enzyme action in mammalian tissue culture systems *(6, 7, 11–13)* preceded its in vivo applications. The first report illustrating that the recombinase, produced from a transgene, can mediate a very specific alteration in vivo in the mouse was published in 1992 *(14)*. Shortly after, we learned that it also works in ES cells *(15)*. These experiments ignited a burst of ideas and applications using the combination of gene-targeting and site-specific recombination (for review, *see (3)*). In this chapter we introduce the major trends through representative examples.

3. Combination with Transgenesis and Homologous Recombination

3.1. Conditional Genome Alterations

3.1.1. Conditional Transgenics

Transgenesis of mammals started with the pioneering work of Gordon and co-workers *(16)* who demonstrated that DNA when injected into the pronucleus of a fertilized egg can integrate into the genome, and can even express a delivered gene *(17, 18)* if designed to do so. During the past 30 years, this technology has brought us a broad spectrum of information about gene functions, regulation, and the consequence of mutations.

In the past, expression of mutant proteins has been particularly fruitful in determining the function of a normal gene *(19)*. However, mutations in many genes cause early developmental arrest, and, consequently, one cannot adequately address any later roles. Present and future objectives include expressing mutant proteins only in specific cell lineages and restricted to developmental periods of interest. Such approaches are invaluable, for instance, to better understand the role of oncogenes in certain organs, or to study the effects of somatic mutations.

One potential strategy takes advantage of the Cre/*loxP* system. Here, a tissue or temporal specific promoter is placed in front of two consecutive genes in such a way that it initially directs only the expression of the upstream (5′-most) gene (*see* **Fig. 19.2A**). This gene is flanked by *loxP* sites, so that it can be removed by Cre-mediated excision. Consequently, the downstream gene will now be expressed instead, and the expression will mimic that of the first gene. If a *lacZ* reporter gene is placed in the upstream position, it

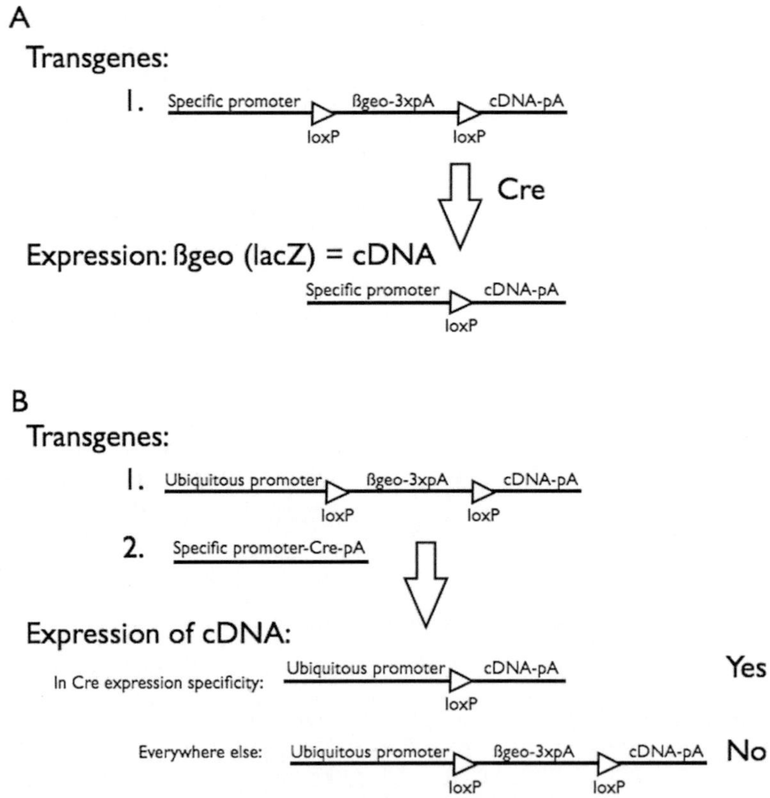

Fig. 19.2. **Typical use of Cre conditional transgenesis. (A)** The conditional transgene is expressing ßgeo, which allows for an easy characterization of gene expression specificity by simple staining of lacZ activity. When the loxP-flanked sequence is removed by Cre recombinase, the cDNA is moved under the transcriptional control of the promoter and expressed in the same manner as previously lacZ.

could facilitate the initial high-resolution characterization of the spatiotemporal expression pattern of a specific promoter by simple lacZ staining techniques. Another variation of this strategy is an arrangement shown in **Fig. 19.2B**. First, transgenic animals with a conditional transgene similar to the above, but under the control of a ubiquitous promoter, are generated. They are then crossed to Cre transgenic lines each displaying different expression patterns. The advantage here is that only the conditional transgenic lines need to be established *(20)*.

An important issue arising when applying such a conditional transgenic strategy is the concern about multiple copy and/or multiple sites of integration of the transgene. In this frequently occurring event, several copies of the *loxP*-containing plasmid integrate into the same and/or different regions of the genome with random orientations. In such an event, inverted *loxP* sites are easily created. These can cause severe chromosomal instability during Cre excision *(21)*, leading to aneuploidity and a potential phenotype that is unrelated to the transgene.

3.1.2. Conditional Knockouts

There are several solutions for creating conditional gene knockouts *(3)*. One of them combines gene-targeting and site-specific recombination. The general strategy here is to introduce *loxP* sites flanking a critical exon (CE) of the gene of interest. These intron-inserted sites should not disturb the coding regions, regulatory elements or proper splicing and therefore functioning of the gene. There is a concern, however, about the positive selectable marker that is also introduced to the modified locus. As a consequence, strategies have been developed for the removal of the marker after the targeted allele has been identified (*see* **Fig. 19.3**). The most common method to do so is to flank the selectable marker cassette with FRT site (*see* **Fig. 19.3A**) and then remove the intervening region with FLP recombinase *(22)*. Another successfully applied idea is to place the selectable marker right beside one of the loxP sites and put a third one on the other side of the marker (*see* **Fig. 19.3B**). A partial Cre excision could then result in removal of the selectable marker, leaving the loxP-flanked CE in place *(23)*.

The conditional alleles are expected to be functionally equivalent to wild type so that the homozygous animals should be normal. If, however, Cre recombinase is expressed in any of the cells, it specifically removes the essential exon from the modified alleles and creates a Cre-excision-dependent knockout. If a cell type-specific Cre transgene is utilized, only the lineage expressing Cre will be deficient for the gene *(24)*.

Almost all the ongoing high-throughput gene-targeting projects (EUCOMM, KOMP (Sanger component) and NorCOMM) are generating conditional alleles. The EUCOMM and KOMP (Sanger) is using the strategy shown in **Fig. 19.3A** for genes that

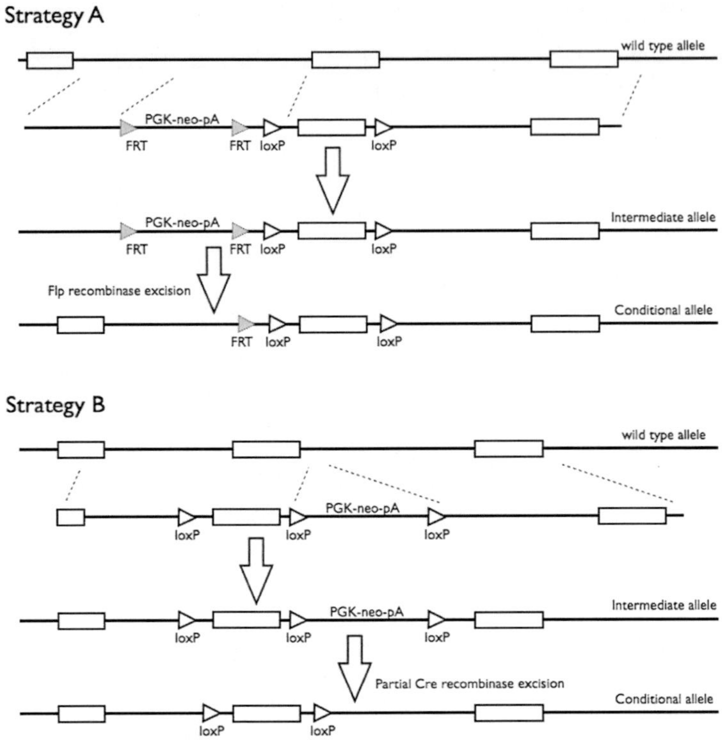

Fig. 19.3. **Two basic strategies to generate Cre recombinase conditional alleles from a targeted gene**. Strategy **A** shows the combination of Cre/loxP and Flp/FRT systems and Strategy **B** depicts the use of three loxP sites to flank a critical exon.

are not expressed in ES cells. Their target vector is slightly different for genes which are expressed in ES cell. Here they use a splice acceptor instead of the PGK promoter to drive a lacZ-neo fusion (ßgeo *(25)*) gene with the endogenous gene expression. The Canadian NorCOMM project is generating a rather unique allele, which is null at the first place. The exogenous DNA segment replacing the CE in the targeted allele, however, is fully replaceable by any sequence. This replacement is mediated by PhiC31 integrase and Flp recombinase. This exchange has high fidelity and no background random integrations can be detected. Therefore no subline cloning and genotyping are necessary to check the correctness of the replacement. The reintroduction of the CE with flanking loxP sites generates a Cre conditional allele (*see* **Fig. 19.4**). With the same simplicity, the primary null allele can be modified in any imaginable way, for example, the insertion of the Cre gene into the targeted locus will "borrow" the endogenous spatial and temporal gene expression for the recombinase.

3.1.3. Conditional Gene Repair

The idea of conditional gene repair is similar to that of the conditional knockout. However, the logic here is complementary. The aim is to create a gene knockout with an insertion into a gene of

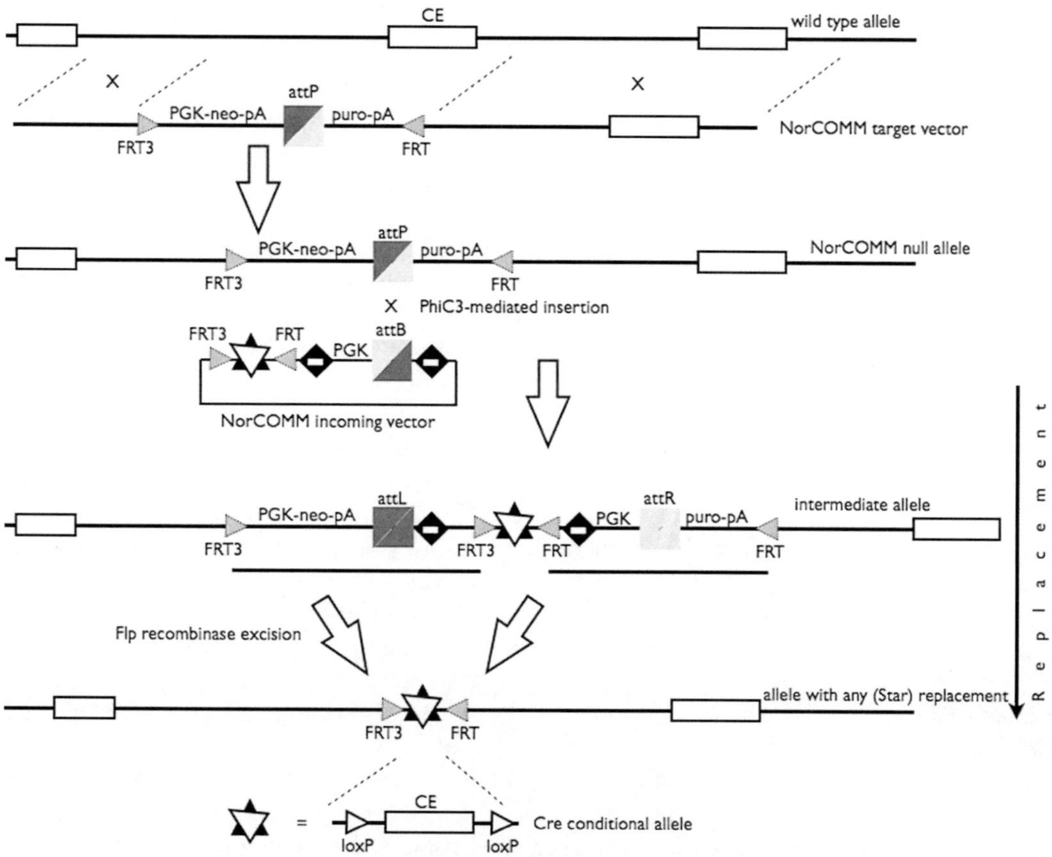

Fig. 19.4. **The NorCOMM versatile gene-targeting strategy.** This approach provides an easy replacement of the null allele with any imaginable sequence by combining PhiC31 integrase-mediated insertion followed by Flp recombinase-mediated removal of the no-longer-needed selectable markers.

interest. However, the insertion has three special features: (1) it interferes with the gene's expression; (2) it is flanked by *loxP* or FRT sites; and (3) after Cre-mediated excision, normal gene function is restored. When such a gene knockout is introduced into the germline, the consequence of the gene deficiency can be characterized, and the primary lineage affected determined. If there is an embryonic lethality related to the primary defect, a Cre transgenic line with specific expression in this particular lineage is crossed with the knockout. Due to the excision of the interfering sequence, the gene is repaired in situ in the primarily affected lineage where the Cre is expressed, resulting in rescue of the primary phenotypic defect. The embryo can thus survive longer, until the chronologically second phenotype is manifest. In theory, this methodology should provide an automatic guide through the multiple functions of a gene. Such an in situ gene repair has been demonstrated for an *N-myc* hypomorphic allele *(26)*. There are several ways to achieve this modification of a gene by targeting. One of these is to replace a NorCOMM allele with a loxP-STOP-loxP CE (*see* **Fig. 19.4**) sequence. Here, the STOP region could be a splice acceptor – reporter (lacZ, GFP, etc.) followed by a strong pA. In this case, transcription of the targeted gene is disrupted until the point when Cre excision of the *loxP*-flanked marker allows for transcriptional and/or translational read-through.

An important difference between conditional gene knockout and repair is their different requirement for the fidelity of the Cre-mediated excision. If, for example, a Cre recombinase transgene excises only in 70% of the cells of the desired lineage, the remaining 30% wild-type equivalent could make analysis of the phenotype extremely complex. Even a weakly mosaic lineage (with respect to wild type) could be sufficient to rescue any phenotypic anomaly. On the other hand, if 70% of cells are repaired in a lineage, it will almost certainly be enough to rescue the phenotype. We expect that numerous Cre transgenic lines will be produced with incomplete Cre-mediated excision in the target lineage. Therefore, many of the lines, although not suitable for most conditional knockout studies, can be useful for conditional repairs.

3.2. Inducible Genome Alterations

The real power of Cre-mediated genome alterations will be realized when temporal regulation of the recombinase becomes possible. Efforts are now moving toward this direction. Both the tetracycline-inducible *(27)*, and the mutant estrogen receptor, tamoxifen-inducible systems *(28, 29)* have been combined with Cre-mediated, site-specific recombination. These experiments have already demonstrated the possibility of inducible Cre recombinase activity and, as a consequence, inducible Cre-mediated excision. The goal for general applicability is to reach minimal or preferably no excision in the noninduced state and the possibility of complete excision in the induced state. An interesting allelic combination allows temporally controlled "self" knockout, when a Cre conditional (floxed) allele is combined with a tamoxifen-inducible CreER recombinase knock-in allele to the homologous locus. This makes "trans" self-knockout possible at any developmental stage *(30)*.

3.3. Introduction of Subtle Changes

Introduction of a subtle change, such as a point mutation, into a gene by homologous recombination is possible by embedding the particular mutation into the target vector. In this case, however, removal of the positive selectable marker is required. The marker can be removed either by transient expression of Cre in the targeted ES cells or after germline transmission of the marker-containing allele. In the latter, crossing the germline chimera to Cre transgenic animals is less labor intensive *(26–29, 31–33)*. The Cre transgene should subsequently be segregated from the targeted allele to maintain the clarity of the experiment. A third way to introduce Cre recombinase is to inject its expression vector as a circular plasmid into a heterozygous F1 zygote. This was reported as a very efficient way of removing *loxP*-flanked segments in the few kilobase range *(34)*. The advantage of this latter method is that the Cre is only transiently expressed during preimplantation stages and does not leave a transgene behind.

3.4. Large Deletions/Duplications and Chromosomal Translocations

The Cre recombinase can also work on two *loxP* sites located at long distances from each other. If two sites with the same orientation are placed on the same chromosome but a few megabases apart, and Cre recombinase is expressed in the cell, excision occurs. Since the frequency is low, a properly designed selection system is required to identify such an event. The most commonly used marker for this technology is a split hypoxanthine phosphoribosyltransferase (HPRT) minigene in combination with an HPRT-negative cell line. Neither of the two halves of the HPRT selectable marker renders cells HPRT positive. The 5′ part of HPRT is placed at one end and the 3′ part at the other of the planned deletion on the same chromosome. The deletion brings the two HPRT pieces *(35)* together and allows the cells to survive in hypoxanthine, aminopterin and thymidine (HAT) medium (*see* **Fig. 19.5A**). If the two *loxP* sites happen to be on homologous chromosomes, the recombination results in a duplication in one chromosome and a deletion on the other (*see* **Fig. 19.5B**).

If two *loxP* sites are placed on different and nonhomologous chromosomes, and their orientation is the same relative to their centromere, a reciprocal translocation occurs (*see* **Fig. 19.5C**). Again, the HPRT selection system can identify this rare event *(36)*.

3.5. A Database of Cre Recombinase Transgenic Mouse Lines

As the site-specific recombinase technology developed further in the mouse system, it became evident that this tool was going to have a significant impact on the power of mammalian genetics. The number of transgenic mouse lines expressing Cre recombinase with different specificities has steadily increased in the past 15 years and has now surpassed 500. Efficient utilization of this community-generated resource calls for a user-friendly database with all necessary information available about the properties of the Cre transgenic lines.

The "CreXmice" database (publicly available online at http://www.mshri.on.ca/nagy/) was created to meet these needs and has evolved over the past 4 years from flat file listings of transgenic lines into its current form, a professionally hosted SQL-driven web application. With hundreds of transgenic mouse lines, CreXmice is enriched by its presence on the World Wide Web allowing visitors the opportunity to search or contribute to the global effort by submitting the specific lines being developed by their laboratories.

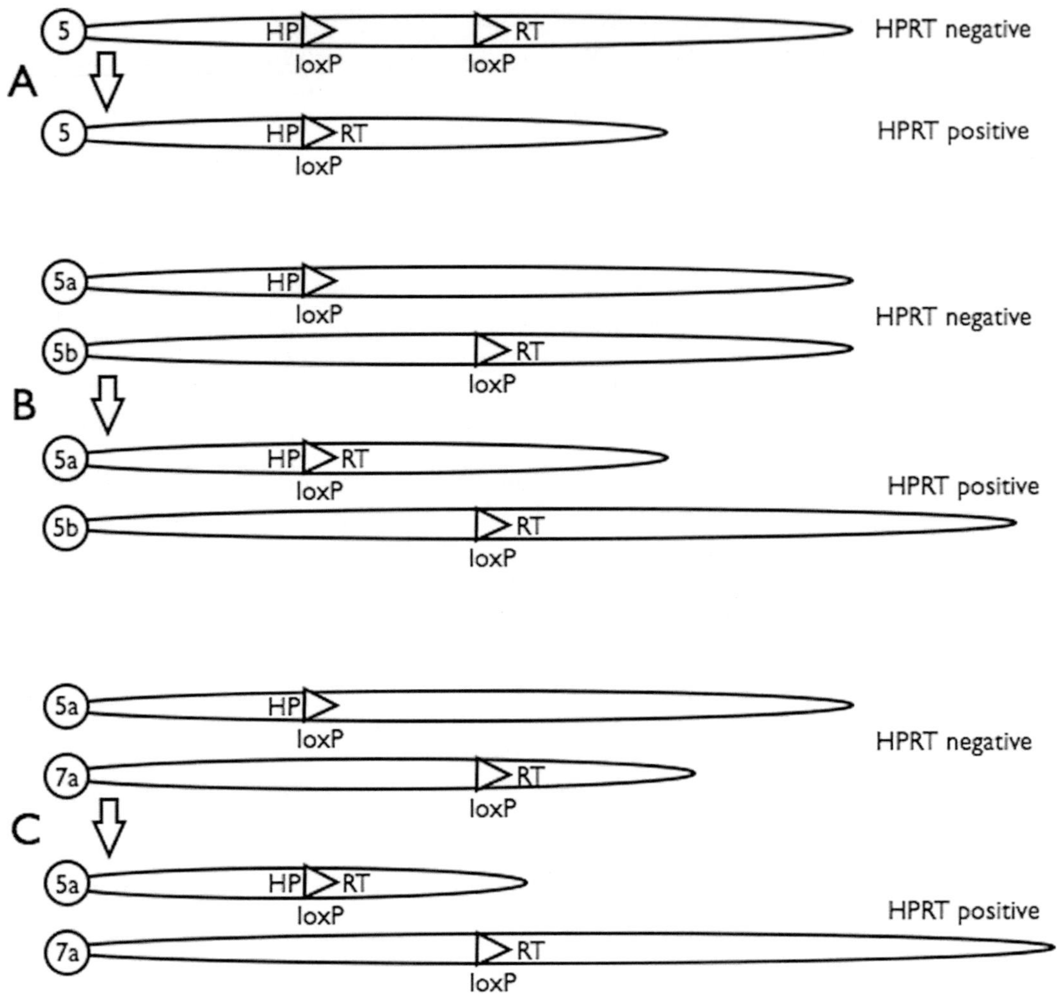

Fig. 19.5. **Creation of large deletions, duplications, and chromosomal translocations by Cre/loxP recombination.** See text for explanations.

Submissions are accepted through a three-step process where users are first asked to detail overall characteristics of the transgenic line including the transgene type, promoter, inducibility, genetic background, and publication information.

The transgenic line information submission is then continued by the user entering the specificities of Cre expression or excision. This second step is a tired approach that utilizes the Edinburgh Mouse Atlas Project's (EMAP) descriptions of anatomical areas at different developmental stages. To describe the specificity, the embryonic or postnatal stage where specificity characterization was performed is selected and further described by cell type, and where Cre excision/expression occurred. Given the previously selected embryonic stage, users then search for an anatomical area that is known to be present during this time of development.

This helps in providing a precise and accurate definition of the specificity. However, if an anatomical area does not exist in the database, the option for manual specification is available. Manually specified areas (if approved by the database curator) are remembered by the application to be used in future submissions, thus growing the EMAP database effortlessly. In fact, this concept of growing not just the number of transgenic lines but also specification data is used throughout the CreXmice web application and is necessary for such a resource to remain current.

The third and final step requests simple name and e-mail information about the authoritative contact for the transgenic line. This, along with the submitter's comments and information on holding sites is then attached to the submission for public viewing.

The quality and depth of characterization of Cre transgenic mouse lines – considering the published papers – are very diverse. Therefore, the effort to standardize the representation of this data has been challenging, and despite a streamlined approach, the database curator holds every submission for review before making it available for visitors on the site.

Visitors to the CreXmice database are invited to use the advanced search functions to identify and locate a transgenic mouse line that is of interest to them. Search results display the most relevant data of a transgenic mouse line, some of which are represented through keyed icons. The use of icons in place of text descriptions provides a clean, intuitive and user-friendly interface. Complete information of a search result is available by clicking the relative link.

Having efficiently standardized the representation of Cre transgenic mouse lines and their respective publications, CreXmice intends to take a further step by promoting an even more interactive and collaborative environment. Further updates to CreXmice will include the feature for registered users to post comments of their experiences with a particular transgenic line and even modify existing entries (*see* **Fig. 19.6**). This open philosophy can

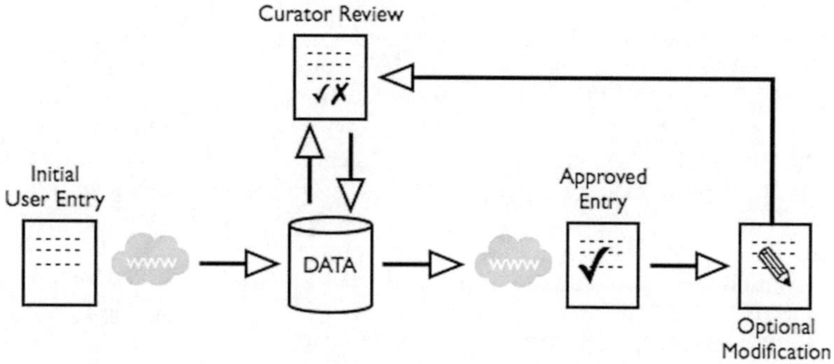

Fig. 19.6. **Block diagram representation of the CreXmice database structure.**

be compared to that of modern-day collaboration tools on the Internet such as Wikis. An opt-in mailing list will be available to update subscribers of new or updated submissions or user comments through e-mail.

4. Prologue

There are an increasing number of publications reporting the successful combination of site-specific recombinase and ES cell-mediated genetics, clearly demonstrating that our initial enthusiasm was well founded. Almost any imaginable genomic alteration can now be produced in mice. For example, we can now create virtually any phenocopy of defects associated with human diseases; from chromosomal aberrations to missense mutations.

The critical tools for reaching this goal are the ones needed for the production of conditional transgenes or conditional mutations, and the generation of Cre transgenic lines with specificity of increasing resolution. The first aspect will soon be achieved by the high-throughput gene knockout projects (*see* **Chapter 1**), and the second will be created by the worldwide research community. The need for user-friendly databases with priority on optimized information presentation will be vitally important for this process. The availability of these databases, such as CreXmice, as shared resources for a wide range of applications will provide far-reaching insights in understanding gene functions and diseases.

Acknowledgments

A.N. is Tier 1, Canadian Research Chair.

References

1. Nagy A, Rossant J. Targeted mutagenesis: analysis of phenotype without germ line transmission. J Clin Invest 1996; 97:1360–5.
2. Rossant J, Nagy A. Genome engineering: the new mouse genetics. Nat Med 1995; 1:592–4.
3. Lobe CG, Nagy A. Conditional genome alteration in mice. Bioessays 1998; 20:200–8.
4. Kilby NJ, Snaith MR, Murray JA. Site-specific recombinases: tools for genome engineering. Trends Genet 1993; 9:413–21.
5. Dymecki SM. Flp recombinase promotes site-specific DNA recombination in embryonic stem cells and transgenic mice. Proc Natl Acad Sci USA 1996; 93:6191–6.
6. Sauer B, Henderson N. Site-specific DNA recombination in mammalian cells by the Cre recombinase of bacteriophage P1. Proc Natl Acad Sci USA 1988; 85:5166–70.
7. Argos P, Landy A, Abremski K, et al. The integrase family of site-specific recombinases: regional similarities and global diversity. Embo J 1986; 5:433–40.

8. Andreas S, Schwenk F, Kuter-Luks B, Faust N, Kuhn R. Enhanced efficiency through nuclear localization signal fusion on phage PhiC31-integrase: activity comparison with Cre and FLPe recombinase in mammalian cells. Nucleic Acids Res 2002; 30:2299–306.
9. Belteki G, Gertsenstein M, Ow DW, Nagy A. Site-specific cassette exchange and germline transmission with mouse ES cells expressing PhiC31 integrase. Nat Biotechnol 2003; 21:321–4.
10. Bertoni C, Jarrahian S, Wheeler TM, et al. Enhancement of plasmid-mediated gene therapy for muscular dystrophy by directed plasmid integration. Proc Natl Acad Sci USA 2006; 103:419–24.
11. Hoess RH, Abremski K. Mechanism of strand cleavage and exchange in the Cre-lox site-specific recombination system. J Mol Biol 1985; 181:351–62.
12. Sauer B, Henderson N. Cre-stimulated recombination at loxP-containing DNA sequences placed into the mammalian genome. Nucleic Acids Res 1989; 17:147–61.
13. Sauer B, Henderson N. Targeted insertion of exogenous DNA into the eukaryotic genome by the Cre recombinase. New Biol 1990; 2:441–9.
14. Lakso M, Sauer B, Mosinger B, Jr., et al. Targeted oncogene activation by site-specific recombination in transgenic mice. Proc Natl Acad Sci USA 1992; 89:6232–6.
15. Gu H, Zou YR, Rajewsky K. Independent control of immunoglobulin switch recombination at individual switch regions evidenced through Cre-loxP-mediated gene targeting. Cell 1993; 73:1155–64.
16. Gordon JW, Scangos GA, Plotkin DJ, Barbosa JA, Ruddle FH. Genetic transformation of mouse embryos by microinjection of purified DNA. Proc Natl Acad Sci USA 1980; 77:7380–4.
17. Brinster RL, Chen HY, Trumbauer M, Senear AW, Warren R, Palmiter RD. Somatic expression of herpes thymidine kinase in mice following injection of a fusion gene into eggs. Cell 1981; 27:223–31.
18. Brinster RL, Chen HY, Trumbauer ME. Mouse oocytes transcribe injected *Xenopus* 5S RNA gene. Science 1981; 211:396–8.
19. Palmiter RD, Brinster RL. Germ-line transformation of mice. Annu Rev Genet 1986; 20:465–99.
20. Grieshammer U, Lewandoski M, Prevette D, Oppenheim RW, Martin GR. Muscle-specific cell ablation conditional upon Cre-mediated DNA recombination in transgenic mice leads to massive spinal and cranial motoneuron loss. Dev Biol 1998; 197:234–47.
21. Lewandoski M, Martin GR. Cre-mediated chromosome loss in mice. Nat Genet 1997; 17:223–5.
22. Meyers EN, Lewandoski M, Martin GR. An Fgf8 mutant allelic series generated by Cre- and Flp-mediated recombination. Nat Genet 1998; 18:136–41.
23. Gu H, Marth JD, Orban PC, Mossmann H, Rajewsky K. Deletion of a DNA polymerase beta gene segment in T cells using cell type-specific gene targeting. Science 1994; 265:103–6.
24. Tsien JZ, Huerta PT, Tonegawa S. The essential role of hippocampal CA1 NMDA receptor-dependent synaptic plasticity in spatial memory. Cell 1996; 87:1327–38.
25. Friedrich G, Soriano P. Promoter traps in embryonic stem cells: a genetic screen to identify and mutate developmental genes in mice. Genes Dev 1991; 5:1513–23.
26. Nagy A, Moens C, Ivanyi E, et al. Dissecting the role of N-myc in development using a single targeting vector to generate a series of alleles. Curr Biol 1998; 8:661–4.
27. St-Onge L, Furth PA, Gruss P. Temporal control of the Cre recombinase in transgenic mice by a tetracycline responsive promoter. Nucleic Acids Res 1996; 24:3875–7.
28. Brocard J, Warot X, Wendling O, et al. Spatio-temporally controlled site-specific somatic mutagenesis in the mouse. Proc Natl Acad Sci USA 1997; 94:14559–63.
29. Schwenk F, Kuhn R, Angrand PO, Rajewsky K, Stewart AF. Temporally and spatially regulated somatic mutagenesis in mice. Nucleic Acids Res 1998; 26:1427–32.
30. Fossat N, Chatelain G, Brun G, Lamonerie T. Temporal and spatial delineation of mouse Otx2 functions by conditional self-knockout. EMBO Rep 2006; 7:824–30.
31. Jacks T, Shih TS, Schmitt EM, Bronson RT, Bernards A, Weinberg RA. Tumour predisposition in mice heterozygous for a targeted mutation in Nf1. Nat Genet 1994; 7:353–61.
32. Moens CB, Auerbach AB, Conlon RA, Joyner AL, Rossant J. A targeted mutation reveals a role for N-myc in branching morphogenesis in the embryonic mouse lung. Genes Dev 1992; 6:691–704.
33. Sakai K, Miyazaki J. A transgenic mouse line that retains Cre recombinase activity in

mature oocytes irrespective of the cre transgene transmission. Biochem Biophys Res Commun 1997; 237:318–24.

34. Araki K, Araki M, Miyazaki J, Vassalli P. Site-specific recombination of a transgene in fertilized eggs by transient expression of Cre recombinase. Proc Natl Acad Sci USA 1995; 92:160–4.

35. Ramirez-Solis R, Liu P, Bradley A. Chromosome engineering in mice. Nature 1995; 378:720–4.

36. Smith AJ, De Sousa MA, Kwabi-Addo B, Heppell-Parton A, Impey H, Rabbitts P. A site-directed chromosomal translocation induced in embryonic stem cells by Cre-loxP recombination. Nat Genet 1995; 9:376–85.

Chapter 20

Transposon Mutagenesis in Mice

David A. Largaespada

Abstract

Understanding the functional landscape of the mammalian genome is the next big challenge of biomedical research. The completion of the first phases of the mouse and human genome projects, and expression analyses using microarray hybridization, generate critically important questions about the functional landscape and structure of the mammalian genome: how many genes, and of what type, are there; what kind of functional elements make up a properly functioning gene? One step in this process will be to create mutations in every identifiable mouse gene and analyze the resultant phenotypes. Transposons are being considered as tools to further initiatives to create a comprehensive resource of mutant mouse strains. Also, it may be possible to use transposons in true forward genetic screens in the mouse. The "Sleeping Beauty" (SB) transposon system is one such tool. Moreover, due to its tendency for local hopping, SB has been proposed as a method for regional saturation mutagenesis of the mouse genome. In this chapter, we review the tools and methods currently available to create mutant mice using in vivo, germline transposition in mice.

Key words: Sleeping Beauty, transposon, mouse transgenesis, insertional mutagenesis, germline mutagenesis.

1. Introduction

1.1. Transposable Elements

There are two major classes of transposable elements. Class I transposons, also called retrotransposons, are mobilized through an RNA intermediate in a "copy-and-paste" manner. Thus, the donor site remains intact during the retrotransposition process. The LINE1 class of transposable elements is representative of this group and is active in many mammalian species including the mouse and human *(1)*. Class II transposons move without going through an RNA intermediate. Instead, these transposons are mobilized by a "cut-and-paste" transposition process. Characteristic of class II

Ralf Kühn, Wolfgang Wurst (eds.), *Gene Knockout Protocols: Second Edition, vol. 530*

DOI 10.1007/978-1-59745-471-1_20 Springerprotocols.com

transposons is an encoded transposase gene required for the transposition process. These transposase proteins usually recognize sequences within inverted terminal repeats that flank the transposon DNA and which are required for transposition. One large family of class II transposons is composed of Tc1/mariner elements *(2)*. Tc1/mariner family elements are flanked by inverted terminal repeats and contain an encoded transposase gene with a paired-like DNA-binding domain and "DDE" catalytic domain *(2)*. This family of transposable elements has been a source of several transposon systems that can be used to engineer vertebrate genomes.

1.2. The Sleeping Beauty (SB) Transposon System

The "Sleeping Beauty" (SB) transposon system, derived from inactive Tc1/mariner family transposable elements is currently the most well-studied "cut-and-paste" transposon available for use as a general gene transfer and insertional mutagenesis tool. Two parts of the SB system must be supplied to a cell for transposition to occur – the transposon vector DNA and the enzyme that mobilizes the vector DNA, the transposase. The SB transposase gene, the first version of which is called SB10, was created by derivation of a functional enzyme gene from multiple disabled, mutant copies of Tc1/mariner family transposase genes. These Tc1/mariner transposase genes had been cloned from the genomes of various species of salmonid fish, where they lay dormant for at least 10 million years *(3)*. In order for the SB transposase to catalyze transposition, a sequence must be flanked by special sequences called inverted repeat/direct repeat (IR/DR) elements. The left and right IR/DRs consist of very similar inverted terminal sequences but they are distinct functionally *(4)*. Moreover each IR/DR contains an inner and an outer DR element (hence IR/DR). These DRs are the binding sites for the transposase protein. Transposon vectors that contain the original IR/DRs built from sequences from salmonid fish are designated "pT" vectors *(3)*. However, sequence changes have been introduced into the IR/DRs that increase transposition rates somewhat. Transposon vectors that are based on this second generation of IR/DRs are designated "pT2" vectors *(5)*. Also, new versions of the SB transposase protein with improved catalytic activity have been developed in which one or more amino acid substitutions are introduced *(6, 7)*.

A variety of insertional mutagenesis applications have been reported for SB including germline mutagenesis in the mouse and zebrafish *(8)* and most recently somatic mutagenesis for cancer gene discovery in the mouse *(9, 10)*. Several other transposable element systems have now been shown to be active in the mouse germline, including *Minos (11)* and *PiggyBac (12, 13)*. However, germline mutagenesis using SB has been most thoroughly described in the literature *(14–22)*. The lessons learned using

SB might in large part apply to these other transposon systems. Therefore, in this chapter we will focus on the use of SB for germline mutagenesis.

This chapter will detail the general methods used to create germline transposon insertion mutations using SB in the laboratory mouse. It is worth pointing out that it has recently been shown that SB germline mutagenesis is also very efficient in the laboratory rat *(23, 24)*. Hence, the methods in this chapter apply also to the rat. Included in the description are the transgenic lines that have been described and that might be made for such a purpose. In addition, special concerns for isolation and analysis of the most useful mutant alleles will be described. This is a complex process from the conceptual point of view, yet in practice is easy to carry out. Using transposons for germline mutagenesis requires no special cell culture technologies or the use of chemicals. In fact, the techniques are similar in concept to paths that are well-worn in other model genetic organisms such as *Drosophila melanogaster (25)*. Because this technology utilizes random insertion of transposon vectors to create mutant alleles, the choice of the internal components of the vector is especially important. Many possible insertion mutations might be created and so all these are not discussed in depth in this chapter. Instead, the reader is referred to published reviews on insertional mutagenesis strategies for in-depth discussion of vector construction *(26, 27)*.

1.3. Creating Transposon Insertion Mutations Using Sleeping Beauty

Transposon design. The first component needed to achieve transposition using SB is a transgenic line of mice carrying the transposon vector to be used as the insertional mutagen. In order to create germline mutations using SB it is by far most efficient to use multi-copy transgene arrays carrying the transposon vector as donors for transposition. Single-copy transposon vectors transpose infrequently in the germline of mice, ranging from once in five gametes to less than once in 100 gametes *(16, 20)*. In contrast, most multi-copy transgene arrays yield 0.5–3 new transposon insertions per gamete *(14, 16, 18)*. This increased rate of transposition seems to be due to more than just an increase in the amount of substrate for transposition *(18)* and may be due to methylation of transposon DNA in a multi-copy array *(28)*. Thus, we and others have used transposon vector transgenes produced using standard pronuclear injection into FVB/n or C57BL/6 J mice, although other strain backgrounds should also work.

The rate of germline transposition achieved in these experiments is critical. This rate determines the number of new insertion mutations in genes that can be induced per offspring. SB transposon vectors have been shown to insert into new sequences at TA dinucleotides *(3)*. Although the sequences immediately adjacent to the TA dinucleotide do influence the likelihood of insertion, on a genomic scale SB insertion is essentially random *(29)*. The one exception to this randomness is the tendency of SB transposon vectors to insert near the donor locus after transposition. This phenomenon is called local hopping *(16, 22)*. Local hopping, and associated genomic damage due to deletions, translocations, and other events will be discussed in more detail below.

The transposon vector used for mutagenesis must be chosen with the desired downstream applications in mind. All described SB vectors, used for germline mutagenesis, include a so-called "5′ gene trap" including splice acceptors and polyadenylation sequences in one or both orientations so that genes can be disrupted upon insertion into an intron *(22)*. In addition, some vectors include a reporter gene (GFP or LacZ) expressed from the chimeric mRNA generated by splicing of the endogenous transcript into the transposon vector *(22)*. These reporters have generally been preceded by an internal ribosome entry site (IRES) so that the reporter can be translated and expressed regardless of the reading frame of the disrupted gene. In one report from our lab, the tTA transcription factor was used as a reporter so that a tet-regulated element (TRE) reporter could be used in conjunction with gene-trap alleles produced with this SB vector *(18)*. A second component of some SB transposon vectors is a so-called "poly-A trap" immediately downstream of the 5′ gene trap. The poly-A trap is composed of an internal promoter, a reporter gene, and a splice donor. In the absence of nearby downstream exons and a polyadenylation signal, the poly-A trap produces an unspliced and non-polyadenylated RNA that is not exported from the nucleus efficiently and is unstable. If the transposon vector has landed in a transcription unit, in the same orientation as that transcription unit, then the RNA transcript initiated by the internal promoter splices to endogenous exons, is processed, and polyadenylated. Thus, the message is stabilized and the reporter is expressed. For SB vectors, the internal promoter that has been shown to be useful for this purpose is the CAGGS promoter *(22)*, which is composed of sequences from the chicken beta actin and human cytomegalovirus immediate early promoters and can be expressed ubiquitously in some transgenic mice *(30)*. By using the GFP gene as a the reporter for the poly-A trap the Takeda lab at Osaka University has succeeded in developing a system that can be used to identify generation 1 (G1) mice with a new transposon insertion into a gene just by visual inspection of neonatal pups for GFP expression *(22)*. This system can thus be used to identify those offspring most likely to have an SB-induced gene mutation. A generalized SB transposon vector, which combines both 5′ and poly-A gene traps is shown in **Fig. 20.1**.

Transposase transgenes. Three different sources of SB transposase have been used in published work on SB-induced germline insertion mutations. A CAGGS-SB10 transgene has been used by several labs to catalyze germline transposition *(14, 21)*. Transgenes

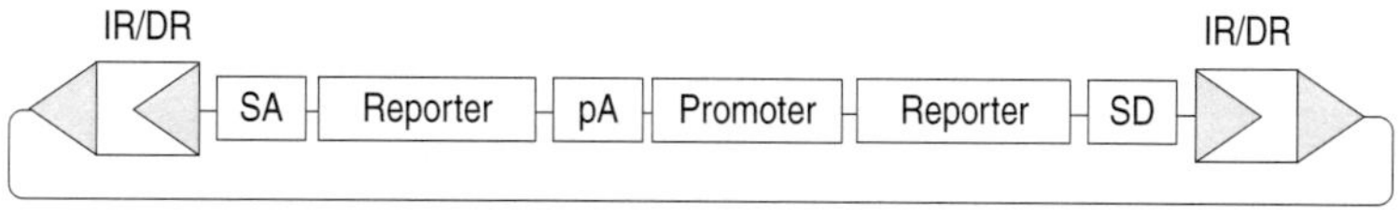

Fig. 20.1 A generic SB transposon vector is shown. The left and right inverted terminal repeat/direct repeat (IR/DR) sequences must flank the transposon vector. Inside a 5′ gene trap consists of a splice acceptor (SA), internal ribosome entry site (IRES), reporter transgene, and polyadenylation (pA) site. This portion of the vector will interrupt the endogenous gene's mRNA processing and result in expression of the reporter in the same spatiotemporal pattern as the disrupted gene. A second gene-trapping cassette, called the poly-A trap is included. The poly-A trap includes an internal promoter (CAGGS has been shown to work *(22)*), a reporter, and splice donor (SD), but no polyadenylation site. The poly-A trap reporter is expressed, and is expressed ubiquitously, only if the transposon lands in a transcription unit in the same orientation as that transcription unit. Thus, G1 mice likely to have a gene disruption can be identified by looking for expression of the poly-A trap reporter. GFP is useful as a reporter for this purpose *(22)*.

created using the CAGGS promoter are reported to be ubiquitously expressed *(31)*. We have found that the CAGGS promoter is not ubiquitously expressed in all transgenic lines, but is expressed at very high levels in the developing germ cells of male mice in some transgenic lines (unpublished data). In one report, the mouse protamine 1 gene promoter (*Prm1*) has been used to drive SB10 expression and catalyze male germline transposition at good rates *(20)*. Finally, a catalytically improved SB11 transposase cDNA under the control of the endogenous *Rosa26* promoter *(32)* can be used to drive germline SB transposition at good rates (our unpublished data). In our experience both CAGGS-SB10 and Rosa26-SB11 transgenes catalyze SB transposon vectors in the male germline at similar rates, with a slight advantage to the CAGGS-SB10 transgene. The Prm1-SB10 transgene seems to work at rates similar to the CAGGS-SB10 transgene *(20)*. At present, the CAGGS-SB10 transgene is the standard for germline mutagenesis. It should be mentioned that improved SB transposases, generated by amino acid substitutions, are being generated by several labs *(7)*. These improved SB transposases promise to allow germline transposition rates much higher than three new insertions per gamete.

Generation of doubly transgenic "seed males" for SB mutagenesis. Once suitable transposon and transposase transgenic lines have been selected the next step is to breed mice together to obtain males that carry both transgenes. These doubly transgenic males are referred to as "seed males" in the literature because they are the source of sperm carrying new SB transposon vector insertions due to ongoing transposition in the developing sperm of these mice *(18, 22)*. The seed males are bred to wild-type females to generate G1 offspring carrying new transposon insertions. The general outline of this procedure is shown in **Fig. 20.2**. We have

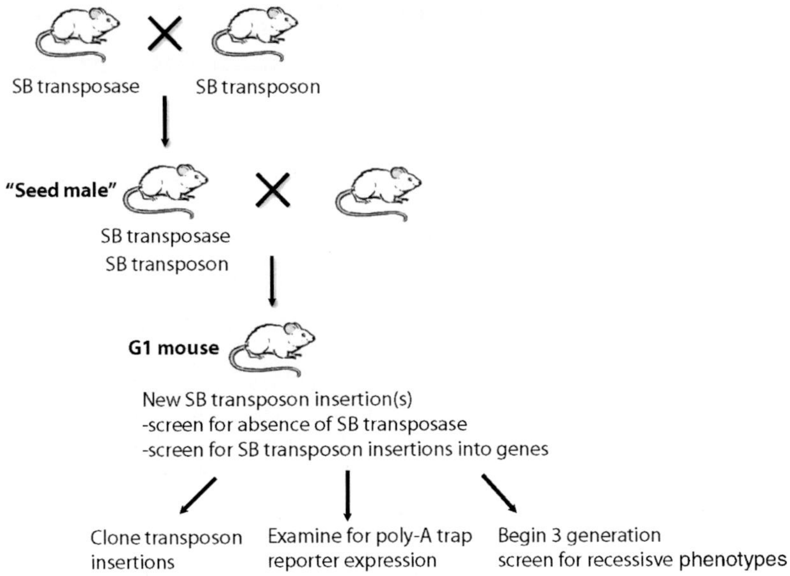

Fig. 20.2. Outline of the crosses and steps used to create and analyze SB transposon insertion mutations. Doubly transgenic seed males, carrying both the transposase and transposon transgenes are generated. These males are bred to wild-type females to generate G1 offspring. The G1 offspring are screened for the absence of the SB transposase transgene and the presence of gene mutations by (1) cloning transposon insertions, (2) expression of 5′or poly-A trap reporters, or (3) entering the G1 mice into a three-generation screen for recessive mutant phenotypes.

found that seed males carrying identical transgenes can vary in the rate of observed germline transposition *(18)*. A single male may vary in germline transposition rate during its lifetime as well (unpublished observations). For this reason, it is advised that multiple seed males should be generated and test litters generated from each male should be screened by Southern blot to identify those males with the highest germline transposition rate *(18)*. Any one male is estimated to produce among its sperm at least 10,000 different transposon insertion mutations *(22)*.

The analysis of G1 offspring including cloning and sequence analysis of transposon insertion sites. In general, the G1 generated from crossing a seed male to a wild-type female are screened for the presence of new transposon insertions into genes by any of several methods. If a CAGGS-GFP-SD poly-A trap was included in the vector, then GFP+ offspring are sought *(22)*. If a visible marker was not included, then new transposon insertions can be sought via Southern blotting and/or cloned using inverse polymerase chain reaction (PCR) or linker-mediated PCR *(16)*. An analysis of the transposon insertion site can be done, by comparison to the draft mouse genome sequence (http://www.ensembl.org) to determine if the insertion is likely to have impaired gene function.

Finally, in one project from my own laboratory, we have bred G1 mice to create G2 mice, which were then intercrossed to determine if a mutant phenotype could be uncovered in mice homozygous for an SB mutagenized chromosome *(17)*. It should be emphasized that any G1 mouse may harbor multiple, independent SB transposon insertion mutations. As mentioned above, roughly 50%–80% of these insertions will occur on the same chromosome as the donor locus, with many being within 10–20 Mbp of the donor locus *(17)*. Thus, several SB transposon insertions may be linked to each other. The tendency of SB for local hopping has been proposed as a method for achieving regional saturation mutagenesis in the mouse *(19)*. However, we discovered that about 40% of all G1 mice carry a deletion, inversion, or translocation of sequences adjacent to the donor locus *(17)*. These chromosomal rearrangements affect hundreds of kilobase pairs and most often are deletions extending from one end of the donor locus. Since these deletions can remove essential genes, it means that the true homozygous phenotype of linked SB transposon insertions cannot be determined unless and until said insertion has been separated from the donor locus by meiotic recombination. Therefore, it is imperative to both identify all the SB insertions present within a G1 animal and separate those of interest from all other transposon insertions and the donor locus. The inheritance of individual transposon insertions can be followed by Southern blotting. All the new transposon insertions can be cloned using PCR-based strategies.

In order to process a large number of tail genomic DNAs and to approach saturation for insertion site recovery, we use PCR-based methods for amplifying insertion sites. Linker-mediated PCR is the most common technique currently used to clone transposon integration sites from mice carrying new SB transposon insertion sites *(14, 16, 17)*. In this method, specially designed linkers are ligated onto restricted tumor genomic DNA, then subjected to two rounds of PCR using transposable element-specific and linker-specific primers before cloning into a plasmid vector for sequencing. The use of linker-mediated PCR amplification and cloning of transposon insertion sites is described in detail in another methods chapter in this series (Collier and Largaespada, Methods in Molecular Biology).

After G1 mice have been obtained and the transposon insertions of interest have been cloned and sequenced, the insertion mutations should be separated from other transposons by meiotic recombination *(16)*. It is relatively simple to design primers useful for three-primer PCR reactions that can distinguish the unmodified, wild-type insertion site from the insertion allele *(16)*. This process can be used to genotype mice and determine whether they are homozygous wild-type, heterozygous for the insertion mutation, or homozygous insertion mutants.

2. Materials

2.1. Transposon Transgenic Lines

Once an SB transposon vector has been generated, it should be used to create transgenic mice by standard pronuclear injection. The reader is referred to previous publications on this technology *(33)*. Several SB transposon vectors have been described in the literature including 5′ gene and poly-A traps *(14, 18, 22)*.

2.2. Transposase Transgenic Line

1. CAGGS-SB10/+ mice *(14)*. These mice are available via the Mouse Models of Human Cancer Consortium (http://mouse.ncifcrf.gov/).

   or

2. Prm1-SB10/+ mice *(20)*

   or

3. Rosa26-SB11 mice *(10)*.

3. Methods

3.1. Generation of Doubly Transgenic "Seed Mice"

1. A transposon vector should be created using recombinant DNA technology. Typically, gene-trapping components would include, at a minimum, splice acceptor and polyadenylation sequences to disrupt genes upon intronic insertion (*see* **Notes 1** and **2**).
2. Transposon transgenic mice are produced by standard pronuclear injection with linearized SB vector DNA (*see* **Note 3**). Alternatively, previously used SB transposon transgenic mice may be available (see above).
3. Several transgene positive founders should be characterized by Southern blotting for copy number. Several medium- and high-copy transposon transgenic lines should be tested for germline transposition rate (*see* **Note 4**).
4. Each transposon transgenic line is bred to SB transposase transgenic mice to create doubly transgenic male "seed mice" (*see* **Note 5**). Several of these seed mice should be tested for germline transposition by crossing to wild-type females and screening tail biopsy genomic DNA from the offspring by Southern blotting for new transposon insertions (*see* **Note 4**).
5. Once a transposon transgenic line has been chosen for further analysis as many G1 offspring carrying new transposon insertions as desired may be generated. Indeed, it would be theoretically possible to mutate most genes in the genome, by transposon insertion, if enough G1 mice are generated.

3.2. Generation and Analysis of G1 Offspring from Doubly Transgenic "Seed Mice"

1. G1 offspring can be screened for new transposon insertions using expression of the 5′ or poly-A trap reporter or by simply cloning the new transposon insertions from the G1 mouse. If the transposition rate is high enough it may be possible to simply screen G1 mice and derived G3 mice for new phenotypes without cloning the transposon insertions first. This "forward genetics" approach will probably require higher rates of germline transposition than are currently possible using SB. In any case, the SB transposon insertion mutations of interest must be identified by molecular cloning eventually.
2. SB transposon insertions are cloned by linker-mediated PCR (see Collier and Largaespada, Methods Mol Biol., 2008, for a description).
3. Preferably at the G1 generation, or at least the G2 generation, it is important to eliminate the SB transposase transgene from mutant mice (*see* **Note 6**). The presence of the SB transposase transgene could cause transposon remobilization, further transposon-induced insertion mutations or genomic rearrangements, most often deletions, at the transposon donor locus *(17)*.

3.3. Propagation of Mutant Mice

1. Specific SB-induced transposon insertion mutations are propagated in mice simply by breeding.
2. Three-primer PCR reactions are used to distinguish the wild-type from the insertion mutant alleles *(16)*.

4. Notes

1. Careful consideration should be given to the SB transposon vector design before mice are created. One critical issue is how causality, linking a specific transposon insertion to a specific mutant phenotype, will be established. It is possible to revert SB transposon insertion mutations, and the attendant phenotype, by remobilization out of the affected gene *(16)*. However, as mentioned above, SB transposon remobilization is inefficient using SB10 or SB11 transposase transgenes currently available. Thus, gene rescue or recreation of the mutation using embryonic stem cell technology could be considered for this purpose. However, these are technically demanding and slow approaches. Thus, it is advised that LoxP or Frt sites flank the gene-trapping portion of the transposon vector as in Geurts et al. *(18)*. Thus, a Cre or Flp transgenic mouse could be used to "revert" the insertion mutation and for most intronic insertions a functional allele.

2. SB transposition is sensitive to the length of the transposon vector *(34)*. While ~10-kbp transposons have been mobilized in the germline of mice *(22)*, it is likely that smaller transposons will yield higher rates of transposition.
3. When the transposon vector DNA is used to create transgenic mice by pronuclear injection the vector should be first linearized or cut out of the plasmid in which it resides. However, it is important that adjacent plasmid sequences still linked to the transposon vector are retained in the DNA fragment that is injected. This plasmid DNA sequence should be incorporated into the transposon array that is generated for several reasons. Data from the Takeda lab suggest that plasmid sequences, and adjacent transposon sequences, may become methylated in the transgenic mouse which promotes SB transposition *(28)*. Secondly, the plasmid sequences provide a good tag for the donor locus in genotyping reactions. Third, if the transposon sequences harbor one or more 6-bp restriction enzyme sites, then they can be used to digest linker-ligated genomic DNA during the transposon insertion site cloning protocol, thus eliminating donor transposons from those that are cloned in this reaction *(16)*.
4. Several different transposon vector transgenic lines should be characterized for any vector. Those with the highest copy number tend to result in the most transposition activity when bred to SB transposase transgenic mice, but line-to-line variation is high due to position effects *(18)*. Screening them for the typical transposition rate that is achieved in seed males is advised. New transposon insertions are detected by Southern blotting using a transposon vector-specific probe and cutting the genomic DNA with an enzyme that cuts once or not at all in the transposon vector. Thus, the transposon vectors that remain in the concatemers will result in a band of defined size while new transposon insertions will result in bands of a different size.
5. SB transposition is low in the female germline and so seed mice should be males *(18)*.
6. When G1 animals are screened for new transposon insertions, it is helpful to also screen them for the presence of the transposase transgene. The transposase transgene will segregate independently of the transposon donor array if it resides on another chromosome. By eliminating G1 mice that carry the transposase transgene, one can ensure that no further SB transposon insertion events will occur and that the transposon insertion(s) of interest will remain stable in subsequent generations.

Acknowledgments

This work was supported by grants R21 CA118600, R01 CA113636, and RO1 DA014764 from the National Institutes of Health (to DAL).

References

1. Ostertag EM, Kazazian HH, Jr. Biology of mammalian L1 retrotransposons. Annu Rev Genet 2001; 35:501–38.
2. Plasterk RH, Izsvak Z, Ivics Z. Resident aliens: the Tc1/mariner superfamily of transposable elements. Trends Genet 1999; 15:326–32.
3. Ivics Z, Hackett PB, Plasterk RH, Izsvak Z. Molecular reconstruction of Sleeping Beauty, a Tc1-like transposon from fish, and its transposition in human cells. Cell 1997; 91:501–10.
4. Cui Z, Geurts AM, Liu G, Kaufman CD, Hackett PB. Structure–function analysis of the inverted terminal repeats of the Sleeping Beauty transposon. J Mol Biol 2002; 318:1221–35.
5. Geurts AM, Yang Y, Clark KJ, et al. Gene transfer into genomes of human cells by the Sleeping Beauty transposon system. Mol Ther 2003; 8:108–17.
6. Yant SR, Park J, Huang Y, Mikkelsen JG, Kay MA. Mutational analysis of the N-terminal DNA-binding domain of Sleeping Beauty transposase: critical residues for DNA binding and hyperactivity in mammalian cells. Mol Cell Biol 2004; 24:9239–47.
7. Zayed H, Izsvak Z, Walisko O, Ivics Z. Development of hyperactive Sleeping Beauty transposon vectors by mutational analysis. Mol Ther 2004; 9:292–304.
8. Miskey C, Izsvak Z, Kawakami K, Ivics Z. DNA transposons in vertebrate functional genomics. Cell Mol Life Sci 2005; 62:629–41.
9. Collier LS, Carlson CM, Ravimohan S, Dupuy AJ, Largaespada DA. Cancer gene discovery in solid tumours using transposon-based somatic mutagenesis in the mouse. Nature 2005; 436:272–6.
10. Dupuy AJ, Akagi K, Largaespada DA, Copeland NG, Jenkins NA. Mammalian mutagenesis using a highly mobile somatic Sleeping Beauty transposon system. Nature 2005; 436:221–6.
11. Drabek D, Zagoraiou L, deWit T, et al. Transposition of the *Drosophila hydei* Minos transposon in the mouse germ line. Genomics 2003; 81:108–11.
12. Ding S, Wu X, Li G, Han M, Zhuang Y, Xu T. Efficient transposition of the piggyBac (PB) transposon in mammalian cells and mice. Cell 2005; 122:473–83.
13. Wu S, Ying G, Wu Q, Capecchi MR. Toward simpler and faster genome-wide mutagenesis in mice. Nat Genet 2007; 39:922–30.
14. Dupuy AJ, Fritz S, Largaespada DA. Transposition and gene disruption in the male germline of the mouse. Genesis 2001; 30:82–8.
15. Roberg-Perez K, Carlson CM, Largaespada DA. MTID: a database of Sleeping Beauty transposon insertions in mice. Nucleic Acids Res 2003; 31:78–81.
16. Carlson CM, Dupuy AJ, Fritz S, Roberg-Perez KJ, Fletcher CF, Largaespada DA. Transposon mutagenesis of the mouse germline. Genetics 2003; 165:243–56.
17. Geurts AM, Collier LS, Geurts JL, et al. Gene mutations and genomic rearrangements in the mouse as a result of transposon mobilization from chromosomal concatemers. PLoS Genet 2006; 2:e156.
18. Geurts AM, Wilber A, Carlson CM, et al. Conditional gene expression in the mouse using a Sleeping Beauty gene-trap transposon. BMC Biotechnol 2006; 6:30.
19. Keng VW, Yae K, Hayakawa T, et al. Region-specific saturation germline mutagenesis in mice using the Sleeping Beauty transposon system. Nat Methods 2005; 2:763–9.
20. Fischer SE, Wienholds E, Plasterk RH. Regulated transposition of a fish transposon in the mouse germ line. Proc Natl Acad Sci USA 2001; 98:6759–64.
21. Horie K, Kuroiwa A, Ikawa M, et al. Efficient chromosomal transposition of a Tc1/mariner-like transposon Sleeping

Beauty in mice. Proc Natl Acad Sci USA 2001; 98:9191–6.

22. Horie K, Yusa K, Yae K, et al. Characterization of Sleeping Beauty transposition and its application to genetic screening in mice. Mol Cell Biol 2003; 23:9189–207.
23. Kitada K, Ishishita S, Tosaka K, et al. Transposon-tagged mutagenesis in the rat. Nat Methods 2007; 4:131–3.
24. Lu B, Geurts AM, Poirier C, et al. Generation of rat mutants using a coat color-tagged Sleeping Beauty transposon system. Mamm Genome2007; 18:338–46.
25. Ryder E, Russell S. Transposable elements as tools for genomics and genetics in *Drosophila*. Brief Funct. Genomic Proteomic 2003; 2:57–71.
26. Stanford WL, Cohn JB, Cordes SP. Gene-trap mutagenesis: past, present and beyond. Nat Rev Genet 2001; 2:756–68.
27. Abuin A, Hansen GM, Zambrowicz B. Gene trap mutagenesis. Handb Exp Pharmacol 2007: 129–47.
28. Yusa K, Takeda J, Horie K. Enhancement of Sleeping Beauty transposition by CpG methylation: possible role of heterochromatin formation. Mol Cell Biol 2004; 24:4004–18.
29. Vigdal TJ, Kaufman CD, Izsvak Z, Voytas DF, Ivics Z. Common physical properties of DNA affecting target site selection of Sleeping Beauty and other Tc1/mariner transposable elements. J Mol Biol 2002; 323:441–52.
30. Okabe M, Ikawa M, Kominami K, Nakanishi T, Nishimune Y. “Green mice” as a source of ubiquitous green cells. FEBS Lett 1997; 407:313–9.
31. Okabe M, Ikawa M, Kominami K, Nakanishi T, Nishimune Y. “Green mice” as a source of ubiquitous green cells. FEBS Lett 1997; 407:313–9.
32. Dupuy AJ, Akagi K, Largaespada DA, Copeland NG, Jenkins NA. Mammalian mutagenesis using a highly mobile somatic Sleeping Beauty transposon system. Nature 2005; 436:221–6.
33. Roths JB, Foxworth WB, McArthur MJ, Montgomery CA, Kier AB. Spontaneous and engineered mutant mice as models for experimental and comparative pathology: history, comparison, and developmental technology. Lab Anim Sci 1999; 49:12–34.
34. Geurts AM, Yang Y, Clark KJ, et al. Gene transfer into genomes of human cells by the Sleeping Beauty transposon system. Mol Ther 2003; 8:108–17.

Chapter 21

Lentiviral Transgenesis

Alexander Pfeifer and Andreas Hofmann

Abstract

Lentiviral vectors efficiently transfer genes into a broad spectrum of cells and tissues, including non-dividing cells and stem cells. Lentiviruses integrate their viral genome into the host chromosome, which is the basis for virus latency as well as stable transgene expression. A rather novel development is the use of lentivectors to transfer transgenes in oocytes and early embryos to generate transgenic animals, a technology also known as lentiviral transgenesis. Lentiviral transgenesis has been shown to be highly efficient in many different species, including mouse, rat, pig, bovine, and even birds. Thus, lentiviral transgenesis has the potential to become a versatile and widespread transgenic technology. The aim of this chapter is to cover important practical aspects of lentiviral transgenesis, including vector preparation, gene delivery into the early embryos and lentiviral RNA interference.

Key words: Lentiviral transgenesis, subzonal injection, preimplantation embryos, transgenic animals, RNA interference.

1. Introduction

The first transgenic animals were generated by injection of simian virus 40 (SV40) DNA into the cavity of mouse blastocysts *(1)*. The demonstration that the SV40 DNA persisted from early embryonic stages to adult life indicated that viral infection might be a way to generate transgenic animals *(1)*. However, a basic requirement for the generation of transgenic lines is vertical transmission of the transgene through the germline. One way to obtain an animal carrying foreign genes in every cell (including the germline) would be to infect embryos at very early stages of development – shortly after fertilization before differentiation has taken place – with viruses that integrate their DNA into the host genome *(2)*.

Ralf Kühn, Wolfgang Wurst (eds.), *Gene Knockout Protocols: Second Edition, vol. 530*

DOI 10.1007/978-1-59745-471-1_21 Springerprotocols.com

Several features made retroviruses attractive candidates for viral transgenesis: retroviruses are enveloped and integrate their genome – after reverse transcription – into host chromosomes to form so-called proviruses *(3)*. Indeed, first experiments with murine leukemia virus (MLV) – a gammaretrovirus (that is considered as a simple, prototypic retrovirus) – demonstrated integration of the MLV genome after infection of early embryos *(2)*. The provirus was transmitted through the germline; however, it was not expressed in the newborn mice due to epigenetic silencing (hypermethylation) of the viral DNA *(4)*.

Because of retroviral silencing, DNA-microinjection (DNA-MI) into the pronuclei of fertilized mouse embryos was developed in the early 1980s as an alternative transgenic technology *(5)*. Although DNA-MI is still the most widespread technology for the production of transgenic mice, it is hampered by very low transgenesis rates (number of transgenic animals per treated embryo) and its use is by and large restricted to a few mouse lines. The generation of transgenic animals by DNA-MI in other biomedical relevant species like rat, pig and bovine is extremely labor- and cost-intensive. This, together with the failure of retroviral vectors to stably express transgenes in other species, for example, cattle *(6)*, has led to a restriction of transgenic technology basically to mouse.

A major advance of transgenic technology was the introduction of lentiviral transgenesis as a novel approach to transfer transgenes to early embryos (for a detailed review see *(7)*). Lentiviruses are members of the large retrovirus family *(3)*. Like other retroviruses they integrate their genome and form proviruses *(3)*. The best-studied lentiviruses are human immune deficiency viruses (HIVs). In contrast to simple gammaretroviruses, such as MLV, lentiviruses carry a more complex genome and virus integration does not require cell division *(3)*. Therefore, lentiviral vectors are able to infect non- or slowly dividing cells. The most advanced vector systems consist of self-inactivating vectors and split-genome packaging plasmids (*see* **Section 2.1.1**), which result in high-titer lentivector preparations (*see* **Section 2.1.2**).

In the initial pioneering studies, lentiviral vectors were used to generate transgenic mice either by subzonal injection of vector particles *(8)* or by direct incubation of embryos with lentiviral vectors *(9)*. The latter approach requires the removal of the zona pellucida (ZP). Another way to produce transgenic mice is the transduction of embryonic stem (ES) cells followed by their injection into the blastocyst cavity of preimplantation embryos *(9)*. An intriguing feature of lentiviral transgenesis is the staggering gain in efficiencies (4- to 20-fold increase depending on the species and the lentivirus) combined with its broad applicability. Meanwhile, lentiviral transgenesis has been developed for rat *(8)*, pig *(10, 11)*, cattle *(12)*, chicken *(13)*, and quail *(14)*.

2. Materials

Unless stated otherwise, all chemicals and biochemicals used in this study are from highest purity (tested for cell culture or for use in embryo culture).

2.1. Preparation of Concentrated Lentiviruses

2.1.1. Virus Plasmids

The original virus plasmids are derived from the lab of Inder Verma (The Salk Institute for Biological Studies, Laboratory of Genetics, La Jolla, CA, USA). Virus plasmids to generate lentiviral vectors of the third generation can be ordered at Addgene (http://www.addgene.org). Detailed vector maps can be downloaded there. In addition, lentivectors and packaging systems are commercially available (e.g., from Invitrogen). The lentivector system consists of two major parts: the vector and the packaging constructs (**Fig. 21.1**).

1. pMDLg/pRRE: The packaging construct pMDLg/pRRE (**Fig. 21.1**) expresses the *gag* and *pol* genes of HIV driven by a Cytomegalovirus (CMV) promoter. Since the transcripts of the *gag* and *pol* genes contain cis-repressive sequences, they are only expressed through binding of *rev* to RRE (*rev* response element), which enables the export of the viral RNA from the nucleus *(15)*.
2. RSV-*rev*: The packaging construct RSV-*rev* (**Fig. 21.1**) expresses the *rev* gene *(15)*.
3. pMD.G: The packaging construct pMD.G (**Fig. 21.1**) encodes a heterologous envelope protein to pseudotype the viral vector. It expresses *VSV.G* driven by the CMV promoter *(15)*.

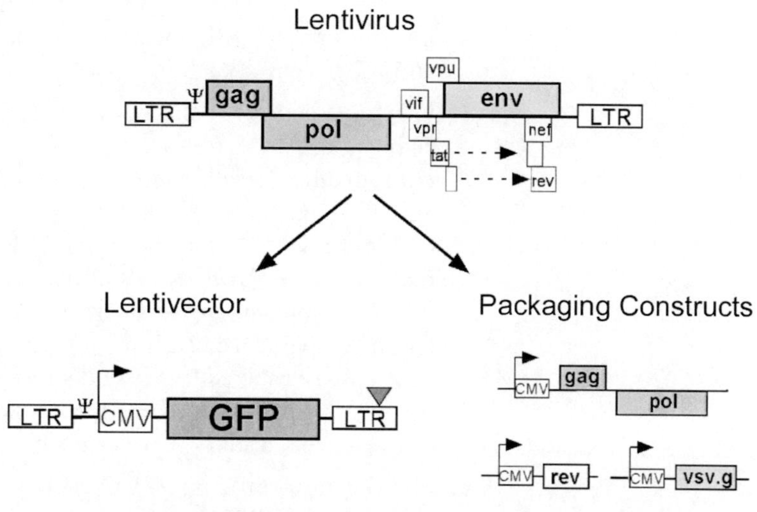

Fig. 21.1. Schematic drawing of the HIV genome (*top*) and the four constructs used to make a lentiviral vector of the third generation (*bottom*). Graphic modified after (*36*).

4. The lentivector (**Fig. 21.1**) contains the sequences of HIV-1 required for packaging, reverse transcription and integration, as well as the transgene expression cassette. The 5′ long terminal repeat (LTR) contains enhancer and promoter either of the Respiratory syncytial virus *(15)* or CMV *(16)* as a substitute of the wild-type U3-region to drive expression of the vector RNA in the packaging cells. In addition, the promoter/enhancer sequences in the U3 region of the 3′ LTR were deleted to generate self-inactivating (SIN, black triangle in **Fig. 21.1**) vectors *(15, 16)*. The deletion of the promoter/enhancer elements of the 3′ LTR are carried over to the 5′ LTR during reverse transcription, which will lead to transcriptional inactivation of the provirus *(15)*.

Transgene expression can easily be achieved by incorporation of promoters in the vector construct (internal promoters). The individual components of the lentiviral vector, the promoter, and/or the transgene, can easily be exchanged, which makes this vector system a versatile tool for gene transfer. **Figure 21.1** shows a lentivector with a CMV-driven eGFP expression cassette (*see* **Note 1** for *Escherichia coli* strain and plasmid preparation).

2.1.2. Production of High-Titer Recombinant Lentivirus

1. High-titer lentivector preparations are essential to achieve high transgenesis rates. For a standard lentiviral vector preparation, twelve 15-cm cell culture dishes (#353025, Falcon) are coated with poly-L-lysine solution (50 ml solved in 500 ml PBS; #P4832, Sigma) for 15 min.
2. HEK 293T cells (#CRL-11268, ATCC) are seeded in 20 ml DMEM (Dulbecco's Modified Eagle Medium, #61965, Gibco; supplemented with fetal calf serum (FCS, #S0115, Biochrom AG), 100 U/ml Penicillin G/100 μg/ml Streptomycin (#A2213, Biochrom AG)), and incubated at 10% CO_2 and 37°C (day 1).
3. For transfection of the plasmids (24 h after plating, day 2), cells should be ~60% confluent. The transfection mix containing all plasmids (lentivector, pMDLg/pRRE, RSV-*rev* and pMD.G) is prepared in a sterile reaction tube as shown in **Table 21.1**. Subsequently, 14 ml of the transfection reagent (2 × BBS, see **Table 21.1**) is added and gently mixed by inversion (5–10 times). After 15 min incubation at room temperature, 2.25 ml of the transfection mix is applied dropwise to each cell culture dish and carefully mixed with the medium. *Important*: For transfection, cells are cultured at only 3% CO_2 and 37°C overnight.
4. On the next morning (day 3), the medium is replaced by 16 ml fresh DMEM per 15-cm dish and cells are cultured again at 10% CO_2 and 37°C.

Table 21.1
Transfection mix (for 12 × 15 cm plates) and protocol for the transfection reagent (2×BBS).

Transfection mix	2 × BBS
270 μg lentivector	4.26 g *N,N*-Bis(hydroxyethyl)-2-aminoethanesulfonic acid
175 μg pMDLg/pRRE	6.54 g NaCl
68 μg RSV-*rev*	0.085 g Na2HPO4
95 μg pMD.G	pH 6.95
1.4 ml $CaCl_2$ (2.5 M) add 14 ml H_2O	add H_2O to a final volume of 400 ml

5. For the first virus collection (24 h after medium change, day 4), the supernatants are harvested and processed through a tensid-free cellulose acetate bottle-top filter (SFCA, 0.45 μm, Nalgene) to remove cell debris. Again, cells are cultured with 16 ml of fresh DMEM at 10% CO_2 and 37°C until the next morning.
6. On the same day (day 4), the filtered supernatant is pre-concentrated in centrifugation tubes (#358126, Beckman Coulter) by an ultra-centrifuge (Optima L-100 XP, Beckman Coulter) with SW32Ti rotor (Beckman Coulter) for 2 h at 19,400 rpm (61,700*g*) and 17°C. Subsequently, each virus pellet is resuspended in 50 μl HBSS (Hanks' Balanced Salt Solution, #14175-046, Gibco). The suspensions are combined and stored until the second virus-harvest in a sterile screw-cap reaction tube (Sarstedt) at 4°C.
7. Second virus-harvest (day 5): Twenty-four hours after the first virus-harvest, supernatants are collected for a second time and pre-concentrated as described above.
8. For the final ultra-concentration, suspensions from days 4 and 5 are combined, transferred into a centrifuge tube (#326819, Beckman Coulter) and layered on top of a 2 ml 20% (w/v) sucrose cushion (#4621.1, Roth). The mixture is centrifuged using a SW55 rotor (Beckman Coulter) for 2 h at 21,000 rpm (53,500*g*) at 17°C. The pellet is resuspended in ~150 μl HBSS and transferred to a sterile screw-cap reaction tube.
9. The tube is vortexed for 45 min at 1,400–1,500 rpm and 17°C, followed by a short centrifugation step (3 s, 16,000*g*) to pellet debris.
10. Finally, the opaque supernatant is aliquoted in sterile screw-caps and stored at –80°C.

2.1.3. Analysis of Virus Titer

1. Using commercially available enzyme-linked immunosorbent assay (ELISA) assays, the content of reverse transcriptase (e.g., Roche, #11468120910) or the lentiviral p24 protein (e.g., Perkin Elmer, #NEK050) can be measured (physical virus titer).
2. In addition, lentiviral vectors that carry an eGFP reporter can be titered by infection of cells followed by flow cytometry analyses. In contrast to ELISA analyses, this method detects only active and infectious viral particles (biological virus titer; infectious units (IUs)).
3. To measure the amount of biological active virus, 100,000 HEK 293T cells (#CRL-11268, ATCC) per well are seeded onto a 24-well plate and adhered for 4–6 h (*see* **Note 2**).
4. One μl of concentrated lentivirus is solved in 330 μl DMEM (dilution 1). Then, 30 μl of dilution 1 is added to 300μl of fresh DMEM (dilution 2, 1:10 of dilution 1) and mixed again. Normally, a total of 3–4 dilution steps (as described above) are sufficient to quantify the virus titer. The virus dilutions are added to the HEK 293T cells (after complete removal of medium) and incubated overnight On the next morning, the medium is filled up to a volume of 1 ml.
5. Seventy-two hours after infection, cells are trypsinized and centrifuged at 300*g* for 5 min. The cell pellet is resuspended in 1 ml of paraformaldehyde solution (4% in PBS) and kept on ice for 15 min. After fixation, cells are centrifuged again at 300*g* for 5 min and the cell pellet is resuspended in 0.5–1 ml of PBS. At 4°C, cells can be stored for several weeks.
6. At least 10,000 cells should be analyzed at λem 530 nm (FL-1) for their mean fluorescence intensity (MFI). Uninfected cells are used as negative control. The cut-off between eGFP-negative ($eGFP^-$) and eGFP-positive ($eGFP^+$) cells is dependent on the control cells and should be defined at a value that separates the negative control into ~90% $eGFP^-$ and ~10% $eGFP^+$ cells (**Fig. 21.2**). Thus, all measured cells that exhibit a MFI higher than this cut-off value are estimated as $eGFP^+$ (**Fig. 21.2**).
7. To calculate the biological virus titer, the percentage of $eGFP^+$ and $eGFP^-$ cells in the different dilutions has to be determined. Cells that have been transduced with a rather low multiplicity of infection (MOI), that exhibit a ~50%–50% distribution of $eGFP^-$ versus $eGFP^+$ cells, should be used for calculation (see example in **Fig. 21.2**): MOI = – ln (percentage of $eGFP^-/100$).
   Virus titer (IU/ml) = (No. of infected cells) × (MOI) × (dilution factor) × 1,000.

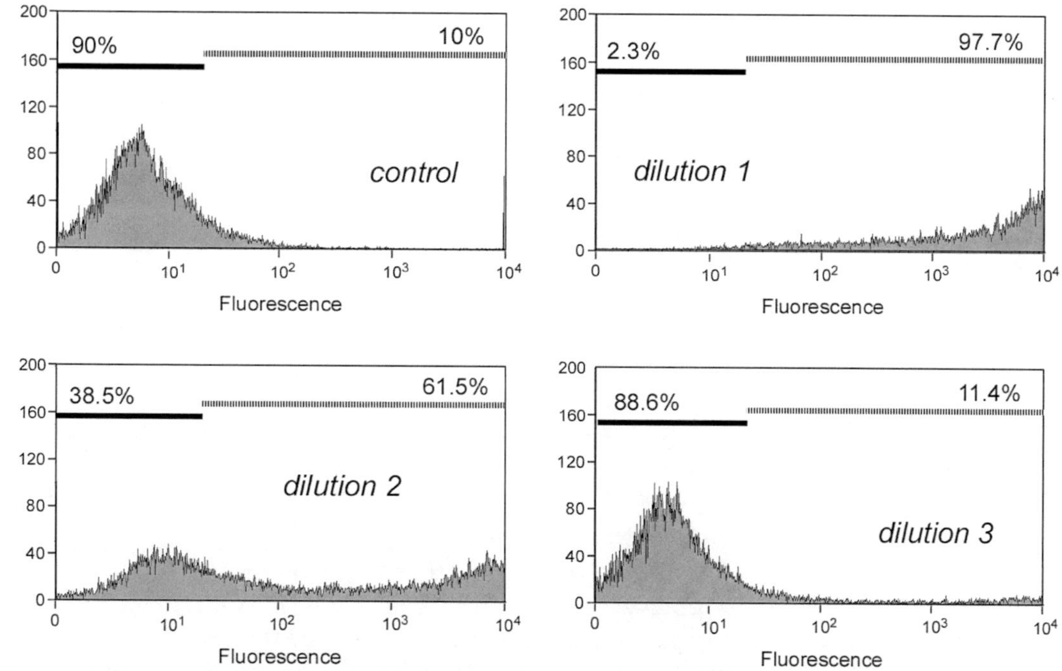

Fig. 21.2. Representative flow cytometry analyses of eGFP expression in HEK 293T cells transduced with different dilutions of a virus preparation for measurement of virus titer. Control cells are used to differentiate between $eGFP^-$ (*solid line*) and $eGFP^+$ (*dotted line*) cells. Then, all dilutions are analyzed using the same histogram markers for their content of $eGFP^-$cells. Finally, dilution 2 is used for calculation (e.g., MOI $= -\ln(0.385) = 0.95$; dilution 2 = dilution factor 10; infected cells = 160,000;→ virus titer $= 1.5 \times 10^9$ IU/ml).

2.2. Media for Embryo Culture

For the culture of embryos, two different media are used.

1. KSOM (*see* **Table 21.2**) embryo culture medium is used for culture in a CO_2 incubator (37°C and 5% CO_2).
2. KSOM-HEPES is used for micromanipulation. The media should be freshly prepared and stored at 4°C not longer than 2 weeks.
3. Embryos are cultured in 60-mm-diameter dishes with centre well (#353653, Falcon) or four-wells for embryo culture (# 353654, Falcon).
4. Removal of the ZP is achieved by treatment with acidic Tyrode's solution (#T1788, Sigma), drops are covered by mineral oil (#M8410, Sigma).

3. Methods

Mammalian preimplantation embryos are surrounded by the ZP. The ZP is a physical barrier against viral infection. Therefore, this hurdle has to be overcome to successfully transduce an embryo (for review see *(7)*). Different routes to overcome the ZP for the generation of lentiviral transgenic mice have been used in the past years: subzonal virus injection *(8)* and removal of the ZP and incubation in virus solution *(9)* will be described in detail in the following sections.

3.1. Subzonal Virus Injection

Subzonal virus injection into the perivitelline space of preimplantation embryos is a quick way to directly transduce embryos. However, complex and cost-intensive manipulation equipment and a well-trained operator are needed to successfully transduce embryos.

1. Preimplantation embryos (zygotes) are collected in KSOM-HEPES (*see* **Section 2.2** and **Table 21.2**) using standard protocols *(17)*. Thereafter, for the in vitro culture, embryos are cultured in KSOM medium (*see* **Section 2.2** and **Table 21.2**) at 37°C and 5% CO_2.
2. For the subzonal injection of recombinant lentivirus into zygotes, an inverted microscope (e.g., DMIL, Leica) with integrated modulation contrast (IMC) combined with a micromanipulation and injection system (e.g., Eppendorf) are required.
3. The micromanipulation system used by our lab consists of following components: The injection capillary is controlled by a TransferMan NK2 micromanipulator (Eppendorf), while a Patchman NP2 micromanipulator (Eppendorf) is used for the holding pipette.
4. Embryos are transferred to a microscope slide (Superfrost PLUS, Menzel-Gläser, Germany) into a 200 μl drop of KSOM-HEPES (*see* **Section 2.2** and **Table 21.2**) with a Transferpettor (#701853, Brand) and a 5-μl glass capillary (#701900, Brand).
5. The virus injection into the perivitelline space of murine zygotes is preferentially done at 400× magnification (e.g., Obj. C PLAN L 40×/0.50 PH2 1.1/-2.0, Leica) using the IMC.
6. Embryos are held in place with a holding pipette (inner diameter = 20–25 μm, angle = 20°, fire-polished ends; #GC100T-15, BioMedical Instruments, Germany) connected to a Cell Tram Oil (Eppendorf).
7. Concentrated lentivirus (>10^9 IU/ml) is filled into an injection capillary (type pronucleus after Zimmermann, #GC100TF-10, BioMedical Instruments, Germany) using Tip Fillers (Eppendorf) and viral particles are injected under the ZP using a Transjector 5246 (Eppendorf). *See* **Note 3** for recommended virus amount and titer.

Table 21.2
Embryo culture media

KSOM Medium		KSOM-HEPES
0.01 mM	EDTA (tetrasodium salt)	0.01 mM
95 mM	NaCl	95 mM
2.5 mM	KCl	2.5 mM
0.35 mM	KH_2PO_4	0.35 mM
0.2 mM	$MgSO_4 \cdot 7H_2O$	0.20 mM
0.2 mM	D-Glucose	0.20 mM
0.2 mM	Na-Pyruvate	0.20 mM
25 mM	$NaHCO_3$	4 mM
1.71 mM	$CaCl_2 \cdot 2H_2O$	1.71 mM
10 mM	Na-Lactate (60% Syrup)	10 mM
1 mM	Glutamine (Glutamax I)	1 mM
1 mg/ml	BSA	4 mg/ml
–	HEPES	20 mM
1 μl/ml	Phenol Red	1 μl/ml
100 U/ml	Penicillin G Na-salt	100 U/ml
5 mg/ml	Streptomycin sulfate	5 mg/ml
1 ×	Amino acids (NEAA)	–
1 ×	Amino acids (EAA)	–

KSOM Medium: Pass through a sterile-filter (0.2 μm bottle top vacuum filtration unit; PES membrane).
Incubate at 37°C and 5% CO_2 over night.

KSOM-HEPES: Pass through a sterile-filter (0.2 μm bottle top vacuum filtration unit; PES membrane).
The pH should be between 7.2 and 7.4.

8. After subzonal injection, the embryos are transferred back into KSOM medium and cultured at 37°C and 5% CO_2 until they reach the blastocyst stage. If a reporter gene (e.g., eGFP) is used, the transduction efficiency can be determined by simple fluorescence microscopy (*see* **Fig. 21.3**).
9. Finally, blastocysts are transferred into synchronized female recipients according to standard protocols for mouse manipulation *(17)*.

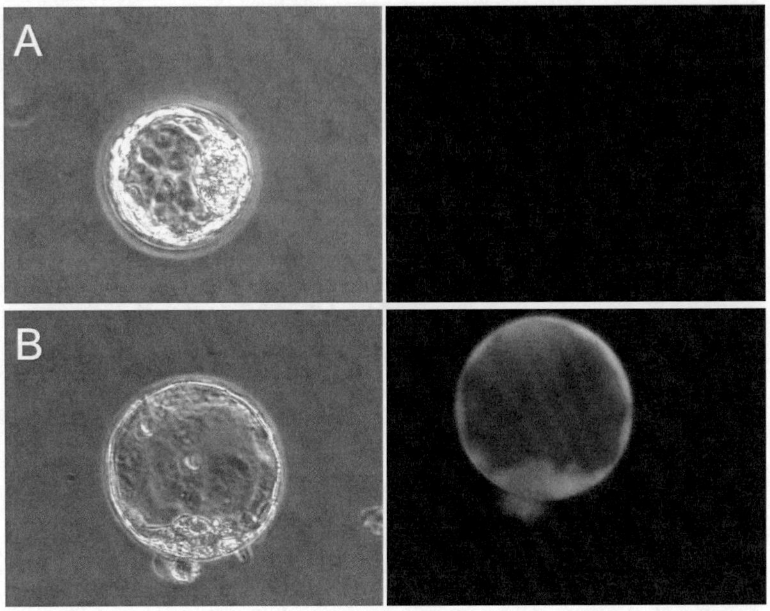

Fig. 21.3. Transgene expression in blastocysts after subzonal injection of lentivectors. (**A**) Wild-type blastocysts and (**B**) blastocysts derived from subzonal injection of lentiviral vectors carrying a phosphoglycerate kinase 1 (PGK) gene promoter-driven eGFP reporter (vector described in (*9*)). Bright-field image, *left*; fluorescence image, *right*.

3.2. Lentiviral Infection of Denuded Preimplantation Embryos

An alternative route for transduction is the co-incubation of ZP-free preimplantation embryos in virus-containing medium. In contrast to subzonal virus injection (*see* **Section 3.1**), this method does not require specialized, expensive mechanical equipment. However, the enzymatic denudation of embryos is a significant intervention that negatively affects embryonic viability.

1. Preimplantation embryos (e.g., zygotes or morulae) are collected using standard protocols for mouse manipulation *(17)*.
2. Removal of the ZP is achieved by acidic Tyrode's treatment. Therefore, 50-μl drops of acidic Tyrode's solution are spotted on a cell culture dish and covered by mineral oil.
3. Embryos are pipetted up and down in acidic Tyrode drops for a few seconds (up to 30 s). Dissolving of the ZP should be followed under a stereo microscope (e.g., SV6, Zeiss). Subsequently, embryos are washed several times in drops of KSOM-HEPES for outside embryo culture *see* **Note 4** for details).
4. For transduction, lentiviral vectors should be diluted in KSOM medium. The rate of dilution highly depends on the desired transgenesis rates (number of viral integrants per cell). The final virus concentration should be between 10^7 and 10^8 IU/ml. Drops (20–50 μl) of virus-containing medium (depending on the number of embryos per drop) should be

prepared on a suitable cell culture dish and covered with mineral oil. Then, embryos are transferred into the drops and incubated at 37°C and 5% CO_2 for 4–6 h.

5. After incubation, embryos are washed several times by transferring them to fresh drops of KSOM. Thereafter, the embryos are transferred to suitable embryo culture dishes with fresh KSOM medium.
6. Embryos are cultured at 37°C and 5% CO_2 until they reach the blastocyst stage and transferred into synchronized female recipients after standard protocols for mouse manipulation *(17)*.

3.3. RNA Interference by Lentiviral Vectors

RNA interference (RNAi) is a powerful cellular process that leads to gene silencing at a post-transcriptional level (for reviews see *(18, 19)*). RNAi is induced by small interfering RNAs (siRNAs). Although siRNAs can be directly used to silence specific genes, this requires chemical or enzymatic synthesis of the desired oligonucleotides. An alternative way to RNAi is to produce siRNAs – or short hairpin RNAs (shRNAs) that can be converted to siRNAs – in the target cell. The most widely used siRNA expression constructs are based on RNA polymerase III transcription units (e.g., nuclear RNA U6, *(20)*). The combination of RNAi and lentiviral transgenesis has been demonstrated to result in efficient and stable knockdown of target genes in vivo *(21–24)*.

3.3.1. Target Site Selection

1. The selection of effective target sites is the prerequisite for efficient RNAi. Several online shRNA-design programs are available on the World Wide Web (e.g., http://www.genelink.com/sirna/shRNAi.asp).

   Furthermore, several companies offer design services for shRNAs: for example, Invitrogen (http://www.invitrogen.com); BBridge (http://www.bbridge.com).
2. The manual selection of target sites is difficult and time-consuming. Several publications describe characteristics for the ideal siRNA: Reynolds et al. describe eight criterias for rational siRNA design *(25)*. In addition, Schwarz et al. point out that "stability of the base pairs at the 5′-end of the two siRNA strands determines the degree to which each strand participates in the RNAi pathway" *(26)*. The group of J.J. Rossi found that the mRNA secondary structure has a major influence on the efficacy of RNAi *(27)* (*see* **Note 5** for additional information).
3. Several sequences for short hairpin loops have been published until now *(28)*. The loop of choice is dependent on the RNAi target sequence and should not generate additional stop signals (i.e., TTTT) and repetitive sequences in the shRNAs.

4. Taken together, presently there is no optimum formula to find an effective shRNA and sometimes a large number of sequences have to be tested by trial and error until an efficient knockdown of the desired gene can be achieved.

3.3.2. Vector Design

1. We design shRNAs by either sense-loop-antisense or antisense-loop-sense orientation and add "TTTTT" as termination signal.
2. Then, these sequences are synthesized commercially as forward and reverse oligonucleotides with restriction sites at the 5′ and 3′ end. After annealing of the oligonucleotides, we ligate the double-stranded fragments into an expression vector containing a suitable polymerase III promoter, like the H1 *(28)* or the U6 *(20)* promoter. In accordance with several recently published studies *(29)*, we have found that the U6 promoter gives higher levels of knockdown than the H1 promoter.
3. Finally, we transfer the complete shRNA expression cassette into a lentiviral vector *(24, 30, 31)* and generate high-titer lentiviruses (*see* **Section 2.1.2**).

3.3.3. Preventing Off-Target Effects

1. Every target sequence for RNAi has to be checked for unintended targeting (i.e., off-target effects), by blasting the sequence of the genome in which the experiments will be performed.
2. shRNA vectors may trigger an interferon (IFN) response. Hornung et al. show IFN-I response after uptake of siRNA molecules by immune cells – specifically plasmacytoid dendritic cells *(32)*. Interestingly, the authors identify a 9-bp consensus motif that induces high IFN-I production in the dendritic cells. Therefore, if siRNA molecules will enter the blood the siRNA design should avoid such motifs.
3. A recent study from the lab of J.J. Rossi described that induction of IFN response (in human CD34+ progenitor cells) to a siRNA can be avoided by switching to a shRNA design in vitro *(33)*.
4. Nevertheless, we recommend to screen lentiviral shRNA vectors for possible induction of IFN by quantitative PCR of IFN target sequences like OAS1 *(34)*.

3.3.4. Analysis of Knockdown Levels

1. We measure knockdown of the desired mRNA by Real-Time PCR analyses. We prefer the TaqMan® method rather than the SYBR® Green method. This method allows the exact measurement of two different genes (the gene of interest and a gene used to determine RNA loading) in one and the same PCR reaction *(24)*.
2. In addition, Western blot analyses can be performed if suitable antibodies for the desired gene product are available.

3.3.5. Controls

1. Scrambled shRNAs – which have no target in the corresponding genome – can be used as a control. Thus, one can control for the effects of transduction (transfection, viral infection, etc.) and the shRNA expression system (expression plasmid, virus, etc.).
2. In addition, several labs also use "nonsense" shRNAs that target, for example, the LacZ gene *(35)*.

4. Notes

1. All virus plasmids are cultured in the *E. coli* strain XL-1 blue MRF (Stratagene) and prepared with an endotoxin-free Maxi kit from MACHEREY-Nagel (NucleoBond® PC 500 EF).
2. As an alternative, HEK 293T cells can be grown overnight to ensure complete attachment. However, since the growth rate of HEK 293T cells is rather high, one has to count the cells at the time point of infection.
3. Virus amount for subzonal injection: The amount of virus that can be administered to the embryos is limited by volume of the perivitelline space. Therefore, only high-titer preparations of lentiviral vectors should be used (a titer of $>10^9$ IU/ml or higher is recommended). As a consequence, the transgenesis rates (number of viral integrants per cell) are directly related to the amount and titer of virus that is used.
4. Prolonged incubation of embryos in acidic Tyrode's solution will significantly harm embryo survival rates. In addition, after denudation embryos tend to adhere to the culture dishes. Therefore, incubation of the culture dishes with FCS (#S0115, Biochrom AG) for 10 min can prevent or lower gluing of the embryos.
5. Using published siRNA target sequences (e.g., for transfectable siRNA oligonucleotides) for the design of viral shRNA expression systems is no guarantee for high knockdown levels. One should test the functionality of each target sequence within the context of the expression vector.

Acknowledgments

The authors would like to thank Dr. Michael Bösl (Head of Group Transgenic Service, Max Planck Institute of Neurobiology, Martinsried) for providing embryos, culture media and valuable tips for embryo manipulation. This work was supported by the DFG and by Bonfor.

References

1. Jaenisch R, Mintz B. Simian virus 40 DNA sequences in DNA of healthy adult mice derived from preimplantation blastocysts injected with viral DNA. Proc Natl Acad Sci USA 1974; 71:1250–4.
2. Jaenisch R, Fan H, Croker B. Infection of preimplantation mouse embryos and of newborn mice with leukemia virus: tissue distribution of viral DNA and RNA and leukemogenesis in the adult animal. Proc Natl Acad Sci USA 1975; 72:4008–12.
3. Goff SP. Retroviridae: The retroviruses and their replication. In: Howley PM, Knipe DM, Griffin D, et al., eds. Fields Virology. Vol. 2 (4th ed.). Philadelphia: Lippincott-Raven Publishers. 2001; 1871–1939. 2001.
4. Jaenisch R. Germ line integration and Mendelian transmission of the exogenous Moloney leukemia virus. Proc Natl Acad Sci USA 1976; 73:1260–4.
5. Gordon JW, Scangos GA, Plotkin DJ, Barbosa JA, Ruddle FH. Genetic transformation of mouse embryos by microinjection of purified DNA. Proc Natl Acad Sci USA 1980; 77:7380–4.
6. Chan AW, Homan EJ, Ballou LU, Burns JC, Bremel RD. Transgenic cattle produced by reverse-transcribed gene transfer in oocytes. Proc Natl Acad Sci USA 1998; 95:14028–33.
7. Pfeifer A. Lentiviral transgenesis. Transgenic Res 2004; 13:513–22.
8. Lois C, Hong EJ, Pease S, Brown EJ, Baltimore D. Germline transmission and tissue-specific expression of transgenes delivered by lentiviral vectors. Science 2002; 295:868–72.
9. Pfeifer A, Ikawa M, Dayn Y, Verma IM. Transgenesis by lentiviral vectors: lack of gene silencing in mammalian embryonic stem cells and preimplantation embryos. Proc Natl Acad Sci USA 2002; 99:2140–5.
10. Hofmann A, Kessler B, Ewerling S, et al. Efficient transgenesis in farm animals by lentiviral vectors. EMBO Rep 2003; 4:1054–60.
11. Whitelaw CB, Radcliffe PA, Ritchie WA, et al. Efficient generation of transgenic pigs using equine infectious anaemia virus (EIAV) derived vector. FEBS Lett 2004; 571:233–6.
12. Hofmann A, Zakhartchenko V, Weppert M, et al. Generation of transgenic cattle by lentiviral gene transfer into oocytes. Biol Reprod 2004; 71:405–9.
13. McGrew MJ, Sherman A, Ellard FM, et al. Efficient production of germline transgenic chickens using lentiviral vectors. EMBO Rep 2004; 5:728–33.
14. Scott BB, Lois C. Generation of tissue-specific transgenic birds with lentiviral vectors. Proc Natl Acad Sci USA 2005; 102:16443–7.
15. Dull T, Zufferey R, Kelly M, et al. A third-generation lentivirus vector with a conditional packaging system. J Virol 1998; 72:8463–71.
16. Miyoshi H, Blomer U, Takahashi M, Gage FH, Verma IM. Development of a self-inactivating lentivirus vector. J Virol 1998; 72:8150–7.
17. Nagy A, Gertsenstein M, Vintersten K, Behringer R. Manipulating the Mouse Embryo A Laboratory Manual. 3rd edition, Cold Spring Harbor Press. 2003.
18. Hannon GJ, Rossi JJ. Unlocking the potential of the human genome with RNA interference. Nature 2004; 431:371–8.
19. Novina CD, Sharp PA. The RNAi revolution. Nature 2004; 430:161–4.
20. Lee NS, Dohjima T, Bauer G, et al. Expression of small interfering RNAs targeted against HIV-1 rev transcripts in human cells. Nat Biotechnol 2002; 20:500–5.
21. Tiscornia G, Singer O, Ikawa M, Verma IM. A general method for gene knockdown in mice by using lentiviral vectors expressing small interfering RNA. Proc Natl Acad Sci USA 2003; 100:1844–8.
22. Rubinson DA, Dillon CP, Kwiatkowski AV, et al. A lentivirus-based system to functionally silence genes in primary mammalian cells, stem cells and transgenic mice by RNA interference. Nat Genet 2003; 33:401–6.
23. Singer O, Marr RA, Rockenstein E, et al. Targeting BACE1 with siRNAs ameliorates Alzheimers disease neuropathology in a transgenic model. Nat Neurosci 2005; 8:1343–9.
24. Pfeifer A, Eigenbrod S, Al-Khadra S, et al. Lentivector-mediated RNAi efficiently suppresses prion protein and prolongs survival of scrapie-infected mice. J Clin Invest 2006; 116:3204–10.
25. Reynolds A, Leake D, Boese Q, Scaringe S, Marshall WS, Khvorova A. Rational siRNA design for RNA interference. Nat Biotechnol 2004; 22:326–30.
26. Schwarz DS, Hutvagner G, Du T, Xu Z, Aronin N, Zamore PD. Asymmetry in the assembly of the RNAi enzyme complex. Cell 2003; 115:199–208.

27. Heale BS, Soifer HS, Bowers C, Rossi JJ. siRNA target site secondary structure predictions using local stable substructures. Nucleic Acids Res 2005; 33:e30.
28. Brummelkamp TR, Bernards R, Agami R. A system for stable expression of short interfering RNAs in mammalian cells. Science 2002; 296:550–3.
29. Boden D, Pusch O, Lee F, Tucker L, Shank PR, Ramratnam B. Promoter choice affects the potency of HIV-1 specific RNA interference. Nucleic Acids Res 2003; 31:5033–8.
30. Naldini L, Blomer U, Gage FH, Trono D, Verma IM. Efficient transfer, integration, and sustained long-term expression of the transgene in adult rat brains injected with a lentiviral vector. Proc Natl Acad Sci USA 1996; 93:11382–8.
31. Follenzi A, Ailles LE, Bakovic S, Geuna M, Naldini L. Gene transfer by lentiviral vectors is limited by nuclear translocation and rescued by HIV-1 pol sequences. Nat Genet 2000; 25:217–22.
32. Hornung V, Guenthner-Biller M, Bourquin C, et al. Sequence-specific potent induction of IFN-alpha by short interfering RNA in plasmacytoid dendritic cells through TLR7. Nat Med 2005; 11:263–70.
33. Robbins MA, Li M, Leung I, et al. Stable expression of shRNAs in human CD34+ progenitor cells can avoid induction of interferon responses to siRNAs in vitro. Nat Biotechnol 2006; 24:566–71.
34. Bridge AJ, Pebernard S, Ducraux A, Nicoulaz AL, Iggo R. Induction of an interferon response by RNAi vectors in mammalian cells. Nat Genet 2003; 34:263–4.
35. Furumoto Y, Brooks S, Olivera A, et al. Cutting Edge: Lentiviral short hairpin RNA silencing of PTEN in human mast cells reveals constitutive signals that promote cytokine secretion and cell survival. J Immunol 2006; 176:5167–71.
36. Pfeifer A, Verma IM. Gene therapy: promises and problems. Annu Rev Genomics Hum Genet 2001; 2:177–211.

Chapter 22

Sperm Cryopreservation and In Vitro Fertilization

Susan Marschall, Auke Boersma, and Martin Hrabé de Angelis

Abstract

Since the mouse has become the most profound model system to investigate the genetics and pathogenetics of human diseases, a huge number of new mutant mouse strains has been generated and still a lot effort is being done to increase the number of suitable mouse models. In nearly all animal facilities the maintenance of breeding colonies is limited and the mouse strains have to be archived in a reliable way. Mouse sperm cryopreservation provides an efficient management of these genetic resources by reducing maintenance space and cost and by safeguarding them against, for example, disease, breeding failure, and genetic drift.

The sperm archiving method has been proven extensively in large-scale ENU mutagenesis programs and in mouse repository and resource centers worldwide (Federation of International Mouse Resources, http://www.fimre.org). Nevertheless, it is crucial to select accurately the most suitable archiving procedure, or combination of different archiving procedures, for each individual mouse mutant strain.

Keywords: Cryopreservation, sperm freezing, in vitro fertilization (IVF), archiving, mouse mutant.

1. Introduction

Since the first reports of successful cryopreservation of mouse spermatozoa were published *(1, 2)* a large number of mouse inbred and hybrid strains as well as mutant mouse lines have been used for the cryopreservation of spermatozoa *(3–6)* but the in vitro fertilization (IVF) rate, recovery rate after thawing, and percentage of live offspring may vary among the strains *(7)*. Nevertheless, mouse sperm cryopreservation provides an efficient means for storing valuable genetically modified mice.

Compared to embryo freezing the cryopreservation of spermatozoa offers a number of advantages. It requires fewer donor animals than embryo freezing as sperm from a single male could potentially give rise to as many as 20 times more offspring than embryos from a single female. Moreover males would not be

Ralf Kühn, Wolfgang Wurst (eds.), *Gene Knockout Protocols: Second Edition, vol. 530*

DOI 10.1007/978-1-59745-471-1_22 Springerprotocols.com

treated with gonadotropins and the collection and manipulation of spermatozoa would be faster *(8)*. Additionally this method will facilitate the preservation of strains in which female reproductive problems are characteristic *(9)*. However, the difference between freezing haplotypes (spermatozoa) and full genomes (embryos) has to be kept in mind. Sperm freezing is problematic in situations where the total genome genotype is of interest, for example, congenic or inbred strains. Recovering frozen embryos results in animals with the genetic background of their parents. To reach the latter after recovery of frozen spermatozoa, oocyte donors from the same strain have to be used. This might be important because several mutations lead to altered phenotypes in different genetic backgrounds. Sperm freezing of strains with maternally inherited alterations will be inappropriate if appropriate oocyte donors are not available. When comparing the costs of both methods, sperm cryopreservation is less expensive in terms of collecting and freezing material but is more expensive when recovering mice *(10)*.

In any case the most suitable archiving procedure, or combination of different archiving procedures, has to be selected accurately for each individual mouse mutant strain to ensure a reliable archive for future work.

2. Materials

2.1. Animals

Male mice used for sperm freezing should be at least 8 weeks and optimally 3 to 6 months old (*see* **Note 1**).

2.2. Equipment

1. Dewar flask, height about 30 cm, with liquid nitrogen (e.g., NeoLab Migge Laborbedarf-Vertriebs GmbH, Heidelberg, Germany, Cat. No. 175030291).
2. Dishes, 40 × 12 mm (*see* **Note 2**).
3. Dishes, 60 × 15 mm.
4. Filter 0.22 μm.
5. Filter 0.45 μm.
6. Four-well dishes.
7. Freezing canisters preparation: Insert a piece of styrofoam tightly onto the bottom of a 50-mL syringe, heat-seal the outlet of the syringe, fix syringe to an acrylic bar (length 26 cm) with a cable connector (*see* **Note 3** and **Fig. 22.1**).
8. Freezing rack (e.g., for 100 individual 0.25-mL straws; Minitüb, Tiefenbach, Germany; or custom-made for goblets, *see* **Fig. 22.2**)
9. Goblets (or Cryo-cups; e.g., MTG, Altdorf, Germany).

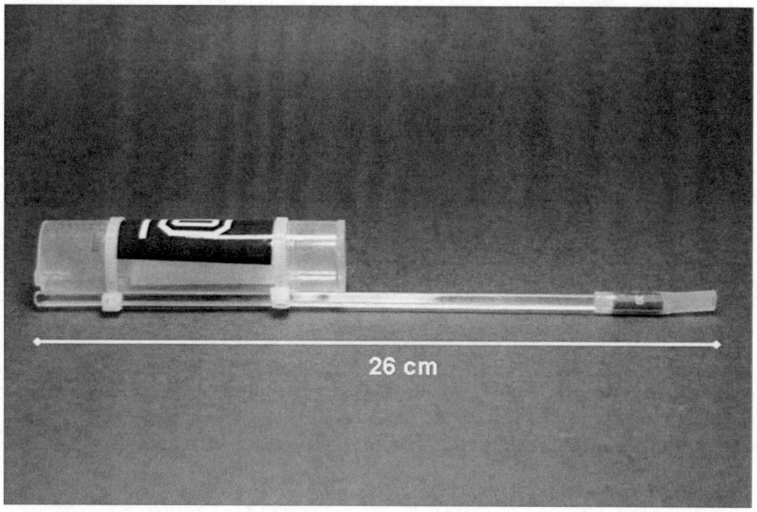

Fig. 22.1. Freezing canister.

10. Incubator (37°C, 5% CO_2 in air).
11. Labels (e.g., CILS International, Worthing, UK).
12. Liquid nitrogen tank.
13. Mouth pipette, consisting of a pipette holder and embryo transfer pipettes (e.g., BioMedical Instruments, Dr. Joachim Gündel, Zöllnitz, Germany; or self-made)
14. Pipetter, 200–1000 μL and corresponding tips.
15. Pipetter, 20–200 μL and corresponding tips.
16. Pipetter, 2–20 μL and corresponding tips (*see* **Note 4**).

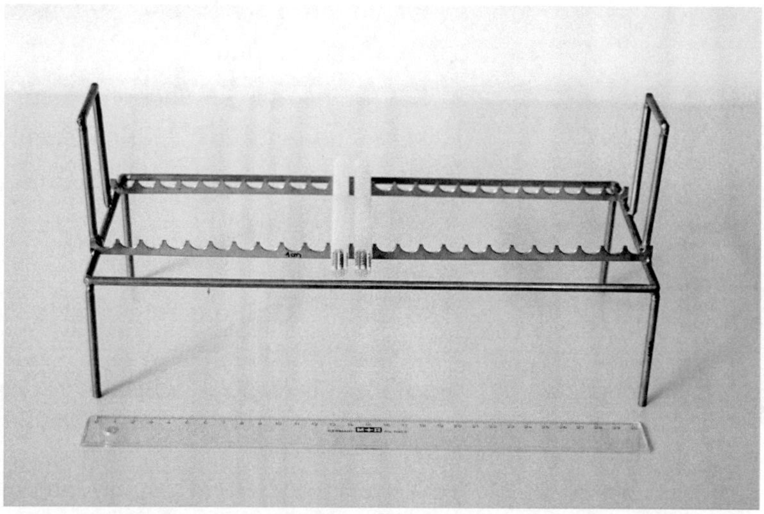

Fig. 22.2. Freezing rack: Custom-made for the freezing of straws (in goblets) in the vapor phase of liquid nitrogen; the length is about 30 cm.

17. Preparation needles.
18. Stereo microscope (ideally with a magnification up to 480 × and a transmitted light base, providing a high resolution and contrast, allowing a reliable evaluation and grading of the embryo quality).
19. Straws, 0.25 mL, transparent (MTG, Altdorf, Germany).
20. Styrofoam box for the freezing procedure, for example, with outer dimensions (width × length × height) of 27 × 43 × 21 cm, with liquid nitrogen.
21. Styrofoam box with ice, for the dissection of the epididymides.
22. Surgical instruments for the dissection of the epididymides (e.g., from FST, Heidelberg, Germany)
    a. Fine forceps
    b. Fine scissors
    c. Spring scissors
    d. Watchmaker forceps
23. Syringe, 1 mL (e.g., Tyco Healthcare Deutschland GmbH, Neustadt a.d. Donau, Germany, *see* **Note 5**).
24. Water bath (37°C).
25. Welding apparatus for plastic films (e.g., Rische + Herfurth GmbH, Hamburg, Germany, Polystar® 100 GE-GS).
26. Optional: Computerized Sperm Motility Analyzer (e.g., IVOS, Hamilton Thorne Biosciences, Inc., Beverly, USA).

2.3. Reagents

1. Cryoprotective agent (CPA); preparation see below.
2. 0.9% NaCl.
3. Liquid nitrogen (L_{N2}, *see* **Note 6**).
4. Human tubal fluid medium (HTF; preparation see below).
5. KSOM medium (Millipore, Billerica, USA).
6. M2 medium (Sigma-Aldrich Chemie GmbH, Germany).
7. Equilibrated mineral oil (Sigma-Aldrich Chemie GmbH, Germany; *see* **Note 7).**

2.4. Preparation of the Media

2.4.1. Cryoprotecting Agent (CPA)

The CPA consists of

a. 18% D(+) Raffinose-Pentahydrate (Sigma-Aldrich Chemie GmbH, Germany),

b. 3% skim milk (Difco™, Becton Dickinson)

in ultrapure water (e.g., Ampuwa, Fresenius Kabi Deutschland GmbH, Germany).

1. Warm approx. 50 mL of ultrapure water up to 60°C.
2. Dissolve 7.2 g raffinose and 1.2 g skim milk in 30 mL warm water (avoid foam production) and fill it up to 40 mL.

3. Centrifuge the solution at 18,000*g* for 60 min at 4°C or room temperature.
4. Take the supernatant, which has to be clear, and sterilize it through a 0.45-μm filter.
5. Measure the osmolarity of the medium (480–500 mOsm).
6. Make aliquots convenient for your usage (0.5–2 mL) and store them at –20°C for no more than 3 months.

2.4.2. HTF

1. Dissolve all reagents as given in **Table 22.1** in the indicated order in 75 mL of ultrapure water.
2. Fill up to exactly 100 mL with ultrapure water.
3. Sterilize the solution through a 0.22-μm filter.
4. Gas the medium using a pipette immersed in the solution with mixture of 5% CO_2 in air for 10 min, or leave it in the incubator (5% CO_2 in air) overnight.
5. Measure the osmolarity of the medium (290–300 mOsm, ideally 296 mOsm, *see* **Note 8**).
6. Store the medium at 37°C for no more than 1 week; leave the lid open.

Table 22.1
Composition of HTF. All components are from Sigma-Aldrich Chemie GmbH, Taufkirchen bei München, Germany, and the Cat. Nos. refer to this company

Component	mg/100 mL	Cat. No.
NaCl	593.75	S-9888
KCl	34.96	P-5405
KH2PO4	5.04	P-5655
MgSO4 × 7H2O	4.93	M-9397
Sodium lactate 60%	342 μL	L-7900
Glucose	50.0	G-6152
NaHCO3	210.0	S-5761
Sodium pyruvate	3.65	P-4562
Penicillin G	7.5	P-4687
Streptomycin	5.0	S-1277
$CaCl_2$ × $2H_2O$	60.0	C-7902
BSA	400.0	A-4378

2.4.3. Hormones for Superovulation

1. Add 20 mL 0.9% NaCl to Intergonan (PMSG) and 30 mL 0.9% NaCl to Ovogest (hCG) and mix well; the final concentrations is 50 IU/mL.
2. Make aliquots convenient for the numbers of females you want to inject; these aliquots can be stored for 1 month at –20°C.

3. Method

3.1. Sperm Cryopreservation

Before freezing valuable mutant mouse strains, a number of prerequisites have to be met. Despite the numerous different published cryopreservation methods, the problem remains that successful methods from one laboratory often prove unsuccessful in other laboratories, indicating that underlying critical factors have not been elucidated. Therefore it is crucial to prove the quality and success of the chosen methods in one's own lab and to demonstrate the reliability of recovering the frozen strains.

In general, spermatozoa show extreme sensitivity to the freezing procedure, and irrespective of the species only a proportion of the spermatozoa survive cryopreservation. Due to freezing injuries descending from the freezing–thawing process a decrease in fertilizing ability or quality between fresh and frozen spermatozoa could be observed *(11)*. Even if the sperm freezing techniques of mouse spermatozoa represent an efficient method of preserving genetically valuable strains, it is not clear whether it is really useful and effective for all known mouse strains.

On principle it has to be noticed that (a) reproductive differences between mouse strains seen in vivo are also evident in vitro and (b) hybrids normally have a better fertility than inbred strains and (c) different mouse strains may have a different freezing sensitivity resulting in varying sperm quality after thawing.

In this chapter a detailed description of the sperm freezing procedure routinely used at the Helmholtz Zentrum München, German Research Center for Environmental Health is given. With some modifications (*see* **Note 9**), the "Nakagata protocol" *(5)* still is successfully used in large-scale mutagenesis programs *(3)* and within the framework of EMMA-The European Mouse Mutant Archive *(12)* (see also www.emmanet.org).

1. Fill 100 μL HTF in a Petri dish (one per male to be frozen), cover it with mineral oil, and put the dish into the incubator (37°C, 5% CO_2 in air). This step is only necessary if the motility of the fresh sperm shall be determined.
2. Put the rack into the freezing box and fill it with LN_2 such that the level is about 3 cm below the straws. Close the box with the lid and wait at least 15 min that the vapor phase can develop.

3. Fill the Dewar vessel three-fourths with LN_2 and let one freezing canister for two males float in the LN_2 to precool them.
4. Label 12 straws per male with the strain name, male ID, and/ or sample ID.
5. Thaw an aliquot of CPA in the incubator. Fill for every male to be frozen one well of the four-well dish with 200 μL 0.9% NaCl, another well with 215 μL CPA. Number the wells clearly according to the number of males to be frozen to avoid mix-up of the samples.
6. Sacrifice the male mouse by cervical dislocation and dissect the two caudae epididymides including a piece of about 0.5 cm of the deferent duct. Place the epididymides in one well with 0.9% NaCl (on ice, alternatively at 37°C, *see* **Note 10**).
7. Remove under the stereo microscope all remaining fat, big blood vessels, and the part of the empty part of the deferent duct with a watchmaker forceps and a spring scissor.
8. Place both epididymides in the well with CPA. Cut the epididymides and the deferent duct several times with a spring scissor and let the spermatozoa swim out for 3–5 min (on ice or at 37°C). Shake the dish carefully until the suspension is homogenous. Do not pipette the sperm to mix it.
9. Take a sample of 2 μL sperm and give it into the preincubated 100 μL HTF drop in the incubator. After an incubation period of 10–20 min you can assess the motility of the sperm by visual-subjective inspection (*see* **Note 11**) using a microscope or, if available, a computerized sperm motility analyzer.
10. Pipette 12 × 17 μL aliquots of the sperm suspension on the lid of a four-well dish.
11. Connect the 1-mL syringe to the cotton plug side of a 0.25-mL straw. Aspirate successively: 100–150 μL HTF (about up to the half of the straw), about 1 cm air, one 15-μL-drop sperm. Then aspirate air until the HTF reaches the cotton-PVA plug and thereby seals the straw. Then weld the other end of the straw with the welding apparatus (*see* **Fig. 22.3**).
12. Repeat the last step for all sperm drops.
13. Put six straws each into one goblet, place goblets horizontally for 15 min on the rack in the vapor phase of the LN_2.
14. After this time, place goblets with straws in the precooled freezing canisters and plunge them into liquid nitrogen.
15. Transport the samples to the LN_2 tank (*see* **Note 12**).

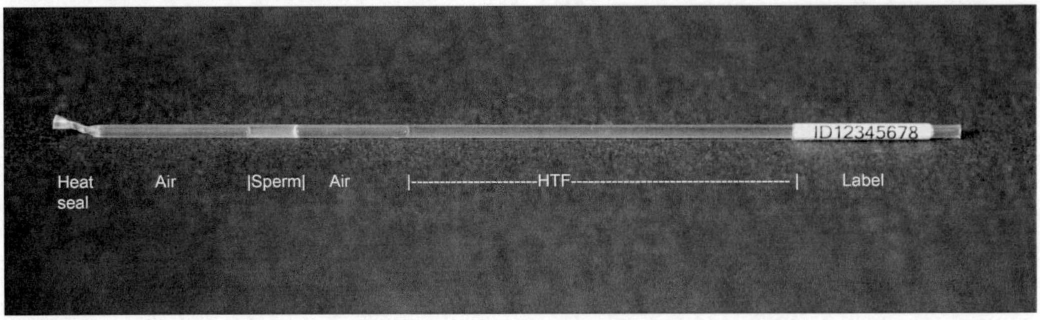

Fig. 22.3. Straw filled with the sperm suspension ready for freezing. The label covers the cotton-PVA plug which seals the straw. The HTF acts as a weight to prevent the straws from floating when plunging them into LN_2. Note the distance between sperm drop and heat seal.

3.2. In Vitro *Fertilization*

The IVF is a routinely used method for the recovery of frozen spermatozoa. For a successful IVF two basic requirements have to be accomplished: (a) capacitated spermatozoa and (b) mature, unaged oocytes. However, due to their inherent genetic differences inbred mutant mouse strains respond variedly to assisted reproductive technologies like superovulation, IVF, and cryopreservation *(13)*.

To capacitate spermatozoa in vitro, freshly collected spermatozoa have to be incubated in a suitable medium (similar to the environment of the female genital tract) for about 90 min at 37°C. For frozen–thawed spermatozoa 45 min (or even less) are sufficient because low temperatures induce capacitation-like changes *(14)* (*see* **Section 3.2.2**).

To acquire a larger number of oocytes a superovulation has to be performed. This treatment with exogenous hormones increases the yield of naturally produced oocytes and allows a specific timing of oocyte collection. Follicle growth is stimulated with pregnant mare serum gonadotropin (PMSG) and a simultaneous ovulation is induced with human chorionic gonadotropin (hCG). Several parameters, for example, age, strain, dose, and injection time of hormones can influence the efficiency of superovulation and should be considered.

Generally F1 hybrid females are recommended for superovulation because of their high yield of oocytes in contrast to inbred strains. Nevertheless, it is often desirable to have a homogeneous genetic background for spermatozoa and oocytes. In this case the appropriate inbred strain has to be used even if it is a low ovulator.

The number of females to be superovulated, which is necessary for the desired outcome in terms of embryos, can roughly be estimated if the in vitro reproductive characteristics of the used strain are known. Byers et al. (2005) *(13)* provide the number of oocytes produced in response to superovulation, the proportion of two-cell embryos produced following IVF and the proportion of live pups born following the transfer of fresh and thawed two-cell embryos for ten inbred strains of JAX® mice, listed as high priority on the Mouse Phenome Database: 129S1/SvImJ, A/J, BALB/cJ, BALB/cByJ, C3H/HeJ, C57BL/6 J, DBA/2 J, FVB/NJ, NOD/LtJ, and SJL/J.

The best age for superovulation is strain-dependent and usually lies between 3 and 6 weeks but even females up to 8 or 10 weeks can be used. The amount and timing of injection are also very important. For both hormones 5–7.5 IU per mouse are most commonly used. Nevertheless, each strain can react differently and if nothing is known about a specific strain a dose–response check should be performed. hCG should be injected between 48–50 h after the PMSG injection. A longer or shorter interval may interfere with the response. Oocytes are collected between 14 h and 15 h after hCG injection. If they are collected earlier than 13.5 h after injection, insemination may fail. Collection after 16 h or later is also unsuitable. At this time oocytes already begin to age and can no longer be fertilized. Due to this injection schedule the collection of the spermatozoa (in vitro capacitation) has to be timed as well.

A successful IVF results in two-cell embryos after overnight culture. These embryos have to be transferred in pseudopregnant recipient females to develop them to term.

3.2.1. Superovulation of the Oocyte Donors

1. Three days before IVF, inject an appropriate number of females intraperitoneally with PMSG; 4 PM is recommended as injection time. The hormone dosage is 2.5–10 IU per female depending on the strain.
2. One day before IVF, inject the females i.p. with hCG; 6 PM is recommended as injection time.

3.2.2. Preparation of the Sperm Dishes

1. On the day before the IVF, prepare for every group of females to be dissected at the same time (about 3–10 females per dish; *see* **Note 13**) the following dishes and put them overnight in the incubator (37°C, 5% CO_2):
   a. For oviduct collection: 3 mL mineral oil in a Petri dish (40 × 12 mm).
   b. For the preparation of the cumulus–oocyte complexes: 400 μL HTF in a Petri dish (60 × 15 mm), covered with mineral oil.
   c. For the sperm and fertilization: 500 μL HTF medium in a Petri dish (60 × 15 mm), covered with equilibrated mineral oil.
2. On the day of the IVF, starting before the oocyte collection, that is, about 13.5 h after the hCG injection, thaw the sperm: take the frozen sperm straw out of the LN_2 and place it directly into the water bath (37°C) for 10–15 min.

3. Dry the straw with a tissue, keep straw horizontally and cut both ends. Let only the sperm suspension slowly flow out onto the lid of a Petri dish. Add 2–4 μL of the sperm suspension to the prepared medium drops.
4. Place all dishes in the incubator for 45 min to capacitate the spermatozoa. In this time the preparation of the oocytes can take place.
5. After an incubation period of 10–20 min you can assess the motility of the frozen–thawed sperm by visual-subjective inspection (*see* **Note 11**) using a microscope or, if available, a computerized sperm motility analyzer.

3.2.3. Preparation of the Oocytes and IVF

1. Sacrifice the females 14 h after hCG injection in groups corresponding to the number of fertilization dishes.
2. Dissect both oviducts of each female and transfer all oviducts in a dish with mineral oil.
3. Transfer the oviducts in the oil surrounding the 400 μL HTF drop in a 60-mm dish.
4. In the oil open the swollen ampullae of every oviduct with two preparation needles, holding the oviduct with one needle and tearing the ampulla carefully with the other. The oocyte–cumulus complexes will flow out or are drawn out with the tip of the needle.
5. Pull the complexes into the HTF medium drop. Any blood or tissue residues which could be detrimental for the fertilization process should be removed at this step.
6. Transfer the cumulus–oocyte complexes into the fertilization dishes, already containing capacitated spermatozoa, using a yellow pipette tip.
7. Incubate oocytes and spermatozoa for 4 h (37°C, 5% CO_2).
8. Prepare dishes and put them into the incubator (37°C, 5% CO_2):
   a. Washing dishes (one per fertilization dish): 4 × 100 μL KSOM in a 60 × 15 mm Petri dish.
   b. Overnight culture dishes (one per fertilization dish): 100 μL KSOM, covered with equilibrated mineral oil, in a 40 × 12 mm Petri dish.
9. After 4 h, wash the oocytes, or presumptive one-cell embryos, respectively, through the four washing drops KSOM using a mouth pipette with embryo transfer glass capillaries (*see* **Note 14**).
10. After washing transfer the oocytes into the 100 μL KSOM drop under oil and culture overnight (37°C, 5% CO_2).

3.2.4. Recovery of the Two-Cell Embryos

1. About 24 h after the IVF, count the number of two-cell embryos. The cleavage rate can be calculated as number of embryos divided by the total number of cells times 100. Only embryos with two symmetrical blastomeres and an intact zona pellucida should be selected for embryo transfer.
2. For the embryo transfer the two-cell embryos are washed through four 100 μL drops M2 medium using a mouth pipette and embryo transfer glass capillaries.
3. After washing transfer the two-cell embryos in 100 μL M2 medium covered with equilibrated mineral oil and put them on a warming plate (37°C) until you transfer them.

4. Notes

1. The use of males younger than 8 weeks or older than 1 year might result in sperm with reduced concentration and motility. However, these samples can be used successfully for IVF, if no other material is available. If possible, the fertility of the males should be proven by successfully mating them to females for 1 week. The mating should be terminated at least one week before the sperm freezing.
2. Plasticware may have a negative effect on the IVF outcome. In our lab we have good experiences with Nunc products (Thermo Fisher Scientific, Langenselbold, Germany), which have been subjected to a mouse embryo toxicity test by the manufacturer.
3. The "freezing canisters" serve in our protocol just as a handle to plunge the straws in LN_2 and for the transport of a large number of samples from several males in a Dewar vessel to the final storage tank. In the original "Nakagata protocol" *(5)* they are used to cool the straws by floating them in the vapor phase of liquid nitrogen before they are plunged into LN_2.
4. Generally for the handling of fresh and even more for frozen–thawed mouse sperm special caution is recommended. The shear forces originating from a narrow tip orifice can injure the sperm. Therefore many labs use wide-bore pipette tips which are commercially available for larger pipette volumes (Rainin, Mettler-Toledo GmbH, Giessen, Germany, Cat. No. HR-250 W for 250 μL and HR-1000 W for 1000 μL), or can be easily made for any pipette tip by cutting off the first 3–5 mm of the tip with a scissors.

5. The Monoject syringe Cat. No. 1180100555 fit directly to the 0.25-mL sperm straw. For syringes of other companies it might be necessary to dilate the opening of the syringe manually.
6. Liquid nitrogen can cause severe damage (frostbite) due to the extreme temperature of –196°C or –320 °F; handle therefore with extreme caution following the safety instructions of your facility.
7. The use of equilibrated mineral oil should increase the compatibility of the oil and the embryos. The equilibration is performed by mixing the same volume of oil and HTF into a 50-mL tube. Do not shake, but mix the two phases by carefully turning the tube upside down several times. Place the tube into the incubator overnight.
8. There are findings which stress the significance of the osmolarity for a successful cryopreservation and recovery. An osmolarity of 294–296 ± 2 mOsm was found to be optimal especially for the recovery of C57Bl6/J and 129S6/SvEv strains. In our lab, however, we could not find a significant difference in the IVF outcome in the osmolarity range 290–300 mOsm.
9. The original "Nakagata protocol" is based on the use of semen straws and "freezing canisters" which serve as a float on the liquid nitrogen in a Dewar flask. It could be shown that this setup involves a number of unspecified parameters which can significantly affect the cooling rate *(15)*. We modified therefore the protocol by replacing the "freezing canister" used in this protocol with a freezing rack which allows a better standardization of the cooling rate. Another more fundamental modification of the "Nakagata protocol" is the use of cryovials instead of straws. This protocol has successfully been used on a large scale, for example, at The Jackson Laboratory in Bar Harbour *(7)* and in the MRC Mammalian Genetics unit in Harwell *(6)*.
10. It is possible to collect the sperm at 37°C, 23°C (room temperature) or 0°C without any detrimental effect on the sperm motility *(9)*. Several sperm freezing protocols recommend working at 37°C. However, if the dissection room and the sperm freezing lab are not closed together, or if several males are processed at the same time, it seems to be advantageous to work on ice to slow down the metabolism of the spermatozoa.
11. Try to evaluate (a) the concentration of the sperm; (b) the ratio of living:dead spermatozoa; (c) the straightness of the swimming spermatozoa (can be seen at the border of the medium drop); (d) the velocity of the spermatozoa.

12. Preferably split the samples of one male to two seperate LN_2 tanks, if possible at two different locations, thus building a backup system.
13. The number of females to be dissected at the same time is among others dependent (a) on the infrastructure of the lab, that is, the distance between the dissection room and the IVF lab; (b) the time required for the dissection and preparation of the females, oviducts, and cumulus–oocyte complexes, that is, eventually the practice and skill of the technician. Make sure that the time between killing the females and adding the oocytes to the sperm suspension is about 5–10 min, but no more than 20 min. Furthermore, oocytes are sensible to chilling. A "cool pack" warmed to 37°C and placed below the dishes can help to maintain a suitable temperature.
14. Again, time is a critical factor during this step, and should not exceed 10 min per dish. If available, warming plates which can be used on a microscope are helpful to reduce heat loss of the embryos during washing. If the oocytes are still surrounded by or even packed within many cumulus cells, this is a hint that the vitality of the spermatozoa is strongly reduced or the sperm concentration was much too low. Degenerated oocytes can already be sorted out.

Acknowledgments

This work was supported by NGFN2, 01GR0430 and by the European Commission's FP6 Research Infrastructures Programme. We thank our technicians Steffie Dunst, Monika Beschorner, Alex Huber and Bernhard Rey for their invaluable work.

References

1. Tada N, Sato M, Yamanoi J, Mizorogi T, Kasai K, Ogawa S. Cryopreservation of mouse spermatozoa in the presence of raffinose and glycerol. J Reprod Fertil 1990; 89(2):511–6.
2. Yokoyama M, Akiba H, Katsuki M, Nomura T. [Production of normal young following transfer of mouse embryos obtained by in vitro fertilization using cryopreserved spermatozoa]. Jikken Dobutsu 1990; 39(1):125–8.
3. Marschall S, Huffstadt U, Balling R, Hrabé de Angelis M. Reliable recovery of inbred mouse lines using cryopreserved spermatozoa. Mamm Genome 1999; 10:773–6.
4. Nakagata N. Use of cryopreservation techniques of embryos and spermatozoa for production of transgenic (Tg) mice and for maintenance of Tg mouse lines. Lab Anim Sci 1996; 46(2):236–8.
5. Nakagata N. Cryopreservation of mouse spermatozoa. Mamm. Genome 2000; 11(7):572–6.
6. Thornton CE, Brown SD, Glenister PH. Large numbers of mice established by in vitro fertilization with cryopreserved

spermatozoa: implications and applications for genetic resource banks, mutagenesis screens, and mouse backcrosses. Mamm Genome 1999; 10(10):987–92.

7. Sztein JM, Farley JS, Mobraaten LE. In vitro fertilization with cryopreserved inbred mouse sperm. Biol Reprod 2000; 63(6):1774–80.
8. Tao J, Du J, Kleinhans FW, Critser ES, Mazur P, Critser JK. The effect of collection temperature, cooling rate and warming rate on chilling injury and cryopreservation of mouse spermatozoa. J Reprod Fertil 1995; 104(2):231–6.
9. Sztein JM, Farley JS, Young AF, Mobraaten LE. Motility of cryopreserved mouse spermatozoa affected by temperature of collection and rate of thawing. Cryobiology 1997; 35(1):46–52.
10. Landel CP. Archiving mouse strains by cryopreservation. Lab. Anim. (NY) 2005; 34(4):50–7.
11. Nishizono H, Shioda M, Takeo T, Irie T, Nakagata N. Decrease of fertilizing ability of mouse spermatozoa after freezing and thawing is related to cellular injury. Biol Reprod 2004; 71(3):973–8.
12. Hagn M, Marschall S, Hrabe de Angelis MH. EMMA: The European Mouse Mutant Archive. Brief Funct Genomic Proteomic 2007; 6(3):186–92.
13. Byers SL, Payson SJ, Taft RA. Performance of ten inbred mouse strains following assisted reproductive technologies (ARTs). Theriogenology 2006; 65(9):1716–26.
14. Fuller SJ, Whittingham DG. Capacitation-like changes occur in mouse spermatozoa cooled to low temperatures. Mol Reprod Dev 1997; 46(3):318–24.
15. Stacy R, Eroglu A, Fowler A, Biggers J, Toner M. Thermal characterization of Nakagata's mouse sperm freezing protocol. Cryobiology 2006; 52(1):99–107.

Part IV

Phenotype Analysis

Chapter 23

Influence of Genetic Background on Genetically Engineered Mouse Phenotypes

Thomas Doetschman

Abstract

The history of mouse genetics, which involves the study of strain-dependent phenotype variability, makes it clear that the genetic background onto which a gene-targeted allele is placed can cause considerable variation in genetically engineered mouse (GEM) phenotype. This variation can present itself as completely different phenotypes, as variations in penetrance of phenotype, or as variable expressivity of phenotype. In this chapter we provide examples from gene-targeting literature showing each of these types of phenotype variation. We discuss ways in which modifier genes can affect the phenotype of a mouse with a mutant gene, and we give examples of modifier locus identification. We also review approaches to minimize gene polymorphism and flanking gene differences between experimental animals, and between them and their controls. In addition, we discuss the advantages and disadvantages of performing the first analysis of a knockout mouse on a mixed genetic background. We conclude that a mixed background provides the quickest preview of possible strain-dependent phenotypes *(1, 2)*. Finally, we review recent approaches to improving genetic diversity by generating new inbred strains that encompass a broader range of alleles within the mouse species.

Key words: Knockout, mouse, genetic background, genetic engineering, penetrance, expressivity, modifier gene.

1. Introduction

The theoretical basis for an understanding of Mendelian inheritance of complex traits and the importance of genetic background in mouse studies was presented in a note to *Science* by C. C. Little in 1914 *(3)*, an argument he later used to explain the Mendelian nature of the apparent non-Mendelian inheritance of tumor transplantation susceptibility *(4)*. It was this understanding that led him

Ralf Kühn, Wolfgang Wurst (eds.), *Gene Knockout Protocols: Second Edition, vol. 530*

DOI 10.1007/978-1-59745-471-1_23 Springerprotocols.com

to believe in the importance of developing inbred mouse strains as tools for investigating complex human disease. A demonstration of the power of utilizing the genetic diversity of mouse to generate models for a complex trait was provided by Gunther Schlager *(5)* who used an eight-way cross of common inbred strains to develop advanced intercross lines of mice with differential blood pressures. More recently, striking phenotypic differences have been found in closely related strains, for example, in response to proteoglycan-induced arthritis among ten different C3H substrains *(6)*. These and other studies make it clear that genetically engineered mouse (GEM) phenotype can be background dependent.

The influence of genetic background on GEM phenotype became apparent in some of the early knockout mice. Hynes reported that fibronectin knockouts had considerable variation in phenotype which he attributed to the analyses being done on embryos of a 129 and C57BL/6 background (Hynes George 1993). Baribault and colleagues *(7)* showed that a keratin-8 deficiency on different backgrounds leads to quite different phenotypes of midgestational lethality or adult colorectal hyperplasia *(8)*. Other notable background-dependent differential GEM phenotypes have been found in EGFR-deficient mice with phenotypes ranging from a peri-implantation lethality to a weaning-age lethality due to abnormalities in multiple organs *(9)*, TGFβ1-deficient mice with phenotypes ranging from preimplantation to weaning-age lethalities (reviewed in *(2, 10)*), and GEM models for cystic fibrosis with the presence *(11)* or absence *(12)* of the lung disease. Another cystic fibrosis model was shown to have background-dependent differences in the severity of intestinal obstruction *(13)*.

2. Methods

2.1. Background-Dependent Variability in Penetrance and Expressivity

The presence of phenotypic variability in penetrance and/or expressivity is nearly always due to the knockout allele being present on a mixed genetic background. This was the case for the fibronectin, keratin-8, cystic fibrosis, and TGFβ1 examples mentioned above. Generation of congenic strains for the mutant allele usually leads to a more consistent phenotype. Incomplete penetrance and variable expressivity in GEMs can also result from environmental influences. Our experience with TGFβ- and SMAD3-deficient mice will primarily be used here to illustrate these points.

We have maintained our *Tgfb1* knockout strain on a mixed genetic background of 129/SvJ and CF-1. On this mixed background only half of the homozygous mutant animals are born, and they subsequently die of a weaning-age autoimmune disease *(14)*. Maintenance on a mixed background is required to prevent loss of the autoimmune phenotype because on several inbred backgrounds nearly all *Tgfb1*$^{-/-}$ animals die of preimplantation *(10)* lethalities. Similarly, on a different mixed genetic background of 129/Sv, C57BL/6, and NIH/Olac, Akhurst's group found that half of *Tgfb1*$^{-/-}$ animals die of a yolk sac developmental defect *(15)*, and that survival to birth ranges from nearly 0% on a C57BL/6 J/Ola background to about 80% on an NIH/Ola background *(16)*. In both *Tgfb1* knockout strains, there is clearly an incomplete penetrance of phenotype.

When rescued by rendering the mice immunocompetent (*Tgfb1*$^{-/-}$ *Rag2*$^{-/-}$ mice) we have found that a TGFβ1 deficiency leads to a colitis-associated colon cancer at 100% penetrance when the mice are of mixed but primarily 129/Sv background (approximately 85% 129 and 15% CF-1) *(17)*; whereas, on a primarily C3H background (approximately 85% C3H and 15% 129 plus CF-1) no colon cancer is detectable in *Tgfb1*$^{-/-}$ *Prkdc*$^{scid/scid}$ immunodeficient mice, even though they have colitis *(18)*. Consequently, there is considerable genetic background influence on TGFβ1-deficient GEM phenotypes, most but not all of which express variable penetrance and expressivity.

The *Tgfb2* knockout strain has been maintained on a mixed genetic background of 129 and Black Swiss, and these mice die from midgestation stage to birth and have severe heart, skeletal, ear, and eye defects *(19–22)*. In general, nearly all of the many congenital defects in these mice present with incomplete penetrance and variable expressivity. Although at first glance it might seem that this variability would be problematic for determining the mechanisms underlying specific defects, it has been useful for correlating the extent of changes in pathways upstream and downstream of TGFβ2 with the variation in penetrance and/or expressivity of the defect. This type of analysis can therefore provide a degree of built-in experimental control.

The *Tgfb3* knockout strain is also maintained on a mixed genetic background of 129 and Black Swiss. These mice die within the first 18 h after birth due to a completely penetrant cleft palate with widely varying expressivity; whereas, in C57BL/6 congenics the expressivity of the cleft palate is very high with low variability. Hence, the developmental progression of the defect could more fully be characterized on the mixed background *(23)*.

Environmental influences can also contribute significantly to incomplete penetrance and variable expressivity of GEM phenotype. Although *Tgfb1*$^{-/-}$ *Rag2*$^{-/-}$ mice can develop colon cancer at 100% penetrance *(17)*, a drop to 0% penetrance occurs in the absence of all enteric flora (germ-free mice), and this drop in penetrance is maintained if the gut flora of the mice are reconstituted with *Helicobacter hepaticus*-free flora *(18)*. Similarly, *Smad3*$^{-/-}$ mice on a 129 background have recently been found to develop colon cancer if *Helicobacter hepaticus* is present *(24)*.

2.2. Modifier Genes

Genetic background-dependent differences in the expressivity of the intestinal obstruction of a GEM model for cystic fibrosis has been used to screen for modifier loci that support increased longevity *(13)*. About 30% of F2 progeny from 129/Sv × CD1 F1 intercrosses were found to live at least 6 weeks while the rest died of bowel obstruction by 2 weeks of age. Polymorphic markers for the two backgrounds were screened for association with increased survival age. A locus on proximal chromosome 7 was identified. Complementary studies on chloride conductance indicated that the increased longevity correlated with upregulation of a compensating chloride current. Guilbault et al. *(25)* has reviewed cystic fibrosis GEMs.

Several studies by Akhurst have identified loci for genetic modifiers of the embryonic yolk sac lethality in TGFβ1-deficient mice. As mentioned above, the penetrance of this lethality varies considerably between the C57BL/6 J/Ola and NIH/Ola backgrounds *(16)*. Their first genetic screen was based upon survival of F2 $Tgfb1^{-/-}$ animals to birth and yielded a locus on NIH/Ola chromosome 5 ($Tgfbkm1^{NIH}$) that accounts for three-fourths of the survival effect. In a follow-up study it was determined that a second locus on chromosome 12 ($Tgfbkm3^{NIH}$) modified the first locus to increase its ability to support survival to birth *(26)*. A third study was based upon data that the TGFβ1 deficiency on a 129S2/SvHsd background had a higher incidence of survival to birth (30%) than when on a C57BL/6NTac (0%) *(27)*. Analysis of survival to birth in F1 progeny of reciprocal crosses revealed differential maternal imprinting effects in which F1s of C57 mothers have a much higher survival-to-birth rate. In addition, an F1 intercross genome scan revealed a chromosome 1 modifier ($Tgfbkm2^{129}$) which accounts for 90% of the survival, independent from the maternal effect.

2.3. Elimination of Gene Polymorphism and Flanking Gene Problems

The flanking gene allele problem was originally addressed by Smithies and Maeda *(28)*. They were concerned about phenotyping GEMs for complex genetic diseases (atherosclerosis and essential hypertension in their case) when there could be allelism in trait-modifying genes flanking the targeted gene. This is especially problematic when one is using GEMs to prove causality for a candidate gene drawn from human linkage studies that show disease association with a particular chromosomal region. In this case one does not know whether it is the targeted candidate gene, a particular allele of another gene in that chromosomal region, or a complex interaction between the two that phenocopies the human disease trait. They suggested that if heterozygous F1 offspring of germline chimeras were crossed with their wild-type F1 littermates, rather than being selfed, then at the F2 generation, wild-type control animals could be screened for those that are non-allelic (129) at the flanking regions (*see* **Fig. 23.1**). This scheme solves the flanking gene problem, but not potential polymorphic differences in unlinked regions of the genome.

Wolfer et al. *(29)* reviewed breeding schemes designed to control for the widely recognized problem of genetic background differences in general. They discussed the Banbury Conference on Genetic Background *(30)* recommendation of producing co-isogenic strains (e.g., co-isogenic 129 and congenic B6) followed by phenotype analysis on F1 hybrids of the two strains (*see* **Fig. 23.2**). This basically solves all polymorphism problems except for the flanking gene problem, but is quite expensive as it requires maintenance of two strains for each targeted gene. However, they discuss breeding schemes to screen for flanking gene effects that may contribute significantly to a knockout phenotype (*see* **Fig. 23.2**). Finally, they point out that another solution to these

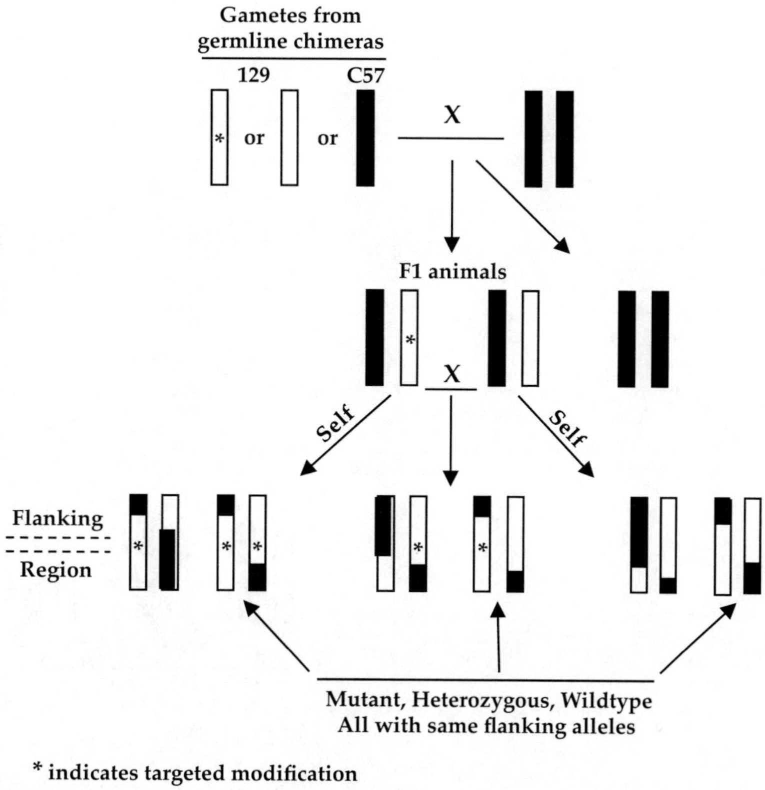

Fig. 23.1 Breeding scheme to eliminate flanking allele differences between experimental and control animals. F1 offspring from both 129 gametes are genetically identical except for the genetically modified allele. Selfing the wild-type F1 animals (second row, right) yields F2 animals, some of which have only 129 alleles in the flanking region (bottom row, far right animal). Selfing the F1 animals that carry the genetic modification will yield homozygous mutant animals which will always be non-allelic in the flanking region. Finally, crossing the wild-type and heterozygous F1 animals will yield heterozygous animals nearly half of which will be non-allelic in the flanking region. Arrows (third row) indicate experimental, heterozygous, and control animals that do not have the flanking gene problem, though each animal will have a different allelic mixture in non-flanking genome regions.

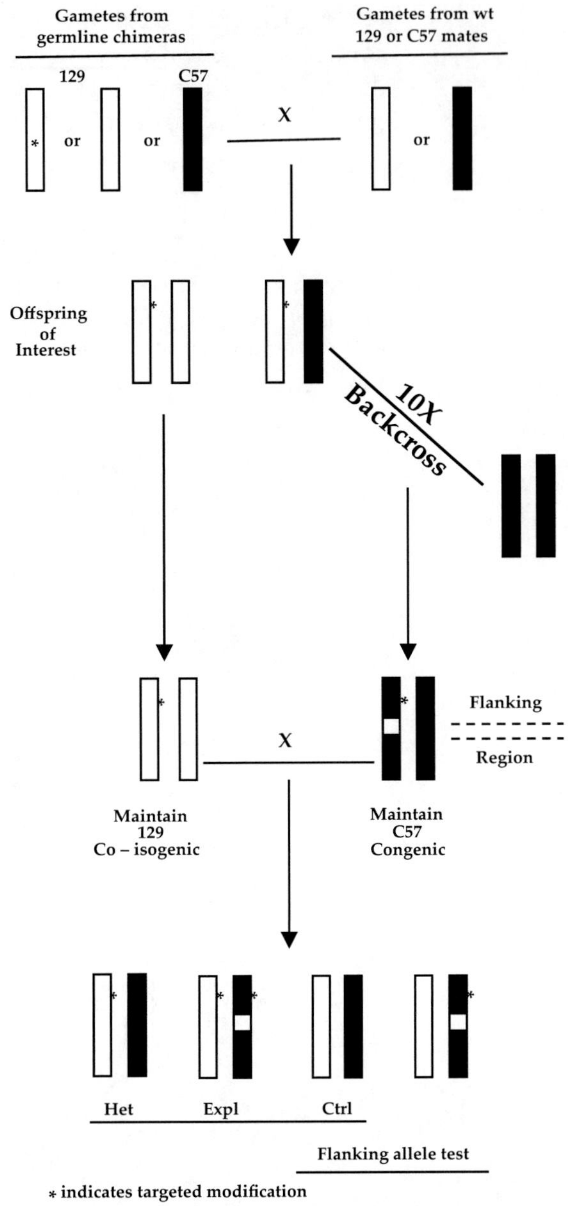

Fig. 23.2. Breeding scheme to eliminate all genetic background differences except in the flanking region. Congenic and co-isogenic strains are generated either by crossing germline chimera to C57 or 129 mates, respectively. Congenic strain can be generated through multiple backcrosses or by speed congenics procedures. Crossing the co-isogenic and congenic strains will yield animals of genetic backgrounds similar to that of F1 hybrids except for the region flanking the modified gene. Hence, experimental, control, and heterozygous animals can be compared with identical (except for flanking region) genetic backgrounds. Flanking allele effects can be tested by phenotypic comparison of the wild-type and heterozygous animal with the modified gene on the C57 congenic chromosome.

problems is to use conditional knockout alleles which allow for comparison between the "on" and "off" states in animals in which the genetic backgrounds are completely identical.

In general, keeping in mind the recommendations of these two papers *(28, 30)* during phenotype analysis of genetically modified genes should allow the investigator to identify any major roles played by differences in genetic background.

2.4. Value of Initially Analyzing Null Phenotypes on a Mixed Genetic Background

Knockout mice are usually generated by crossing a germline chimera, in which the knockout allele is on a 129 background, with an animal of any desired background. With speed congenic techniques congenic animals can be generated within 1–2 years *(31, 32)*. This is the case even though there are breeding schemes to test for or eliminate flanking gene effects. The resulting offspring are then intercrossed to generate homozygous mutant animals that will either be inbred 129 strain or F2 generation mice with a 50/50 mixture of 129 and the other desired background. With the exception of doing the gene targeting in an ES cell of another background, putting the targeted allele on a background other than 129 requires a standard backcrossing scheme. For convenience, the first homozygous mutant animals can be produced both on a 129 inbred background and on a mixed background. If one produces experimental and control mice from Fn generations derived from the mixed background strain, each experimental and control animal will have a different mixture of the two original backgrounds, assuming that the mixed strain is maintained as an advanced intercross line *(33)*. The question arises as to which background is better for phenotype analysis. Obviously, the more backgrounds the better; however, with limited resources, choices must be made.

We suggest that the background most likely to provide the widest range of phenotypes is the mixed background. This is due to the considerable background dependence of knockout phenotypes discussed previously. On a mixed background this phenotype variation could often play itself out as incomplete penetrance and variable expressivity. These, in turn, would likely decrease as the targeted allele were moved to a more inbred state. Consequently, the mixed-background knockouts potentially display a wide range of phenotypes, and those phenotypes with incomplete penetrance and variable expressivity would be candidate phenotypes upon which a modifier gene search could be based. If resources allow phenotype analysis of a second knockout strain, the 129 strain knockout would be most appropriate because it would reveal which of those mixed-strain phenotypes may have 129 strain modifiers, and, by elimination, which phenotypes may have modifiers on the other strain of the original mixture. It is for these reasons that we always make our first phenotype screen on a mixed strain. The TGFβ1 knockout mouse provides an important case in point. Had the decision been made to put the knockout allele on the C57BL/6 or 129 background, embryonic lethalities would have precluded discovery of the important roles played by TGFβ1 in autoimmunity *(34)*, platelet aggregation *(35)*, colon cancer *(17)*, and cardiac hypertrophy *(36)*.

2.5. Inbred Strain-Specific GEMs: Present and Future

As we have seen, there is significant background-dependent phenotypic variability found in GEM strains. Nonetheless, the power of combining the inherent phenotypic differences between inbred strains with GEMs is underutilized. Ideally one should be able to choose ES cell lines from the inbred strain(s) that are most appropriate for the phenotype/disease/defect to be investigated. Given advances in the understanding of embryonic "stemness" *(37–39)* procedures will likely be developed that improve the ability to generate strain-specific ES cell lines. Such cell lines could then be used to put "disease" genes, for example, into inbred strains with differences in susceptibility to that disease. This would be akin to studies that have used congenic lines for the *ApoE* KO gene to model and investigate aspects of hyperlipidemia *(40–42)*. This direct ES cell approach would save time by avoiding backcrossing and it would eliminate the "flanking gene" problem.

With completion of the mouse genome sequencing project *(43)* it is now known that the nearly 500 traditional strains of inbred mice (http://www.informatics.jax.org/external/festing/search_form.cgi) represent from only one-third *(44)* to one-half *(45)* of the genetic diversity present in the *Mus musculus* species. Of the ancestral subspecies groups, *domesticus*, *castaneus*, and *musculus*, there is a disproportionately low representation of *castaneus* in the traditional strains. Hence, the value of mouse genetics for investigating complex disease would be greatly improved were a new set of inbred strains developed that more fully represented the full potential of mouse genetic diversity. To this end the Complex Trait Consortium (http://www.complextrait.org/) has initiated the Collaborative Cross *(46)* in which 1000 new recombinant inbred strains will be derived from a randomized cross of eight inbred strains that more fully represent the genetic diversity of *Mus musculus* (http://lsd.ornl.gov/mgg/projects/collabcross.html) than that presently available. The addition of these new RI strains will expand the availability of genetic background choices upon which to make GEMs.

3. Conclusion

The study of mouse genetics has taken an exciting step forward with the advent of gene targeting via homologous recombination in ES cells. The knockout mice are wonderfully informative because of the unexpected and wide-ranging phenotypes that can result. Cognizance of the importance of genetic background on differences in knockout phenotype and of approaches for analyzing the genetics of those differences will broaden our understanding of the complexities of gene function.

References

(1) Doetschman T. Interpretation of phenotype in genetically engineered mice. Lab Anim Sci 1999; 49(2):137–143.

(2) Sanford LP, Kallapur S, Ormsby I, Doetschman T. Influence of genetic background on knockout mouse phenotypes. Methods Mol Biol 2001; 158:217–225.

(3) Little CC. A possible Mendelian explanation for a type of inheritance apparently non-Mendelian in nature. Science 1914; 40(1042):904–906.

(4) Little CC, Tyzzer EE. Further experimental studies on the inheritance of susceptibility to a transplantable carcinoma (JA) of the Japanese Waltzing Mouse. J Med Res 1916; 33:393–427.

(5) Schlager G. Selection for blood pressure levels in mice. Genetics 1974; 76(3):537–549.

(6) Glant TT, Bardos T, Vermes C, Chandrasekaran R, Valdez JC, Otto JM et al. Variations in susceptibility to proteoglycan-induced arthritis and spondylitis among C3H substrains of mice: evidence of genetically acquired resistance to autoimmune disease. Arthritis Rheum 2001; 44(3):682–692.

(7) Baribault H, Penner J, Iozzo RV, Wilson-Heiner M. Colorectal hyperplasia and inflammation in keratin 8-deficient FVB/N mice. Genes Dev 1994; 8(24):2964–2973.

(8) Baribault H, Price J, Miyai K, Oshima RG. Mid-gestational lethality in mice lacking keratin 8. Genes Dev 1993; 7(7A):1191–1202.

(9) Threadgill DW, Dlugosz AA, Hansen LA, Tennenbaum T, Lichti U, Yee D et al. Targeted disruption of mouse EGF receptor: effect of genetic background on mutant phenotype. Science 1995; 269(5221):230–234.

(10) Kallapur S, Ormsby I, Doetschman T. Strain dependency of TGFbeta1 function during embryogenesis. Mol Reprod Dev 1999; 52(4):341–349.

(11) Kent G, Iles R, Bear CE, Huan LJ, Griesenbach U, McKerlie C et al. Lung disease in mice with cystic fibrosis. J Clin Invest 1997; 100(12):3060–3069.

(12) Clarke LL, Grubb BR, Gabriel SE, Smithies O, Koller BH, Boucher RC. Defective epithelial chloride transport in a gene-targeted mouse model of cystic fibrosis. Science 1992; 257(5073):1125–1128.

(13) Rozmahel R, Wilschanski M, Matin A, Plyte S, Oliver M, Auerbach W et al. Modulation of disease severity in cystic fibrosis transmembrane conductance regulator deficient mice by a secondary genetic factor. Nat Genet 1996; 12(3):280–287.

(14) Shull MM, Ormsby I, Kier AB, Pawlowski S, Diebold RJ, Yin M et al. Targeted disruption of the mouse transforming growth factor-beta 1 gene results in multifocal inflammatory disease. Nature 1992; 359(6397):693–699.

(15) Dickson MC, Martin JS, Cousins FM, Kulkarni AB, Karlsson S, Akhurst RJ. Defective haematopoiesis and vasculogenesis in transforming growth factor-beta 1 knock out mice. Development 1995; 121(6):1845–1854.

(16) Bonyadi M, Rusholme SA, Cousins FM, Su HC, Biron CA, Farrall M et al. Mapping of a major genetic modifier of embryonic lethality in TGF beta 1 knockout mice. Nat Genet 1997; 15(2):207–211.

(17) Engle SJ, Hoying JB, Boivin GP, Ormsby I, Gartside PS, Doetschman T. Transforming growth factor beta1 suppresses nonmetastatic colon cancer at an early stage of tumorigenesis. Cancer Res 1999; 59(14):3379–3386.

(18) Engle SJ, Ormsby I, Pawlowski S, Boivin GP, Croft J, Balish E et al. Elimination of colon cancer in germ-free Transforming Growth Factor *beta*1-deficient mice. Cancer Res 2002; 62(22):6362–6366.

(19) Sanford LP, Ormsby I, Gittenberger-de GA, Sariola H, Friedman R, Boivin GP et al. TGFbeta2 knockout mice have multiple developmental defects that are non-overlapping with other TGFbeta knockout phenotypes. Development 1997; 124(13):2659–2670.

(20) Bartram U, Molin DG, Wisse LJ, Mohamad A, Sanford LP, Doetschman T et al. Double-outlet right ventricle and overriding tricuspid valve reflect disturbances of looping, myocardialization, endocardial cushion differentiation, and apoptosis in *Tgfb2*knockout mice. Circulation 2001; 103(22):2745–2752.
(21) Poelmann RE, Jongbloed MR, Molin DG, Fekkes ML, Wang Z, Fishman GI et al. The neural crest is contiguous with the cardiac conduction system in the mouse embryo: a role in induction? Anat Embryol (Berl) 2004; 208(5):389–393.
(22) Molin DG, Poelmann RE, DeRuiter MC, Azhar M, Doetschman T, Gittenberger-de Groot AC. Transforming growth factor beta-SMAD2 signaling regulates aortic arch innervation and development. Circ Res 2004; 95(11):1109–1117.
(23) Proetzel G, Pawlowski SA, Wiles MV, Yin M, Boivin GP, Howles PN et al. Transforming growth factor-beta 3 is required for secondary palate fusion. Nat Genet 1995; 11(4):409–414.
(24) Maggio-Price L, Treuting P, Zeng W, Tsang M, Bielefeldt-Ohmann H, Iritani BM. Helicobacter infection is required for inflammation and colon cancer in SMAD3-deficient mice. Cancer Res 2006; 66:828–838.
(25) Guilbault C, Saeed Z, Downey GP, Radzioch D. Cystic fibrosis mouse models. Am J Respir Cell Mol Biol 2007; 36(1):1–7.
(26) Tang Y, McKinnon ML, Leong LM, Rusholme SA, Wang S, Akhurst RJ. Genetic modifiers interact with maternal determinants in vascular development of Tgfb1(–/–) mice. Hum Mol Genet 2003; 12(13):1579–1589.
(27) Tang Y, Lee KS, Yang H, Logan DW, Wang S, McKinnon ML et al. Epistatic interactions between modifier genes confer strain-specific redundancy for Tgfb1 in developmental angiogenesis. Genomics 2005; 85(1):60–70.
(28) Smithies O, Maeda N. Gene targeting approaches to complex genetic diseases: atherosclerosis and essential hypertension. Proc Natl Acad Sci USA 1995; 92(12):5266–5272.
(29) Wolfer DP, Crusio WE, Lipp HP. Knockout mice: simple solutions to the problems of genetic background and flanking genes. Trends Neurosci 2002; 25(7):336–340.
(30) Banbury Conference. Mutant mice and neuroscience: recommendations concerning genetic background. Banbury Conference on genetic background in mice [see comments]. Neuron 1997; 19(4):755–759.
(31) Markel P, Shu P, Ebeling C, Carlson GA, Nagle DL, Smutko JS et al. Theoretical and empirical issues for marker-assisted breeding of congenic mouse strains. Nat Genet 1997; 17(3):280–284.
(32) Wakeland E, Morel L, Achey K, Yui M, Longmate J. Speed congenics: a classic technique in the fast lane (relatively speaking). Immunol Today 1997; 18(10):472–477.
(33) Darvasi A, Soller M. Advanced intercross lines, an experimental population for fine genetic mapping. Genetics 1995; 141(3):1199–1207.
(34) Bommireddy R, Ormsby I, Yin M, Boivin GP, Babcock GF, Doetschman T. TGFbeta1 inhibits Ca^{2+}-Calcineurin-mediated activation in thymocytes. J Immunol 2003; 170(7):3645–3652.
(35) Hoying JB, Yin M, Diebold R, Ormsby I, Becker A, Doetschman T. Transforming growth factor beta1 enhances platelet aggregation through a non-transcriptional effect on the fibrinogen receptor. J Biol Chem 1999; 274(43):31008–31013.
(36) Schultz JJ, Witt SA, Glascock BJ, Nieman ML, Reiser PJ, Nix SL et al. TGF-beta1 mediates the hypertrophic cardiomyocyte growth induced by angiotensin II. J Clin Invest 2002; 109(6):787–796.
(37) Okita K, Ichisaka T, Yamanaka S. Generation of germline-competent induced pluripotent stem cells. Nature 2007; 448(7151):313–317.
(38) Wernig M, Meissner A, Foreman R, Brambrink T, Ku M, Hochedlinger K et al. In vitro reprogramming of fibroblasts into a pluripotent ES-cell-like state. Nature 2007; 448(7151):318–324.
(39) Takahashi K, Yamanaka S. Induction of pluripotent stem cells from mouse embryonic and adult fibroblast cultures by defined factors. Cell 2006; 126(4):663–676.
(40) Matsushima Y, Sakurai T, Ohoka A, Ohnuki T, Tada N, Asoh Y et al. Four strains of spontaneously hyperlipidemic (SHL) mice: phenotypic distinctions determined by genetic backgrounds. J Atheroscler Thromb 2001; 8(3):71–79.
(41) Shi W, Wang NJ, Shih DM, Sun VZ, Wang X, Lusis AJ. Determinants of

atherosclerosis susceptibility in the C3H and C57BL/6 mouse model: evidence for involvement of endothelial cells but not blood cells or cholesterol metabolism. Circ Res 2000; 86(10):1078–1084.

(42) Shi W, Pei H, Fischer JJ, James JC, Angle JF, Matsumoto AH et al. Neointimal formation in two apolipoprotein E-deficient mouse strains with different atherosclerosis susceptibility. J Lipid Res 2004; 45(11):2008–2014.

(43) Waterston RH, Lindblad-Toh K, Birney E, Rogers J, Abril JF, Agarwal P et al. Initial sequencing and comparative analysis of the mouse genome. Nature 2002; 420(6915):520–562.

(44) Wade CM, Kulbokas EJ, III, Kirby AW, Zody MC, Mullikin JC, Lander ES et al. The mosaic structure of variation in the laboratory mouse genome. Nature 2002; 420(6915):574–578.

(45) Ideraabdullah FY, Casa-Esperon E, Bell TA, Detwiler DA, Magnuson T, Sapienza C et al. Genetic and haplotype diversity among wild-derived mouse inbred strains. Genome Res 2004; 14(10A):1880–1887.

(46) Churchill GA, Airey DC, Allayee H, Angel JM, Attie AD, Beatty J et al. The Collaborative Cross, a community resource for the genetic analysis of complex traits. Nat Genet 2004; 36(11):1133–1137.

Chapter 24

Pathologic Phenotyping of Mutant Mice

Roderick T. Bronson

Abstract

The easiest and cheapest way to analyze the phenotype of most knockout mice is to do a comprehensive necropsy and histopathologic examination of slides of all tissues. Once any lesion is found in a knockout mouse a vast contemporary and traditional literature can be searched for occurrences of similar lesions in other species, including human beings. This may provide further insights into the molecular and cellular pathogenesis of the lesion. In this chapter we will focus on the best way to turn a mouse into a set of slides which can thereafter be studied by investigators and pathologists. Some techniques suggested are not generally used in conventional histology laboratories. Most are decidedly old fashioned. They have all been used successfully in many diverse studies of mice of all ages with all kinds of lesions.

Key words: Mouse, pathology, histotechnique, disease, lesion, gene, artifact, phenotype, embryos, immunohistochemistry, experimental design, mouse pathologists, statistics.

1. Introduction

Hundreds of mouse mutants are being discovered or engineered each year and many more are on the way. For many years now everyone has complained that there are not enough mouse pathologists to study these mutants. It seems that there are even fewer at this time in mid-2007 than there were a few years back. This chapter cannot address that issue. What can be done is to help scientists who may not be acquainted with the methods of histological slide preparation to process tissues appropriately so that slides are of high quality and free of confusing artifacts. Mouse histology, like mouse pathology, has fallen so low that most histology labs prepare slides of very poor quality. Even experienced histologists and pathologists may find some helpful hints in this

Ralf Kühn, Wolfgang Wurst (eds.), *Gene Knockout Protocols: Second Edition, vol. 530*

DOI 10.1007/978-1-59745-471-1_24 Springerprotocols.com

chapter. These methods have been perfected through many a frustrating trial and embarrassing error, mouse by mouse, tissue by tissue, and slide by slide over many years. Believe nothing you read, but do not stand around arguing that one procedure is better than another, or more "standard" than another, or is how your lab has always done it. Do a test and compare your procedure with any new one recommended to you.

2. Materials

2.1. General Approach

When most people knock genes out they assume that they are engaging in hypothesis-driven research. Not true. When you knock out a gene anything or nothing can happen. You have to be prepared to spend quality time with your mice to find out what is wrong with them. It is your job to do so. Do not hope that a "phenotyping service" can be of any help to you. Do not rely on expensive small-animal imaging machines like magnetic resonance imaging (MRI) or ultrasound sonography to help you. Much mouse data derived from those machines have never been correlated with reality. Do not assume that anyone else can figure out what is wrong with your mouse. There are no experts in this area. You will be the expert, but you will have to learn about things that you never wanted to know. Open your eyes, observe, observe some more, and finally do a little reading. Look for normal numbers of mutant and control pups in litters to check for possible embryonic lethality. Monitor growth of pups until weaning. Check to see if both males and females can breed. Weigh adults and observe them at various times of the day to see if their behavior is normal. Play with them. Take a pencil, put the mouse on it, and gently twirl and shake it around. This is a wonderful test for strength and agility. Watch for fighting, which can result in alopecia or wounds. Check for malocclusion of incisor teeth, a very common cause of wasting in mice. Follow adults for as long as you can afford to, up to 2 years of age, to see if there is an age-related phenotype. Make sure that you maintain equal numbers of male and female mutant and control mice for these aging studies, since both control and mutant mice will start to die from diseases unrelated to the mutated gene. Some age-related diseases like lymphoma might occur more often in males than in females.

If you detect any phenotype (defect, anomaly, disease, lesion, whatever you want to call it) in mice of a certain age, follow the abnormal mice as long as you can so that their abnormality becomes most pronounced. Watch mice that are going to die for as long as possible and collect them for pathologic analysis as close to death as possible. Even if they suddenly die they may provide useful post-mortem data unless they are so decomposed (autolyzed) that you gag when opening them up for fixation.

2.2. Pathogenesis

For all phenotypes the key to understanding is "pathogenesis", a word seldom used today. Pathogenesis refers to how a disease process evolves over time from the earliest visible abnormality to the fully developed disease. To understand pathogenesis start with the most evolved, severe disease and then progress to earlier and milder stages of the disease in other mice. Eventually you can hope to find the disease at a stage when lesions are tiny and just beginning to be abnormal. For phenotypes that do not start at the same age in all mice, pathogenesis studies can be frustrating. To find mice in an early stage of disease some mice have to be killed before they look sick and before lesions are large enough to be detected clinically or by various imaging devices like CT and MRI, even if these are available and offer sufficient resolution to detect your mouse's lesions. However, some mice killed early may have no lesions because the disease has not started in them yet. As the search for earlier and earlier time points progresses, more and more mice might have to be killed that have no lesions. On the positive side only a few mice with early lesions need to be found. You do not need large, statistically significantly numbers of early-stage mice. You only need a few to study how lesions begin and then evolve.

2.3. Etiology

In knockout mouse pathology it is usually assumed that any phenotype is entirely determined by the knocked-out gene. That is not always a safe assumption. First the phenotype may be as much determined by a modifying gene as by the knocked-out gene. Thus the knockout mouse may show an embryonic lethality phenotype on the B6 genetic background and no phenotype on the 129 background. One has to wonder in cases like this which is the gene and which is the modifier. Environmental factors can also play roles in determining phenotypes. For example, it has happened several times that KO mice and not heterozygous or wild-type controls develop dermatitis, for unknown reasons. (Affected mice scratch themselves incessantly and excoriate themselves. The etiopathogenesis of this very common disease is not known.) Better known is that mice with knocked-out immune genes may develop colitis, but only when infected with *Helicobacter*. In fact, that some infectious agent is responsible for a phenotype is often suggested. Lesions can be cultured and special stains for microorganisms like *Pneumocystis carinii* can be performed to rule out infectious etiologies. Seldom are such tests positive. In general, infectious agents are not responsible for phenotypes.

2.4. Comparative Pathology

Once a phenotype is discovered and its pathogenesis elucidated, it is necessary to find all other situations in which the same phenotype occurs in mice or in any species. If the defect is ventricular septal defect, for example, you will want to find all other cases of that defect in the mouse, human or veterinary literature. For other phenotypes the invertebrate and fish literature could be usefully searched. When each report of a similar phenotype in mice or other species is found, its genetics and molecular biology should be closely examined to see how the reported phenotype and your phenotype compare and contrast in pathogenesis. Beware that the reported phenotype might be a "phenocopy" of yours, representing the final common pathway of different molecular or pathogenic pathways. What must also be avoided is claiming that a nonspecific phenotype like "hepatic necrosis" in a mouse is the same as a human phenotype also called "hepatic necrosis". Many things cause necrosis in liver and there are many kinds of liver necrosis. This can be said of many pathologic diagnoses; comparisons must always be made with circumspect. Names can be dangerous. If you call a mouse lesion by a human name everyone assumes that the mouse lesion is the same as the human one even if it is only vaguely similar. For that matter even if it the mouse and human lesions are morphologically identical they may be caused by different pathologic mechanisms and genetic lesions.

2.5. Histology

Histologic technique was invented and essentially perfected a century ago. In this century we have much better equipment to perform many of the techniques that used to require painstaking handiwork. Automatic cassette labelers, tissue processors, slide labellers, slide stainers, and cover slippers are found in most modern laboratories. Most of the materials used in histology are standard, though each laboratory will have its favorite brands of solvents, paraffins, and stains to use. It makes little difference which brands are used. However, choice of fixative is very important and will be discussed in Section 3.

Despite automation of some the most tedious steps in slide preparation, four steps in turning a mouse into slides cannot be replaced by machines. They will be discussed in detail in Section 3 and are mouse dissection, tissue trimming and cassetting, tissue embedding, paraffin block sectioning, slide staining, interpretation, and reporting of results. Mouse dissection requires knowledge of mouse anatomy and a sharp eye for any abnormality. If the dissector does not find and sample lesions they will not end up on slides. Trimming and cassetting involves removing tissues from fixative, slicing them up into pieces, and putting them into small plastic containers, called cassettes, which will carry the tissues through the processor. If the trimmer does not slice up a tissue with its cut surface passing through the lesion in the correct orientation, the slide will be difficult to interpret or the lesion might not even show up on the slide. Embedding can be critical. If a tissue is not placed with the correct orientation at the bottom of the embedding mold as the paraffin solidifies around it, the tissue will not show up with the correct orientation on the slide. Sectioning, while fairly standard, requires the efforts of a technical expert. Slide staining using hematoxylin and eosin (H&E) is routine but often people fail to change the stains, particularly hematoxylin, often enough. How pathologists should report diagnoses and how they should interact with investigators is very important and will be discussed in Section 3.

2.6. Special Stains

All conventional histology labs do special stains. These have been used for many decades and are reliable, though some are quirky. Nervous tissue is usefully stained with luxol fast blue for myelin and cresyl violet for cells. Luxol fast blue is a so-called "regressive" stain in which the tissue is overstained and then the stain is removed by dipping in ethanol. The sections have to be checked as the destaining proceeds to make sure that all of the luxol is removed from gray matter. Slides from young mice are tricky to destain since they have little myelin, and it can be tricky to find an appropriate end point for destaining. After destaining, little or no stain is left. Silver stains for axons like Bodians and Beilchowski are very pretty if done well. They are not very useful for diagnosis, as it happens. A wonderful old staining technique that has gone almost completely out of fashion is the Golgi technique. Famously, this silver staining technique impregnates the dendritic arborization of neurons completely so they can be studied in three dimensions in thick sections. Some immunostains like that for Calbindin, specific for Purkinje cells, have somewhat the same effect. Many people use trichrome stains for collagen. Commercial staining kits are handy. Tissue Gram stains for bacteria are very useful but it is essential to have a control with both Gram-positive and Gram-negative bacteria. Incubating a piece of meat with feces will assure contamination by a variety of bacteria. Congo red stains for amyloid are useful. Amyloid famously glows apple green when examined microscopically under crossed polarizing lenses.

2.7. Clinical Pathology

Many people studying mice have not been trained in medicine. That may explain why they sometimes have a misconception about clinical chemistry parameters like blood urea nitrogen (BUN). They think that BUN is regulated like any other molecule, and when it is elevated it represents something similar to overexpression of a transgene. But BUN is no more than a useful but entirely non-specific marker of kidney disease. The same can be said for most clinical chemistry parameters. In studying mice it is seldom worth the trouble and expense to do clinical chemistry, although conventional drug and chemical toxicity studies have always included clinical pathology. It is not clear that that ever was cost effective. You are going to be looking at the kidneys anyway. If there is kidney disease you will know it by looking at the kidney slide. Moreover, kidneys can develop a lot of disease before going into failure, which is when BUN becomes elevated. Of course once you know that your mice of interest have kidney disease then you could use BUN to monitor the progress of the disease. That would be useful. Clinical chemistry is at best a marginally useful adjunct to histopathology.

2.8. Role of Immunohistochemistry (IHC) in Phenotyping

Flourescent pictures are quite decorative. "Decorate" is a lovely word that IHC enthusiasts have used, not realizing the irony that IHC is more for show than for science. Look in any journal and you will see many fluorescent images, usually with some merged two-color images showing co-localization of two proteins in the same cells. The one antibody says what is being expressed in the cells or what is happening, like apoptosis, in the cells, and the other says what kind of cells they are. As long as some fluorescent objects are double-stained it is assumed that they are the cells of interest even if they look nothing like real cells, are too few or too numerous to be the cells of interest in that tissue, and are located where no such cells can be located. It has become conventional (canonical, as Christians and experts in signal transduction say) that you are not permitted to identify any cell using H&E only. Even immunology journals, for example, will not let you call a cell a neutrophil without labeling it with an antibody, even though no cell, other than an eosinophil, could possibly be confused with a neutrophil *(1)*. (Many investigators, even some trained in pathology, do not know that eosinophils have eosinophilic (i.e., red) granules in H&E sections.) Few molecular biologists know that apoptosis is definitively and readily diagnosed in H&E sections if nuclear fragmentation is present. They insist on using fluorescent techniques to diagnose and count apoptotic cells, even though nuclear fragmentation may be entirely absent.

Generally those using IHC seem unaware that antibodies are almost never specific, even though Western blots almost always show multiple bands, and IHC almost always stains many different kinds of cells. This kind of staining is usually called "non-specific". Actually it is not necessarily non-specific at all. The antibody made to a particular peptide which folds in a particular way recognizes other peptides with similar folding patterns. The antibody may have a reduced affinity to the other peptides, but its binding is just as specific to the other peptides as it is to the peptide of interest. Affinity purification will not eliminate these antigens that mimic the antigen of interest, though it might eliminate low-affinity antigens. The definitive negative control for the specificity of an antibody is to use it on tissues in which the gene has been knocked out. This control is often not used even if the knockout mouse tissues are available. There is also little awareness that most genes are expressed in most cells, as we know from countless microarrays in which even minor perturbations to cells cause thousands of genes to be more or less upregulated or downregulated. The Allen Brain atlas (www.brainatlas.com), wonderful as it is, is definitive proof that most genes are expressed everywhere. You can prove this by looking up your favorite immune gene, or podocyte gene, or hair follicle gene. Sure enough, there it is in the brain, usually expressed in some anatomically specific part of the brain. In any case, the way to assure that what you see in IHC slides correlates with H&E-stained paraffin or frozen slides is to stain replicate slides using both techniques, or divide the tissue at necropsy, taking some for conventional sections and some for IHC. Once you have both H&E and IHC slides, you should study them together, going back and forth from the ICH slides to the H&E slides to make sure which cells are staining with the antibody and where they are in the tissue. This is a lot easier if you can use immunoperoxidase on paraffin sections rather than fluorescence on frozen sections. In the end, if there is not good concordance between H&E and IHC results, throw out the IHC results. This is how to keep your feet firmly attached to the ground so as not to float off into a colorful fluorescent miasma.

2.9. Morphometrics and Statistics

A popular abuse of statistics is to stain slides with an antibody specific for apoptosis or proliferation, for example, then use a computer program to count tagged cells, and finally to conclude that the mutants have significantly more or fewer apoptotic or proliferative cells than the controls. Significance here could be a 7.75832% greater number of cells, significant at the p, 0.05 level using Student's T test or some other test which sounds more sophisticated and scientific. But often investigators do not tell you whether an 8% difference is biologically significant, and what the evidence for that significance is? The rule should be that if a 3-year old cannot tell the difference at a glance between controls and test subjects then there is no difference worth studying. A very simple way of evaluating whether mutants and controls have more or less of any phenotype in a certain organ is to take all of the slides of the organ from the two test groups and shuffle them. Then examine a few of them until you have an idea of the range of severity of lesions. Then, without looking at the slide labels, examine each slide at relatively low magnification, determine how severe the lesion is and put the slide in one of two stacks: good and bad. The entire enterprise should take no more than a few minutes. Afterwards, look at the labels. If nearly all of the mutant slides are in one pile and nearly all of the controls are in the other, you may have publishable results. If not, forget it. Later you can reshuffle the slides and pile them up into four stacks: normal, mild, moderate, and severe. Once you know that you have real differences that pass the 3-year-old test, you can think about spending time getting data to please the reviewer "scientists" using some more scientific-appearing morphometric analysis. If you score lesions for severity, 0–4, it is not considered very good statistical practice to use a T test, since the data are ordinal and not normally distributed; but go ahead anyway. Everyone does it. When you publish your work show a typical random picture of what you scored as a 1, 2, 3, and 4. Do not lie by labeling the severe lesions as "untreated control mouse" and the mild lesions as "treated mouse". After all, you evaluated the slides blindly and there were, in fact, a few treated mice with +3 and +4 lesions, now weren't there?

2.10. Mouse Neurology and Behavior

Behavioral abnormalities can be grouped into two broad categories: those that affect motor behavior and those that do not. Motor abnormalities can be detected just by looking at the mouse movement. The non-motor abnormalities involving learning, memory, and emotional behavior require special equipment and techniques for evaluation. In general, motor abnormalities are associated with neuropathologic lesions located in hind brain and spinal cord or muscle. (In mice and other animals the corticospinal tract and basal nuclei, so important in human neurology, play only a minor role in controlling motor behavior). Some motor abnormalities are not associated with lesions, much to the chagrin of neuropathologists. In general, abnormalities of learning, memory, and emotion are associated with forebrain defects and may have no neuropathologic lesions. If lesions are found anywhere in the nervous system a large literature is available to help investigators establish clinical–anatomic correlations. An example of a clinical–anatomic correlation that should be widely known, but is not, is the distinction between spastic paralysis and flaccid paralysis. In both situations the patient or mouse cannot walk. But in flaccid paralysis the affected limbs are limp like cooked spaghetti. When you see that, there must be a defect in the motor neuron cell body, in its axon or in the myoneural junction. Histologically in muscle there will be neurogenic atrophy: groups of tiny muscle fibers. By contrast, in spastic paralysis the limbs are springy, like uncooked spaghetti, when they are pushed back and forth by the investigator. When you see that, the lesions are in the upper motor neurons that descend from the brain down the spinal cord and finally tell lower motor neurons what to do. In many mouse diseases with hind-limb paralysis the lesions responsible for the paralysis are in the upper, not lower motor neuron *(2)*. Other clinical–pathologic correlations are useful but inexact. Whirling and bobbing mice often have vestibular disease and often are deaf as well, which can often be detected by clapping the hands to see if the mouse startles. Shaking mice often have abnormalities of myelin. Mice that stumble and fall from side to side often have cerebellar abnormalities. A simple and not very precise test for blindness is to lower a mouse by the tail downwards past a tabletop. The mouse might tend to reach out to grab the tabletop if it can see it.

2.11. Small-Animal Imaging

Everyone wants to have the latest million dollar imaging machine, like MRI, to be used to phenotype their mice. They are confused. Human patients are imaged by various kinds of scanners so that the doctor can know what is going on inside them. It would be much more efficient to simply kill the patient, do an autopsy, and analyze the cellular basis of their diseases using conventional histopathology. Patients do not like this approach, however. Mice probably do not either, but necropsy and histopathology are still much better than in vivo imaging, no matter how good the resolution of the machine used. The only rational for in vivo imaging is when therapies are being used to reduce the size of lesions like tumors. Even then it is good practice to study the histopathology of lesions in some mice as the treatment progresses so that the cellular mechanisms of changes induced by the therapy can be understood. Thus, small-animal imaging is an adjunct to, not a replacement for, histopathology.

2.12. Reporting Results to Investigators

If you are the pathologist on a project, never ever write a report. The truth is on the slide, not in your head. If the results are published anyone reviewing the slides later must be able to see what you saw. You must have a multiheaded microscope and you must review slides with investigators. It is their experiment, not yours. They will write the paper. They must understand the pathology. Your job is to find the lesion, name it, explain what you know about it, and what is known about its pathogenesis. The diagnosis and early discussions are, of course, only the beginning. After that, many more experiments will usually be necessary before anyone understands how the genetic lesion causes the phenotypic lesion.

2.13. Words that Pathologists Use

The name you give a lesion or disease can be the key that unlocks a whole new world of science to the post doc, graduate student, or Principal Investigator (P.I.). For example, if there is too much bone and you call it "osteopetrosis" you and the scientist can find all of the papers on that subject in Pubmed. The richer your vocabulary in pathology the more useful you will be. In this example it would help if you, unlike some who have published in this field, knew that osteosclerosis is the same as osteopetrosis. Both words have to be searched. Beware of medical words. Like any other words, those used by pathologists have histories and have accumulated meanings that go way beyond and are even contrary to their literal meaning. Consider "retinitis pigmentosa" of humans, which is nothing other than retinal degeneration, the term used in mouse pathology. There is no inflammation ("itis") and nothing wrong with pigment. "Achondroplasia", the term for one kind of human dwarf, literally translated means that these short-statured people have no cartilage ("a" = absent, chondro = cartilage, and "plasia" = growth). That is nonsense. Dogs like basset hounds with a similar short stature have a disease which veterinarians call "chondrodysplasia", ("dys" = abnormal) which is more accurate but just as pompous and confusing as the human term. More confusion comes in when you realize that G.I. pathologists use the word "dysplasia" to mean early cancer. All of this Latinizing and Greekizing of medical terms has never clarified anything. Even simple words in the vernacular mean different things to different kinds of investigators coming from different traditions. For example "hepatitis" in human medicine means viral hepatitis. In veterinary medicine "hepatitis" means any inflammation of liver ("hepat" = liver and "itis" = inflammation). The issue of nomenclature in pathology has always been controversial and recent bioinformatic concepts being used like "controlled vocabulary" and "heuristics" only add to the confusion. The disease is what is the reality; what you call it should be as descriptive, precise, specific, simple, and honest as possible.

What is critical is that everyone studying the disease understands to what extent any lesion or disease is non-specific and part of some common process, or is specific or unique. Many lesions/diseases are unusual but not unique; one can find other examples of the same disease in mice or other species. However, when searching using the disease/lesion name as the key word be aware that the papers you find are likely to have misused the word. A paper you find that describes cerebral infarction might be describing cerebral hemorrhage instead. You have to check the figures in the papers describing the lesion. Even that can often be misleading, given the low quality of most published pictures of lesions. In any event, in science lesions and diseases should have as precise names as possible; but everyone must understand that any name is always provisional and imprecise. A conflict of interest arises when investigators ascribe words to diseases as they write grants and papers, when the truth inevitably becomes compromised by funding needs. Grantsmanship demands that mouse disease models be designated by human disease names because tax payers want cures for disease and are not interested in basic science. Pharmaceutical and biotech companies are interested in neither, but only in profits. It would be nice if everyone understood that no mouse disease will ever be exactly like any human disease, no matter what it is called until humans grow tails and start squeaking.

2.14. Publications, Careers, and Promotions

For publication, you as the pathologist can help the investigators with photomicroscopy, which they usually need, and can provide nice words for the paper when it finally emerges. Whether your name as the pathologist is put on the paper is the decision of the P.I. who paid for the work. Some, including a Nobelist, think that mouse pathologists should not be co-authors unless the histology costs are provided free of charge. Experience has shown that when it comes to publications a "don't ask – don't tell" policy is best. It is humiliating to ask to be on a paper and even more humiliating to get into arguments over it. The job of the histopathology lab is to provide high-quality slides in a timely fashion with expert slide interpretation and consultation by the pathologist if requested. For that service the investigator pays a fee. End of story. If the service provided helps the investigators to publish a paper, at their discretion they may want to include the pathologist or even a technician. The pathologist's name will usually be buried somewhere in the middle of a dozen names, including the name of the P.I.'s mother. It would be nice if the investigators at least would acknowledge the histopathology lab at the end of their papers, since the lab often must compete for funding and needs to cite papers that have resulted from use of its services. That seldom happens. Whether mouse pathologists will ever be promoted for providing services that are often critical to the success of other faculty members and their graduate students and post docs is very unclear. Work in a mouse histopathology core, no matter how good, will never lead to tenure. This is one of many complex reasons why it will be more and more difficult everywhere to find an experienced mouse pathologist.

3. Methods

3.1. Fixation

3.1.1. Bouin's Fixative and Demineralization

When commercially available 10% neutral buffered formalin (NBF, a buffered 10% solution of commercial 37%–40% formaldehyde) or paraformaldehyde (PFA) are used for fixation, mouse tissues and cells shrink even more than tissues from large species and become very brittle and difficult to section. Commercially available Bouin's fixative (16 ml of 40% formaldehyde, 4 ml glacial acetic acid, and 170 ml of saturated aqueous picric acid) is unquestionably the fixative of choice since tissues shrink but nuclei of cells do not. Particularly for tumor diagnosis it is essential to have a good look at nuclear detail, which is impossible to do if the tissues have been fixed in aldehydes. Tissues can be left in Bouin's fixative for weeks without ill consequences. Since Bouin's fixative contains picric acid, mouse bones and even teeth are demineralized in a week's time. Picric acid is a weak acid, and tissues exposed to weak acids can always be stained with hematoxylin for nuclei (basophilia) and with metachromatic stains like cresyl violet, used to stain neurons. However, when tissues are demineralized with dilute strong acids like 5% nitric acid, they lose their ability to be stained with these two stains. Why the dissociation constant of an acid should make a difference in basophilia is not clear. Of course, demineralization can also be accomplished using EDTA but that takes a long time and is not suitable for routine use.

3.1.2. Aldehyde Fixatives, Cell Shrinkage, and IHC

Most labs believe as an article of faith that PFA (paraformaldehyde) is best for everything including IHC. This is simply wrong. PFA is difficult to make up from powder before use. It is supposed to be purer than commercial 10% NBF, but the difference is not significant for light microscopic work. Chemically it is nearly identical to NBF. Whether anyone has ever tested this is not clear. It is certain that more than just a few hours of fixation of mouse tissues in aldehydes causes cells and nuclei to shrink so severely that nuclear detail cannot be evaluated. This can be a huge problem if careful analysis of tumor cells is required, since abnormalities in nuclear details determine the malignancy grade of a tumor. Amazingly it can be very difficult to distinguish nuclei of neutrophils from those of lymphocytes fixed in aldehydes, even though everyone knows that nuclei from the two kinds of cells are entirely different. PFA is supposed to be the universal fixative for all antibodies used in IHC. This is a fable. Some antibodies work on PFA-fixed tissue, some do not. It is quite likely that antibodies that work on PFA-fixed tissue will also work on NBF-fixed tissue. Some antibodies work only on fresh frozen tissue slides. A comparative study of 83 antibodies used on mouse tissues fixed using a variety of fixatives has shown that Bouin's fixative is more effective for many antibodies than aldehydes *(3)*. On the other hand, one school of thought claims that only freshly made PFA has no cross-linked aldehyde molecules, and as a consequence there is less background staining than when old PFA or formalin is used. It is not clear that there is much evidence for this. No matter which fixative is used, it is best to fix small mouse tissues or slices of larger tissues like liver for no longer than 4 h and then transfer them to very dilute fixative to prevent bacteria from growing. It should be added that formalin should be used for LacZ staining or for in situ hybridization. For those procedures Bouin's fixative is not good. Bouin's fixative is also not good for electron microscopy (EM). For EM a mixture of PFA or NBF and glutaraldehyde are the necessary fixatives.

3.1.3. White Matter Holes Due to Exposure to Ethanol

Traditionally in many research labs tissues fixed in Bouin's fixative, PFA or formalin have been post-fixed and stored in 70% ethanol (EtOH). This has caused a problem that continues to be unrecognized. Prolonged exposure of myelinated nervous tissue to ethanol produces holes in white matter. Holes always occur in nervous tissue fixed in Telly's fixative (16 ml 40% formaldehyde, 8 ml glacial acidic acid, and 160 ml 70% EtOH). This is true if brain or spinal cord tissue is put in a tissue processor in which 70% ethanol is the first solution and left there for several days. That is what happens when the processor is loaded on Friday night and set to switch on Sunday evening for processing overnight. That 48-h exposure to ethanol will cause white matter holes. Since holes are definitive lesions in prion diseases like scrapie, and some neurodegenerative diseases like that of the mahogany mutation *(4)*, a zero-holes policy is essential for mouse histopathology labs. Fixed nervous tissue should be stored in very dilute fixative to prevent bacteria from growing. Tissues should also be shipped in a minimum of dilute fixative. Shipping companies definitely have a no-leak policy.

3.2. Dissection and Cassetting

3.2.1. Strategies for Collecting Tissues

Few graduate students, post docs, professors, or doctors know anything about mouse anatomy. Many prefer not to learn. A mouse pathology lab should offer dissection as a service. Technicians can learn how to dissect a whole fixed mouse and will provide excellent, uniform results. In a well-designed pathology lab the pathologist should have his or her office near to the dissection area, which should be part of the histology lab, so that help can be provided at a moment's notice. Laboratory users often request that some particular organ like brain be dissected in a certain way, to show the hippocampus, for example, and the dissector might need help interpreting the request. Since it is usually a poor idea to transport live mice from another lab to the histology lab, given the possible transmission of mouse pathogens, it is best in practice for the investigator to kill and fix the mouse in his or her lab or in the procedure room in the mouse facility.

3.2.2. Dissection or Fixation of Freshly Killed or Dead Mice

Traditionally people perform necropsies and dissections on freshly dead, unfixed mice, just as they perform necropsies on large animals and autopsies on people. These larger animals cannot readily be fixed whole as can mice. Organs are weighed, sliced up, and placed in large containers of fixative. Later, at the time of trimming, each organ is removed from the fixative, sliced up, and the pieces of tissue are placed in cassettes. It is bad policy to put unfixed tissues directly into cassettes. Fixation is poor under those conditions because fixative cannot flow freely around tissues. Necropsies of freshly killed mice are traditionally performed in diagnostic laboratories and in pharmaceutical companies where sets of mice are necropsied by specially trained prosectors according to fixed protocols. However in small-scale laboratory settings complete necropsies of fresh tissues are so impractical that they are seldom, if ever performed. Dead mice are found by mouse room technicians and usually placed in the refrigerator. By the time the investigator finds out that the mouse has died it has become badly autolyzed. Mice that are killed by investigators are usually killed either in procedure rooms in the mouse facility or back in their research labs. Usually the investigator needs to remove specific tissue samples from the mice for molecular biology procedures, and there is no time for a complete dissection. Often investigators do not want learn how to do a complete dissection. They are willing at most to remove a few large tissues like spleen and liver, which are fixed or frozen for future processing. Because of these practical limitations complete necropsies of whole fixed mice are best done in the histology lab by trained technicians. This service can be performed profitably for as little as $25.00. Admittedly, if a mouse is fixed before dissection, fresh organ weights cannot be measured, but fixed organ weights from mutants and controls are comparatively just as good as fresh organ weights. Organ weights are of little values anyway. It is sufficient to note at the time of necropsy if any organ appears unusually enlarged or shrunken.

3.2.3. Berry Picking Tissues for Pathology

Many investigators dissect out and fix bits of fresh tissue of interest to them. Often a number of tissues from the same mouse are put in one vial, in which case the histotechnicians will cassette as many of them as possible in the same cassette and they will appear in the same block and on the same slide. This saves a lot of money as compared with another common practice of putting a single tissue in a single vial and asking for single tissues on one slide. Sometimes investigators make the double mistake of putting only a tiny piece of a large tissue like liver alone in a vial. The slide made from this is 99% blank and 1% tissue for the $8.00 charged for a single slide. Another common expensive mistake is to ask for two or more slides from the same block. Investigators seem unaware that a 6-μm section is thinner than a red blood cell. There cannot be new information on a second slide that is not present on the first. Berry picking, of course, reflects only what the investigator knows. Since he or she may never have heard of the adrenal gland: for example, any adrenal pathology will always be missed.

3.2.4. Fixation by Immersion and Dark Neuron Artifact

Mice are killed with carbon dioxide, preferably from a compressed tank and not as vapor rising from dry ice. The head should not be cut off. That destroys important tissue and allows blood to enter the lungs. After being killed the mouse's chest and abdomen are sliced open completely and the intestines allowed to protrude outside the body. The skin over the head and back are sliced open and reflected to the sides to expose the spine and vertebral muscles to the fixative. If the fixative is Bouin's the skull need not be opened. Bouin's fixative penetrates the skull rapidly and fixes the brain. For formalin fixation the skull should be opened. Try not to touch the brain or you will cause neurons to shrink, so called "dark neuron artifact". Never remove the fresh brain from the skull before it is fixed. You will squash it, distort it, and cause dark neuron artifact. Also do not try to dissect out the spinal cord. It is far better to make sections of spinal cord in situ within the vertebrae. After opening up the mouse put it in a generous amount of Bouin's fixative. Do not cram the mouse into a 50-ml Falcon tube, though it will just fit, with its nose in the pointed base of the tube. After the mouse has been in fixative for at least 4–5 days bones will be soft and dissection will be simple. To speed things up change the Bouin's fixative every day.

3.2.5. Fixation by Perfusion

A variant of whole-body fixation by immersion is fixation by perfusion. This procedure has been used for many decades to fix nervous tissues for ultrastructural examination. It is also an excellent way to fix nervous tissue for light microscopic examination, since it provides instant complete fixation. It eliminates the nasty shrinkage artifact around blood vessels and neurons in brains fixed by immersion. Importantly it is impossible to get good sections of spinal cord without perfusion. By the time a spinal cord has fixed by immersion, even using Bouin's fixative, motor neurons have become artifactually shrunken and even vacuolated. The perfusion procedure is made to look very difficult by the experts but is actually very simple. The mouse is deeply anesthetized until the hind leg does not retract when the toes are pinched with forceps. Alternatively, the mouse is killed with carbon dioxide. Then within a leisurely few minutes the mouse is pinned down on its back and the chest is opened by cutting along the rib cage on both sides about midway along the ribs. The flap of sternum is gently pinned back over the head but not squashed down. Then the right auricle of the heart is snipped off, causing blood to flow into the chest cavity. This is a good time to collect blood if you want it. The heart is lifted up gently with forceps and a 20-gauge needle on a butterfly tube is inserted straight down into the tip of the heart through the wall, but not deeper. If there is a concern about going too deep, the plastic sleeve provided with the needle can be cut off so that only 4 mm of bare needle sticks out of the sleeve. The tip of the needle will be in the left ventricle. Next, infuse 5–10 ml of PBS into the heart to wash out the blood. After that infuse around 20–30 ml of fixative, preferably Bouin's fixative, into the heart using a 20–30-ml syringe. The mouse will stiffen and turn yellow. You can push the full 30 ml of fixative through in around 5 min. You cannot do much damage by pushing too much fluid too fast, but do not rush it. Sometimes the fixative runs out of the nose, in which case the perfusion will not be very good. However, this is not for EM but for light microscopy, so it does not make a lot of difference. For EM a mixture of glutaraldehyde and PFA (Karnovsky's fixative) or NBF (Trump's fixative) has to be used. After perfusion follow the procedures in the next paragraph.

3.3. An Optimized Adult Mouse Necropsy and Special Techniques for Each Organ

A full set of mouse tissues can be dissected into enough parts to fill ten cassettes resulting in ten slides. The key to dissection and cassetting is that the cost of the complete necropsy will depend on the efficiency of selecting and slicing up the tissue. Here is a useful comprehensive and cost-effective way to do it. Note that many tissues are put into each cassette. This contradicts one of the silly old rules required by so-called "Good Laboratory Practice" and used by drug companies, that each cassette, paraffin block, and slide should have only one tissue. That may be necessary for larger animals, but is certainly not for mice. Ten cassettes, blocks, and slides will suffice to represent the entire set of mouse tissues from a standard necropsy.

3.3.1. Heart

It is sliced longitudinally so that four chambers are seen on the slide. Slice it through the center of the atria on each side. Put both pieces, along with lung slices, into cassette #1. For those interested in atherosclerosis the most efficient way is to slice a cross-section through the whole heart as close to the root of the aorta as you dare go. The heart is embedded flat side down and serially or step sectioned towards the great vessels. Since in mice the most severe atherosclerosis is at the base of the aortic valve leaflets, sectioning has to proceed until that level is reached. Traditionally people do this sectioning on frozen sections since they want to stain with oil-red-O for lipid. This is poor practice. Serial frozen sections are time-consuming and difficult to cut and have poor histologic detail. In easily cut paraffin sections lipid shows perfectly well as vacuoles in macrophages and cholesterol shows well as clefts. In good-quality paraffin sections other important cellular details such as inflammatory cells in addition to macrophages can be studied.

3.3.2. Lung

Slice fixed right and left lung lobes longitudinally from the rostral (snoutward) or proximal part of the lungs, where the bronchi are, to the caudal (tailward), peripheral part of the lobes. For the most careful examination of lung architecture it is often preferable to vary the fixation procedure by opening the chest of the freshly killed mouse and then placing a needle into the trachea after finding it by dissecting the neck. The lung is then filled with fixative via the trachea until the lungs fill the chest comfortably. Do not overfill. Some people tie off the trachea before removing the needle so the lungs do not empty again, but this is not necessary. Lung research labs make this procedure very complicated by removing the lungs in toto and then filling them with a fixed head of infusion pressure, but routinely this is a waste of time. In any case never slice lungs before fixation. The razor blade will only crush the tissue. If you want to avoid any crushing at all, separate the lung lobes at the hilus after complete fixation by infusion, process them into paraffin and then slice up the lungs at the time of cassetting after they have hardened in paraffin. It should be added that if you perfuse-fix mice the lungs always fix very well with appropriately dilated alveoli. Finally be aware that in filling lungs for optimal fixation any cells or debris in alveoli and bronchi will be washed out. For example, if the lungs have edema, which on slides looks like homogenous eosinophic fluid in alveoli, filling the lungs will wash out the fluid making the diagnosis of edema impossible. In some experiments, then, it is best to fill some lungs for fixation and to leave others in other mice unfilled.

3.3.3. Salivary Glands and Submandibular Lymph Nodes

These are arranged in a chain of organs extending from below one ear around under the throat to the other ear. Parotid, submandibular, and sublingual are the three kinds of salivary glands in the mouse. Many minor salivary glands are present around the back of the mouth. Take as much of this chain as convenient and put it flat into a cassette. Salivary glands, thyroids, and thymus go into cassette #2.

3.3.4. Thyroids

These are located on each side of the trachea just below the larynx. Dissect out the larynx and trachea in toto without the muscle that runs along the ventral side of the trachea and without the esophagus, which is dorsal to the trachea. Cut the trachea across below the thyroid and include half the larynx. The little tube should be embedded on its side, not on its end, so that the trachea is longitudinal with the thyroids on each side. If you dissect the thyroids off of the trachea and try to process them separately you are likely to lose them. It is a matter of chance and luck whether the parathyroids will be present within or adjacent to the thyroids. To find them you may have to serial section the entire piece of trachea plus thyroids. Unlike other animals mice have parathyroids that often are found adjacent to thyroids and not embedded in them.

3.3.5. Thymus

This is the pale tissue in the chest in front of the heart. It is small in old mice, immunodeficient mice, and in stressed baby mice in which it undergoes extensive apoptosis. It is worth hunting for.

3.3.6. Liver

Slice a few lobes into 4-mm thick slices and place into a cassette. It makes little difference where you take the slices from since the liver is homogenous, but include a slice of a lobe that includes the gall bladder. Liver, spleen and pancreas and kidney plus adrenal go into cassette #3.

3.3.7. Spleen

Many people would like to analyze spleens by flow cytometry (FACS). Beg them to leave half the spleen for you, which they can include in a separate container. Some object that FACS spleen data must use whole spleen as the denominator value as in "number of CD8+ cells/spleen. They do not want to divide the spleen. If this problem arises you can ask for spleen from another mouse of the same genotype. Remember that the red pulp of the spleen is very much like bone marrow. To understand the pathology of the immune system or of the blood or of the bone marrow of a mouse you need to do pathology on the spleen. It is best to slice the spleen in such a way that the cross-section on the slide will represent its largest average diameter. Spleens in mice are very labile in size, usually because of increased or decreased hematopoiesis. You want to be able to say if any spleen was enlarged or not just by looking at the slide. The average normal spleen is around 50% white pulp and 50% red pulp.

3.3.8. Lymph Nodes

The submandibular lymph nodes will appear in section with the salivary chain. Mesenteric lymph nodes can be difficult to find but can be included with the intestines as described below. If investigators are specifically interested in lymph nodes careful examination of inguinal and axillary regions is necessary. The inguinal and axillary nodes are always present but usually are tiny.

3.3.9. Pancreas

Pancreas is dissected as completely as possible and put in the liver cassette. Many people are interested in the pancreatic islets of the mouse. They are not scattered randomly in the spleen but tend to be clumped. Exactly where it is is never clear. To make sure you get the entire pancreas on the slide to maximize your chances of finding islets you should remove the entire fresh pancreas after killing and opening the mouse. It is a flat tissue that extends between the spleen, stomach, and duodenum. After removal, spread the pancreas flat onto a piece of paper towel and put pancreas and paper into the fixative. It will stick to the paper and will fix flat. If you do not do this and simply drop the pancreas into fixative it will coil up. If flat-fixed a single section will include many islets. Otherwise, the block will have to be step sectioned to find enough islets to evaluate. If a mouse is dissected after fixation by immersion the pancreas will not be flat.

3.3.10. Kidneys and Adrenals

Kidneys should be removed from the body with adrenal still attached. The kidneys should be cut sagitally on the midline so that the cut section passes through the renal pelvis and also passes close to the adrenal. To make sure the pelvis and renal papilla are included in the section it is best to make the sagittal slice from the ventral aspect of the kidney where the ureter protrudes. The sagittal slices with adrenals still attached should be put into the liver cassette.

3.3.11. Small Intestine

The small intestine varies quite markedly along its length with villi generally longest proximally in duodenum and shortest in ileum. Lesion can be randomly distributed along its length. The most important thing to remember about gut is that superficial epithelium is very easily disturbed by handling, scraping, or washing. It is best not to touch the mucosal surface of gut tissue. For routine necropsies on fixed mice it is best to slice out 4–5 segments of the intestine at various places along its length that will lie flat in cassette #4. The sections will be longitudinal. Some people seem to think that cross-sections are preferable to longitudinal ones. This is true only in one sense. A cross-section will always show villi in longitudinal section from crypt to tip of villus. Longitudinal sections have to be sectioned deeply to get close to the center of the lumen or many villi will be cross-sectioned or sectioned on a bias. In any case, in addition to routine samples any abnormalities are sampled. Peyer's patches occur randomly anywhere along the length of both small and large intestine and can be seen as slight elevations of the serosal surface. It is worthwhile trying to find the mesenteric lymph nodes at the base of the mesentery and including them if you can.

3.3.12. Large Intestine and Stomach

Slices of cecum and stomach can be removed preferably with the knife blade passing through the greater curvature of each organ. Segments of proximal and distal colon should be removed. There is nothing wrong in including pieces of feces. Colonic disease like colitis or cancer can be randomly distributed, so the more colon one takes the better. The pieces of stomach and large bowel are put into cassette #5.

3.3.13. Swiss Rolls of Guts

An alternative approach to fixing intestines, if it is necessary to rule out any gut pathology anywhere along its length, is to make "Swiss rolls". No one has consulted with the Swiss to see if they agree with this name. This is done in either of two ways. The first, used by those who hunt for intestinal tumors, is to open up the gut along its full length and examine the opened intestine under a dissecting microscope. The gut is then rolled up into one or two coils, the way one might roll up ones belt, which are then placed in cassettes. This tends to cause considerable distortion and fixation is not very good. A better way, which precludes examination of the opened gut first, is to remove the gut and then coil it down flat onto a sheet of paper. It will stick to the paper. So that the coils are small enough to fit flat in a cassette, 2–3/per intestine must be made. The paper and gut is then lowered down into a jar of Bouin's fixative. Bouin's fixative fixes gut rolls very well; formalin does not penetrate as well and tips of villi can slough due to poor fixation. For the colon the entire colon can be removed and coiled down or laid out in two or three straight pieces on paper for flat fixation. Note that the gut is not washed. Washing removes important pathologic findings like blood, sloughed cells, and pathogens. Again note also that the mucosa is very fragile and should not be touched.

3.3.14. Whole Head with Brain

For a routine necropsy of a mouse that has been fixed by immersion or perfusion it is best to leave the brain in the whole skull and make sections of brain and demineralized skull together. A midline sagittal razor blade slice is made first. It includes everything from the tip nose to the caudal medulla of brain. Next a parasagittal slice through the right eye and ear is made. With luck the knife blade goes right through the eye and mid-mediolar part of cochlear as described below. The sagittal and parasagittal sections of head are put into cassette #6. Next the left half of skull and brain is sliced into six cross-sections, one through left eye, one through left ear and cerebellum, two between these two slices, one in front of the eye slice, and a sixth caudal to the ear slice. The coronal slices go into cassette #7. Slides of cassettes #6 and #7 will include a wide variety of tissues including eyes, ears, Zimbal's gland, respiratory and olfactory epithelia, nasal glands, vomeronasal organ, and salivary ducts, minor salivary glands, Harderian and exorbital lacrymal glands, tongue, molar, and incisor teeth with constantly growing roots, skin, and vibrissae. The midline sagittal section of head will include the pituitary. This is the optimal way to get sections of that gland. With luck this section will include a section of the pineal gland too.

3.3.15. Pituitary

If you try to dissect out fresh pituitary you will ruin it every time. The best way to see the pituitary is in a midsagittal slice of whole head with brain. If, however, you want to sample the pituitary alone, carefully remove the brain, being careful that the pars nervosa does not pull away from the pars intermedia. Then fix the pituitary in the skull without removing it until after fixation.

3.3.16. Perfusion Fixed Brains

For optimal pathologic examination of brain, perfusion with Bouin's fixation is essential. After perfusion do not open the skull. Simply put the whole head into Bouin's fixative to continue fixation. After a few days open the softened demineralized skull widely so that the brain can be removed without damage. After removal, sagittal or coronal sections can be sliced and cassetted. Slicing is a critical step. It is best to use one of the disposable microtome blades used in the histology lab for this. Where you slice will determine what structures you see on the slide. It is a good idea to consult a brain atlas to decide where you will make the slice. The cut should be exactly perpendicular and exactly at right angles to the perpendicular so that the brain sections are precisely symmetrical. You will have to make a second slice parallel to the first to have a 4 mm-thick slab of brain that will fit into the cassette. You will also have to make sure that the correct face of the slab is sectioned. For this place a dot of India ink on the face you do not want to look at. When the slab of brain is embedded the marked face is placed upwards in the embedding mold. Be aware that commercially available steel or plastic brain molds used to slice whole brain are designed for slicing brains that have been lightly fixed and cryopreserved by soaking in sucrose. Slices from such brains are frozen and cryosectioned. This is the way that most neuroscientists process brains for immunostaining and morphometrics. Brains that have been fixed for several days or longer for conventional paraffin sections are so shrunken that they occupy only two-thirds of the mold. The molds are therefore of no use for slicing such brains. To get a good look at cerebellar folia, precise midline sagittal sections are essential. The optimum midline section of the cerebellum should include no deep cerebellar nuclei and no peduncles so that the cerebellum is not attached to the medulla. To get a good look at the commissures, frontal sections are beautiful. For everything else cross-sections (coronal) are best. Sometimes serial sections of parts of the brain are required. Experience has shown that nothing is gained by making serial cross-sections of several slices of brain embedded together. Rather a single slab of brain including the level of interest is cut from the brain using a razor blade, embedded, and serially sectioned.

3.3.17. Spine with Spinal Cord and Leg

The vertebra, paraspinal muscles and spinal cord in situ are sliced into six equidistant cross-sectional segments representing two cervical, two thoracic and two lumbar spines. These slices go into cassette #8. The slides from these cassettes will include spinal cord, meninges, dorsal root ganglia and spinal roots, vertebra, skeletal muscle, and nerves. Thoracic levels will show sympathetic trunk. A midsagittal slice is made through the entire hind leg and placed in cassette #8. The slide will show bones, joints, muscles, and nerves. Note that the three cassettes with brain and spinal cord have to be placed into a separate container from the other cassettes. These will be washed for a few hours in running water to wash out the Bouin's fixative and will not be placed in 70% ethanol. Holes in white matter due to prolonged exposure to ethanol must be avoided. Other cassettes with non-neural tissues can be washed in 70% ethanol or in running water and can be stored in 70% ethanol until time for processing. If you can avoid soaking of any tissues in ethanol it is best since in a busy lab some brain tissue always seems to get into ethanol and to develop holes.

3.3.18. Bone and Muscle

Bone experts have insisted for a long time on using a lot of very difficult techniques such as cutting whole undemineralized bone using a sledge microtome. They like to determine how well mineralized bone is using vonKossa stain. Various imaging techniques like dual energy X-ray absorptiometry (DEXA) are used for determining how dense bones are. Bouin's fixation and demineralization and H&E sections are, however, totally adequate to analyse all aspects of bone. One can see how thick cortical bone is, how much medullary bone there is in metasphyses and epiphyses (none in diaphysis of long bones). How well bone is mineralized is evaluated by looking for slightly blue staining cement lines in H&E sections. Unmineralized bone (osteoid) in osteomalacea stains a homogenous pink color. If bones have true osteoporosis one can see holes in bones. Osteoclasts and osteoblasts are easily observed in young bone. Muscle is best fixed while still attached to bone. Otherwise the fibers contract abnormally. It has to be admitted, however, that most paraffin sections of fixed muscle have many small condensed fibers scattered among larger less dense fibers giving a "checkerboard pattern". This pattern is artifactual and does not represent staining of fast and slow twitch fibers. With the exception of the soleus and quadriceps muscles mouse muscle is composed only of fast twitch fibers. In making the diagnosis of neurogenic atrophy one cannot rely on finding individual small or tiny fibers. These are found in most mice. Grouped atrophy is the only reliable diagnostic finding in neurogenic atrophy in mice.

3.3.19. Reproductive Tracts

Reproductive tissues are placed in cassette #9. The female tracts should be removed in toto with vagina, cervix, both horns of the uterus and both ovaries. For males one or both testes plus epididymus are removed and cassetted. The seminal vesicles, coagulating gland (sometimes called anterior prostate), prostate at the bladder's neck and urethra are dissected together in toto. This tissue is embedded dorsal or ventral side down and pressed flat in the embedding mold as the paraffin hardens. The preputial glands on each side of the penis are often inflamed and should be removed separately. The bulbourethral gland, distal urethra, and penis, though often ignored, should be included. Mouse prostates present problems because they are composed of four biologically distinct lobes on each side, with differential susceptibility to various kinds of cancer. For careful examination of the four lobes of the prostate some people dissect them individually under a dissecting microscope. This is very time consuming. A quicker way is to remove bladder and seminal vesicles and to make a cross-sectional slice through the middle of the prostate at the bladder neck. Then process the slices of prostate in a single block and serial or step section the block. The sections from this slice will include dorsal prostate with red homogenous secretion, lateral prostate with red granular secretion and ventral prostate with pale blue secretion. If the block from cassette #9 has only dorsal prostate and one wants to see ventral prostate a simple way to avoid step or serial sectioning is to melt the paraffin block, flip the tissue over, re-embed, and section what is now the ventral side. Re-embedding and flipping tissues in blocks is generally a quick and inexpensive way of finding tissues that do not show up on slides but are assumed to be deeper in the block.

3.3.20. Skin

Skin samples can be removed from abdomen and back of the fixed whole body and placed in cassette #10. It is important that slices of skin be sliced so that the sections will not be twisted and that the sections will run in a straight midline sagittal orientation. A more comprehensive study of skin requires study of whole pelt. The fresh skin is removed from much of the body in toto, placed flat on a piece of paper, sticky side down, and put in fixative. It will fix flat. Then midline sagittal slices and coronal slices are cut and cassetted. It might be useful to take plucks of hair if the hair looks abnormal. The hair is put on the slide, covered with embedding compound, and cover slipped. The four kinds of hair – guard, awl, auchene, and zigzag – can then be examined. Skin samples with wounds or implants have to be handled in a special way. A rectangular slice of skin is removed with the wound in the center and fixed flat on paper. After fixation a single slice of the skin is made lengthwise exactly through the center of the wound or implant. That face is embedded down, so that the section will include the maximum central extent of the lesion. Only one slice is required. It will include normal skin at one end, progressing through the lesion in the center, and then extending to normal tissue again on the other side. If many slices of the lesion are made it is never clear which represents the center. For wound-healing experiments, for example, one can tell how completely the healing has advanced if the sole section on the slide represents the center of the lesion where closure is least complete.

3.3.21. Mammary Gland

Mammary tissue is located in subcutis from under the ears, down the ventral neck to the chest and back up behind the forelegs. Another patch begins in the inguinal area extends dorsally up the sides and medially down between the legs to the back of the hind legs. In virgin mice the tissue is composed only of small ducts running in the subcutaneous fat, which is a thin layer in younger lean mice. Mammary tissue is best sampled by removing inguinal skin with all attached subcutaneous fat right down to the abdominal muscle. The skin is fixed flat on paper, sliced into strips including skin and fat together and placed in cassette #10. For special studies of mammary biology the gland at all stages of gestation, lactation, and involution are examined. A nice adjunct to histopathology is study of whole mounts of mammary tissue made by clearing the whole fat pad in xylene or acetone, staining with methylene blue and mounting on a slide. The same tissue can later be processed for paraffin slides.

3.3.22. Tumors

Any large tumors found under the skin or elsewhere at the time of necropsy should be left in place but sliced on the midline to assure proper fixation. At the time of dissection tumors should be removed, leaving surrounding tissues still attached. Histological sections should always include the tumor tissue margins. The margins are where invasion and early metastasis into lymphatic and blood vessels will be seen on slides. It is best to make a single slice through the largest diameter of a tumor. The slide will then be a permanent record of the tumor's size. Tumor slices can go into cassette #10.

3.3.23. Eyes

For best results eyes should be dissected out of the orbit by inserting curved scissors behind the eye and snipping the optic nerve. Do not just yank the eye out. That causes the optic nerve to snap back into the optic nerve head. It is difficult to get good sections of eyes in which the retina has not detached and become distorted. To avoid that, eyes should be fixed in fixative of choice, including Bouin's fixative, for 4 h, and then transferred to 70% ethanol. This will minimize shrinkage of the retina which causes it to pull away from the retinal pigment epithelial layer and sclera which shrink less than the retina. Eyes should never be sliced unless one wishes to dissect out the retina for whole retinal mounts. Eyes should be embedded on their sides so that slides will show the eye from front to back – cornea to optic nerve. Some people like to embed eyes in plastic. The sections are very elegant, but expensive.

3.3.24. Ears

Ears have to be handled specially for best results. The petrous temporal bones are quite easily removed from the skull since the bone is so much harder than surrounding bones of the skull. It is then fixed in formalin and demineralized in EDTA or fixed in Bouin's fixative. Traditionally ears of mice are embedded ventral side down so that sections have a frontal (also called horizontal) orientation. They are not traditionally embedded so that sections are coronal (also called transverse and cross-sections). The embedded ears are then serially sectioned. Appropriate levels will have the optimal "mid-modiolar" plane so favored by ear experts. It turns out that an even better mid-modiolar plane is seen when the ears are embedded for serial sagittal sections. Such sections show the whole 2.5 turns of the cochlea. (This orientation is the same as that seen when the whole skull is sliced in the parasagittal plane and the slice fortuitously goes through the center of the cochlea). The vestibular ear is also particularly important for mice that are spinning and tail chasing. The otoconia of the maculae of utricle and saccule are easily visualized if the fixed whole ear bone is cleared in methyl salicylate (oil of wintergreen) and examined under polarized light using a dissecting microscope. The otoconia can be seen in demineralized paraffin embedded tissue sections since the matrix remains, even though the calcium carbonate is removed.

3.3.25. Blood

A blood smear taken from a heart blood sample at the time of necropsy right after the mouse has been killed is a simple and very powerful tool to assess the health of the blood. Leukemias, various kinds of anemias, and blood "dyscrasias" all can be tentatively diagnosed from one drop of blood. It is fairly safe to say that if a blood smear is normal the mouse has no blood disease. Remember to put only a tiny drop on the slide. The smear should end half way down the slide in at tongue-shaped smear. If you put too much blood on the slide you will drag the white blood cells off the end of the slide. Remember to fix the slide after air-drying in 100% methanol. Otherwise, the cells will disintegrate over time before you get around to staining them with Wright-Geimsa stain. For future studies in other mice, after you have determined that there is a blood phenotype, you will have to do complete blood counts and other assays to understand the disease. Remember that the red pulp of spleen is like bone marrow so it must be analyzed along with bone marrow taken from the legs and vertebrae to understand blood disorders. A foolish convention from human and large animal pathology is to sample the sternum for bone marrow. Unlike large animals and people the medullary cavities of all mouse bones, except those of tail and distal limbs, are full of bone marrow and have almost no adipose tissue at all.

3.4. Newborn and Baby Mice

Baby mice should be killed with CO_2. This can be frustrating since CO_2 seems to take forever to kill them. You can put the baby mice in the refrigerator to cool them down to anesthetize them and then put them in CO_2. If the pups are newborn it is best not to dissect them. Just put them in Bouin's fixative whole. Do not use PFA or formalin. Immature mouse tissues shrink and become very brittle when fixed in aldehydes and slides are very poor or not interpretable. The nuclei shrink so much that nuclear detail such as nucleoli cannot be studied. If you have to use aldehydes, fix no longer than overnight and then transfer to 70% ethanol. For whole-body fixation of mice older than 2–3 days open the skull, abdomen, and chest before putting them in Bouin's fixative. This is especially important if you use aldehydes which do not penetrate well. After fixation slice the head sagitally or coronally and slice the body into four cross-sections including the legs. You might miss one or another organ like the adrenals when only one slide is made from the block; deeper sectioning into the block will reveal any overlooked organs. It might be necessary to rule out heart defects such as abnormal septae, great vessels or valves. Random slices of heart will not serve for careful analysis of heart. Instead, the fixed body should be sliced through the chest above the shoulders and just below the level of the diaphragm and embedded rostral face down. Complete serial sections through the entire chest are then cut. Start making slides with as many sections/slide as possible. For larger postnatal mice only 4–5 sections/slide will be possible and to get through most of the heart 30–40 slides will be necessary. For smaller mice or for embryonic mice fewer slides with more sections/slide will get through the entire heart. If you step section or throw sections away to save money and time you will inevitably throw away the exact slide with the defect you are trying to find.

3.5. Embryos

Bouin's fixation is essential for embryos or tissues become brittle and good sections without tearing and shattering cannot be cut. If embryonic death is suspected, start with a time-mated dam carrying pups at e18.5. Kill the dam and check each placental site. Open it, remove the pup plus placenta, brush off the nose of the mouse to start it breathing, and lay it out in a warm place. Normal mice will turn pink, breathe, and squirm around for hours. Mutants may gasp, turn blue, and die. This failure to breath is a common lethal phenoytpe, unexplained more often than not. Cut off a piece of leg for genotyping and put pup plus placenta in Bouin's fixative. You can select mutants and their placentas for future histological analysis if any are present at that age. It may be, however, that you find no mutants at e18.5 but there are some placental sites without any living embryo. In that case you have to go earlier to, say, e16.5 days. Look to see if all embryos are present and accounted for; if not, step pack to an earlier time point. Eventually you will find a point at which mutants are present but perhaps look different from wild-type mice and heterozygotes. You want to focus your phenotyping time at the age when mutants are as close to death as possible. Next you will want to go earlier and earlier to find the earliest time when the mutants deviate phenotypically from controls. In this way you will be able to study the pathogenesis of the phenotype. Placenta must always be evaluated. At 13.5 days or earlier yolk sac must also be evaluated. Really small embryos at 10.5 days or earlier must not be handled with forceps at any time. Processing for histology must be done by hand. Embryos are moved from solution to solution by drawing them up into the barrel of a plastic pipette, the tip of which has been cut off. It is seldom useful to make random sections of an embryo, and sagittal sections are difficult to assess. Rather, the embryos should be embedded on the top of their heads and serially cross-sectioned. Pictures with this orientation can be seen in Kaufman's atlas *(5)*. Placentas should be embedded so that sections pass through the labyrinth. That is where nutrient exchange occurs and it is where significant pathology should be sought. When you get back earlier than e9.5 days you will have difficultly removing, fixing, and embedding tiny embryos. You may have to resort to serially sectioning the whole uterus with embryos still in their placentas, hoping to find mutants with clearly observable phenotypes. Confirmation of the genotype would require removing DNA from the embryos using laser capture or by picking the embryo section off the slide with a fine needle.

References

1. Matsumoto, H., Kumon, Y., Watanabe, H. et al. Antibodies to CD11b, CD68, and lectin label neutrophils rather than microglia in traumatic and ischemic brain lesions. J Neurosci Res. 2007; 85:994–1009.
2. Bronson, R.T., Lake, B.D., Cook, S. et al. Motor neuron degeneration of mice is a model of neuronal ceroid lipofuscinosis (Batten disease). Ann Neurol 1993; 33:381–85.
3. Mikaelian, I., Nanney, L.B., Parman, K.S. et al. Antibodies that label paraffin-embedded mouse tissues: a collaborative endeavor. Toxicol Pathol 2004; 32:181–91.
4. Bronson, R.T., Donohue, L.R., Samples, R. et al. Mice with mutations in the mahogany gene, Atrn, have cerebral spongiform changes. J Neuropath Exp Neurol 2001; 60:7245–730.
5. Kaufman M.H. 1992 The Atlas of Mouse Development, Academic Press, London.

Chapter 25

Systemic First-Line Phenotyping

Valérie Gailus-Durner*, Helmut Fuchs*, Thure Adler, Antonio Aguilar Pimentel, Lore Becker, Ines Bolle, Julia Calzada-Wack, Claudia Dalke, Nicole Ehrhardt, Barbara Ferwagner, Wolfgang Hans, Sabine M. Hölter, Gabriele Hölzlwimmer, Marion Horsch, Anahita Javaheri, Magdalena Kallnik, Eva Kling, Christoph Lengger, Corinna Mörth, Ilona Mossbrugger, Beatrix Naton, Cornelia Prehn, Oliver Puk, Birgit Rathkolb, Jan Rozman, Anja Schrewe, Frank Thiele, Jerzy Adamski, Bernhard Aigner, Heidrun Behrendt, Dirk H. Busch, Jack Favor, Jochen Graw, Gerhard Heldmaier, Boris Ivandic, Hugo Katus, Martin Klingenspor, Thomas Klopstock, Elisabeth Kremmer, Markus Ollert, Leticia Quintanilla-Martinez, Holger Schulz, Eckhard Wolf, Wolfgang Wurst, and Martin Hrabé de Angelis

Abstract

With the completion of the mouse genome sequence an essential task for biomedical sciences in the twenty-first century will be the generation and functional analysis of mouse models for every gene in the mammalian genome. More than 30,000 mutations in ES cells will be engineered and thousands of mouse disease models will become available over the coming years by the collaborative effort of the International Mouse Knockout Consortium. In order to realize the full value of the mouse models proper characterization, archiving and dissemination of mouse disease models to the research community have to be performed. Phenotyping centers (mouse clinics) provide the necessary capacity, broad expertise, equipment, and infrastructure to carry out large-scale systemic first-line phenotyping. Using the example of the German Mouse Clinic (GMC) we will introduce the reader to the different aspects of the organization of a mouse clinic and present selected methods used in first-line phenotyping.

Key words: Mouse clinic, phenotyping, skeletomuscular diseases, osteoporosis, bone metabolism, cardiovascular disease, blood pressure, atrial natriuretic peptide, obesity, cachexia, diabetes energy expenditure, energy intake, lung function, neurological disease, motor coordination, muscle

*Equal contribution

Ralf Kühn, Wolfgang Wurst (eds.), *Gene Knockout Protocols: Second Edition, vol. 530*

DOI 10.1007/978-1-59745-471-1_25 Springerprotocols.com

function, eye, blood, clinical chemistry, hematology, urine, flow cytometry, bead-array, lymphocyte, antibody, infection, IgE, allergy, asthma, airway inflammation, bronchoalveolar lavage, steroid metabolism, pathology.

1. Introduction

The German Mouse Clinic (GMC) was established as the first mouse phenotyping platform worldwide with the logistics of systemic, standardized phenotypic analysis with open access for the scientific community (www.mouseclinic.de). Three additional mouse clinics – at the MRC Harwell (UK), ICS (France) and Sanger Institute (UK) – have established the same rationale for systemic phenotyping of mouse models used in the GMC. These large-scale phenotyping platforms collaborate for the standardization of phenotyping protocols in the EUMODIC consortium (arose from the EUMORPHIA consortium; *(5)* http://www. eumodic.org/) and are embedded in the collaborative effort of the International Mouse Knockout Consortium (the European EUCOMM, the US KOMP and the Canadian NorCOMM projects) *(1–4)*.

In a systemic first-line phenotypic analysis a series of tests are performed on a large cohort of animals. Basic parameters of different physiological areas can be measured relatively easily and faster allowing the screening of large numbers of mice. At the GMC this "primary" screen comprises the areas of behavior, bone and cartilage, neurology, clinical chemistry and hematology, eye development, immunology, allergy, steroid metabolism, cardiovascular function, energy metabolism, lung function, vision and pain perception, gene expression, and pathology. Part of the standardized primary phenotyping screen is in common with the first-line analysis (so-called EMPReSSslim) of the other European mouse clinics, ensuring sharing and comparability of data.

Mouse lines showing an interesting phenotype are subjected to a more in-depth assessment using more time- and cost-intensive measurements. These "secondary" and "tertiary" screens include tests with training cycles, non-invasive in vivo imaging methods, and analysis requiring surgery like telemetry or electro encephalography (EEG).

2. Aspects of Systemic First-Line Phenotyping – Organization of the Phenotyping Facility and Workflow

Comprehensive phenotyping of a larger number of mice requires expertise in many different disease areas, equipment and infrastructure – prerequisites that single laboratories cannot provide. Therefore, the concept of a phenotyping platform for the scientific community has evolved. The GMC provides the phenotypic analysis for mouse models generated by various methods (knockout, knockin, knockdown, gene trap, ENU) on the basis of scientific collaboration. Here we would like to summarize the most important aspects of systemic first-line phenotyping by means of the GMC. *Expertise*: At the GMC experts from various fields of mouse physiology and pathology work side-by-side in close cooperation with clinicians, computer scientists and engineers at one location. This allows active discussions and exchange of observations during the screening process, serves the systemic view for the interpretation of data, and accumulates an enormous amount of expertise and knowledge. After the screening of a mouse line the data are discussed with the provider of the mouse model, and together with a comprehensive report recommendations for further analyses are given. *The building and equipment*: The building (specific pathogen-free (SPF) area) for first-line phenotyping has a modular structure of 11 units and rooms for shared infrastructure (total GMC capacity of around 10,000 mice). Each unit consists of a laboratory and an adjacent mouse room, which can be run independently with respect to room temperature, air humidity, and air pressure. Animals are exposed to the same environmental conditions keeping variability due to miscellaneous housing/sanitary conditions to a minimum (essential for the primary screening protocol). For the analysis of genome–environment interactions which are not assessed in the primary screening workflow a challenge platform with additional laboratories and attached mouse rooms was established (capacity of an additional 10,000 mice). State-of-the-art technology and equipment (e.g., micro-computed tomography, automated blood analyzer, nine-color FACS, Bioplex) are established and constantly improved. *The animal husbandry*: Mice in the GMC are housed in type II polycarbonate cages in individually ventilated caging (IVC) systems. The main challenge of this multi-user unit is the maintenance of animal health and avoidance of infections by importing mice from other institutions with different health status. Therefore, mice are only accepted for import if a recent health report according to the FELASA recommendations has been approved *(6)*. *Management*: For the organization and management of a mouse phenotyping platform with a capacity for systemic phenotyping of more than 3,000 mice per year and with respect to more than 320 parameters, it is mandatory to establish a distinct management unit. The GMC management team coordinates the different operating levels in the GMC, including animal management, logistics for the import of mice, primary phenotyping workflow, quality control, communication with collaborating partners, and organization of infrastructure and communication platforms for the GMC members. *Database/bioinformatics*: For the central structured storage of any data produced in the animal facility, including ordinary attributes of mice, health reports, phenotyping schedules, files, examination results and metadata, etc., we strongly recommend using a mouse colony management system (MausDB is used in the GMC *(7)*). It facilitates daily work as well as generation of reports, and is a prerequisite for data mining and data exchange. For making use of the results of a complex phenotypic analysis the data have to be accessible for further correlation and analysis. A process to develop standards for the description of phenotypes and file formats for the description of phenotyping protocols and phenotype data sets has been initiated *(8)*.

In addition, the management of the import and phenotypic analysis of a mouse line every week requires support by a database. For mid-term scheduling of overlapping phenotyping workflows of mouse cohorts we have developed a software tool called CoordDB. It includes a phenotyping request form on the GMC website (http://www.mouseclinic.de/), which stores the information about mutant mouse lines and collaboration partners and manages administrative tasks arising thereof. The web-based, database-driven software tools MausDB and CoordDB have been developed in-house and will soon be made available as open-source systems on our website (http://www.helmholtz-muenchen.de/ieg/).

3. First-Line Phenotyping Scheme

Mouse models are analyzed in separate cohorts of mutant female and male mice and are compared with their respective wild-type littermates or to an established baseline of the corresponding wild-type strain. Genetic background and the choice of the control group play an important role in assessing a phenotype *(3)*. Each cohort consists of at least ten animals and the age range is supposed to be within 7 days. To allow for adaptation of the animals to the new environment the animals are moved into the animal facility 2 weeks before the start of the analysis (which is especially important for the behavioral tests). Before the screening starts all mice are examined for gross morphological alterations or anomalies.

The analysis of a mouse model in both pipelines allows a systemic view of the phenotype (*see* **Fig. 25.1**). The screening period from the age of 9–17 weeks also allows the detection of the development of phenotypes: for example, the body weight is continuously documented, the mice are constantly under observation for behavioral and physical changes, and mice that are dying during the screening period are immediately transferred to the pathology for macroscopic and microscopic histopathological examination.

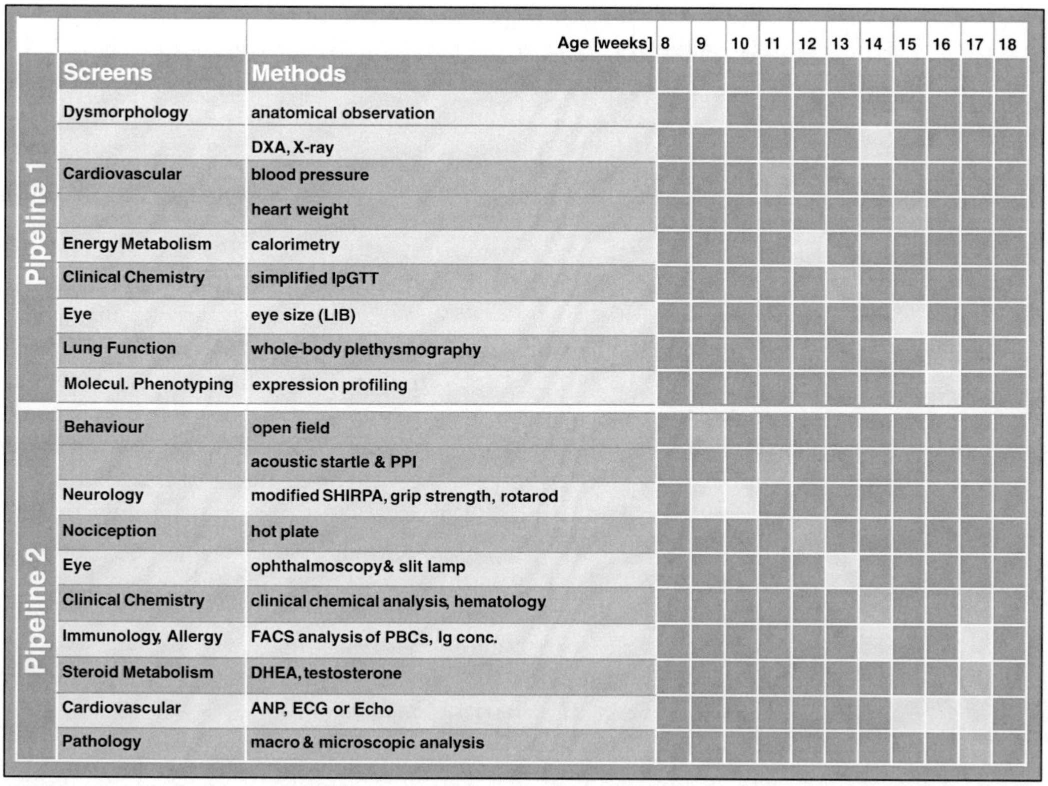

Fig. 25.1. Systemic primary phenotyping in the German Mouse Clinic. Mouse models are analyzed in two different pipelines.

In the following chapters we want to give an overview of selected methods and protocols used in first-line phenotyping with additional information about secondary and tertiary phenotyping methods that are applied in the specific areas. Since more than 320 parameters are measured in both pipelines the methods are not described in detail. Rather we intend to give an example how phenotypic analyses can be performed in a consecutive order taking into account the different influences the tests might have on one another.

3.1. Bone and Cartilage Phenotyping of Mice

3.1.1. Introduction

The skeleton is the third largest organ system in the human body and consists of bone and cartilage. In humans, skeletomuscular diseases encompass a series of defects including developmental defects and more chronic and progressive disorders, such as osteoporosis or rheumatoid arthritis. With an incidence of 20% in the adult population and a higher prevalence in women and in older age groups, they represent a major cause of physical disabilities in Canada, US, and Europe according to the World Health Report 2001 *(9)*. A direct consequence is a huge burden corresponding to direct and indirect long-term disability and morbidity costs. Studies of naturally occurring and engineered mutant mice have led to a dramatic increase in the understanding of bone biology *(10)* and have in many cases been essential for the discovery and understanding of bone-related human diseases like osteoporosis *(11)*, osteoarthritis *(12)*, osteogenesis imperfecta *(13)*, and scoliosis *(14)* or limb defects *(15)*.

3.1.2. Methods

For a first-line screen, a broad spectrum of parameters involving bone development, metabolism, and homeostasis should be covered (*see* **Table 25.1**). Morphological analysis of the mouse is the easiest way to obtain first-line information about skeletal malformations. There are many protocols available, like the SHIRPA protocol or a 54-parameter protocol according to Fuchs and co-workers *(16)*. For the description of the phenotypes we recommend using the nomenclature of the Mammalian Phenotype Ontology wording (www.informatics.jax.org/searches/MP_form.shtml). Subsequently animals should be analyzed by dual-energy X-ray absorption (DXA) and X-ray imaging since both techniques enable a high-throughput non-invasive first-line phenotyping for bone and cartilage abnormalities. X-ray analysis is a reasonable technique to get information about developmental disorders. X-ray applications can diagnose various conditions like congenital bone diseases, joint dislocations, degenerative conditions, or metabolic bone diseases.

DXA analysis is one of the most commonly used techniques for the diagnostics of osteoporosis and bone mineralization defects. It is based on the different absorption of X-rays by different organ systems and is able to accurately discriminate between the skeleton and fatty and lean tissues *(17)*. DXA provides measures of total body bone mineral density (BMD), total body bone mineral content (BMC), fat mass (FM), and bone-free lean tissue mass (LTM). The major disadvantage of DXA is the area-based measurement of BMD data.

As a secondary test, peripheral quantitative computed tomography (pQCT) technology provides volumetric bone density information and it separates cortical and trabecular bone compartments and thus can monitor metabolic changes very quickly and precisely *(18)*. For in vivo analysis, the pQCT analysis is restricted to locations of the appendicular skeleton and tail vertebra. Proximal tibia metaphysis and distal femur metaphysis are preferred sites for measurements. For a whole-body analysis micro-computed tomography (μCT) *(19)* can accurately determine bone parameters like tissue volume (TV), bone volume (BV), percent bone volume (BV/TV), bone surface (BS), bone specific surface (BS/BV), bone surface density (BS/TV), trabecular thickness (Tb.Th), trabecular number (Tb.N), and trabecular separation (Tb.Sp). The purpose of using bone markers within the skeletal screen is to identify phenodeviants of systemic metabolic bone

Table 25.1
Methods for bone and cartilage screening

Method	Application	Indicated defects
Morphological analysis	Morphological anomalies	Skeletal malformations
X-ray imaging	Bone and cartilage abnormalities, developmental disorders	Congenital bone diseases, joint dislocations, degenerative conditions or metabolic bone diseases (fusions, fractures, translocations, etc.)
DXA (dual-energy X-ray absorption)	Area-based measurement of bone mineral density (BMD) and body composition for the evaluation of bone mineralization defects	Alterations in total-body BMD, bone mineral content (BMC) and body composition parameters (fat mass, lean mass), metabolic bone changes
pQCT (peripheral quantitative computed tomography)	Volumetric bone density distribution for the evaluation of bone mineralization defects, in vivo restricted to locations of the appendicular skeleton and tail vertebra	Alterations in volumetric bone density, mass and area of the cortical and trabecular bone, metabolic bone changes
μCT (micro-computed tomography)	3D-visualization of the whole skeleton, bone microanalysis for the evaluation of bone mineralization defects	Alterations in volumetric trabecular bone density, mass, area and many other bone parameters, metabolic bone changes
Bone marker measurement	Quantification of the formative and resorptive activities of bone cells in near real time, assessing disease progression or response to therapy	Alterations in bone formation, resorption and hormonal regulation, systemic metabolic bone diseases
Three-point bend test	Biomechanical and geometrical properties of single bones	Metabolic bone changes, mineralization defects
Histomorphometry	Analysis of cellular dynamics, localized bone turnover and remodeling, structure on the cellular level	Alterations in bone osteogenesis, metabolic bone changes, defects on the cellular level, mineralization defects
Skeletal preparation	3-D information of the whole skeleton, observation of the endochondral ossification process	Cartilage and bone malformations
In-vitro analysis of bone cells	Cellular causes of bone diseases	Defects on the cellular level (proliferation, apoptosis, metabolic activity, mineralization)

diseases quickly and non-invasively *(20)*. The rate of bone formation or resorption can be assessed either by measuring an enzymatic activity of the osteoblastic or osteoclastic cells or by measuring components of the bone matrix released into the circulation. Bone formation markers include bone-specific alkaline phosphatase (BSAP), osteocalcin (OC), type I procollagen C-terminal propeptide (PICP), and type I procollagen N-terminal propeptide (PINP), whereas bone resorption markers include collagen metabolites resulting from type I collagen degradation like C-terminal (CTX) and N-terminal (NTX) telopeptide, deoxypyridinoline (DPD) and pyridinoline (PYD), helical peptide, lysosomal enzymes like tartrate-resistant isoenzyme of acid phosphatase (TRAP5b), cysteine-proteinases such as Cathepsin K, soluble receptor activator of NF-κB ligand (sRANKL) which is the main stimulatory factor for the formation of mature osteoclasts. Bone turnover regulators include calcitonin, parathyroid hormone (PTH), 1.25 di-hydroxy vitamin D3 (1.25$(OH)_2$D3), and fibroblast growth factor 23 (FGF-23) which is an important regulator of phosphate homeostasis.

Biomechanical and geometrical properties of single bones can be analyzed by three-point bending technique. In the three-point bend test, the whole long bone is loaded in bending until failure. The test should be performed on intact mouse femora using a servo-hydraulic test machine. Each femur is placed on two lower supports that are 8 mm apart. Force is applied at the mid-diaphysis on the anterior surface such that the anterior surface is in compression and the posterior surface in tension. From the load-deformation curve, for example, the load to failure (F_u), stiffness (S), and work to failure (U) can be calculated *(21)*. The dynamics of bone can be analyzed by histomorphological–histometrical and immunohistochemical techniques *(22)*. Using these techniques, information on localized bone turnover and remodeling, as well as structure on the cellular level can be collected. Skeletal preparation uses alizarin red and alcian blue stain for bone and cartilage, respectively. The disadvantages of this experiment are that the animals have to be sacrificed for this analysis and the fact that it is very time consuming.

To broaden bone phenotyping to the cellular level in vitro analysis of bone cells can be applied to unravel and describe potential cellular causes of an observed bone alteration. During a 3-week cultivation period, primary osteoblasts are investigated and characterized on the RNA, protein, and functional level by measuring several cellular and molecular parameters – each at multiple defined time points to receive kinetic results. For example, determination of proliferation, apoptosis, and metabolic activity of the cells, application of qRT-PCR for bone-related genes, and investigation of the matrix mineralization and nodule formation among others provide a deeper insight into the cellular phenotype *(23)*.

As the cells are uncoupled and investigated independently from systemic influences of the whole organism and under identical, well-known conditions, it is possible to assess if a cell-autonomous or a secondary effect is responsible for the observed phenotype.

The skeleton is influenced by environmental factors, as bone shape, bone mass, density, and structure adapt to the forces that act as load. To test the dynamic interplay of genetic and environmental factors, challenge experiments can be applied. Reaction to the challenge strongly depends on the genetic background of the experimental animals. Examples for challenge experiments are: loading experiments like jumping *(24)*, ulna loading *(25)*, vibration loading *(26)*, treadmill or running-wheel *(27)*, as well as disuse experiments by hind-paw unloading *(28)*.

3.2. Cardiovascular Phenotyping of Mice

3.2.1. Introduction

Being the major cause of morbidity and mortality, cardiovascular diseases continue to be a large socioeconomic burden for the next 20 years. Mouse models are essential tools for the investigation of the underlying disease mechanisms, environmental influences, and genetic risk factors. They provide a comparable mammalian heart physiology and many advantages for controlled and standardized experiments. The cardiovascular screen of the GMC offers a comprehensive phenotyping program for cardiovascular function. Primary screening includes tail-cuff blood pressure (BP) analysis and the determination of the N-terminal propeptide of the atrial natriuretic peptide (Nt-proANP) by immunoassay. More sophisticated, so-called secondary and tertiary examinations include electrocardiography (ECG) and echocardiography. Furthermore, specialized applications and analyses can be performed in cooperation with cardiology experts, who are members of the CardioVascular Disease Network in the National Genome Research Program.

3.2.2. Methods

BP analysis provides insights into functions of the vascular system including the regulation of vascular tone and left ventricular pump function. BP is strongly influenced by defects in many organ systems (heart, kidney, lung, liver) and metabolic or (neuro)endocrine pathways. Imbalances in one or, usually several organs and pathways, result in changes of this sensitive global parameter *(29, 30, 31)*. For the primary screening the non-invasive BP analysis measures the peripheral BP in conscious mice using a tail-cuff inflation system (MC4000 Blood Pressure Analysis Systems, Hatteras Instruments Inc., Cary, North Carolina, USA). Systolic, diastolic, and mean arterial pressures are determined over a 4-day period to accommodate the biological variability of BP. Prior to examination, the animals are habituated to the complete procedure to minimize stress and consequently the effect of handling on BP analysis. During measurement the animals are slightly immobilized on a pre-warmed platform in small metal boxes serving as restrainers (*see* **Fig. 25.2**).

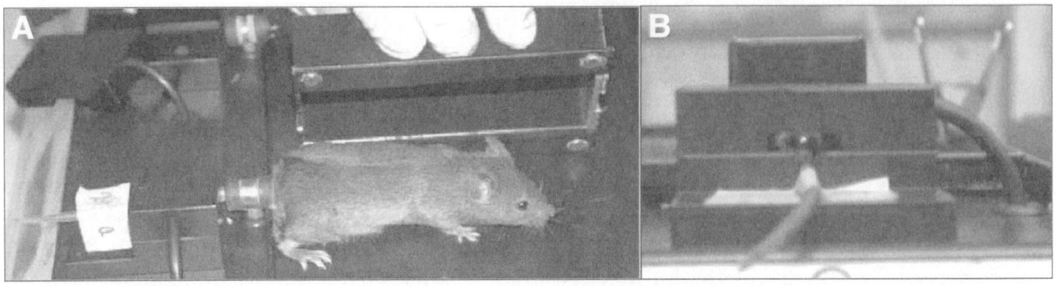

Fig. 25.2. Blood pressure set-up with a mouse fixed in the tail-cuff while the restrain box is lifted (**A**). View of the mouse tail in optical path (**B**).

The tails are looped through a tail-cuff and fixed in a notch containing an optical path with an LED light source and a photo sensor. The blood pulse wave in the tail artery can be detected by light extinction and is transformed into a pulse amplitude signal. During measurement this pulse amplitude is reduced by the rising air-pressure applied on the tail through the inflating tail-cuff and will finally appear leveled if the air-pressure exceeds the systolic BP.

Atrial natriuretic peptide (ANP) is a cardiac hormone predominantly secreted by atrial myocytes in response to atrial wall stretch occurring if cardiac filling pressures are increased *(32)*. ANP has diuretic, natriuretic, and vasodilating properties antagonizing the renin-angiotensin-aldesterone system. Recent studies suggest an additional cardioprotective effect by antihypertrophic and antifibrotic actions which may inhibit abnormal remodeling *(33, 34)*. Plasma concentrations of natriuretic peptides are increased in patients with cardiovascular disorders and heart failure and are therefore used as diagnostic and prognostic markers (for review see *35* and *36*). Since ANP is rapidly cleared from the circulation (half-life 3–4 min), the more stable N-terminal propeptide (Nt-proANP, half-life 60–120 min), which is cleaved from the released prohormone, is used because it better reflects chronic levels of ANP secretion *(37, 38)*. Nt-proANP concentrations are determined in lithium-heparin plasma using an Nt-proANP enzyme-linked immunosorbent assay (ELISA) (Biomedica Medizinprodukte, Austria). Recently, we were able to show that Nt-proANP may distinguish wild-type and mutant strains with different cardiomyopathy phenotypes and myocardial function. Moreover, Nt-proANP concentrations were correlated with left ventricular function which was determined by echocardiography *(39)*. In summary, this biomarker is a valuable tool to detect cardiovascular abnormalities in mice to be applied as a primary screen in large-scale phenotyping projects as in the GMC and the EUMODIC (www.eumodic.org).

The ECG measures the electrical activity, rate, and rhythm of the heart beat, supplying information about the conductive properties (a function of ion channels), the excitable myocardial mass, and the propagation of excitation within the heart tissue. Almost all types of cardiac pathologies will eventually cause also distinct ECG changes. Therefore, the ECG provides also insight into cardiac function *(40, 41, 42)*. A six-lead limb surface ECG is performed in sedated mice (isoflurane inhalation) using electrode gel and metal bracelets fixed at the joints of both front-paws and the left hind-paw. Wiring and mouse platform are located in a Faraday cage. The bipolar standard limb leads I, II, and III and the augmented unipolar leads AVF, AVR, AVL trace recordings are sampled and digitized for several minutes. A semi-automated shape analysis is performed on the ECG traces in lead II (ECG-auto, EMKA technologies, Paris, France) to detect characteristic trace elements, such as P-, Q-, R-, S-, and T-waves, for measurement of their duration and peak heights (*see* **Fig. 25.3**). In addition, the recordings are screened for arrhythmias including supraventricular and ventricular extrasystoles and conduction blockages.

Transthoracic ECG is the gold standard to determine ventricular dimensions and to identify a reduced cardiac pump function. In order to visualize the small and extremely fast-beating hearts of mice (typically approx. 600 bpm) high-frequency ultrasound biomicroscopy is used. A dedicated small-animal ultrasound machine, the Vevo 660 (Visalsonics, Canada), provides high resolution and a sufficiently large frame rate for imaging *(43)*. B-mode imaging generates a two-dimensional picture of the anatomical structures such as ventricle and large blood vessels. M-mode imaging displays the motion of tissue structures over time providing information about contractility, ventricular diameters, and wall thicknesses (*see* **Fig. 25.4**). Doppler-mode imaging examines the blood flow velocity and direction to assess hemodynamics and valvular function *(see* **Fig. 25.5**; for review see *44* and *45*).

Fractional shortening and ejection fraction are calculated using left ventricular end-diastolic (LVEDD) and end-systolic internal diameters (LVESD). These echocardiographic parameters describe left ventricular pump function. For faster turnaround conscious mice can be examined after 2 days of "training" to accustom the mice to the procedure. For more specific analysis ECG is performed in sedated (isoflurane inhalation) mice under temperature and ECG control (*see* **Table 25.2**).

In summary, invasive and non-invasive methods as well as dedicated small-animal equipment have reached a level of sophistication and performance that enables even the most informative and detailed physiological studies in mice. Future high-throughput screens will also include more and more blood-borne biomarkers to "fish" for novel mouse models with interesting phenotypes.

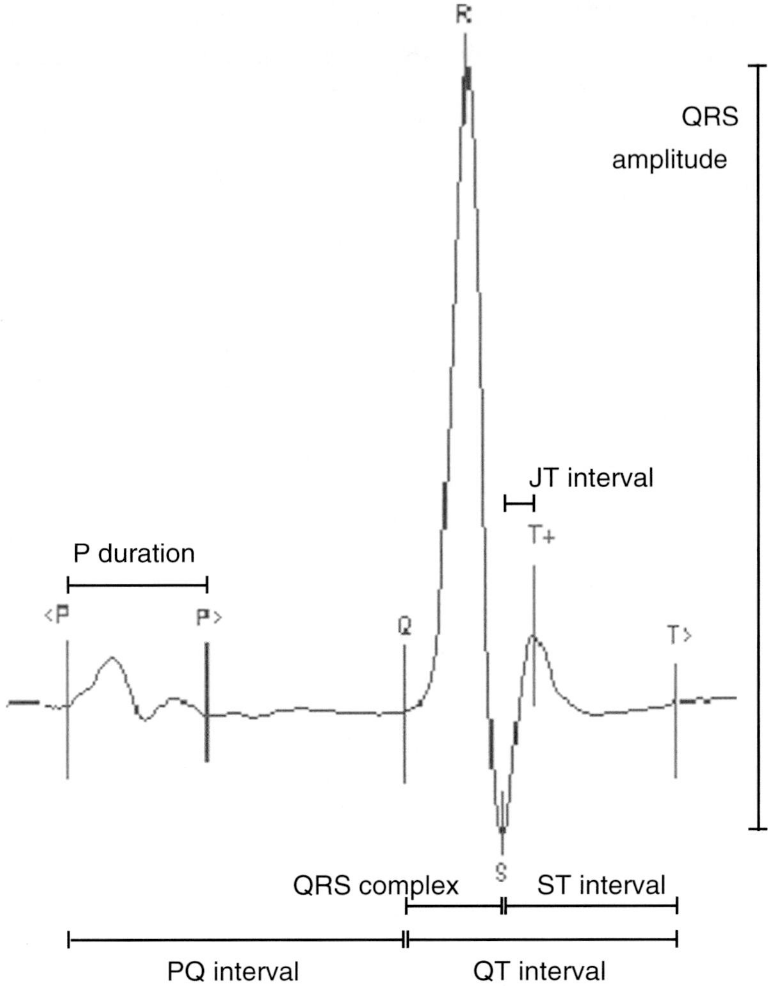

Fig. 25.3. Example of electrocardiogram trace with analyzed parameters.

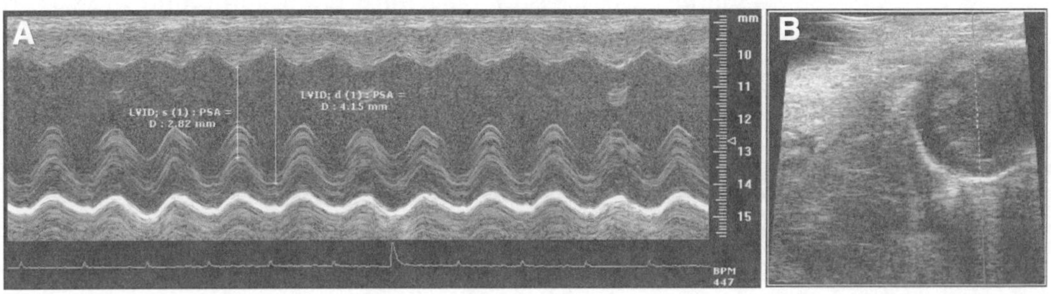

Fig. 25.4. M-Mode (time motion) imaging of the left ventricle (short axis view) (A) and the corresponding 2D picture (B).

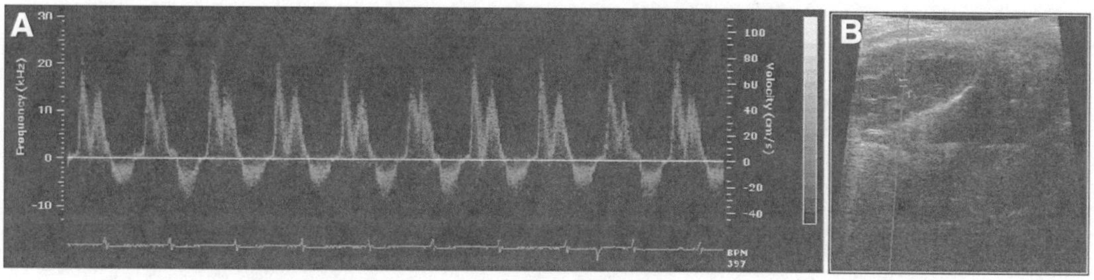

Fig. 25.5. Doppler flow spectrum of the mitral valve (A) and the corresponding 2D picture (B).

Table 25.2
Cardiovascular phenotyping of mice

Method	Parameters
Blood pressure analysis	Vascular tone and heart rate
Nt-proANP ELISA	Level of biomarker for cardiovascular disorders
Electrocardiography	Electrical activity, rate and rhythm of the heart beat
Echocardiography – B-Mode	Two-dimensional picture of anatomical structures
Echocardiography – M-Mode	Motion of tissue structures
Echocardiography – Doppler-Mode	Blood flow velocity and direction

3.3. Metabolic Phenotyping of Mice

3.3.1. Introduction

Major health problems are caused by disturbances in energy balance due to the dysfunctional regulation of either energy intake or energy expenditure. Long-term maintenance of a positive energy balance, for example, results in the development of obesity, which is frequently associated with the metabolic syndrome (including diabetes, cardiovascular diseases, and other comorbidities, *46*). The prevalence of obesity is still increasing; therefore, it was identified as a prime public health risk factor in many countries of the world according to the World Health Report 2002 *(47)*. Other serious health problems are associated with a negative energy balance such as certain types of cancer driving patients into cachexia (wasting syndrome; *48*). Against this background, animal disease models are in the center of interest to identify previously unknown genes related to the regulation of energy balance. Understanding the physiological mechanisms that regulate mammalian energy balance, the genetics, and the interactions of genetic predisposition and environmental factors is therefore of high clinical and societal importance.

3.3.2. Methods

Metabolic phenotypes can be detected, for example, regarding abnormalities in body mass and body composition, energy assimilation and gastrointestinal functions (nutrient absorption and transport), energy expenditure (metabolic rate), and body temperature regulation. On the basis of standardized procedures key parameters of energy balance are analyzed. Food intake, energy assimilation and the efficiency of energy turnover are related to the intake part of the equation. Metabolic rate, body temperature, and locomotor activity are the parameters that are measured to investigate energy expenditure. A set of laboratory methods needs to be implemented and evaluated, for example, indirect calorimetry, bomb calorimetry, monitoring of locomotor activity, body temperature registration, and software tools that facilitate the sound interpretation of a considerable data volume collected in a high-throughput screen (*see* **Table 25.3**) *(49, 50, 51)*. Mice can routinely be exposed to moderate metabolic challenges to discover mutations that do not show clear effects under *ad libitum* conditions (e.g. *52*). These challenge tests comprise mild food reduction for up to 5 days, a 2-day food restriction, or moderate cold resulting in an acute negative energy balance or feeding a high caloric diet to stimulate, for example, diet-induced obesity or diabetes.

Standard operating protocols need to be developed for all implemented methods to allow for the comparability of data sets, for example, for monitoring of body mass at defined time points, daily food intake (corrected for spillage and water content), and rectal or core body temperature obtained by thermosensitive sensors or implanted transponders. Total samples of excreted feces and reference samples of the diet are collected, dried, and combusted in a bomb calorimeter to calculate assimilated energy and

Table 25.3
List of methods and addressed target variables

Indirect calorimetry, activity and body temperature monitoring	Standard and resting metabolic rate Metabolic fuel utilization Spontaneous locomotor activity Thermoregulation
Food intake monitoring	Energy assimilation Efficiency of energy turnover Gastrointestinal functions
Body composition (destructive chemical carcass analysis, desiccation, Soxhlet fat extraction)	Disturbed allometric relations of body components Long-term disturbances of energy balance Dysfunctional metabolic pathways (e.g., fat metabolism)

assimilation efficiency. Regarding energy expenditure, indirect calorimetry is a powerful tool to monitor energy expenditure by measuring O_2 consumption and CO_2 production. As part of the high-throughput screen, metabolic rate is determined during night time (nocturnal energy expenditure from lights off to lights on) and additionally for the first four hours after lights on as an estimate for resting metabolism (www.eumodic.eu). Measurements are conducted using a multi-channel open flow indirect calorimetry system equipped with analyzers for O_2 and CO_2 gas concentrations, mass flow meters, multiplex unit, and motion detectors that register spontaneous activity. This set up can be combined with an automated drinking and feeding monitoring system. In addition to energy expenditure estimates, the respiratory exchange ratio which is calculated from gas analysis data can be used to detect differences in preferred metabolic fuel utilization.

In summary, the continuous measurement of metabolic rate, body temperature, locomotor activity, and food intake provides data that can be used to decide which part of the energy balance equation is affected by the mutation, that is, energy expenditure or energy assimilation. In addition, the determination of energy content of the feces and the calculation of assimilation efficiency indicates possible abnormalities in energy uptake. In collaboration with other modules of the phenotyping center, for example, blood chemistry screen or dysmorphology, complementary tests can be conducted (e.g., glucose and insulin tolerance test, or DXA scans to determine body composition). Finally, mutant mouse lines with elevated or lower body mass compared to control mice can be subjected to chemical carcass analysis to clarify whether changes in structural body size or disorders in allometrical relations of body composition caused the effect of the mutation.

3.4. Lung Function Screen

3.4.1. Introduction

According to the World Health Organization *(53, 54, 55)*, lung diseases are among the leading causes of death worldwide. Morbidity and mortality from lung disease have steadily increased in recent years, major contributors being chronic obstructive pulmonary disease (COPD) and asthma, both of which are known to be determined by a complex interplay of inherited predispositional and environmental factors. Despite considerable research efforts and the development of various animal models, a substantial gap of knowledge remains about gene–phenotype interactions presumed to play a role for asthma and COPD, and available mouse models for testing the significance and interplay of genetic variants and for developing new therapeutic strategies in biomedical research *(56–60)*. The Respiratory Function Screen provides a comprehensive overview on spontaneous breathing patterns, respiratory mechanics, gas exchange, and airway responsiveness. The screen aims at the detection of as yet unknown phenotypes in mutant mice, with a focus on the association with lung diseases and disorders of central respiratory control. These mutant lines may serve as model organisms for testing the interplay of genetic variants and for developing new therapeutic strategies in biomedical research.

3.4.2. Methods

Spontaneous Breathing Pattern: This screen assesses alterations of breathing patterns in unrestrained mutant mouse lines at different levels of activity (sleep, rest, activity). Spontaneous respiration in unrestrained animals is studied by whole-body plethysmography (Fa Buxco®). Pressure changes due to inspiratory and expiratory temperature and humidity fluctuations during breathing are measured over 40 min and are transformed into flow and volume signals, so that automatic data analysis provides tidal volumes, respiratory rates, minute ventilations, inspiratory and expiratory times, as well as peak inspiratory and peak expiratory flow rates (*see* **Figs. 25.6** and **25.7**).

Specific tidal volumes and minute ventilations are calculated from the respective absolute values in relation to the body weight of the animals. Primary parameters of interest are alterations in tidal volume, respiratory rate, and timing of breathing as well as impaired adaptation to different levels of activity.

Lung Function Testing: Our unit provides lung function testing comparable to that available in pulmonary divisions. The unit allows the measurement of lung size (total lung capacity) and all volumetric subunits of the lungs including the volume of the conducting airways (dead space). Respiratory system resistance, static and dynamic lung compliances, as well as intrapulmonary gas mixing and alveolar-capillary gas transfer are other parameters delivered by the unit. All measurements were performed using a computer-controlled piston-type servo ventilator (*see* **Fig. 25.8**

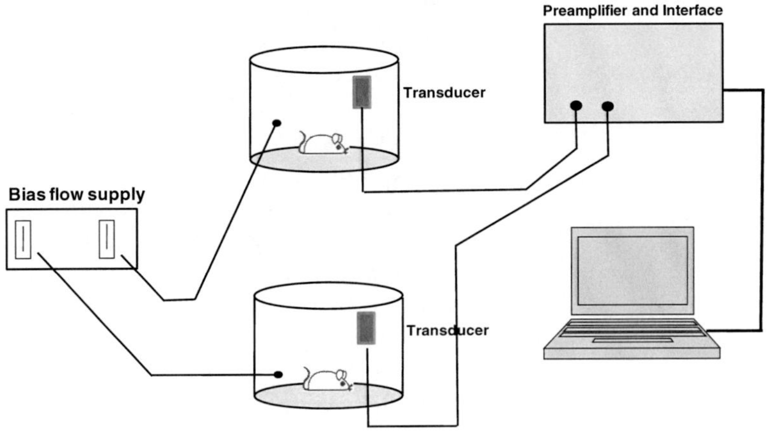

Fig. 25.6. Setup for whole-body plethysmography.

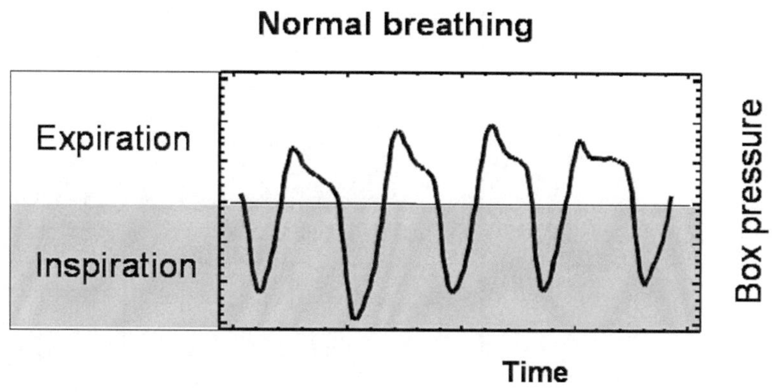

Fig. 25.7. Typical pressure swings during breathing.

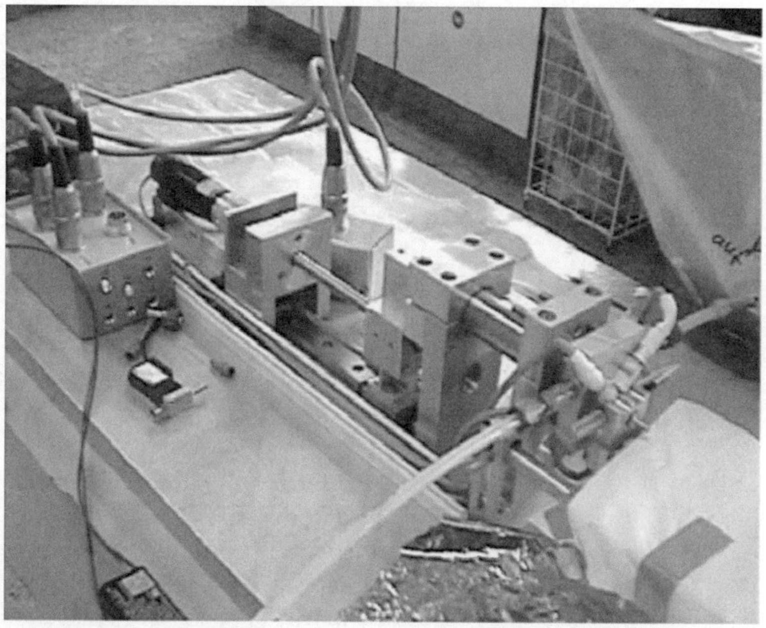

Fig. 25.8. Working place lung function.

and **25.9**) *(56, 61)*. The ventilator provided for positive pressure ventilation of intubated mice and for defined respiratory maneuvers for lung function testing using either air or specific gas mixtures, for example, for measurement of diffusing capacity. Respiratory volume and flow settings can be adapted in steps of 10 μl to fit the physiological limitations of mice. Concentrations of the respiratory gases oxygen (O_2) and carbon dioxide (CO_2) and the test gases helium (He) and carbon monoxide ($C^{18}O$) were measured by a magnetic sector field mass spectrometer (modified M3 Varian, Finnigan MAT, Bremen, Germany) via a specially modified capillary through which a continuous flow of gas was

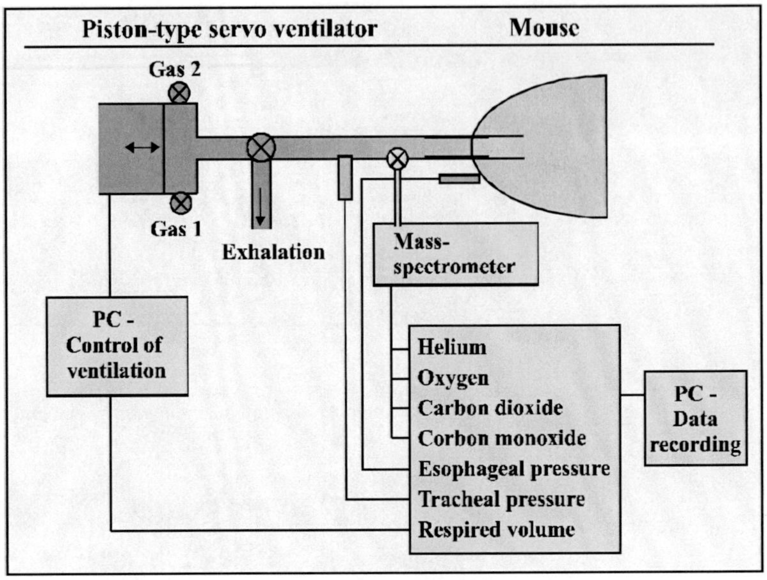

Fig. 25.9. Schematic diagram of a lung function working place.

sucked into a sample chamber. Our unit has been applied to study sex and inter-strain differences of lung functions in various inbred mouse strains. This study was embedded in the "Phenome Project" from Jackson Laboratories (*56*; http://aretha.jax.org/pub-cgi/phenome/mpdcgi?rtn=projects/details&sym=Schulzl).

Airway Responsiveness: To characterize the degree of airway reactivity SOPs, for bronchial challenge with methacholine, are established *(62)*. For measurement in restrained animals, the above-mentioned lung function unit is applied and changes in respiratory system resistance as a function of intravenously injected metacholine determined. In unrestrained animals, whole-body plethysmography is applied and changes in the parameter "enhanced pause" (Penh) are used to assess the level of airway reactivity. The protocol starts with a determination of baseline values and a negative control with buffered PBS. Then, five progressively increasing concentrations of a methacholine aerosol are delivered to the animals to set up a dose–response curve.

3.5. Behavioral Phenotyping of Mice

3.5.1. Introduction

The aim of the behavioral screen in the GMC is the investigation of the molecular and genetic basis of emotional and cognitive dysfunctions that are relevant for human neuropsychiatric disorders like anxiety disorders, post-traumatic stress disorder, depression, schizophrenia, autism, attention-deficit hyperactivity disorder, Parkinson's, and Alzheimer's Disease. To this end we perform a careful behavioral characterization of mouse mutants and their control littermates of both sexes in a hierarchical manner.

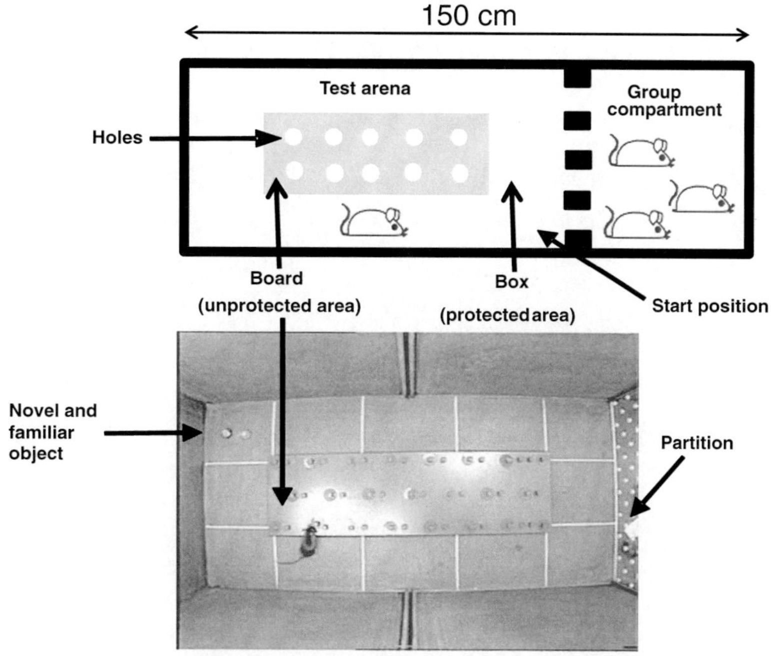

Fig. 25.10. Arena for the modified hole board test.

3.5.2. Methods

The modified hole board test *(63)* has proven useful as introductory examination of spontaneous behavior in the GMC primary screen (*see* **Fig. 25.10**). This test allows the comprehensive analysis of a range of parameters known to be indicative of behavioral dimensions such as locomotor activity, exploratory behavior, arousal, emotionality, memory, and social affinity in a single short test (**Table 25.4**).

Individual mice freely explore the test arena for 5 min in the presence of their cage mates in the adjacent group compartment behind a transparent, perforated partition. Behavioral measures are recorded by a trained observer blind to the genotype of the tested mouse by the use of a handheld computer. In addition, a camera is mounted above the center of the test arena and the locomotor path is analyzed by the use of a video-tracking system.

A clear advantage of this test procedure is the efficient generation of an overview over behavioral parameters that, in the absence of other behaviorally relevant pathological processes which are clarified in the other GMC screens, suggest alterations in CNS function. And the reliable detection of CNS-specific dysfunctions is the goal of the behavioral screen. However, because the modified hole board test was only recently developed (2001), it is not yet a widely accepted method for primary behavioral phenotyping. Since the manual behavioral observation is hardly amenable to automatization, large-scale phenotyping projects like EUMORPHIA

Table 25.4
Parameters measured in the modified hole board and related behaviors

Behavior	*Parameters*
Forward locomotion	Distance moved, line crossings (latency, frequency)
Speed of movement	Velocity (mean, maximum, angular)
Exploration	*Vertical*: Rearings (box, board; latency, frequency) *Horizontal*: Holes (latency, frequency), objects (novel, familiar; latency, frequency, duration)
Risk assessment	Stretched attends (latency, frequency)
Anxiety-related	Board entries (latency, frequency, duration); distance to the wall
Grooming	Grooming (latency, frequency, duration)
Defecation	Boli (latency, frequency)
Social affinity	Exploration of the partition (latency, frequency, duration)
Object memory	Object recognition index

(www.eumorphia.org) and EUMODIC (www.eumodic.org) prefer the open field test to assess spontaneous behavior. The open field test can be automated, but assesses less behaviors.

To evaluate the utility of our modified hole board testing procedure as primary behavioral screen, we analyzed the results obtained since the beginning of its usage in the GMC. Out of 62 mutant mouse lines analyzed in the modified hole board, 97% showed a significant genotype effect in comparison to their concurrently tested control littermates in at least one parameter. In 23% of these cases the alterations were subtle and could not be allocated to a behavioral phenotype, so they were classified as of unclear relevance. The remaining 77% showed behavioral patterns that could be allocated to the following phenotypic alterations: general activity level (29%), forward locomotion (22%), exploration (25%), social affinity (13%), anxiety-related behavior (6%), grooming (2%) and object memory (2%). In these remaining mutant lines with a behavioral phenotype 55% of the phenotypes were found in both sexes, 15% in males and 30% in females. Sixty-three percent of these phenotypes were considered interesting enough for further detailed, in-depth analysis. In a couple of years, through EUMODIC we will be able to compare the detection efficiency of interesting phenotypes of the modified hole board with that of the open field.

Interesting mouse lines detected in the primary screen are characterized in detail by the use of more specific behavioral tests. In hearing mice, sensorimotor behavior and sensorimotor gating as an endophenotype associated with schizophrenia can effectively be analyzed by the acoustic startle reflex and its prepulse inhibition *(64)* in a procedure cross-validated by EUMORPHIA (www.eumorphia.org). To further evaluate a possible anxiety-related phenotype we apply the light–dark box test, the elevated plus-maze, and the social interaction test (for review see *65*). To evaluate depression-related behavior we apply the forced swim test and the tail suspension test (for review see *66*). In addition, aspects of learning and memory are assessed by testing social, olfactory recognition in a social discrimination paradigm *(67, 68)*, fear memory in a fear-potentiated startle procedure *(69)*, and object recognition in a procedure based on the work by Genoux and co-workers *(70)*. For detailed analysis of appetitively motivated learning and memory processes, operant procedures can be applied *(71)*. Once a working hypothesis is formed about brain areas or neuronal circuits involved in the observed phenotype, the analysis can ideally be complemented by functional neuroanatomical investigations and appropriate pharmacological challenges for further clarification of the mechanism.

3.6. Neurological Phenotyping of Mice

3.6.1. Introduction

Neurological dysfunction results in a wide variety of disorders ranging from impaired movement to severe mental illness. Most of these disorders appear to be caused by a complex combination of genetic and environmental factors, but these are largely unknown and there is an urgent need for causal therapies. Studying the neurobehavioral phenotype of mutant mice is a powerful tool to understand the neural basis of behavior and the pathophysiology of neurological and psychiatric disorders *(72, 73)*. The neurological screen of the GMC aims to provide well-characterized mouse models for neurological diseases, to investigate the in vivo consequences of the mutations, and to allow for therapeutic trials *(4)*.

3.6.2. Methods

The neurological screening comprises numerous tests from a first fast snapshot to more comprehensive analyses dependent on the precedent results *(74)*. **Table 25.5**. Some of the tests are also performed as part of the European phenotyping platforms EUMORPHIA (www.eumorphia.org) and EUMODIC (www.eumodic.org) and can be used for large-scale phenotyping projects. For the primary screening an overall evaluation of basic neurological functions is recommended using a "modified SHIRPA-protocol" *(75)*. This is a series of close observations and simple tests to examine various reflexes, to rate, for example, gait and posture as well as to assess muscle, motor neuron, spinocerebellar, sensory and autonomic functions and apparent behavioral alterations, which might also interfere with further analyses.

Table 25.5
Methods for neurological screening

Method	Application	Indicated defects
Modified SHIRPA	General appearance, basic neurological functions	Basic muscle, motor neuron, spino-cerebellar, sensory and autonomic functions and apparent behavioral alterations
Grip strength testing	Evaluation of muscle function	Motor pathways, neuromuscular defects, muscle dysfunction
Pole test	Basic motor functions	Motor pathways
Swim ability	Balance/motor coordination	Motor pathways and vestibular defects
Rotarod	Motor coordination/ balance and motor learning	Sensorimotor impairment, especially basal ganglia and cerebellar defects and motor cortex/striatum, respectively
Footprint analysis	Gait abnormalities	Ataxias
Staircase test	Evaluation of skilled reaching	Motor pathways of the brain, for example, in motor cortex, striatum, nigrostriatal tract and subthalamic nucleus
Telemetric EEG	Analysis of paroxysmal disorders	Epileptic seizures

Quantification of forelimb and combined "forelimb and hindlimb grip strength" generates information about muscle function but also indicates defects in motor pathways as well as at neuromuscular connections (*see* **Fig. 25.11**) *(76)*.

Fig. 25.11. Measurement of grip strength.

The strength is measured as the maximum force a mouse applies to a transducer-coupled bar or grid (TSE, Bad Homburg, Germany) when slowly pulled away before it releases the grip. For the evaluation of the results of this test as well as some other tests where body movements are implemented, the effects of body weight have to be considered *(77)*.

A simple "pole test" allows for the detection of motor deficits. Mice are positioned at the top of a vertical pole of about 50 cm height *(78, 79)*. The time needed for the mice to descend the pole is measured and the strategy they use is classified using a defined scale. Another basic test for the detection of motor restrictions and especially for balance problems is the "swim test" since this motor task is also very sensitive towards inner ear defects *(80)*. Mice are placed in tepid water and the swimming performance is rated. The "accelerating rotarod" is used for the evaluation of coordination and/or balance *(81)*. This method is more elaborate than the basic test mentioned above but also more sensitive to smaller alterations, since motor performance can be quantified more exactly. For this test mice are placed on a slowly accelerating rotating rod (TSE, Bad Homburg, Germany) so they continuously have to adjust their walking opposite the rotation direction to stay on the rod. The latency time on the rod is recorded. A total of 3–4 trials are conducted with an intertrial interval of 15–20 min. For analysis of motor learning the rotarod can be used over several days to compare respective improvements of naïve mice against the trained ones *(82)*.

Ataxic gait is a common symptom of various neurological diseases and "footprint analyses" can be used for depiction of gait abnormalities observed, for example, in the modified SHIRPA analysis. Mice feet are painted with non-toxic ink, ideally front and hind legs with different colors, and the mice are then launched over a paper-covered runway to get a respective footprint pattern. From these different parameters like stride length and base dimensions can be measured and differences between mutants and controls calculated *(83)*. Evaluation of fine-tuning of movement is done with a "staircase test" *(84)*. Food pellets are placed on a narrow stairway and the mice have to reach for them. Different difficulty levels of accessible steps allow the evaluation of reaching skills. This test provides a valuable tool for the detection of extrapyramidal motor dysfunction as well as unilateral neurological deficits, for example, brain lesions in stroke models. "Muscle biopsies" are carried out to fathom possible reasons for muscle function defects. Beneath muscle morphology a series of histological and histochemical stains are used that are also useful for the detection of mitochondrial abnormalities. Telemetric electroencephalography (EEG) is performed to record electrical activity of the cerebral cortex for the characterization of paroxysmal disorders. The employment of a telemetric device allows the investigation of awake mice not compromised by outgoing wires and independent of narcosis side effects *(74)*. Changes in nociceptive as well as analgesic pathways can be analyzed using carefully selected experimental procedures covering different aspects of pain signaling *(85)*.

3.7. Eye Screen

3.7.1. Introduction

Visual impairment and blindness are widespread diseases; worldwide 50 million people are affected by blindness *(86)*. To understand the genetic and biochemical mechanisms of ocular disorders, mouse models are used as an appropriate tool *(87, 88)*. The Eye Screen of the GMC offers a comprehensive phenotyping of the eye.

3.7.2. Methods

In a basic examination, the eyes are characterized in vivo by noninvasive methods for morphological alterations **Table 25.6**.

At first, the "dysmorphological examination" of the eye is performed for apparent anophthalmia (absent eyes), microphthalmia (small eyes), eyelid closure, abnormal secretion, and buphthalmos (bulging eyes). The light responsiveness of the pupil can be easily observed with a slit lamp or a funduscope, without medicamentous dilation of the pupil. "Slit-lamp biomicroscopy" is used to examine the mice for abnormalities of the anterior part of the eye, especially the cornea, iris, and lens *(89)*. The main focuses are opacities of the cornea and the lens (cataract), but also the vascularization of the cornea, the iris pigmentation, and the position and shape of the pupil are observed. At least 10 min prior to examination pupils are dilated with a 1% atropine solution applied to the

Table 25.6
Screening for eye and vision phenotypes

Parameters	Observations
Dysmorphology of the eye	Obvious morphological alterations
Funduscopy	(Qualitative) abnormalities of the retinal fundus and optic disc, vessel alterations and development disorders
Slit-lamp biomicroscopy	(Qualitative) abnormalities of lens and cornea, like opacity and development disorders
Laser interference biometry (LIB)	Abnormalities of eye size parameters: axial eye length, cornea and lens thickness, anterior chamber depth
Optokinetic drum	Vision test
Electroretinography (ERG)	Test for retinal function
Pupillometry	Functional test for lesions of the anterior visual pathway
Histology	Morphological details

eyes. Both eyes of the mice are examined by slit-lamp biomicroscopy (Zeiss SLM30) at 48 × magnification with a narrow-beam slit-lamp illumination at 25–30°angle from the direction of observation. "Funduscopy (Ophthalmoscopy)" is applied to examine the posterior parts of both eyes for retinal degeneration and abnormalities of the optic disc and the pigmentation; moreover, the retinal vasculature is observed for abnormal vascular patterning and vessel appearance *(90)*. After pupil dilation with one drop of atropine (1%), the mouse is grasped firmly in one hand and clinically evaluated using a head-worn indirect ophthalmoscope (Sigma 150 K, Heine Optotechnik, Herrsching, Germany) in conjunction with a condensing lens (Volk 90D lens) mounted between the ophthalmoscope and the eye. Digital fundus images are taken with a Heine Video Omega 2C indirect Ophthalmoscope connected to a VRmAVC Video Grabber (Dieter Mann GmbH, Mainaschaff, Germany) and a Volk 40D or 60D lens.

An expanded basic examination is the determination of different eye size parameters, like axial eye length, cornea and lens thickness, and anterior chamber depth, by "Laser interference biometry" (LIB). The measurement is performed with the "ACMaster" (Meditec, Carl Zeiss, Jena, Germany) equipped with the new technique of optical low-coherence interferometry (OLCI), adapted for short measurement distances *(92, 93)*. Mice are anesthetized with an intraperitoneal (i.p.) injection of 137 mg Ketamine and 6.6 mg Xylazine per kg body weight and placed on a platform in front of the ACMaster. For the appropriate orientation of the mice, the reflection of six infrared LEDs arranged in a circle must be placed in the centre of the pupil. A modified software (Zeiss Meditec) optimized for the dimensions in the mouse eye was used to enable peak recognition automatically. For each mouse, the left and the right eyes are measured independently; the value for each eye is the mean of 10–30 individual measurements in an interval of 30–60 s.

However, these morphological examinations often do not allow conclusions about visual acuity or functionality of the central visual system, which is an important information for other screens in the GMC, for example, the behavior screen. For this purpose, the "optokinetic drum" is used as a fast and effective vision test *(95, 96)*. A mouse with a regular visual system will follow a rotating vertical stripe pattern with its head (head-tracking). The tested mouse is freely moving in a transparent acrylic glass cylinder (diameter, 15 cm; height, 18 cm), which is placed in the center of the optokinetic drum (diameter, 63 cm; height, 35 cm; *see* **Fig. 25.12**). Movement behavior and head-tracking are analyzed under room light, a spatial frequency of 0.1 cyc/deg and a rotation speed of 10 rpm *(97)*. Prior to the analysis, the mice were allowed to adapt to the environment of the non-rotating drum for

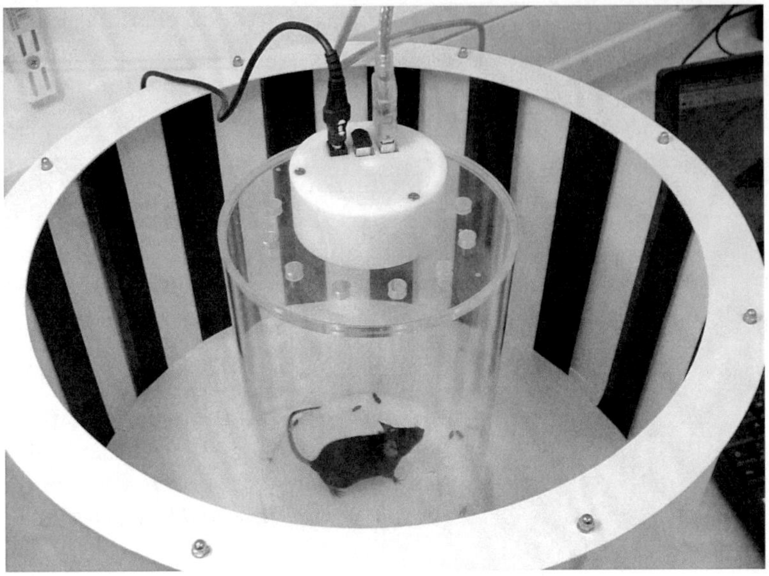

Fig. 25.12. A mouse performing the vision test in the optokinetic drum.

3 minutes. The moving direction of the drum is changed after 30 s and 3–5 30-s intervals per direction are monitored for the head-tracking behavior.

Electroretinography (ERG) is used to examine the retinal function as described *(94)*. Mice are dark-adapted for at least 12 h and anesthetized with 137 mg Ketamine and 6.6 mg Xylazine per kg body weight. After pupil dilation (1 drop Atropine 1%), individual mice are fixed on a sled with Velcro straps. Gold wires (as active electrodes) are placed on the cornea; care was taken not to obstruct the pupillary opening. The ground electrode is a subcutaneous needle in the tail; a reference electrode is placed subcutaneously between the eyes. The mice are introduced into an ESPION ColorBurst Handheld Ganzfeld LED stimulator (Diagnosys LLC, Littleton, MA, USA) on a rail to guide the sled (High-Throughput Mouse-ERG, STZ for Biomedical Optics and Function Testing, Tübingen, Germany). To minimize temperature influences on the ERG, body temperature is kept at 37°C using a warming plate. Ten milliseconds light pulses are delivered at a frequency of 0.48 Hz in two steps at 500 cd/m^2 and 12,500 cd/m^2. Bandpass filter is set ranging from 0.15 Hz to 1,000 Hz. Responses are recorded simultaneously from both eyes with an ESPION Console (Diagnosys LLC, Littleton, MA, USA) and stored for offline analysis after averaging 10–40 individual measurements at each step.

The "Pupillometry" will be implemented as a functional test for cognition and to diagnose lesion of the anterior visual pathway. In this test the pupillary light reflex (PLR) is evoked by a light stimulus, which is conveyed over a reflex arc, including retinal ganglion cells and their axonal projection toward pupillomotor centers in the central nervous system *(91)*.

At last, the morphological analysis can be completed by a "histological analysis". For this purpose, the eyes are extracted immediately after death of the mouse and fixed 24 h in Davidson solution, dehydrated, and embedded in plastic medium (JB4-Plus; Polysciences, Inc. Eppelheim, Germany). Transverse 2-μm sections are cut with an ultramicrotome (Ultratom OMU3; Reichert, Walldorf, Germany) and stained with methylene blue and basic fuchsin. Sections are evaluated with a light microscope (Axioplan; Carl Zeiss Jena GmbH, Göttingen, Germany) and images are taken with a scanning camera (Axiocam; Carl Zeiss) equipped with a screen-capture program (KS100; Carl Zeiss Vision, Hallbergmoos, Germany) and imported in an image-processing program (Photoshop 7.0; Adobe, Unterschleissheim, Germany).

3.8. Clinical–Chemical and Hematological Phenotyping

3.8.1. Introduction

Mutant mouse lines produced by gene-driven or phenotype-driven methods provide important animal models for biomedical research. Since many genetic changes lead directly or indirectly, via altered organ functions, to changes of diagnostic laboratory parameters, the clinical–chemical and hematological phenotyping covering a large range of organ functions is an indispensable tool for the comprehensive characterization of any kind of mutant mice. It permits the detection of hematological changes, defects of various organ systems, and changes in metabolic pathways as well as electrolyte homeostasis. Suitable laboratory diagnostic procedures allow the accurate and efficient examination of expected effects as well as the discovery of additional, more subtle consequences of particular genetic modifications, for example, in knockout mice without obvious phenotypic alterations. Thus, in-depth clinical–chemical examinations contribute to the understanding of pathomechanisms in mouse models and to the development of systems biology of complex organisms *(98–100)*.

3.8.2. Methods

The basic clinical–chemical and hematological phenotyping is carried out by automated routine procedures allowing the high-throughput screening of a large number of mice for a broad spectrum of clinical–chemical blood parameters including substrates, proteins, electrolytes, and enzymes, as well as hematological parameters such as red and white blood cell counts *(101, 102)*.

Blood collection: The limiting factors of the analysis in mice are the volume of the blood sample and the time intervals between the examinations. The total blood volume corresponds to 7.5% of the body weight of the mouse. Ten percent of the total blood volume can be removed once without causing significant alterations, and up to 15% of the blood volume can be collected if fluid replacement is carried out. For repeated sampling, 10%, 7.5%, and

1% of the total blood volume can be removed every two weeks, every week, and each day, respectively. Blood collection is carried out by retroorbital sinus puncture under short-term general anesthesia (e.g., with isoflurane) using non-heparinized glass capillaries less than 0.9 mm in diameter. The blood is collected in EDTA-coated tubes for hematological measurements and in Li-heparin-coated tubes for plasma chemistry.

Hematological analysis: The basic hematological analysis consists of ten parameters (**Table 25.7**) measured with a blood analyzer (ABC-Animal Blood Counter, Scil), which has been validated for the analysis of mouse blood. The number and cell volume of red blood cells, and the numbers of white blood cells and platelets are measured by electrical impedance. The hemoglobin concentration is determined by spectrophotometry. Mean corpuscular volume (MCV), mean platelet volume (MPV), and red blood cell distribution width (RDW) are directly calculated from the cell volume measurements. Mean corpuscular hemoglobin (MCH) and mean corpuscular hemoglobin concentration (MCHC) are calculated from hemoglobin/red blood cell count (MCH), and hemoglobin/hematocrit (MCHC), respectively.

Clinical–chemical analysis of blood plasma or serum: The clinical–chemical parameters of blood plasma or serum are automatically analyzed using an Olympus AU400 autoanalyzer (Olympus, Hamburg, Germany) and the adapted reagents from Olympus (Hamburg, Germany), Biomed (Oberschleißheim, Germany), or Wako (Neuss, Germany). The basic phenotyping set of 21 parameters including the concentrations of various substrates, proteins, and electrolytes, as well as several enzyme activities (**Table 25.7**) are analyzed using 130 μl plasma diluted with the same volume of deionized water. This analysis provides a good overview and is the base for subsequent in-depth functional analyses of specific organ systems *(101–104)*. If significant deviations from the baseline levels are found, specific profiles for affected organs are performed (**Table 25.8**).

Table 25.7
Basic blood-based screen established for the mouse

Blood plasma	*Substrates*: Cholesterol, creatinine free fatty acids glucose, triglycerides, urea; *Proteins*: Albumin, ferritin, total protein, transferrin *Electrolytes*: Calcium, chloride iron, phosphorus, potassium, sodium; *Enzyme activities*: Alanine aminotransferase (EC 2.6.1.2), alkaline phosphatase (EC 3.1.3.1), α-amylase (EC 3.2.1.1), aspartate aminotransferase (EC 2.6.1.1), lactate dehydrogenase (EC 1.1.127)
Hematology	*Red blood cells*: Hematocrit, hemoglobin, mean corpuscular hemoglobin, mean corpuscular hemoglobin concentration, mean corpuscular volume, red blood cell count, red blood cell distribution width; *White blood cells*: Mean platelet volume, platelet count, white blood cell count

Table 25.8
Examples for organ profile analysis of the mouse blood

Tissue	Parameter
Bone/muscle	Calcium, phosphorus, alkaline phosphatase, creatine kinase, lactate dehydrogenase
Iron pathway/ anemia	Ferritin, transferrin, iron, lactate dehydrogenase, unsaturated iron-binding capacity pottassium, total bilirubin
Kidney	Urea, creatinine, total protein, albumin, uric acid, chloride, potassium, sodium
Liver	Direct bilirubin, total bilirubin, cholesterol, high-density lipoprotein cholesterol, low-density lipoprotein cholesterol, alkaline phosphatase, alanine aminotransferase, aspartate aminotransferase
Pancreas	Cholesterol, high-density lipoprotein cholesterol, low-density lipoprotein cholesterol, glucose, triglycerides, α-amylase, lipase

Urine analysis: For the in-depth investigation of kidney function a broad spectrum of parameters (urea, uric acid, creatinine, total protein, glucose, electrolytes, amylase, albumin) is determined in urine samples using the AU400 autoanalyzer. For these analyses either single spot or 24 h urine collected in metabolic cages is used. Proteinuria can be analyzed qualitatively by SDS-polyacrylamide gel electrophoresis *(105)*. The weight density displaying the concentration of soluble substances in the urine is determined with a refractometer. Aberrant production of urinary sediment is investigated by microscopic evaluation of host cells, casts, and crystals in the urine samples after concentration and staining procedures.

Additional tests: Additional tests routinely established in our analysis include oral (oGTT) or intraperitoneal (ipGTT) glucose tolerance tests *(100)*, blood gas measurement, differential white blood cell counts, and reticulocyte counts. For the glucose tolerance tests the fasted glucose level is determined after 16 h of food deprivation. Then the mice are challenged by the administration of a defined amount of glucose (2 g of glucose per kg body weight) and blood glucose is measured at defined time points, for example, 15 min, 30 min, 60 min, and 120 min after glucose administration in a simplified ipGTT protocol we use to screen mutant mouse lines for impaired glucose tolerance.

Blood gas values such as pH, pO_2, pCO_2, and bicarbonate concentrations are evaluated in cooperation with the metabolic screen of the GMC using an ABL 5 blood gas analyzer (Radiometer GmbH, Willich, Germany).

Differential white blood cell counts and reticulocyte counts are determined by light-microscopic evaluation of appropriately stained blood smears.

Standardization

Many factors including individual features, mouse husbandry, and experimental procedures affect the results of the clinical–chemical and hematological analyses *(104)*. Execution and outcome of the analysis are optimized by implementing comprehensive automation, standardization, and quality control protocols. Obtaining valid and reproducible results as well as the comparability of the results in clinical chemistry analyses of different projects increase with the use of standardized protocols (http://www.interphenome.org) in phenotyping centers like the GMC (http://www.mouseclinic.de) *(4)*.

3.9. Immunological Phenotyping

3.9.1. Introduction

Mouse models have been a primary source of information for understanding the intricate mechanisms of the immune system. The Immunology Screen at the GMC was set up to conduct a broad immunological phenotyping of mouse mutant lines with the intention of identifying distinct gene functions, which play key roles in the immune defense of the organism through a complex network of cellular and soluble components. We developed a screen sensitive enough to detect minor alterations under steady-state (non-challenge) conditions, using a limited amount of samples. In the basic screen we measure leukocyte populations in peripheral blood and immunoglobulin levels in blood plasma (**Table 25.9, 25.10**).

Table 25.9
Main parameters determined for primary screening of mouse blood samples

Flow cytometry *Main lineages* Staining1: T cells ($CD3^+$), $CD4^+$ T cells, $CD8^+$ T cells, g/d T cells ($CD3^+$g/d TCR+), T reg cells ($CD4^+CD25^+$) Staining2: B cells ($CD19^+$), B1 B cells ($CD19^+CD5^+$), mature B cells ($CD19^+IgD^+$), granulocytes (CD11b $^+$Gr1 $^{++}$), NK cells (NKp46$^+$ CD5$^-$) and NK T cells (NKp46$^+$ CD5$^+$) *Subpopulations* Further subpopulations are identified by bi-variate and boolean gating with the following markers Staining 1: CD25, CD62L, Ly-6C, CD44. Staining 2: CD19+ cells: IgD, B220, CD11b, MHC-II(I-A, I-E), CD5, Gr1; CD19- cells: Gr1, B220, CD5, MHCII, CD11b.
Bioplex/ELISA IgM, IgG_1, IgG_{2a}(resp., IgG_{2c}) , IgG_{2b}, IgG_3, IgA; anti-DNA antibodies, rheumatoid factor
Flow cytometry data are acquired on a LSR II flow cytometer (Becton Dickinson, USA) and frequencies of populations are analyzed using FlowJo software (TreeStar Inc., USA). The plasma levels of immunoglobulin isotypes (IgG1, IgG2a (IgG2c), IgG2b, IgG3, IgGM, IgGA) are determined using a bead-based array with monoclonal anti-mouse antibodies conjugated to beads of different regions (Biorad, USA), and acquired on a Bioplex reader (Biorad). The presence of rheumatoid factor and anti-DNA antibodies is evaluated by indirect ELISA.

Table 25.10
Simultaneous analysis of cell surface marker molecules on blood leukocytes

Laser	P	Longpass dichroic mirror	Bandpass filter	Fluochrome	Blood staining 1	Blood staining 2
Blue: 488 nm	G	–	488/10		Side scatter	Side scatter
	F	505LP	530/30	FITC	CD44	IgD
	E	550LP	575/26	PE	gdTCR	NKp46
	D	600LP	610/20		Propidium iodide	Propidium iodide
	B	685LP	695/40	PerCPCy55	CD4	Gr1
	A	755LP	780/60	PECy7	CD62L	CD19
Violet: 405 nm	B	–	450/50	Pacific Blue	CD3	CD11b
	A	505LP	525/50	Cascade Yellow	Ly6C	I-A/E
Red: 633 nm	C	–	660/20	APC	CD25	CD5
	B	710LP	730/45	Alexa Fluor 700	CD45	CD45
	A	755LP	780/60	APC-Alexa750	CD8a	B220

The proportions of leukocyte populations in peripheral blood are genetically regulated *(106)*. As a consequence, inbred strains differ in frequencies of leukocyte subsets in lymphoid organs and in peripheral blood. Moreover, several CD antigens are restricted to specific mouse strains or interstrain differences occur concerning the level of expression of certain CD antigens. In individual mice, the number of circulating leukocytes and the proportions of subpopulations show daily rhythmic variations *(107)* and depend on homeostatic proliferation and/or retraction *(108)*, as well as on activation through environmental and/or microbial factors (e.g.,*109*). Furthermore, sex-dependent factors are documented to influence the immune status *(110)*.

3.9.2. Methods

Multi-color flow cytometry has proven successful for the discovery of mutants based on the simultaneous analysis of a number of cell surface marker molecules on blood leukocytes (**Table 25.10**). Reliable values of frequencies of leukocyte clusters are very much dependent on the appropriate preparation, acquisition and gating of the leukocytes (*see* **Fig. 25.13**).

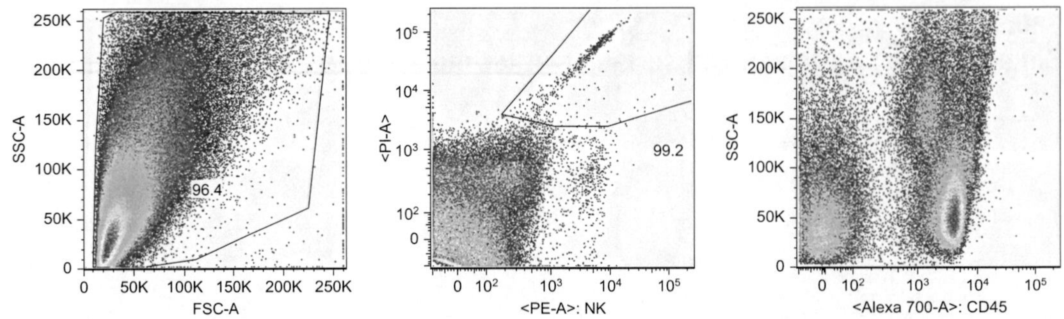

Fig. 25.13. Gating strategy for high-throughput flow cytometry data analysis. *Left dot plot:* Intact cells are first identified by their FSC/SSC profile. *Middle dot plot:* Subsequently, dead cells are eliminated on the basis of their propidium iodide signal. *Dot plot to the right:* In a last step, living cells are gated for CD45+ leukocytes over their SSC/CD45 signal pattern. CD45+ cells are subsequently analyzed by software-based semi-automatical multi-parameter analysis. Fluorescence minus one (FMO) controls are used to define "positive" and "negative" regions *(121)*.

Our method includes ammonium chloride-mediated erythrocyte lysis, thus preventing interference by larger amounts of erythrocytes. We further use for all stainings a monoclonal antibody to the cell surface glycoprotein CD45 to be able to create a CD45+ gate, allowing us to discriminate leukocytes from debris, erythrocytes, and thrombocytes *(111)*. Furthermore, we stain with propidium iodide in order to gate out dead cells, which loose their specific antigen expression profile or might non-specifically bind antibodies. Samples are acquired automatically from 96-well plates with an HTS on a LSRII flow cytometer until a number of 30,000 living CD45+ cells is reached for each well. The profile of Ig subclasses is of special interest, because it reflects the general direction of T helper cell differentiation (e.g.,*112*); for example, Th1-directed switching correlates with higher IgG2a levels, respectively, IgG2c in C57BL/6 mice, whereas Th2-directed responses bias towards IgG1. Under baseline conditions, the levels of Ig classes and IgG isotypes are characteristic of a special inbred mouse strain and under genetic control *(113)*. For the analysis of immunoglobulin isotypes, the usage of bead array-based systems like the Luminex X-map® technology *(114)* has challenged traditional methods like ELISA, mostly due to its power, combining the simultaneous analysis of many different analytes in a single and relatively small amount of sample, with high sensitivity.

Additional tests (secondary and tertiary screening) include the flow cytometrical analyses of primary (bone marrow, thymus) and secondary (spleen, mesenteric lymphnodes) lymphoid organs. A high purity of organ cell suspensions is normally achieved after erythrocyte lysis, allowing the acquisition of a high number of cells (>50,000), enabling even to analyze clusters of minor subpopulations. The antibody panels for these stainings are specifically adapted to each organ. Stages during T-cell development in the thymus can be identified by CD4 and CD8 co-staining. Double-negative cells are "classically" further subdivided by their CD44 and CD25 surface expression *(115)*. For the bone marrow, lineage markers, including markers for B cells and their developmental stage (e.g., CD43, IgM, IgD), and for hemopoietic stem cells and precursors (e.g., c-kit) are markers of choice (for B cells: *116*). For spleen and lymph nodes, a broad panel for B-cell *(116)* and T-cell markers can be completed with specific markers for dendritic cells (CD11c).

Following the sanitary status of mice is crucial for all immunological assays as infections can severely alter many immunological parameters. On the other hand, immunological phenotyping can also focus on infection as in vivo challenge. This approach allows intensive investigation of innate and/or adaptive immune mechanisms and their capability to deal with a defined dose of antigen or infection.

We have decided to implement infection with *Listeria monocytogenes* into the screening protocols for analysis of the integrity of the immune system under challenge conditions, since this infection fulfils most of the specific requirements for such an approach: for example, *L. monocytogenes* infection can easily be treated with antibiotics to rescue mice with an increased susceptibility for this facultative intracellular bacterium *(117)*. *L. monocytogenes* infection is a well-described infection model (e.g.,*118*). Monitoring the immune response of animals within the first days after infection can be achieved under high-throughput conditions by measuring serum-liver enzyme activities and/or cytokine levels, mainly of IFNγ. MHC-I tetramers allowing the detection of antigen-specific CD8+ T cells *(119)* enable direct monitoring of pathogen-specific T cells in the host organism, and their additional characterization for other differentiation and memory markers, like CD62L or CD127 *(120)*.

Such a precise determination of resistance patterns under standardized infection challenge conditions is a promising attempt to identify novel functions for defined genes within the complex regulatory network of immune response. Hopefully, other immunological challenge models for high-throughput applications will follow in the near future (e.g., allergy, tumor rejection, tissue/organ transplantation, chronic inflammatory diseases, autoimmunity, other infectious agents), since the identification of resistance- or susceptibility-associated genes is likely to provide interesting new target structures for new therapeutic strategies fighting the multitude of human diseases, which are under strong control of innate and/or adaptive immune responses.

3.10. Allergy Screen

3.10.1. Introduction

The increasing prevalence of allergic asthma and other allergic airway diseases in the industrialized world over the last decades seems to be part of a generalized trend of an increasing prevalence of immune-mediated disorders concomitant with a decline in the prevalence of some of the most common infectious diseases. Although these findings provide clear evidence that environmental factors play a major role in the cause of asthma and allergy, findings that helped to formulate the so-called hygiene hypothesis *(122)*, family and twin studies clearly showed that a strong component of asthma and atopic allergy risk is genetically determined *(123)*. Thus, IgE-mediated atopic disorders such as allergic asthma, allergic rhinitis, and atopic dermatitis are now considered as environmentally and exposure-driven immune disorders leading to the expression of various clinical phenotypes in individuals with defined genetic risk profiles *(124)*. Genome screens with classical linkage and fine mapping approaches suggest that susceptibility to asthma is determined by multiple genes with each gene having a moderate dose effect. However, the precise function and linkage to allergic disease has only been provided for a limited number of individual genes. Understanding the complex genetic program that leads to the precipitation of such phenotypically distinct diseases in atopy will be essential for the development of new diagnostic and therapeutic approaches in IgE-mediated allergy. In this respect, the development of phenotypically and genotypically defined animal models will be an important step. To detect allergy-prone mouse mutants in systematic screening efforts, total plasma IgE was established as a powerful screening parameter *(125, 62)*. However, both the clinical experience in patients suffering from allergic diseases and data from murine models clearly indicate that total IgE has its limits as screening parameter for allergy. IgE-mediated allergies often become evident only under exposure conditions (e.g., pollen allergy) and can be either aggravated or ameliorated upon long-term exposure to distinct environmental factors. Thus, allergy phenotyping in the laboratory mouse should cover the whole spectrum from measuring baseline total IgE levels to allergen exposure-oriented challenges. Identifying mutant mouse lines with elevated total IgE or enhanced allergen-specific IgE sensitization and allergic airway inflammation on allergen exposure and challenge is likely to provide important advances with regard to the pathophysiology, the diagnosis, and the preventive as well as therapeutic treatment of IgE-mediated allergic diseases *(126)*.

3.10.2. Screening Procedures in IgE-Mediated Allergy

The various methods used in allergy phenotyping of mutant mouse lines are compiled in **Fig. 25.14** and **Table 25.11**.

Classical immunoassay techniques like ELISA *(62)* or more innovative multiplex approaches such as the bead-based assays (Luminex) *(127)* are available as ready-made or custom-use assays

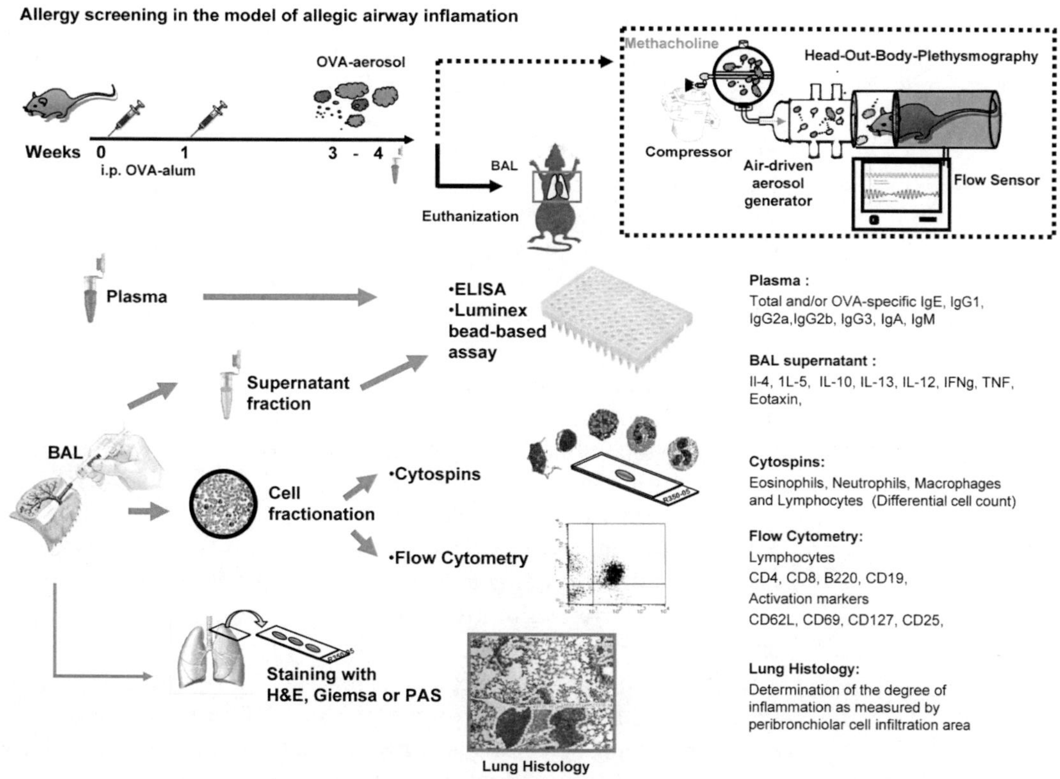

Fig. 25.14. Allergy screening in the model of allergic airway inflammation.

Table 25.11
The most important factors that influence the allergic screen under challenge condition

Allergy phenotype	Severe	Less severe
Age	Young	Old
Sex	Female	Male
Paradigmatic strain	BALB/c	C57BL/6
Antigen dose	Low dose	High dose

to measure different parameters in mice. Of particular interest for initial screening procedures are plasma immunoglobulin levels including IgE and IgG1. These parameters help to initially define a potential allergic phenotype in mice by providing information on a possible allergic predisposition. Based on such initial information, challenge screens with an allergen can be performed which will lead to more substantial data on the allergic phenotype in a particular mouse mutant line.

A classical symptom of IgE-mediated type I hypersensitivity in men and mice is allergen-induced allergic airway inflammation, which is characterized by periodic airflow limitation, inflammation and airway hyperresponsiveness (AHR). Testing AHR in mice is an established procedure for modeling challenge conditions in IgE-mediated allergy. AHR can be induced by exposure to a model allergen such as ovalbumin (OVA) aerosol in OVA-sensitized mice. Several methodologies have been employed to screen for AHR in murine allergy models after sensitization with allergen and allergen aerosol challenge. We have successfully established head-out body plethysmography to evaluate AHR in murine allergen challenge models *(62)*. This technique records breathing patterns in allergen-challenged mice exposed to increasing doses of inhaled methacholine, which provokes narrowing of the airways (bronchoconstriction). As a consequence, mice with severe allergic airway inflammation will react to lower doses on inhaled methacholine. To evaluate the underlying inflammatory and immune mechanisms in AHR, bronchoalveolar lavage (BAL) is the most widely used and preferred technique *(128)*. BAL allows the quantification and differentiation of both the cellular infiltration in the airways and the various inflammatory markers in the cell-free BAL fluid (e.g., cytokines and chemokines). Cell fractions can either be evaluated using cytospins and differential microscopy or by flow cytometric analysis. Finally, histologic and immunohistologic staining of lung tissue sections can be used to characterize and quantitate the inflammatory infiltrate under the microscope. Total and allergen-specific immunoglobulin values from plasma or BAL samples can be analyzed and correlated to the AHR response and the degree of airway inflammation in the BAL read-outs.

More refined immunological techniques can be included into the in-depth allergy phenotyping of mutant mouse lines. Among them are the use of MHC-multimers (e.g., tetramers) to directly visualize allergen-specific T-cell reactivities *(129)* and, as an in vivo technique, the adoptive transfer of immunologically competent cells from mutant to wild-type lines and vice versa, the latter being a powerful technique to delineate the functional role of certain gene products on cellular fractions and subtypes of immunocompetent cells in the allergic airway inflammation model *(130)*. Many of these techniques cannot be applied to human subjects. Therefore, murine models are essential for understanding of the genetics and pathogenesis of human IgE-mediated allergic diseases. Atopic IgE-mediated allergy can express itself in susceptible individuals through various clinical phenotypes. Allergic asthma is one of the most important clinical phenotypes of IgE-mediated allergy which can be mimicked in mice by the murine AHR model after allergen sensitization and challenge (see above). Murine models have also been successfully established for other clinical phenotypes of IgE-mediated allergy. Among them are models for allergic rhinoconjunctivitis *(131)*, atopic dermatitis/eczema *(132)*, and food allergy which can be induced by using epicutaneous sensitization followed by an intragastric challenge resulting in the development of diarrhea *(133)*. All these murine models are associated with allergen-specific IgE levels, which underline the essential role of IgE as a first allergy screening parameter. Other methods like passive cutaneous anaphylaxis (PCA) or active cutaneous anaphylaxis (ACA) were originally developed in the guinea pig or rat and were subsequently adapted to the murine system. PCA and ACA are important immunological tools in the analysis of murine mast cell function.

Screening methods

1. A total of 7–10 mutant and 7–10 control mice of the same background (significant differences between strains of mice have been described *(134)*).
2. Female animals (females are more susceptible to develop allergic airway inflammation *(135)*).
3. Mice aged 6–8 weeks old (age of mice is an important factor in allergy *(136)*).
4. Total IgE measurement from 15–30 μl of plasma samples (ELISA or bead-based assay).

Challenge methods

1. Sensitization of mice with allergen: Intraperitoneal sensitization of 6–8-week-old mice with of 10 μg of OVA-alum at day 0 and day 7.
2. OVA-aerosol challenge at week 3.
3. Body plethysmography at week 4.
4. Euthanization soon after body plethysmography.
5. Bronchoalveolar lavage (BAL).
6. Collection of lungs for cell isolation and histology.

3.11. Steroid Metabolism

3.11.1. Introduction

Steroids control differentiation and proliferation processes of cells and tissues. They participate in regulation of apoptosis *(137)*, bone remodeling *(138)*, and neuroregeneration *(139)*. Severe diseases are caused by monogenic mutations with loss of function of steroid pathway proteins (e.g., pseudohermaphroditism *(140)*). Defects in steroid metabolism contribute as well to the pathogenesis of many different multifactorial diseases like cancer (e.g., breast or prostate cancer), polycystic ovary syndrome, diseases of cartilage and bone (e.g., osteoporosis) or neurological diseases (e.g., Alzheimer's) *(141, 142)*. Other hereditary diseases are caused by disorders in biosynthesis of the steroid precursor cholesterol (e.g., CHILD syndrome, chondrodysplasia punctata, and Smith–Lemli–Opitz syndrome) *(143)*. Altered relations of steroid concentrations are typical signs for steroid-related disorders. Therefore, screening for aberrations in the steroid concentrations in mouse plasma should be of interest. Only a few animal models for the study of steroid-related disorders exist. Mice exhibiting dwarfism or cataracts (where changes in glucocorticoids are indicative), skin defects (androgens), limb malformations (cholesterol), and distorted reproduction biology or behavior (testosterone, estradiol, progesterone) are the most interesting candidates since such phenotypic defects are observed in human steroid-related diseases. For our screening, we adopted commercially available ELISA kits for measurements of samples obtained from rodents.

3.11.2. Methods

For screening of alterations in plasma concentrations of the chosen steroids, blood is collected from the tettootbital sinus of fully mature anesthetized mice (isoflurane) and plasma is prepared by centrifugation and stored at –20°C. Plasma was chosen as the source for steroid assays to facilitate sharing of blood withdrawal with other screeners.

Since no steroid ELISA kits are available for mouse samples, human ELISA kits have to be used, but analysis is disturbed by the mouse plasma matrix. Therefore, the steroids have to be extracted from the matrix by liquid/liquid-extraction. The respective amount of plasma is extracted three times in each case with a tenfold excess of tert-butylmethylether (TBME). After evaporation of the combined organic extracts the dried residues can be stored at –20°C. For the subsequent ELISA, the material is reconstituted for the respective kit, for example, in assay buffer or in steroid-free serum. The steroids are quantified by competitive ELISA according to the manufacturer's protocols. The plates are read in a standard microplate reader at the named wavelength and the concentrations reported on basis of the respective kit standard curve. We use this procedure for the quantification of dehydroepiandrosterone, testosterone, progesterone, and estradiol. The ELISA kits are chosen considering measurement range and costs. We are confident that further steroids could be quantified in the same way. There are ELISA kits available also for estrone, androstenedione, aldosterone, dehydroepiandrosterone-sulfate, dihydrotestosterone, estriol, pregnenolone, and 17-hydroxyprogesterone. Nevertheless, for these steroids the pre-ELISA-sample preparation has to be established with respect to plasma demand and possible matrix effects.

Although steroid concentration determinations are routine in clinical chemistry applied for human health, these approaches represent a major challenge when studying the mouse. This problem can be addressed in different ways, for example, RIA, GC/MS, LC/MS, and ELISA. RIA is a very sensitive method of quantification but associated with radioactive exposure of people and environment. GC/MS is often applied for human samples. When applied to screening measurements of mouse plasma, the main limitations are due to minute sample volume and matrix effects *(144)*. The latest technology developments increase potential of LC/MS methods for steroid quantification. However, the technology is still cost-intensive and not standardized worldwide. ELISA is our method of choice for steroid screening in mouse plasma because it is a worldwide established and comparable method with reasonable costs and sound interpretation. Standardization is reached relatively easyly. The necessity of the plasma extraction is a drawback concerning workload. But on the other hand, the extraction procedure results in more clean samples. It is also useful for further enrichment of the analytes in the sample, especially when the mouse plasma concentration of an analyte is lower than the sensitivity of the kit (e.g., estradiol).

3.12. Pathological Examination of Mouse Models

3.12.1. Introduction

Genetically engineered mice (GEM) have become indispensable tools for cancer research, and the identification of genes involved in human diseases. A complete phenotyping of a mouse model is a challenge for which most research laboratories are not prepared. The GMC was designed to offer a large-scale, standardized and comprehensive phenotype analysis of mouse mutants to the scientific community. The pathology screen within the GMC plays a pivotal role in the final analysis of mouse models. The morphologic phenotype helps to find a correlation between the functional abnormalities found in other screens with the anatomical changes. This helps to understand how these genes influence the development of human diseases. Mouse mutants entering the pathology screen of the GMC are first examined in a primary screen, which includes a macroscopic and microscopic examination of over 30 organs (H&E staining). Interesting mouse models are further analyzed in a secondary and a tertiary screen where more sophisticated analysis is performed in order to achieve a complete phenotype of the corresponding mouse model.

3.12.2. Methods

Basic pathological examination (primary screen).

The macroscopic examination represents a very important step of a pathological analysis and gives a first impression about the organ systems that are likely to be altered due to a genetic manipulation. A final diagnosis is based on objective findings, such as size measurements (weight), evaluation of color, consistency, and form of the organs. X-ray analysis is always performed in mouse models with suspected bone diseases or malformations. All organs are then fixed in 4% buffered formalin and embedded in paraffin. The microscopic examination starts with the evaluation of tissue sections stained with hematoxylin and eosin (H&E). Most diagnoses can be made using this approach *(145–147)*. However, in some instances special stains or immunohistochemical analyses are necessary in order to achieve a precise diagnosis.

Further pathological examination (secondary and tertiary screen)

Special stains (e.g., PAS and Masson's trichrome) are used to demonstrate special structures. The periodic acid Schiff (PAS) staining is used for the demonstration of basement membranes, fungi, or for the differentiation of mucosubstances secreted from the epithelial cells of various organs *(148)*. It is a routine staining for liver and kidney. The Masson's trichrome staining is used for the detection of collagen fibres *(149)*. Several other stains can be performed to demonstrate special structures *(150)* or lipid deposits (Sudan staining *(151)*). Immunohistochemistry is used for the identification of specific and highly selective cellular epitopes in routinely processed paraffin-embedded tissues using an antibody and an appropriate labeling system. Immunohistochemistry allows a diagnosis in cases where the morphological and clinical data alone are not enough to reach a conclusive diagnosis in a tissue section. A very good example is the use of immunohistochemistry for the classification of lymphomas and leukemias. Different antibodies, specific for a special cell type, (myeloperoxidase, B220, CD79a etc.) are available to make a precise diagnosis or classification of hematological neoplasias *(152, 153)*. In addition, specific antibodies can be used to demonstrate cellular processes, like proliferation (e.g., Ki67) and apoptosis (activated caspase 3) or the presence of a genetic event, such as p53 mutations *(154–156)*. Electron microscopy has the fundamental advantage of disclosing the substructure and ultrastructure of individual cells *(157)*. Transmission electron microscopy (TEM) is particularly useful when conventional light microscopy can identify only minor abnormalities despite clear evidence of disease *(158)*. Furthermore, it plays an important role for the investigation of newly recognized diseases. It is also an important diagnostic tool in renal diseases when a glomerulonephritis or an autoimmune disorder is suspected *(159)*. Laser microdissection enables to focus on the cells of interest. Microscopically identified cell populations can be isolated and extracted from tissue sections or cytologic preparations *(160)*. The main goal is the elimination or minimization of contamination by other cell types also present in the tissue or cytologic sample. These extracted cells can then be utilized for analysis of DNA, RNA, or protein expression **Fig. 25.15**.

The pathological examination is essential in the final analysis of mouse models because it is the only screen that can detect the morphological basis of the alterations found in the other screens. This helps to discover gene functions and aims to support the understanding how these genes influence human diseases. According to the International Mouse Knockout Consortium (20,000 genes) *(161)* the number of GEMs to be analyzed, in the near future, is surpassing by far the number of pathologists (human and veterinary) with expertise in comparative pathology, who could effectively characterize and validate these model animals *(162)*. The pathology screen of the GMC was designed to cover this gap, and therefore, is a unique institution not only in Germany but also in all of Europe.

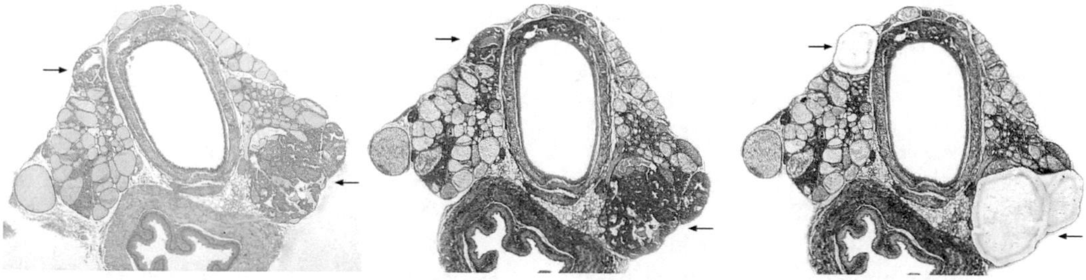

Fig. 25.15. Laser-assisted microdissection (LAM). The picture on the left-hand side shows the H&E-stained cross-section of a murine thyroid gland with nodular hyperplasia of the follicular cells (arrows). In the next picture, the hematoxylin-stained section of this thyroid gland is depicted before the microdissection. On the right-hand side, the hyperplastic nodular lesions are specifically and precisely removed via LAM, for the purpose of minimizing the contamination with normal cells for detailed molecular genetic analysis. The two steps of this non-contact LMPC method (Palm®) include firstly laser-microbeam microdissection, and secondly laser-pressure catapulting (all figures: original magnification 31.25 ×).

Acknowledgement

This work was supported by NGFNplus grants from the Bundesministerium für Bildung und Forschung (01GS0850, 01GS0851, 01GS0852, 01GS0868, 01GS0869, 01GS0854) and by an EU grant (LSHG-2006-037188).

References

1. Collins FS, Rossant J, Wurst W. A mouse for all reasons. Cell 2007; 128:9–13.
2. Rosenthal N, Brown S. The mouse ascending: perspectives for human-disease models. Nat Cell Biol 2007; 9:993–9.
3. Brown SD, Hancock JM, Gates H. Understanding mammalian genetic systems: the challenge of phenotyping in the mouse. PLoS Genet 2006; 2:e118
4. Gailus-Durner V, Fuchs H, Becker L, Bolle I, et al. Introducing the German Mouse Clinic: open access platform for standardized phenotyping. Nat Meth 2005; 2:403–4.
5. Brown SD, Chambon P, Hrabé de Angelis M, Eumorphia Consortium. EMPReSS: standardized phenotype screens for functional annotation of the mouse genome. Nat Genet 2005; 37:1155.
6. Brielmeier M, Mahabir E, Needham JR, Lengger C, Wilhelm P, Schmidt J. Microbiological monitoring of laboratory mice and biocontainment in individually ventilated cages: a field study. Lab Anim 2006; 40:247–60.
7. Maier H, Lengger Ch, Simic B, Fuchs H, Gailus-Durner V, Hrabé de Angelis M. MausDB: an open source application for phenotype data and mouse colony management in large-scale mouse phenotyping projects. BMC Bioinformatics 2008; 9:169 (epub 26 March 2008)
8. The Mouse Phenotype Database Integration Consortium. Integration of mouse phenome data resources. The Mouse Phenotype Database Integration Consortium. Mamm Genome 2007; 18:157–63.
9. WHO. The World Health Report 2001 – Mental Health: New Understanding, New Hope; http://www.who.int/whr/2001/en/index.html.
10. Karsenty G, Wagner EF. Reaching a genetic and molecular understanding of skeletal development. Dev Cell 2002; 2:389–406.
11. Ralston SH, de Crombrugghe B. Genetic regulation of bone mass and susceptibility to osteoporosis. Genes Dev 2006; 20:2492–506.

12. Abe K., Fuchs H., Lisse T., Hans W. and Hrabé de Angelis M. New ENU induced semidominant mutation, Ali18, causes inflammatory arthritis, dermatitis, and osteoporosis in the mouse. Mamm Genome 2006; 17:915–26.
13. Rauch F, Glorieux FH. Osteogenesis imperfecta. Lancet 2004; 363:1377–85.
14. Giampietro PF, Blank RD, Raggio CL, et al. Congenital and idiopathic scoliosis: clinical and genetic aspects. Clin Med Res 2003; 1:125–36.
15. Mariani FV, Martin GR. Deciphering skeletal patterning: clues from the limb. Nature 2003; 423: 19–25.
16. Fuchs H, Schughart K, Wolf E, Balling R, Hrabé de Angelis M. Screening for dysmorphological abnormalities – a powerful tool to isolate new mouse mutants. Mamm Genome 2000; 11:528–30.
17. Sorenson JA, Duke PR, Smith SW. Simulation studies of dual-energy X-ray absorptiometry. Med Phys 1989; 16:75–80
18. Gasser JA. Bone measurements by peripheral quantitative computed tomography in rodents. Methods Mol Med 2003; 80:323–41.
19. Turner CH, Hsieh YF, Muller R et al. Genetic regulation of cortical and trabecular bone strength and microstructure in inbred strains of mice. J Bone Miner Res 2000; 15:1126–31
20. Allen MJ. Biochemical markers of bone metabolism in animals: uses and limitations. Vet. Clinical Path 2003; 32:101–113
21. Turner CH, Burr DB. Basic biomechanical measurements of bone: a tutorial. Bone 1993; 14:595–608.
22. Akhter MP, Iwaniec UT, Covey MA, Cullen DM, Kimmel DB, Recker RR. Genetic variations in bone density, histomorphometry, and strength in mice. Calcif Tissue Int 2000; 67:337–44.
23. Roman-Roman S, Garcia T., Jackson A, et al. Identification of genes regulated during osteoblastic differentiation by genome-wide expression analysis of mouse calvaria primary osteoblasts in vitro. Bone 2003; 32:474–82.
24. Umemura Y, Baylink DJ, Wergedal JE, Mohan S, Srivastava AK. A time course of bone response to jump exercise in C57BL/6 J mice. J Bone Miner Metab 2002; 20:209–15.
25. Robling AG, Turner CH. Mechanotransduction in bone: genetic effects on mechanosensitivity in mice. Bone 2002; 31:562–9.
26. Rubin C, Turner AS, Bain S, Mallinckrodt C, McLeod K. Low mechanical signals strengthen long bones. Nature 2001; 412:603–4.
27. Hoshi A, Watanabe H, Chiba M, Inaba Y. Effects of exercise at different ages on bone density and mechanical properties of femoral bone of aged mice. Tohoku J Exp Med 1998; 185:15–24.
28. Judex S, Donahue LR, Rubin C. Genetic predisposition to low bone mass is paralleled by an enhanced sensitivity to signals anabolic to the skeleton. FASEB J 2002; 16:1280–2.
29. Krege JH, Hodgin JB, Hagaman JR, Smithies O. A noninvasive computerized tail-cuff system for measuring blood pressure in mice. Hypertension 1995; 25:1111–5.
30. Lorenz JN. A practical guide to evaluating cardiovascular, renal, and pulmonary function in mice. Am J Physiol Regul Integr Comp Physiol 2002; 282:R1565–82.
31. Deschepper CF, Olson JL, Otis M, Gallo-Payet N. Characterization of blood pressure and morphological traits in cardiovascular-related organs in 13 different inbred mouse strains. J Appl Physiol 2004; 97:369–76.
32. Lang RE, Unger T, Ganten D. Atrial natriuretic peptide: a new factor in blood pressure control. J Hypertens. 1987; 5:255–71.
33. Nishikimi T, Maeda N, Matsuoka H. The role of natriuretic peptides in cardioprotection. Cardiovasc Res 2006; 69:318–28.
34. Kilic A, Bubikat A, Gassner B, Baba HA, Kuhn M. Local actions of atrial natriuretic peptide counteract angiotensin II stimulated cardiac remodeling. Endocrinology 2007; 148:4162–9.
35. Yap LB, Mukerjee D, Timms PM, Ashrafian H, Coghlan JG. Natriuretic peptides, respiratory disease, and the right heart. Chest 2004; 126:1330–6.
36. Moro C, Berlan M. Cardiovascular and metabolic effects of natriuretic peptides. Fundam Clin Pharmacol 2006; 20:41–9.
37. Lerman A, Gibbons RJ, Rodeheffer RJ, et al. Circulating N-terminal atrial natriuretic peptide as a marker for symptomless left-ventricular dysfunction. Lancet 1993; 341: 1105–9.
38. Kettunen RV, Leppaluoto J, Jounela A, Vuolteenaho O. Plasma N-terminal atrial natriuretic peptide in acute myocardial infarction. Am Heart J 1994; 127:1449–55.

39. Schoensiegel F, Bekeredjian R, Schrewe A, et al. Atrial natriuretic peptide (Nt-proANP) and osteopontin (Opn) are useful tools to screen for cardiac disorders in mice. Comp Med 2007; 57:546–53.
40. Doevendans PA, Daemen MJ, de Muinck ED, Smits JF. Cardiovascular phenotyping in mice. Cardiovasc Res 1998; 39:34–49.
41. Ehmke H. Mouse gene targeting in cardiovascular physiology. Am J Physiol Regul Integr Comp Physiol 2003; 284: R28–30.
42. Royer A, van Veen TA, Le Bouter S, et al. Mouse model of SCN5A-linked hereditary Lenegre's disease: age-related conduction slowing and myocardial fibrosis. Circulation 2005; 111:1738–46.
43. Yang XP, Liu YH, Rhaleb NE, Kurihara N, Kim HE, Carretero OA. Echocardiographic assessment of cardiac function in conscious and anesthetized mice. Am J Physiol 1999; 277:H1967–74.
44. Collins KA, Korcarz CE, Lang RM. Use of echocardiography for the phenotypic assessment of genetically altered mice. Physiol Genomics 2003; 13:227–39.
45. Syed F, Diwan A, Hahn HS. Murine echocardiography: a practical approach for phenotyping genetically manipulated and surgically modeled mice. J Am Soc Echocardiogr 2005; 18:982–90.
46. Grundy SM, Brewer HB, Cleeman JI, Smith SC, Lenfant C. Definition of metabolic syndrome: report of the National Heart, Lung, and Blood Institute/American Heart Association conference on scientific issues related to definition. Circulation 2004; 109:433–8
47. WHO. The World Health Report 2002 – Reducing Risks, Promoting Healthy Life. http://www.who.int/whr/2002/en/index.html.
48. Tisdale MJ. Cachexia in cancer patients. Nat Rev Cancer 2002; 2:862–71.
49. Meyer CW, Neubronner J, Rozman J. et al. Expanding the body mass range: associations between BMR and tissue morphology in wild type and mutant dwarf mice (David mice). J Comp Physiol [B] 2007; 177:183–92.
50. Meyer CW, Elvert R, Scherag A, et al. Power matters in closing the phenotyping gap. Naturwissenschaften 2007; 94:401–6.
51. Kallnik M, Elvert R, Ehrhardt N, et al. Impact of IVC housing on emotionality and fear learning in male C3HeB/FeJ and C57BL/6 J mice. Mamm. Genome 2007; 18:173–86.
52. Haemmerle G, Lass A, Zimmermann R, et al. Defective lipolysis and altered energy metabolism in mice lacking adipose triglyceride lipase. Science 2006; 312:734–7.
53. WHO 2006, Mortality Data, http://www.who.int/whosis/mort/en/
54. WHO 2005, Particulate matter and air pollution: how it harms health, http://www.euro.who.int/document/mediacentre/fs0405e.pdf
55. European Respiratory Society & European Lung Foundation. The European lung white book. 2003 ERSJ Ltd, UK.
56. Reinhard C, Eder G, Fuchs H, Ziesenis A, Heyder J, Schulz H. Inbred strain variation in lung function. Mamm Genome 2002; 13:429–37.
57. Reinhard C, Meyer B, Fuchs H, et al. Genomewide linkage analysis identifies novel genetic loci for lung function in mice. Am J Respir Crit Care Med. 2005; 171:880–8.
58. Ganguly K, Stoeger T, Wesselkamper SC, et al. Candidate genes controlling pulmonary function in mice: transcript profiling and predicted protein structure. Physiol Genomics 2007; 31:410–21.
59. Tankersley CG. Genetic control of ventilation: What are we learning from murine models? Curr Opinion Pul Med 1999; 5:344–8.
60. Tankersley CG, Fitzgerald RS, Levitt RC, Mitzner WA, Ewart SL Kleeberger SR. Genetic control of differential baseline breathing pattern. J Appl Physiol 1997; 82:874–81.
61. Schulz, H., Johner, C., Eder, et al. Respiratory mechanics in mice: strain and sex specific differences. Acta Physiol Scand 2002; 174:367–75.
62. Alessandrini F, Schulz H, Takenaka S, et al. Effects of ultrafine carbon particle inhalation on allergic inflammation of the lung. J Allergy Clin Immunol 2006; 117:824–30.
63. Ohl F, Holsboer F, Landgraf R. The modified hole board as a differential screen for behavior in rodents. Behav Res Meth Instr Comp 2001; 33:392–7.
64. Geyer MA. Assessing prepulse inhibition of startle in wild type and knockout mice. Psychopharmacology 1999; 147:11–3.
65. Belzung C, Griebel G. Measuring normal and pathological anxiety-like behaviour in mice: a review. Behav Brain Res 2001; 125:141–9.
66. Cryan JF, Holmes A. The ascent of mouse: advances in modelling human depression

and anxiety. Nat Rev Drug Discov 2005; 4:775–90.

67. Engelmann M, Wotjak CT, Landgraf R. Social discrimination procedure: An alternative method to investigate juvenile recognition abilities in rats. Physiol Behav 1995; 58:315–21.
68. Richter K, Wolf G, Engelmann M. Social recognition memory requires two stages of protein synthesis in mice. Learn Mem 2005; 12:407–13.
69. Falls, W. Fear-potentiated startle in mice. In Crawley, J. (ed), Current Protocols in Neuroscience 2002. Unit 8.11B.1–8.11B.16.
70. Genoux D, Haditsch U, Knobloch M, Michalon A, Storm D, Mansuy I. Protein phosphatase 1 is a molecular constraint on learning and memory. Nature 2002; 418:970–5.
71. Hölter SM, Kallnik M, Wurst W, Marsicano G, Lutz B, Wotjak CT. Cannabinoid CB1 receptor is dispensable for memory extinction in an appetitively-motivated learning task. Eur J Pharmacol 2005; 510:69–74.
72. Watase K, Zoghbi HY. Modelling brain diseases in mice: the challenges of design and analysis. Nat Rev Genet 2003; 4:296–307.
73. Oliver PL, Davies KE. Analysis of human neurological disorders using mutagenesis in the mouse. Clin Sci (Lond) 2005; 108:385–97.
74. Schneider I, Tirsch WS, Faus-Kessler T, et al. Systematic, standardized and comprehensive neurological phenotyping of inbred mice strains in the German Mouse Clinic. J Neurosci Methods 2006; 157:82–90.
75. Rogers DC, Fisher EM, Brown SD, Peters J, Hunter AJ, Martin JE. Behavioral and functional analysis of mouse phenotype: SHIRPA, a proposed protocol for comprehensive phenotype assessment. Mamm Genome 1997; 8:711–3.
76. Meyer OA, Tilson HA, Byrd WC, Riley MT. A method for the routine assessment of fore- and hindlimb grip strength of rats and mice. Neurobehav Toxicol 1979; 1: 233–6.
77. Maurissen JP, Marable BR, Andrus AK, Stebbins KE. Factors affecting grip strength testing. Neurotoxicol Teratol 2003; 25:543–53.
78. Freitag S, Schachner M, Morellini F. Behavioral alterations in mice deficient for the extracellular matrix glycoprotein tenascin-R. Behav Brain Res 2003; 145:189–207.
79. Karl T, Pabst R, von Horsten S. Behavioral phenotyping of mice in pharmacological and toxicological research. Exp Toxicol Pathol 2003; 55:69–83.
80. Sawada I, Kitahara M Yazawa Y. Swimming test for evaluating vestibular function in Guinea Pigs. Acta Otolaryngol. (Stokh) 1994; Suppl 510:20–3.
81. Carter RJ, Morton AJ, Dunnet SB. Motor coordination and balance in rodents. In: Crawley JN, Gerfen CR, Rogawski MA, Sibley DR, Skolnick P, Wray S (eds), Current Protocols in Neuroscience 2001. John Wiley & Sons, Inc., New York, unit 8.12. pp. 1–14.
82. McFadyen MP, Kusek G, Bolivar VJ, Flaherty L. Differences among eight inbred strains of mice in motor ability and motor learning on a rotorod. Genes Brain Behav 2003; 2:214–9.
83. Brooks SP, Pask T, Jones L, Dunnett SB. Behavioural profiles of inbred mouse strains used as transgenic backgrounds. I: motor tests. Genes Brain Behav 2004; 3:206–15.
84. Baird AL, Meldrum A and Dunnett SB. The staircase test of skilled reaching in mice. Brain Res Bull 2001; 54:243–50
85. Rácz I, Zimmer A. Animals Models of Nociception. In: Hrabé de Angelis M, Chambon P, Brown S (eds), Standards of Mouse Model Phenotyping 2006. Wiley-VCH, Weinheim, pp. 221–235.
86. Johnson GJ, Foster A. Prevalence, incidence and distribution of visual impairment. In: The Epidemiology of Eye Disease 2004 (eds.: Johnson GJ, Minassian DC, Weale RA, West SK). Arnold, London UK, pp. 3–28.
87. Dalke C, Graw J. Mouse mutants as models for congenital retinal disorders. Exp Eye Res 2005; 81:503–12.
88. Graw J. The genetic and molecular basis of congenital eye defects. Nat Rev Genet 2003; 4:876–88.
89. Favor J. A comparison of the dominant cataract and recessive specific-locus mutation rates induced by treatment of male mice with ethylnitrosourea. Mutat Res 1983; 110:367–82.
90. Hawes NL, Smith RS, Chang B, Davisson M, Heckenlively JR, John SW. Mouse fundus photography and angiography: A catalogue of normal and mutant phenotypes. Mol Vis 1999; 5:22.
91. Aleman TS, Jacobson SG, Chico JD, et al. Impairment of the transient pupillary light reflex in Rpe65(−/−) mice and humans with leber congenital amaurosis. Invest Ophthalmos Vis Sci 2004; 45:1259–71.

92. Schmucker C, Schaeffel F. In vivo biometry in the mouse eye with low coherence interferometry. Vision Res 2004; 44:2445–56.
93. Puk O, Dalke C, Favor J, Hrabé de Angelis M, Graw J. Variations of eye size parameters among different strains of mice. Mamm Genome 2006; 17:851–57.
94. Dalke C, Löster J, Fuchs H, et al. Electroretinography as a screening method for mutations causing retinal dysfunction in mice. Invest Ophthalmos Vis Sci 2004; 45:601–9.
95. Thaung C, Arnold K, Jackson IJ and Coffey PJ. Presence of visual head tracking differentiates normal sighted from retinal degenerate mice. Neurosci Lett 2002; 325:21–4.
96. Douglas RM, Alam NM, Silver BD, McGill TJ, Tschetter WW, Prusky GT. Independent visual threshold measurements in the two eyes of freely moving rats and mice using a virtual-reality optokinetic system. Vis Neurosci 2005; 22:677–84.
97. Schmucker C, Seeliger M, Humphries P, Biel M, Schaeffel F. Grating acuity at different luminances in wild-type mice and in mice lacking rod or cone function. Invest Ophthalmos Vis Sci 2005; 46:398–407.
98. Rathkolb B, Decker T, Fuchs E, et al. The clinical–chemical screen in the Munich ENU Mouse Mutagenesis Project: screening for clinically relevant phenotypes. Mamm Genome 2000; 11:543–6.
99. Aigner B, Rathkolb B, Herbach N, et al. Screening for increased plasma urea levels in a large-scale ENU mouse mutagenesis project reveals kidney disease models. Am J Physiol – Renal Physiol 2007; 292:F1560–7.
100. Herbach N, Rathkolb B, Kemter E, et al. Dominant-negative effects of a novel mutated Ins2 allele causes early-onset diabetes and severe beta-cell loss in Munich Ins2C95S mutant mice. Diabetes 2007; 56:1268–76.
101. Loeb WF, Quimby FW. The Clinical Chemistry of Laboratory Animals. Taylor & Francis, 1999, Philadelphia.
102. Klempt M, Rathkolb B, Aigner B, Wolf E. Clinical chemical screen. In: Hrabé de Angelis M, Chambon P, Brown S (eds), Standards of Mouse Model Phenotyping 2006, pp. 87–107. Wiley-VCH, Weinheim.
103. Hough TA, Nolan P, Tsipouri V, et al. Novel phenotypes identified by plasma biochemical screening in the mouse. Mamm Genome 2002; 13:595–602.
104. Klempt M, Rathkolb B, Fuchs E, Hrabé de Angelis M, Wolf E, Aigner B. Genotype-specific environmental impact on the variance of blood values in inbred and F1 hybrid mice. Mamm Genome 2006; 17:93–102.
105. Rathkolb B, Tran TV, Klempt M, et al. Large scale albuminuria screen for nephropathy models in chemically induced mouse mutants. Nephron Exp Nephrol 2005; 100:e143–9.
106. Chen J. and D.E. Harrison. Quantitative trait loci regulating relative lymphocyte proportions in mouse peripheral blood. Blood 2002; 99:561–6.
107. Yellon SM, Tran LT. Photoperiod, reproduction, and immunity in selected strains of inbred mice. J Biol Rhythms 2002; 17:65–75.
108. Freitas AA, Rocha B. Population biology of lymphocytes. The Flight for survival. Annu Rev Immunol 2000; 18:83–111.
109. Grewal IS, Heilig M, Miller A, Sercarz EE. Environmental regulation of T-cell function in mice: group housing of males affects accessory cell function. Immunology 1997; 90:165–8.
110. Krzych U, Strausser HR, Bressler JP, Goldstein AL. Quantitative differences in immune responses during the various stages of the estrous cycle in female BALB/c mice. J Immunol 1978,121:1603–5.
111. Weaver JL, Broud DD. Serial phenotypical analysis of mouse peripheral blood leukocytes. Toxicol Mech Methods 2002; 12:95–118.
112. Stevens TL, Bossie A, Sanders VM, et al. Regulation of antibody isotype secretion by subsets of antigen-specific helper T cells. Nature 1988; 334:255–8
113. Sant'Anna OA, Mouton D, Ibanez OM, et al. Basal immunoglobulin serum concentration and isotype distribution in relation to the polygenic control of antibody responsiveness in mice. Immunogenetics 1985; 22:131–9.
114. Fulton RJ, McDade RL, Smith PL, Kienker LJ, Kettman JR, Jr. Advanced multiplexed analysis with the FlowMetrix system. Clin Chem 1997; 43:1749–56.
115. Godfrey DI, Kennedy J, Suda T, Zlotnik A. A developmental pathway involving four phenotypically and fuctionally distinct subsets of CD3-CD4-CD8-Triple negative adult mouse thymocytes defined by CD44 and CD25 expression. JI 1993; 150:4244–52
116. Tung JW, Parks DR, Moore WA, Herzenberg LA. Identification of B-cell subsets. Methods Mol Biol 2004; 271:37–58.

117. Wong P, and EG Pamer. Feedback regulation of pathogen-specific T cell priming. Immunity 2003; 18:499–511.
118. Busch D H, and E G Pamer. T lymphocyte dynamics during Listeria monocytogenes infection. Immunol Lett 1999; 65: 93–8.
119. Busch DH, I M Pilip, SVijh, and EG Pamer. Coordinate regulation of complex T cell populations responding to bacterial infection. Immunity 1998; 8:353–62.
120. Huster, KM, V Busch, M Schiemann, et al. Selective expression of IL-7 receptor on memory T cells identifies early CD40L-dependent generation of distinct CD8+ memory T cell subsets. Proc Natl Acad Sci USA 2004; 101:5610–5.
121. Baumgarth N, and M Roederer. A practical approach to multicolor flow cytometry for immunophenotyping. J Immunol Methods 2000; 243:77–97.
122. Bach JF. The effect of infections on susceptibility to autoimmune and allergic diseases. N Engl J Med. 2002; 347:911–20.
123. McCunney RJ. Asthma, genes, and air pollution. J Occup Environ Med 2005; 47: 1285–91.
124. Wahn U, von Mutius E, Lau S, Nickel R. The development of atopic phenotypes: genetic and environmental determinants. Nestle Nutr Workshop Ser Pediatr Program 2007; 59:1–11; discussion 11–15.
125. Jakob T, Köllisch GV, Howaldt M. et al. Novel mouse mutants with primary cellular immunodeficiencies generated by genome-wide mutagenesis. J Allergy Clin Immunol 2008; 121:179–84
126. Jakob T., Alessandrini F, Gutermuth J, et al. Phenotyping allergy in the laboratory mouse. In: Hrabé de Angelis M, Chambon P, Brown S (eds), Standards of Mouse Model Phenotyping 2006 Wiley-VCH, Weinheim, pp. 253–81
127. Bauer R, Scheiblhofer S, Kern K, et al. Generation of hypoallergenic DNA vaccines by forced ubiquitination: preventive and therapeutic effects in a mouse model of allergy. J Allergy Clin Immunol 2006; 118:269–76.
128. Lommatzsch M, Julius P, Kuepper M et al. The course of allergen-induced leukocyte infiltration in human and experimental asthma. J Allergy Clin Immunol 2006; 118:91–7.
129. Klenerman P, Cerundolo V, Dunbar PR. Tracking T cells with tetramers: new tales from new tools. Nat Rev Immunol 2002; 2:263–72.
130. Miyahara N, Swanson BJ, Takeda K, et al. Effector CD8+ T cells mediate inflammation and airway hyper-responsiveness. Nat Med 2004; 10:865–9.
131. McCusker CT. Use of mouse models of allergic rhinitis to study the upper and lower airway link. Curr Opin Allergy Clin Immunol 2004; 4:11–6.
132. Gutermuth J, Ollert M, Ring J, Behrendt H, Jakob T. Mouse models of atopic eczema critically evaluated. Int Arch Allergy Immunol 2004; 135:262–76.
133. Dearman RJ, Kimber I. A mouse model for food allergy using intraperitoneal sensitization. Methods 2007; 41:91–8.
134. Shinagawa K, Kojima M. Mouse model of airway remodeling: strain differences. Am J Respir Crit Care Med 2003; 168:959–67.
135. Melgert BN, Postma DS, Kuipers I, et al. Female mice are more susceptible to the development of allergic airway inflammation than male mice. Clin Exp Allergy 2005; 35:1496–1503.
136. Mayuzumi H, Ohki Y, Tokuyama K, et al. Age-related difference in the persistency of allergic airway inflammation and bronchial hyperresponsiveness in a murine model of asthma. Int Arch Allergy Immunol 2007; 143:255–62.
137. Bansal N, Houle A, Melnykovych G. Apoptosis: mode of cell death induced in T cell leukemia lines by dexamethasone and other agents. FASEB J 1991; 5:211–6.
138. Jerome CP. Hormonal therapies and osteoporosis. Ilar J 2004; 45:170–8.
139. Chowen JA, Azcoitia I, Cardona-Gomez GP, Garcia-Segura LM. Sex steroids and the brain: lessons from animal studies. J Pediatr Endocrinol Metab 2000; 13:1045–66.
140. Geissler WM, Davis DL, Wu L, et al. Male pseudohermaphroditism caused by mutations of testicular 17 beta-hydroxysteroid dehydrogenase 3. Nat Genet 1994; 7:34–9.
141. Mindnich R, Adamski J. Functional aspects of 17beta-hydroxysteroid dehydrogenase 1 determined by comparison to a closely related retinol dehydrogenase. J Steroid Biochem Mol Biol 2007; 104:334–9.
142. Möller G, Adamski J. Multifunctionality of human 17beta-hydroxysteroid dehydrogenases. Mol Cell Endocrinol 2006; 248:47–55.
143. Herman GE. Disorders of cholesterol biosynthesis: prototypic metabolic malformation syndromes. Hum Mol Genet 2003; 12 Spec No 1:R75–88.

144. Prehn C, Ströhle F, Haller F, et al. A comparison of methods for assays of steroidogenic enzymes: new GC/MS versus HPLC and TLC. In: Enzymology and Molecular Biology of Carbonyl Metabolism, eds. In Weiner H, Plapp B, Lindhal R, Maser E (eds), pp 277–283, Purdue University Press, 2007, West Lafayette, Indiana, USA.
145. Sorensen KD, Kunder S, Quintanilla-Martinez L, Sorensen J, Schmidt J, Pedersen FS. Enhancer mutations of Akv murine leukemia virus inhibit the induction of mature B-cell lymphomas and shift disease specificity towards the more differentiated plasma cell stage. Virology 2007; 362:179–91.
146. Favor J, Gloeckner CJ, Janik D, et al. Type IV procollagen missense mutations associated with defects of the eye, vascular stability, the brain, kidney function and embryonic or postnatal viability in the mouse, mus musculus: an extension of the Col4a1 allelic series and the identification of the first two Col4a2 mutant alleles. Genetics 2007; 175:725–36.
147. Wilson I, Gamble M. The hematoxylins and eosin. In: Theory and Practice of Histological Techniques 5th Edition (2000, reprint 2006), eds. Bancroft, J.D. and Gamble, M., Churchill Livingstone, Edinburgh, London, New York, Oxford, Philadelphia, St. Louis, Sydney, Toronto.
148. Bancroft JD, Gamble M. Theory and Practice of Histological Techniques 5th Edition, eds. Bancroft, J.D. and Gamble, M., Churchill Livingstone, 2002, Edinburgh
149. Toyota E, Kawaguchi Y, Ogasawara Y, et al. Novel rat model of ischemic cardiomyopathy induced by repetitive myocardial ischemia/reperfusion injury while conscious. Circ J 2007; 71:788–95.
150. Kwok-Kee C, Lowe J. Techniques in Neuropathology in Theory and Practice of Histological Techniques 5th Edition, eds. Bancroft, J.D. and Gamble, M., Churchill Livingstone, 2002, Edinburgh.
151. Jones LM. Lipids in Theory and Practice of Histological Techniques 5th Edition, eds. Bancroft, J.D. and Gamble, M., Churchill Livingstone, 2002, Edinburgh.
152. Kunder S, Calzada-Wack J, Hölzlwimmer G, et al. A Comprehensive Antibody Panel for Immunohistochemical Analysis of Formalin-fixed, Paraffin-embedded Hematopoietic Neoplasms of Mice. Analysis of mouse specific and human antibodies cross-reactive with murine tissue. Toxicol Pathol 2007; 35:366–75.
153. Kremer M, Quintanilla-Martínez L, Nährig J, von Schilling C, Fend F. Immunohistochemistry in bone marrow pathology: a useful adjunct for morphologic diagnosis Virchows Arch 2005; 447:920–37.
154. Otha Y, Ichimura K. Proliferation markers, proliferating cell nuclear antigen, Ki67, 5-bromo-2′-deoxyuridine, and cyclin D1 in mouse olfactory epithelium. Ann Otol Rhinol Laryngol 2000; 109:1046–8.
155. Vyas D, Robertson CM, Stromberg PE, et al. Epithelial apoptosis in mechanistically distinct methods of injury in the murine small intestine. Histol Histopathol 2007; 22:623–30.
156. Skalicky SE, Ow K, Hannan M, Russell PJ, Crowe PJ, Yang JL. P53 expression is associated with malignant potential in xenograft tissues of a fibrosarcoma mouse model. Anticancer Res 2007; 27:973–8.
157. Galeano B, Klootwijk R, Manoli I, et al. Mutation in the key enzyme of sialic acid biosynthesis causes severe glomerular proteinuria and is rescued by N-acetylmannosamine. J Clin Invest 2007; 117:1585–94.
158. Stirling JW, Coleman M, Thomas A, Woods AE. Role of transmission electron microscopy in tissue diagnosis: Diseases of the kidney, skeletal muscle and myocardium. J Cell Pathol 1999; 4: 223–43.
159. Qiao B, Wu J, Chu YW, et al. Induction of systemic lupus erythematosus-like syndrome in syngenic mice by immunization with activated lymphocyte-derived DNA. Rheumatology 2005; 44:1108–14.
160. Potter SS, Hartman HA, Kwan KM, Behringer RR Patterson LT. Laser Capture-Microarray Analysis of Lim1 Mutant Kidney Development. Genesis 2007; 45:432–9.
161. Anon. Mutant mice galore. Nature 2007; 446:469–70
162. Cardiff RD. Pathologists needed to cope with mutant mice. Nature 2007; 447:528

Index

V

W

X

Z

Color Plates

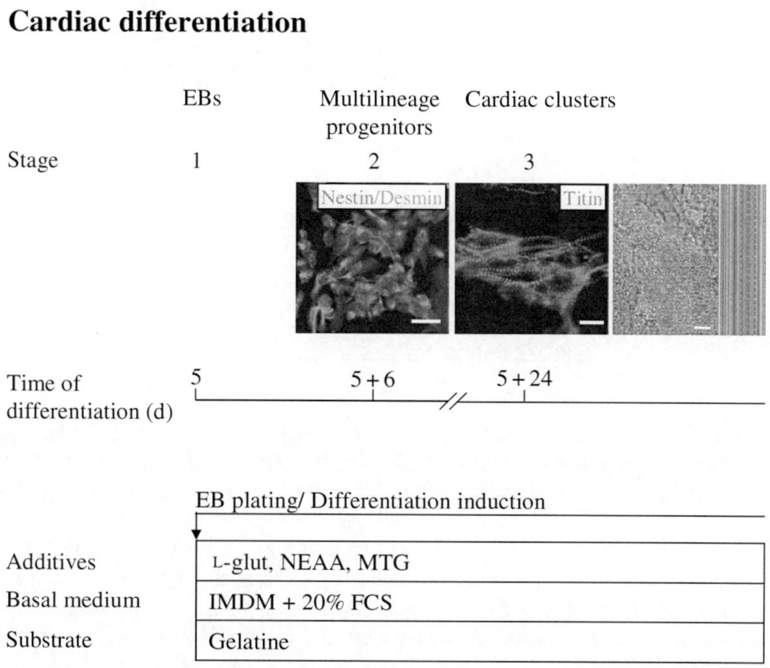

Fig. 12.2. Protocol for mES cell-derived cardiac differentiation. Five-day EBs were plated onto gelatin- or laminin-coated plates and cultured in IMDM+20%FCS supplemented with L-glutamine, NEAA, and MTG for up to 24 days. Multilineage progenitors at the intermediate stage 2 co-express nestin and desmin, while terminally differentiated cardiac clusters (stage 3) show well-organized sarcomeric staining of Z-disk epitopes of titin. Beating frequency measured from a beating cluster (phase contrast) by the LUCIA HEART imaging system is shown at the *right*, bar = 50 μm (*see* discussion on p. 229).

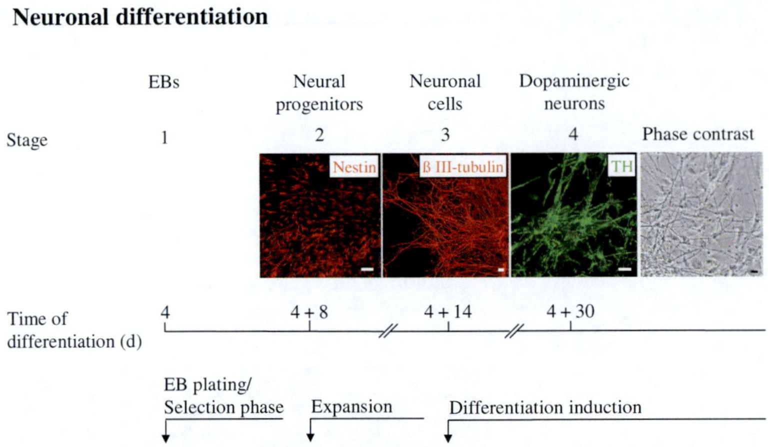

Additives	B1*	B2 + EGF + bFGF	Neurobasal + B27 (2%) + SPFs
Basal medium	DMEM/F12	DMEM/F12	
Substrate	Gelatine	poly-L-ornithine/laminin	

Fig. 12.3. Protocol for mES cell-derived neuronal differentiation. ES cells were cultured as EBs for 4 days. After plating onto gelatin, cells were cultured in B1 supplements and FCS-containing medium for 24 h (*). After medium change (at day 4+1), EB outgrowths were cultured until day 4+8 without FCS to select for neural progenitors. At day 4+8, EBs were dissociated and replated onto poly-L-ornithine/laminin until day 4+14, when differentiation of mature neurons was induced by "Neurobasal" medium, B27 supplement, and SPFs ("survival promoting factors"). The table shows the media, additives, and substrates used with this protocol. Differentiation led to nestin-positive neural progenitors (stage 2) followed by β-III-tubulin-expressing neuronal cells at stage 3 (4+14 d) and dopaminergic neurons expressing tyrosine hydroxylase at stage 4. A phase contrast picture shows the morphology of the ES cell-derived neurons at stage 4 (*right*) (*see* discussion on p. 230).

Pancreatic differentiation

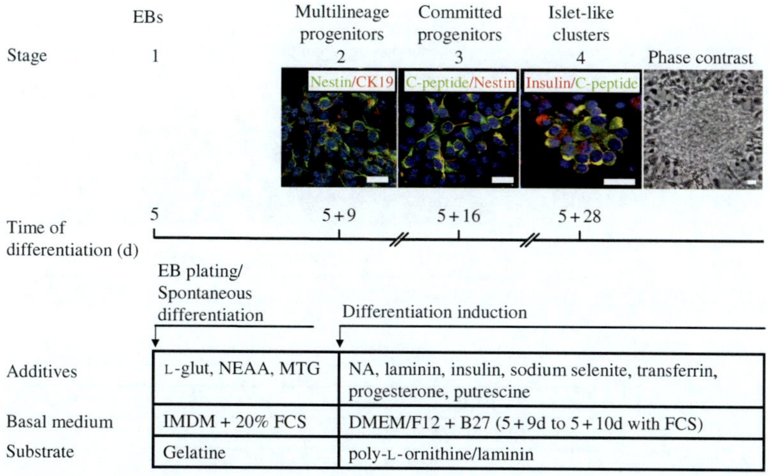

Fig. 12.4. Protocol for mES cell-derived pancreatic differentiation. Scheme displays media, additives, and substrates used during the differentiation process. Five-day EBs were plated onto gelatin for spontaneous differentiation in IMDM containing 20% FCS, L-Glut, NEAA, and MTG. At day 5+9, EBs were dissociated and replated onto poly-L-ornithine/laminin and subjected to differentiation by adding the differentiation factors niacinamide (NA), laminin, insulin, sodium selenite, transferrin, progesterone, and putrescine (and FCS for 24 h after plating). After medium change (at day 5+10), differentiation was continued (without FCS) until day 5+28. During spontaneous differentiation, nestin/CK19 co-expressing multilineage progenitors were formed (stage 2). Directed differentiation resulted in C-peptide/nestin-positive committed progenitors (stage 3) and insulin/C-peptide co-expressing islet-like clusters (stage 4; images from (*68*)). Morphology by phase contrast is shown (*right*) from (*77*). Cell nuclei were visualized by Hoechst 33342 (*blue*). Bars = 20 μm (*see* discussion on p. 231).

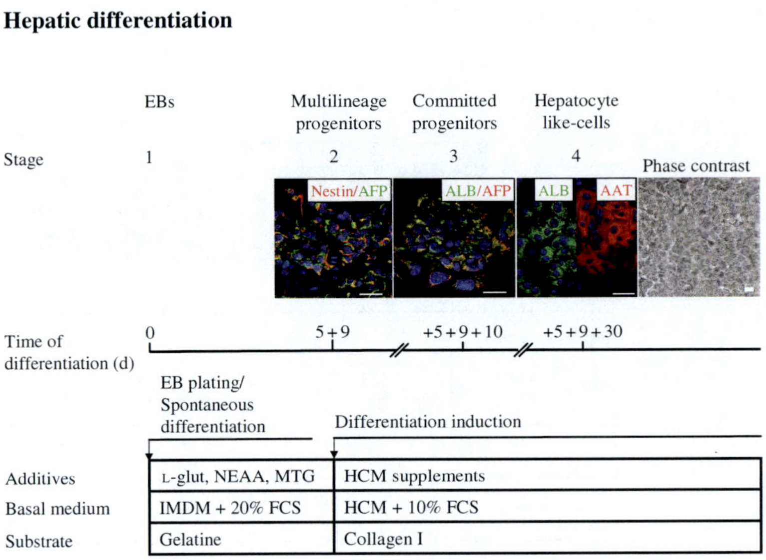

Fig. 12.5. Protocol for mES cell-derived hepatic differentiation. Scheme displays media, additives, and substrates used during the differentiation process. Five-day EBs were plated onto gelatin for spontaneous differentiation in IMDM containing 20% FCS, L-Glut, NEAA, and MTG. At day 5+9, differentiation into the hepatic lineage was induced by dissociation of the EBs and replating onto collagen I. Cells were cultured in differentiation medium (HCM) containing 10% FCS until day 5+9+30. Spontaneous differentiation led to nestin/AFP-positive multilineage progenitors (stage 2). Differentiation resulted in albumin/AFP co-expressing committed progenitors at stage 3, and albumin- and AAT-positive, partially binucleated hepatocyte-like cells (stage 4, images from (*53*)) with cuboidal morphology (phase contrast, *right*, from (*77*)) at day 5+9+30. Cell nuclei were visualized by Hoechst 33342 (*blue*). Bars = 20 μm (*see* discussion on p. 232).

FSC
www.fsc.org
MIX
Papier | Fördert gute Waldnutzung
FSC® C083411

Kolibri 360 GmbH
Ferdinand-Jühlke-Straße 7
99095 Erfurt, Deutschland
produktsicherheit@kolibri360.de

For an old Centralite

Ralph A Raimi
22 May 1985